Quantum Networks

Ming-Xing Luo

Quantum Networks

Introduction and Applications

Springer

Ming-Xing Luo
School of Information Science
and Technology
Southwest Jiaotong University
Chengdu, China

ISBN 978-981-97-6225-5 ISBN 978-981-97-6226-2 (eBook)
https://doi.org/10.1007/978-981-97-6226-2

This Springer imprint is published by the registered company Springer Nature Singapore Pte Ltd.
The registered company address is: 152 Beach Road, #21-01/04 Gateway East, Singapore 189721, Singapore

To my parents, wife and son.

Preface

The history of quantum information is a fascinating one, spanning over a century of scientific discoveries and technological advancements. The origins of quantum information can be traced back to Planck in 1900, who introduced the concept of quantization, which suggested that energy could only be exchanged in discrete units or quanta. This idea laid the foundation for the development of quantum mechanics with Heisenberg's uncertainty principle.

The initiation of quantum information theory depends mainly on the entanglement theory built by Schrodinger, Einstein, Podolsky, & Rosen in the 1930s. The major development in quantum information came in the 1960s when Feynman proposed the idea of using quantum computers to simulate quantum systems. Quantum computers, on the other hand, could simulate quantum systems efficiently by exploiting the properties of quantum mechanics beyond classical computers.

Around ten years later, Wiesner proposed the concept of quantum encoding, which laid the foundation for the development of quantum cryptography such as BB84 protocol in 1984 to secure communication by the principles of quantum mechanics. Meanwhile, the 1980s saw a surge of interest in exploring the potential of quantum computing by Benioff for the quantum Turing machine and Deutsch for quantum circuit algorithms. This intrigues significant progress in the 1990s by Shor who proposed an algorithm for factoring large numbers efficiently on a quantum computer. This algorithm is considered one of the most important breakthroughs in quantum computing and has significant implications for cryptography. Another breakthrough is from Grover for searching an unsorted database using a quantum computer.

The history of the quantum network is relatively short, as the concept of quantum communication and cryptography only emerged in the 1980s. However, the development of the quantum internet may be started by Bennett and Brassard, and then Ekert proposed a new form of communication that relied on the principles of quantum mechanics. This method, known as quantum key distribution (QKD), uses quantum communication to securely transmit a random key between two parties. The key can then be used to encrypt and decrypt messages, ensuring that they cannot be intercepted or read by anyone else. This idea laid the foundation for quantum communication, which is the backbone of the quantum internet.

Over the next few decades, researchers continued to develop and refine the concepts of quantum communication and cryptography. The first wide discussion is from Kimble for the creation of a quantum internet. Several potential strategies are then proposed for building a quantum internet. Meanwhile, several teams of scientists successfully demonstrated the quantum key distribution and quantum-encrypted communication over free space or fiber optic networks.

In 2016, China launched the world's first quantum satellite, called Micius. The satellite was designed to test the feasibility of using quantum communication to transmit information over long distances. This satellite is then used for quantum communication between the satellite and ground stations on Earth, marking a significant step forward in the development of the quantum network.

Today, researchers around the world are working on developing the various components of the quantum network, including both the quantum communication theory of quantum internet protocol, quantum internet standard, quantum infrastructure, and quantum repeaters, quantum routers, and quantum controls. The development of quantum network has the potential to revolutionize the way we communicate and share information. A quantum network could also enable new forms of computing and data processing, as quantum computers are capable of solving problems that are beyond the reach of classical computers.

While some books have focused on special aspects of quantum computing, quantum communication, and quantum information theory. This is not enough for advanced students or researchers, especially the lack of recent results including but not limited to hot topics, such as quantum internet, distributed quantum computation, quantum network communication, and nonlocality of entangled networks. Besides, entanglement as another aspect of quantum physics is far beyond the classical description. Thus the goal of this proposal is to complete theoretical topics related to quantum networks consisting of entangled systems. It will include various fundamental but most recent publications of wide topics on quantum Internet. Our motivation is to provide a reference for both graduated students and experts who are interested in this challenging, vigorous, and fascinating field. It is designed for a reader having some basic knowledge of quantum mechanics, linear algebra, and computer science but not being familiar with quantum networks particularly, large-scale quantum information processing. While we regret that we can only include a limited selection of excellent achievements due to the author's capabilities and research interests, readers are encouraged to explore other relevant results online.

The authors would like to thank Profs. Shao-Ming Fei, Jingyun Fan, Jing-Lin Chen, Xiu-Bo Chen, Yi-Xian Yang, Hong Lai, Li-Ming Zhao, Song-Ya Ma, Xue Yang, Armin Tavakoli, Alejandro Pozas-Kerstjens, and Marc-Olivier Renou for collaborations/comments/discussions/encouragement/supports during the preparation of this book. The authors thank Yan-Han Yang, Ran Zhang, Yan-Jun Xiang, Hui-Ming Cheng, Jun-Li Jiang, Xia Yan, Si-Yi Li, and Si Jin for draft checks. We also would like to express our sincere gratitude to the editor. This work is partially supported by NSF of China under grants 62172341 and 61702427, NSF of Sichuan under grant 2023NSFSC0447, and Fundamental Research Funds for the Central Universities (SWJTU).

Chengdu, China
June 2024

Ming-Xing Luo

Contents

Chapter 1
Mathematical Foundations

In this chapter, we present familiar concepts, definitions, and properties from quantum information theory for self-understanding. These can be found in numerous books on quantum computation or quantum information Dirac (1981), Peres (2006), Nakahara and Ohmi (2008), Nielsen and Chuang (2000). This chapter is on finite-dimensional scenarios of quantum states, operations, and information processing related to the main topic of the book. However, some comparable outcomes may be extended to infinite-dimensional quantum states.

1.1 Linear Algebra

The set $\mathbb{N}$ contains natural numbers, namely $\{1, 2, \cdots\}$. The set $\mathbb{Z}$ consists of integers, that is, $\{\cdots, -1, 0, 1, \cdots\}$. $\mathbb{R}$ and $\mathbb{C}$ represent the sets of real and complex numbers, respectively.

$\mathbb{R}^n$ refers to the n-dimensional real vector space, while $\mathbb{C}^n$ denotes the complex n-dimensional vector space.

In the space $\mathbb{C}^n$, a vector is denoted as $|x\rangle$. This vector is represented as a column of n complex numbers x_i for $1 \leq i \leq n$, which can be written as $|x\rangle = (x_1, x_2, \cdots, x_n)^T$, where T denotes the transpose of the vector. The adjoint of $|x\rangle$ is denoted by $\langle x|$, which is the conjugate transpose of $|x\rangle$.

A normalization orthogonal basis of the space $\mathbb{C}^n$ is a set of vectors denoted by $\mathcal{S} := \{|e_i\rangle\}$ that satisfies the condition:

$$\langle e_i|e_j\rangle = \delta_{ij}, \text{ for all } i, j, \tag{1.1}$$

where δ_{ij} is the Kronecker delta function, which equals 1 if $i = j$ and equals 0 otherwise. The basis $\mathcal{S}$ is a complete basis if it satisfies the following identity relation:

M.-X. Luo, *Quantum Networks*,
https://doi.org/10.1007/978-981-97-6226-2_1

$$\sum_{i=1}^{n} |e_i\rangle\langle e_i| = \mathbb{1}, \tag{1.2}$$

where $\mathbb{1}$ is the identity matrix on $\mathbb{C}^n$. This means that any vector $|x\rangle \in \mathbb{C}^n$ can be linearly represented with complex coefficients $a_i \in \mathbb{C}$ as $|x\rangle = \sum_i a_i |e_i\rangle$.

The symbol $\mathbb{H}$ denotes a finite-dimensional Hilbert space without specifying its dimension. It is a complex vector space equipped with an inner product, given by $\langle u, v\rangle = \sum_i u_i^* v_i$, where $|u\rangle = (u_1, u_2, \cdots)^T$ and $|v\rangle = (v_1, v_2, \cdots)^T$ are two vectors in $\mathbb{H}$.

An $n \times n$ matrix A on the space $\mathbb{C}^n$ is called Hermitian if it satisfies $A^\dagger = A$, where $\dagger$ represents the conjugate transpose operation.

A unitary matrix U on the space $\mathbb{C}^n$ is an $n \times n$ matrix satisfying $U^{-1} = U^\dagger$, where $\dagger$ represents the conjugate transpose operation. If U is unimodular, i.e., $\det(U) = 1$, then it is called a special unitary matrix, denoted by $SU(n)$. One example is Pauli matrices on $\mathbb{C}^2$ defined by

$$\sigma_x = \begin{pmatrix} 0 & 1 \\ 1 & 0 \end{pmatrix}, \sigma_y = \begin{pmatrix} 0 & -i \\ i & 0 \end{pmatrix}, \sigma_z = \begin{pmatrix} 1 & 0 \\ 0 & -1 \end{pmatrix}. \tag{1.3}$$

They are known as the spin matrices and denoted by σ_1, σ_2 and σ_3, respectively.

For any two matrices A and B in the same space, the commutator is a matrix defined as

$$[A, B] = AB - BA \tag{1.4}$$

while the anticommutator is given by

$$\{A, B\} = AB + BA. \tag{1.5}$$

The Pauli matrices satisfy

$$\{\sigma_i, \sigma_j\} = 2\delta_{ij}\mathbb{1}, \tag{1.6}$$

$$[\sigma_i, \sigma_j] = 2i \sum_k \varepsilon_{ijk}\sigma_k, \tag{1.7}$$

where ε_{ijk} are the Levi-Civita symbols defined by

$$\varepsilon_{ijk} = \begin{cases} 1, & (i, j, k) = (1, 2, 3), (2, 3, 1), (3, 1, 2); \\ -1, & (i, j, k) = (2, 1, 3), (1, 3, 2), (3, 2, 1); \\ 0, & \text{Otherwise.} \end{cases} \tag{1.8}$$

Theorem 1.1.1 Strang (2023) *Let A be a complex $m \times n$ matrix. It can be decomposed into*

$$A = U\Sigma V^\dagger, \tag{1.9}$$

where $U \in SU(m)$, $V \in SU(n)$ and Σ is an $m \times n$ matrix whose diagonals are nonnegative real numbers, called the singular values, while all the off diagonal components are zero.

The matrix Σ is known as the singular value matrix. The decomposition given in Equation (1.9) is called the singular value decomposition (SVD). A special case of SVD is the spectral decomposition.

A matrix X is said to be normal if it commutes with its adjoint, which is expressed as $[X, X^\dagger] = 0$, or equivalently, $XX^\dagger = X^\dagger X$.

Theorem 1.1.2 Strang (2023) *Let A be a normal matrix. The spectral decomposition of A is given by*

$$A = \sum_i \lambda_i |\lambda_i\rangle\langle\lambda_i|, \tag{1.10}$$

where $\{\lambda_i\}$ are eigenvalues and $\{|\lambda_i\rangle\}$ are corresponding unit eigenvectors which are orthogonal.

For a normal matrix A the spectral decomposition implies that

$$A^n = \sum_i \lambda_i^n |\lambda_i\rangle\langle\lambda_i| \tag{1.11}$$

for any $n \in \mathbb{N}$. This equality may be extended to any integer $n \in \mathbb{Z}$ if A^{-1} exists.

A special case is for Hermitian matrix, that is, conjugate symmetric matrix with $H = H^\dagger$. This kind of matrices has all real eigenvalues.

The tensor product (Kronecker product) of the $m \times n$ matrix A and any finite matrix B is defined by

$$A \otimes B = \begin{pmatrix} a_{11}B, a_{12}B, \cdots, a_{1n}B \\ a_{21}B, a_{22}B, \cdots, a_{2n}B \\ \cdots \\ a_{m1}B, a_{m2}B, \cdots, a_{mn}B \end{pmatrix}. \tag{1.12}$$

The tensor product satisfies the following equalities

$$(A \otimes B)(C \otimes D) = (AC) \otimes (BD), \tag{1.13}$$
$$A \otimes (B + C) = A \otimes B + A \otimes C, \tag{1.14}$$
$$(A \otimes B)^\dagger = A^\dagger \otimes B^\dagger, \tag{1.15}$$
$$(A \otimes B)^{-1} = A^{-1} \otimes B^{-1}, \tag{1.16}$$
$$\mathrm{Tr}(A \otimes B) = (\mathrm{Tr}A)(\mathrm{Tr}B), \tag{1.17}$$

where $\mathrm{Tr}(\cdot)$ denotes the trace operation, that is, the summation of diagonal components.

Let $\{|e_k\rangle, \forall k\}$ be an orthogonal basis on a Hilbert space $\mathbb{H}$. The matrix $P_k = |e_k\rangle\langle e_k|$ is the projection operator associated with the direction $|e_k\rangle$.

The set of projection operators $\{P_k, \forall k\}$ satisfies the following conditions:

(i) $P_k^2 = P_k$ for any k;
(ii) $P_k P_j = 0$ for all k and j with $k \neq j$;
(iii) $\sum_k P_k = \mathbb{1}$ if $\{P_k, \forall k\}$ is complete.

1.2 Quantum State

1.2.1 Quantum Hypothesis

Quantum mechanics was discovered roughly a century ago. There are different interpretations of the basic ingredient of wave functions Dirac (1981), Peres (2006), Nakahara and Ohmi (2008), Nielsen and Chuang (2000). Here, we explore the Copenhagen interpretation with the following hypotheses.

A1. *Superposition principle.* A pure state in quantum mechanics is represented by a normalized vector $|\psi\rangle$ in Hilbert space $\mathbb{H}$. The linear superposition $c_1|\psi_1\rangle + c_2|\psi_2\rangle$ of any two states $|\psi_1\rangle$ and $|\psi_2\rangle$ is a possible state of the same system for any $c_1, c_2 \in \mathbb{C}$.

A2. *Observable.* For any physical quantity λ, there exists a corresponding Hermitian operator A_λ acting on Hilbert space $\mathbb{H}$. The outcome is one of the eigenvalues of the operator A_λ when measuring λ. Let λ_1 and λ_2 be two eigenvalues of A. If we measure λ in a superposition state $c_1|\lambda_1\rangle + c_2|\lambda_2\rangle$, the wave function collapses as: $c_1|\lambda_1\rangle + c_2|\lambda_2\rangle \to |\lambda_1\rangle(|\lambda_2\rangle)$, and the state after the measurement is $|\lambda_1\rangle(|\lambda_2\rangle)$. The probability of collapsing to the state $|\lambda_i\rangle$ is given by $|c_i|^2 (i = 1, 2)$ under when one repeats the measurements under the independent and identically distributed assumptions of states.

A3. *Evolution.* The time dependence of a state is governed by the *Schrödinger equation* as

$$i\hbar \frac{\partial |\psi\rangle}{\partial t} = H|\psi\rangle, \tag{1.18}$$

where $\hbar$ is the *Planck constant* and H is a Hermitian matrix corresponding to the energy of the system and is called the *Hamiltonian*.

Let A and B be Hermitian matrices and $|\psi\rangle$ be some quantum state on a Hilbert space. It is easy to evaluate that

$$|\langle\psi|[A, B]|\psi\rangle|^2 + |\langle\psi|\{A, B\}|\psi\rangle|^2 = 4|\langle\psi|AB|\psi\rangle|^2. \tag{1.19}$$

The uncertainty principle in terms of standard deviation has been formulated first in Kennard (1927), Robertson (1929) and then improved for different motivations Coles et al. (2017).

1.2.2 Quantum States

In classical contexts, a set of states can be represented by Σ, which is a collection of possible outcomes or events. A probability distribution over the elements of Σ can be described using a probability vector $v \in \mathbb{R}^{\Sigma}$, where $v(x)$ represents the probability of observing the classical state $x \in \Sigma$.

In a quantum setting, we describe a state by a *density matrix* . A density matrix ρ is a matrix that satisfies the following properties:

- It is a *positive semidefinite* Hermitian matrix, that is, all the eigenvalues are nonnegative.
- It is normalized, that is, the trace of ρ equals to 1.

Consider an ensemble of quantum pure states given by $\{|\psi_i\rangle, p_i\}$, where p_i represents the probability of the system is in the pure state $|\psi_i\rangle$. Such a system is named as mixed state. The density matrix of the mixed state is given by:

$$\rho = \sum_i p_i |\psi_i\rangle\langle\psi_i|, \tag{1.20}$$

where $\mathcal{D}(\mathbb{H})$ denotes the set of all density matrices on Hilbert space $\mathbb{H}$ associated with the system. A state ρ is pure if and only if $\rho^2 = \rho$.

The set $\mathcal{D}(\mathbb{H})$ is a convex set since $t\rho_1 + (1-t)\rho_2 \in \mathcal{D}(\mathbb{H})$ for $\rho_1, \rho_2 \in \mathcal{D}(\mathbb{H})$, where $t \in [0, 1]$.

A special case of a mixed state is the uniform mixture of orthogonal basis states $\{|e_i\rangle\}$, given by:

$$\rho = \frac{1}{N} \sum_i |e_i\rangle\langle e_i| = \frac{1}{N} \mathbb{1}, \tag{1.21}$$

which is called the maximally mixed state. Here, N represents the dimension of the state space.

To quantify the distance between two quantum states $\rho_1, \rho_2 \in \mathcal{D}(\mathbb{H})$, a measure such as fidelity can be used. The fidelity, denoted by $F(\rho_1, \rho_2)$, is defined by Jozsa (1994):

$$F(\rho_1, \rho_2) = \mathrm{Tr}(\sqrt{\sqrt{\rho_1}\rho_2\sqrt{\rho_1}}), \tag{1.22}$$

where ρ_1 and ρ_2 are Hermitian matrices on Hilbert space $\mathbb{H}$.

Any state ρ can be represented in terms of its spectral decomposition as $\rho = \sum_i p_i |p_i\rangle\langle p_i|$. This implies that $\sqrt{\rho} = \sum_i \sqrt{p_i} |p_i\rangle\langle p_i|$. Hence, we have $F(\rho_1, \rho_2) = 1$ if and only if $\rho_1 = \rho_2$. The fidelity function is nonnegative and satisfies $F(\rho_1, \rho_2) < 1$ whenever $\rho_1 \neq \rho_2$. The fidelity satisfies the basic properties of a distance measure, namely non-negativity, triangle inequality, and symmetry.

1.2.3 Quantum Operations

In quantum mechanics, quantum operations are known as quantum channels. They describe discrete-time changes in states that can be physically implemented. These operations map density matrices to density matrices, unlike classical operations that map probability vectors to probability vectors.

A quantum operation that transforms a state ρ into ϱ is described by a linear mapping of the form $\mathcal{E} : \mathcal{D}(\mathbb{H}) \to \mathcal{D}(\mathbb{H})$. For a linear mapping $\mathcal{E}$ to be admitted for density matrices, two conditions must be satisfied:

- The mapping $\mathcal{E}$ is positive: for every positive semidefinite matrix $A \in \mathbb{C}^{\Sigma \times \Sigma}$, the matrix $\mathcal{E}(A)$ must also be positive semidefinite.
- The mapping $\mathcal{E}$ is trace-preserving, meaning that $\mathrm{Tr}(\mathcal{E}(A)) = \mathrm{Tr}(A)$ for all states $\rho \in \mathcal{D}(\mathbb{H})$.

Given a state ρ in Hilbert space $\mathbb{H}$, a unitary mapping is described by a unitary operator $U \in SU(\mathbb{H})$. The final state after applying the unitary mapping is given by $U\rho U^\dagger$. This mapping is also called a unitary quantum channel.

Definition 1.2.1 A general quantum channel is the analog of a trace-preserving mapping for mixed states. It is a linear map $\mathcal{E} : \mathcal{D}(\mathbb{H}) \to \mathcal{D}(\mathbb{H})$ that satisfies the conditions of being positive and trace-preserving. A quantum channel can be represented using a Kraus operator sum representation, where the output state ϱ is obtained by applying a set of operators K_j to the input state ρ:

$$\varrho = \sum_{j=1}^{m} K_j \rho K_j^\dagger, \tag{1.23}$$

where K_j satisfy $\sum_{j=1}^{m} K_j^\dagger K_j = \mathbb{1}$ with a finite integer m.

Kraus operators are not unique for a quantum channel. There is no unitary channel between two Hilbert spaces with different dimensions, but a completely positive trace-preserving (CPTP) channel, which is called Stinespring dilation.

Independent quantum operations are described by tensor products. For instance, if $\mathcal{E}_1$ transforms ρ_1 into ϱ_1 on Hilbert space $\mathbb{H}_1$, $\mathcal{E}_2$ transforms ρ_2 into ϱ_2 on Hilbert space $\mathbb{H}_2$, and the two operations are performed independently, then the transformation from the pair (ρ_1, ρ_2) to the pair (ϱ_1, ϱ_2) on Hilbert space $\mathbb{H}_1 \otimes \mathbb{H}_2$ is given by the mapping $\mathcal{E}_1 \otimes \mathcal{E}_2$. This linear mapping is uniquely defined as:

$$\mathcal{E}_1 \otimes \mathcal{E}_2 : \mathcal{D}(\mathbb{H}_1 \otimes \mathbb{H}_2) \to \mathcal{D}(\mathbb{H}_1 \otimes \mathbb{H}_2), \tag{1.24}$$

and it satisfies

$$(\mathcal{E}_1 \otimes \mathcal{E}_2)(\rho_1 \otimes \rho_2) = \mathcal{E}_1(\rho_1) \otimes \mathcal{E}_2(\rho_2) \tag{1.25}$$

for any density matrices $\rho_1 \in \mathcal{D}(\mathbb{H}_1)$ and $\rho_2 \in \mathcal{D}(\mathbb{H}_2)$. This holds for three or more operations.

A well-known result Nielsen and Chuang (2000) provides a way to check whether a given mapping $\mathcal{E}$ is completely positive. Suppose that $\Sigma = \{x_1, \cdots, x_n\}$, we consider the block matrix

$$\begin{pmatrix} \mathcal{E}(|x_1\rangle\langle x_1|) & \cdots & \mathcal{E}(|x_1\rangle\langle x_n|) \\ \vdots & \ddots & \vdots \\ \mathcal{E}(|x_n\rangle\langle x_1|) & \cdots & \mathcal{E}(|x_n\rangle\langle x_n|) \end{pmatrix} \tag{1.26}$$

which is called the *Choi matrix* of $\mathcal{E}$. In the Dirac notation, $|a\rangle\langle b|$ is the matrix whose (a, b) entry is 1 and whose other entries are 0. $\mathcal{E}$ is completely positive if and only if its Choi matrix is positive semidefinite.

Since the quantum transformation is linear, combined with quantum state superposition it follows a general kind of quantum no-go theorems.

Theorem 1.2.1 Wootters and Zurek (1982) *(No-cloning Theorem) There is no quantum transformation to clone any unknown state on Hilbert space, that is, for any $\mathcal{E}$ and any state ρ_A on $\mathbb{H}_A$ we obtain $\mathcal{E}(\rho_A \otimes \varrho_B) \neq \rho_A \otimes \rho_B$, where ϱ_B is an auxiliary state.*

Classical information of bits can be copied faithfully. However, the quantum No-cloning Theorem forbids the same task in quantum scenarios. This provides a fundamental security guarantee for quantum cryptography, such as quantum key distribution.

In contrast to classical network communication, a quantum router cannot copy unknown incoming states and send them to different paths. This implies another no-go theorem in quantum communication, known as the Quantum No-broadcasting Theorem Barnum et al. (1996). This states that no quantum channel exists to broadcast an unknown quantum state to multiple parties, unlike classical information which can be broadcasted.

Theorem 1.2.2 (No-broadcast Theorem) *Given any initial state ρ_1, there is no quantum transformation $\mathcal{E}$ which creates a state ρ_{AB} in a Hilbert space $\mathbb{H}_A \otimes \mathbb{H}_B$ such that* $\mathrm{Tr}_A \rho_{AB} = \mathrm{Tr}_B \rho_{AB} = \rho_1$.

Another no-go theorem is from the superposition of two unknown states. Although the superposition principle is the landmark feature of quantum mechanics, there is no universal quantum transformation to superpose any two unknown quantum states which is known as the Quantum No-superposing Theorem Oszmaniec et al. (2016).

Theorem 1.2.3 (No-Superposing Theorem) *Given any two unknown states ρ_1 and ρ_2 on a Hilbert space $\mathbb{H}$, there is no quantum transformation $\mathcal{E}$ which creates a state $\varrho = \alpha\rho_1 + \beta\rho_2$ with two known parameters α and β.*

Based on these no-go theorems, quantum network communication is different from classical network communication which can make use of faithful copy, broadcast, and generalized encoding for routers. These no-go theorems may be further extended to Quantum No-transformation Theorem Luo et al. (2017), Li et al. (2018), further reading see review Fan et al. (2014).

A nonzero mapping $\mathcal{G} \in \mathcal{M}(\mathbb{H})$ is degenerate if there exists a pure state $|\phi\rangle \in \mathbb{H}$ such that $\mathcal{G}(|\psi\rangle) \propto |\phi\rangle$ for all pure states $|\psi\rangle$ in $\mathbb{H}$. Otherwise, $\mathcal{G}$ is non-degenerate.

Let $\mathcal{G}$ be a non-degenerate mapping in $\mathcal{M}(\mathbb{H})$. From the linearity and superposition of pure states, there does not exist a CP mapping $\mathcal{F} \in \mathcal{CP}(\mathbb{H}, \mathbb{H}^{\otimes 2})$ such that for all pure states $|\psi\rangle$ in complex or real Hilbert $\mathbb{H}$

$$\mathcal{F}(|\psi\rangle) \propto |\psi\rangle|\varphi\rangle, \tag{1.27}$$

where $|\varphi\rangle = r\mathcal{G}(|\psi\rangle)$ with a normalization constant r.

Theorem 1.2.4 *Let α_j and β_j be complex constants satisfying $|\alpha_j|^2 + |\beta_j|^2 = 1$, and $\mathcal{G}_j$ be a mapping on complex or real Hilbert space $\mathbb{H}$ for $j = 1, 2, \cdots, m$. There does not exist a CP mapping $\mathcal{F} \in \mathcal{CP}(\mathbb{H}, \mathbb{H}^{\otimes m})$ such that for all pure states $|\psi\rangle \in \mathbb{H}$*

$$\mathcal{F}(|\psi\rangle) \propto \otimes_{j=1}^{m}|\varphi_j\rangle \tag{1.28}$$

if $\mathcal{G}_j$ satisfy one of the following conditions:

(i) there exist two mappings $\mathcal{G}_i$ and $\mathcal{G}_j$ such that they are not affine mappings in $AF(\mathbb{H})$ or non-identity mapping;
(ii) there exists a nonlinear mapping $\mathcal{G}_j$ when $\alpha_j \neq 0$ and $\beta_j \neq 0$;
(iii) there exists a mapping $\mathcal{G}_j \notin \mathcal{CP}(\mathbb{H})$ when $\alpha_i = 0$ for $i = 1, 2, \cdots, m$;

where the vector representative of $|\varphi_j\rangle$ is given by $|\varphi_j\rangle \propto \alpha_j|\psi\rangle + \beta_j\mathcal{G}_j(|\psi\rangle)$ for $j = 1, 2, \cdots, m$.

Proof It is sufficient to prove the case (i). Assume that there is a CP mapping $\mathcal{F}$ that maps any state $|\psi\rangle$ to a product state proportional to $|\varphi_1\rangle|\varphi_2\rangle$, where $|\varphi_j\rangle$ is proportional to $r_j(\alpha_j|\psi\rangle + \beta_j\mathcal{G}_j(|\psi\rangle))$, and $\mathcal{G}_j$ is a non-degenerate mapping. Note that one of the mappings $\mathcal{G}_j$ is proportional to the identity mapping, i.e., $\mathcal{G}_j \propto \mathbb{1}$. So, the assumption is impossible.

The following proof is sufficient to consider the qubit space $\mathbb{H} := \mathbb{R}^2$ (or $\mathbb{C}^2$) by using the Stinespring Theorem. There are an ancillary state $|\Omega\rangle$ on Hilbert space $\mathbb{H}_A$, two orthogonal states $|P_0\rangle$ and $|P_1\rangle$ in the Hilbert space $\mathbb{H}_P \subset \mathbb{H}_A$, and a unitary mapping $\mathcal{U} \in SU(\mathbb{H} \otimes \mathbb{H}_A)$ such that $\mathcal{U}(|\psi\rangle|\Omega\rangle)$ is proportional to the state $p(a, b)|\varphi_1\rangle|\varphi_2\rangle|P_0\rangle + q(a, b)|\Psi\rangle|P_1\rangle$. Here, $|\psi\rangle$ is proportional to the state $a|0\rangle + b|1\rangle$ with $a^2 + b^2 = 1$, $|\Psi\rangle$ is the vector representative of the failure state in the Hilbert space $\mathbb{H}^{\otimes 2}$, and $p(a,b)$ and $q(a,b)$ are positive probability functions satisfying $p(a,b)^2 + q(a,b)^2 = 1$ for all a and b. Specifically, for classical states $|0\rangle$ and $|1\rangle$, we obtain that

$$\mathcal{U}(|0\rangle|\Omega\rangle) \propto p_0|\varphi_{10}\rangle|\varphi_{20}\rangle|P_0\rangle + q_0|\Psi_0\rangle|P_1\rangle, \tag{1.29}$$

$$\mathcal{U}(|1\rangle|\Omega\rangle) \propto p_1|\varphi_{11}\rangle|\varphi_{21}\rangle|P_0\rangle + q_1|\Psi_1\rangle|P_1\rangle, \tag{1.30}$$

where $|\varphi_{ij}\rangle$ is proportional to the state $r_{ij}(\alpha_j|j\rangle + \beta_j\mathcal{G}_i(|j\rangle))$ with the normalization constant r_{ij} for $i = 1, 2$ and $j = 0, 1$. $|\Psi_0\rangle$ is the vector representative of the failure state to transform the pure state $|0\rangle$ with probability $q_0 > 0$, and $p_0 > 0$ is the corresponding success probability. Similarly, $|\Psi_1\rangle$ is the vector representative of the failure state to transform the pure state $|1\rangle$ with probability $q_1 > 0$, and $p_1 > 0$ is the corresponding success probability.

Using the linearity of unitary mapping, we obtain that $\mathcal{U}(|\psi\rangle|\Omega\rangle)$ is proportional to the state $|\Phi_0\rangle|P_0\rangle + |\Phi_1\rangle|P_1\rangle$, where $|\Phi_0\rangle$ and $|\Phi_2\rangle$ are respectively given by $|\Phi_0\rangle = ap_0|\varphi_{10}\rangle|\varphi_{20}\rangle + bp_1|\varphi_{11}\rangle|\varphi_{21}\rangle$ and $|\Phi_1\rangle = ap_0|\Psi_0\rangle + bp_1|\Psi_1\rangle$. We find that $ap_0|\varphi_{10}\rangle|\varphi_{20}\rangle + bp_1|\varphi_{11}\rangle|\varphi_{21}\rangle \propto p(a,b)|\varphi_1\rangle|\varphi_2\rangle$. As the right side is a product state for all a and b, it follows that $|\varphi_{10}\rangle \propto |\varphi_{11}\rangle$ or $|\varphi_{20}\rangle \propto |\varphi_{21}\rangle$. Otherwise, the left side is an entangled state. Assume that $|\varphi_{10}\rangle \propto |\varphi_{11}\rangle$, it implies $\mathcal{G}_1(|j\rangle) \propto |j\rangle$ for $j = 0$ and 1. We can prove that $\mathcal{G}_1 \propto \mathbb{1} + c$ with a constant c, i.e., $\mathcal{G}_1 \in AF(\mathbb{H})$, and $\mathcal{G}_2 \propto \mathbb{1}$. This contradicts to the assumption of $\mathcal{G}_1$. □

If all mappings $\mathcal{G}_j$ are identity mappings, i.e., $\mathcal{G}_j \propto \mathbb{1}$ for $j = 1, 2, \cdots, m$, Theorem 1.2.4 reduces to the Quantum No-cloning Theorem. If $\alpha_j = 0$ and $\mathcal{G}_j$ is the orthogonal mapping for all $j = 1, 2, \cdots, m$, Theorem 1.2.4 implies the Quantum No-anticloning Theorem Pati (2002). To prove the Quantum No-splitting Theorem Pati and Sanders (2006), we assume that $|\phi_1\rangle$ and $|\phi_2\rangle$ are the split states of an unknown state $|\psi\rangle$, and use the fact that $r_j(\alpha_j|\psi\rangle + \beta_j\mathcal{G}_j(|\psi\rangle)) \propto |\phi_j\rangle$ for $j = 1, 2$. Similarly, Theorem 1.2.4 is useful to prove the No-partial Erasure Theorem Pati and Sanders (2006) or the No-encoding Theorem Luo et al. (2017), which includes the No-superposing Theorem Oszmaniec et al. (2016) as a special case.

1.2.4 Multipartite Systems

Suppose that a composed system consists of two systems, where one is on Hilbert space $\mathbb{H}_1$ and the other is on Hilbert space $\mathbb{H}_2$. The system as a whole is on Hilbert space $\mathbb{H} = \mathbb{H}_1 \otimes \mathbb{H}_2$. For a pure state, the general vector is written as

$$|\psi\rangle = \sum_{i,j} c_{ij}|e_{1,i}\rangle \otimes |e_{2,j}\rangle, \tag{1.31}$$

where $\{|e_{k,i}\rangle\}(k = 1, 2)$ is an orthogonal basis of $\mathbb{H}_k$ and c_{ij} satisfy $\Sigma_{i,j}|c_{ij}|^2 = 1$.

A state $|\psi\rangle \in \mathbb{H}$ is a *separable state* if it can be written as a tensor product of two vectors as $|\psi\rangle = |\psi_1\rangle \otimes |\psi_2\rangle$, $(|\psi_k\rangle \in \mathbb{H}_k)$. Otherwise, the state is *entangled*. One example is the Einstein-Podolsky-Rosen (EPR) state Einstein et al. (1935) given by

$$|\phi\rangle_{AB} = \frac{1}{\sqrt{2}}(|0\rangle \otimes |1\rangle + |1\rangle \otimes |0\rangle) \tag{1.32}$$

on Hilbert space $\mathbb{H}_A \otimes \mathbb{H}_B$. It is easy to prove $|\phi\rangle \neq |\psi_1\rangle|\psi_2\rangle$ for any two pure states $|\psi_1\rangle \in \mathbb{H}_A$ and $|\psi_2\rangle \in \mathbb{H}_B$.

Quantum measurement provides an alternative way to characterize pure entangled systems. A state is not entangled if a measurement of one qubit does not affect the state of the other qubits. However, if the qubit A of the state $|\phi\rangle_{AB}$ is measured to be 0 or 1, then the outcome of the measurement of the qubit B is 1 or 0, respectively. Therefore, local measurement on one qubit affects the outcome of the measurement on the other qubit, indicating the presence of strong correlations between the two qubits. These correlations are essential ingredients in quantum information processing, which will be further extended in this book.

Consider a system of two qubits. A simple orthogonal basis for the combined system is $\{|00\rangle, |01\rangle, |10\rangle, |11\rangle\}$. Another is the Bell basis given by

$$|\phi^{\pm}\rangle = \frac{1}{\sqrt{2}}(|00\rangle \pm |11\rangle), |\psi^{\pm}\rangle = \frac{1}{\sqrt{2}}(|01\rangle \pm |10\rangle). \tag{1.33}$$

All the Bell states are entangled.

From the fact of $\dim(\mathbb{H}_1)\dim(\mathbb{H}_2) > \dim(\mathbb{H}_1) + \dim(\mathbb{H}_2)$ for $\dim(\mathbb{H}_1) > 2$ or $\dim(\mathbb{H}_2) > 2$, most bipartite systems are entangled when the associated Hilbert spaces have higher dimensions Horodecki et al. (2009).

One straightforward method to determine whether a given pure state is entangled or not is to examine its reduced density matrix. Specifically, for a pure state $|\phi\rangle_{AB}$, the reduced density matrix ρ_A is obtained by taking the partial trace over the subsystem B as $\rho_A = \mathrm{Tr}_B(|\phi\rangle\langle\phi|)$, and the state is entangled if and only if ρ_A is a mixed state.

Definition 1.2.2 The entanglement may be extended to mixed systems. In this case, the separable state is defined as any probabilistic mixing of pure separable states, i.e., a separable state can be decomposed into

$$\rho = \sum_i p_i \varrho_A^{(i)} \otimes \varrho_B^{(i)}, \tag{1.34}$$

where $\{p_i\}$ is a probability distribution, $\varrho_A^{(i)}$ and $\varrho_B^{(i)}$ are states of respective system A and B. One combined system is entangled if it cannot be decomposed into probabilistic mixing of pure separable states.

The local mixture is not useful for verifying the entanglement of mixed combined systems. One possible way is using the negativity. Let ρ be a bipartite state and define the *partial transpose* ρ^{pt} of ρ with respect to the second Hilbert space as

$$\rho_{ij,kl} \mapsto \rho_{il,kj}, \tag{1.35}$$

where $\rho_{ij,kl}$ are defined by

$$\rho_{ij,kl} = (\langle e_{1,i}| \otimes \langle e_{2,j}|)\rho((\langle e_{1,k}| \otimes \langle e_{2,l}|). \tag{1.36}$$

Here, $\{|e_{1,k}\rangle\}$ is the basis of the first system and $\{|e_{2,k}\rangle\}$ is the basis of the second system. Suppose one state takes the form (1.34). The partial transpose yields to

$$\rho^{pt} = \sum_i p_i \rho_{1,i} \otimes \rho_{2,i}^T, \tag{1.37}$$

where ρ^T for any density matrix ρ is a density matrix since it is still a positive semi-definite Hermitian with a unit trace. It was conjectured by Peres and subsequently proved by the Hordecki family Horodecki et al. (2009) that the positivity of the partially transposed density matrix is a necessary and sufficient condition for ρ to be separable in the cases of $\mathbb{C}^2 \otimes \mathbb{C}^2$ systems and $\mathbb{C}^2 \otimes \mathbb{C}^3$ systems. Conversely, ρ is inseparable if the partial transpose of ρ of these systems is not a density matrix.

Entangled states are characterized by using the *negativity* as

$$N(\rho) \equiv \frac{\Sigma_i |\lambda_i| - 1}{2}. \tag{1.38}$$

Note that the negativity vanishes if and only if all the eigenvalues of ρ^{pt} are nonnegative. Negativity is one of the so-called entanglement monotones Horodecki et al. (2009), which include other quantities such as the concurrence, entanglement of formation, and entropy of entanglement Horodecki et al. (2009).

Definition 1.2.3 Let $\mathbb{H} = \mathbb{H}_1 \otimes \mathbb{H}_2$ be the Hilbert space of a bipartite system. Then a pure state $|\Psi\rangle \in \mathbb{H}$ admits the following *Schmidt decomposition* as

$$|\Psi\rangle = \sum_{i=1}^{n} \sqrt{s_i}|\phi_{1,i}\rangle \otimes |\phi_{2,i}\rangle, \tag{1.39}$$

where s_i are positive and called the *Schmidt coefficients* satisfying $\sum_i s_i = 1$. $\{|\phi_{j,i}\rangle\}$ is an orthonormal state set of $\mathbb{H}_j$. $N \in \mathbb{N}$ is called the *Schmidt number* of $|\Psi\rangle$.

A similar definition holds for multipartite entangled systems Horodecki et al. (2009). An n-qubit state $|\Phi\rangle$ on Hilbert space $\otimes_{i=1}^n \mathbb{H}_{a_i}$ is entangled if it does not satisfy $|\Phi\rangle = |\psi_1\rangle_A |\psi_2\rangle_B$, where $|\psi_1\rangle_A$ and $|\psi_2\rangle_B$ are pure states on systems in respectively A and B, where A and B is the bipartition of $\{a_1, \cdots, a_n\}$. For mixed states, there are different multipartite entanglement for verifying strong correlations Horodecki et al. (2009), Svetlichny (1987), Mermin (1990). One is to characterize the genuine multipartite entanglement beyond any biseparable states.

Definition 1.2.4 An n-partite state is genuinely n-partite entangled if it cannot be written as a convex combination of biseparable pure states, that is, $\rho = \sum_i p_i |\phi_i\rangle\langle\phi_i|$, where $\{|\phi_i\rangle\}$ are biseparable with respect to different bipartitions.

The genuine multipartite entanglement shows a strong correlation among all the involved subsystems. This will be discussed in the following chapter.

Among three-qubit entangled pure states, the following two states are important for quantum applications. One is Greenberger-Horne-Zeilinger (GHZ) Greenberger and Horne (1989) state defined by

$$|GHZ\rangle_{ABC} = \frac{1}{\sqrt{2}}(|000\rangle + |111\rangle). \tag{1.40}$$

The other is the W state Dür et al. (2000) given by

$$|W\rangle = \frac{1}{\sqrt{3}}(|100\rangle + |010\rangle + |001\rangle). \tag{1.41}$$

All the three-qubit entangled pure states can be transformed into one of two states by stochastic local operations and classical communications (SLOCC) Horodecki et al. (2009). Here, local operations and classical communications (LOCC) may be represented by using local trace-preserving quantum mapping as follows.

Definition 1.2.5 A LOCC quantum channel is a CPTP $\mathcal{E} : \mathcal{D}(\otimes_a \mathbb{H}_a) \to \mathcal{D}(\otimes_b \mathbb{H}_b)$ which can be defined by using the superoperators $\mathcal{E}(\rho) = \sum_i (\otimes_a A_{i,a}) \rho (\otimes_a A_{i,a}^\dagger)$, where $A_{i,a}$ are local Kraus or Kraus-Choi operators satisfying $\sum_i A_{i,a}^\dagger A_{i,a} = \mathbb{1}_a$ with the identity operator $\mathbb{1}_a$ on $\mathbb{H}_a$.

Under local unitary operations, there is a general normal form Dür et al. (2000), Acín et al. (2000) defined as

$$|\Psi\rangle = \lambda_0|000\rangle + \lambda_1 e^{i\varphi}|100\rangle + \lambda_2|101\rangle + \lambda_3|110\rangle + \lambda_4|111\rangle \tag{1.42}$$

where $\lambda_i \geq 0$, $0 \leq \varphi \leq \pi$, and $\sum_{i=0}^4 \lambda_i^2 = 1$.

More generally, for an n-qubit system, one orthogonal basis is defined by $\{|i_0 i_1 \cdots i_{n-1}\rangle\}$, where $i_j \in \{0, 1\}$. One may express the basis in terms of the decimal system as $|x\rangle$, where x is the decimal expression of the binary series $i_0 i_1 \cdots i_{n-1}$. The basis of the two-qubit system is then written into $\{|0\rangle, \cdots, |3\rangle\}$.

1.3 Quantum Information

Information is one of the main motivations for this book. We give a heuristic definition that will be used in our discussions.

1.3.1 Classical Information

Information as an ability to distinguish among possible alternatives has been widely used Shannon (1948), Cover and Thomas (2012). In this book, we distinguish between an abstract message and the physical signal that represents the message. Another important point is that information can be transformed from one physical signal to another so that the same message can be encoded by different physical signals.

In classical settings, an unknown message is mathematically described by a probability distribution $\{p(x), x \in \mathcal{X}\}$. Here, X as a random variable denotes a possible message in $\mathcal{X}$. We generally consider discrete random variables which can take on finite numbers or infinite sequences of values. $p(x)$ denotes probability function which returns the probability of message x and satisfies $p(x) \geq 0$ and $\sum_x p(x) = 1$.

Definition 1.3.1 Shannon (1948) Define the entropy H of an unknown message to be

$$H = \log M, \tag{1.43}$$

where M is the number of possible messages.

The log function denotes the logarithms with base 2. The entropy notion for discrete variables is first proposed by Shannon in his pioneering work Shannon (1948). If $n < H$, n bits are not enough to represent all the messages faithfully. On the other hand, if $n \geq H$, one can distinguish all the messages. The entropy measures the number of bits that is to distinguish or represent the messages.

Another feature related to the entropy is the *uncertainty* of the unknown message, that is, of how much one does not know before getting the message Cover and Thomas (2012). This is a measure of how the uncertainty is reduced when one identifies the right message.

Suppose that X and Y are two unknown messages, having n_X and n_Y possible values respectively. If X and Y are independent of each other, every combination of X and Y values is a possible joint message of XY, and so $n_{XY} = n_X n_Y$. In this case, the joint entropy is given by

$$H(XY) = \log n_{XY} = \log n_X + \log n_Y = H(X) + H(Y), \tag{1.44}$$

which is additive. On the other hand, if the messages are not independent, the joint entropy $H(XY)$ may be less than the summation of two independent entropies, i.e., $H(XY) \leq H(X) + H(Y)$.

The entropy (1.43) or (1.44) denotes the maximal uncertainty of a given random variable with the same number of messages, i.e., the uniform distribution. In general, we may define the entropy of messages with probability distribution $\{p(x)\}$ as

$$H(X) = -\sum_x p(x) \log p(x). \tag{1.45}$$

This quantity may be naturally extended to joint entropy $H(XY)$ by using joint distribution $\{P(x, y)\}$.

The second is a conditional entropy $H(X|Y)$, i.e., the uncertainty of X under the condition of known messages Y. This quantity may be defined by using conditional distribution $\{P(x|y)\}$, or an operational way as

$$H(X|Y) = H(XY) - H(Y). \tag{1.46}$$

Another useful quantity is the common entropy contained in two random variables X and Y, i.e., the mutual information $I(X; Y)$ of X and Y. This information is specially important in the Shannon communication theory Shannon (1948) for a general communication system $(\mathcal{X}, P(y|x), \mathcal{Y})$. Here, X denotes the random variable describing the input messages chosen by the sender from $\mathcal{X}$. Y denotes the random variable describing the output messages obtained by the receiver after the input messages under the specific classical channel. A general classical channel is mathematically represented by a conditional probability function $P(y|x)$. With these notions, $I(X; Y)$ represents how much information related to the input messages is obtained by the receiver for a given channel. This is further externalized as

$$C = \max_{\{p(x)\}} I(X; Y), \tag{1.47}$$

where the maximization is over all the input messages with different distributions. The quantity C characterizes the maximal transfer ability of a given channel and is named channel capacity. The Shannon communication theory is mainly about how to realize reliable communication with the maximal transition efficiency over noise channels Shannon (1948). We present the basic properties of Shannon entropy for the convenience of other chapters. For the further results, one may read the book Cover and Thomas (2012).

Theorem 1.3.1 *There are some properties of Shannon entropy for the general case.*

1. *The Shannon entropy $H(X)$ is a continuous function of $p(x)$. If all $p(x)$ are equal, then H is maximal.*
2. *The Shannon entropy satisfies the Gibbs inequality:*

$$H(X) \leq \log m, \tag{1.48}$$

   where the equality holds if and only if $p(x) = \frac{1}{m}$ for all x.
3. *The Shannon entropy is monotonically increasing with the number m of samples m.*
4. *The information distance of two probability distributions, $P(X)$ and $Q(X)$, of the same random variable X may be measured by the relative entropy or Kullback-Leibler divergence as*

$$D(P(X)||Q(X)) = \sum_x p(x) \log \frac{p(x)}{q(x)}. \tag{1.49}$$

5. *The relative entropy is nonnegative, i.e., $H(P(X)||Q(X)) \geq 0$, with the equality only for the equal distribution of $P(X) = Q(X)$.*

6. *For two random variables, X and Y, the total information satisfies the strong additivity as*

$$H(XY) = H(X) + H(Y) - I(X; Y). \tag{1.50}$$

7. *All unconstrained three or fewer random variables X, Y and Z satisfy Shannon information inequalities as*

$$I(X; Y|Z) \geq 0. \tag{1.51}$$

8. *For a discrete Markov process $X \rightarrow Y \rightarrow Z$, i.e., a random process for which the future state depends only on the present state, the Shannon information satisfies*

$$I(X; Z) \leq \min\{I(X; Y), I(Y; Z)\}. \tag{1.52}$$

This is known as the Data Compression Theorem.

Different from the Shannon entropy there are other formations of entropy. One is Rényi entropy defined by

$$R_\alpha(X) = \frac{1}{1-\alpha} \ln \left(\sum_{i=1}^{n} p_i^\alpha \right), \tag{1.53}$$

where $\alpha > 0$ and $\alpha \neq 1$. As $\alpha \rightarrow 0$, the Rényi entropy is the logarithm of the support size. As $\alpha \rightarrow 1$, the Rényi entropy converges to Shannon entropy. It is then reduced to the collision entropy $R_2(X)$ for $\alpha = 2$ or Min-entropy $R_\infty(X)$ for $\alpha \rightarrow \infty$. It is easy to show that

$$H(X) > R_2(X) > R_\infty(X). \tag{1.54}$$

Different from the encoding length of Shannon entropy, the collision entropy quantifies the collision probability between samples while the min-entropy quantifies the unpredictability. Both entropies are applicable for explaining cryptographic features such as extractable randomness or key derivation.

Another is Tsallis entropy Tsallis (2009) defined by

$$T_q(X) = \frac{1}{q-1} \left(1 - \sum_{i=1}^{n} p_i^q \right), \tag{1.55}$$

where $q > 0$, and recovers the Boltzmann-Gibbs (BG) expression referred to as nonextensive statistical mechanics. Tsallis entropy is proposed as the generalization nonadditive entropy. As $q \rightarrow 1$, the Tsallis entropy converges to Shannon entropy. Tsallis entropy is not additivity but satisfies

$$T_q(XY) = T_q(X) + T_q(Y) - (1-q)T_q(X)T_q(Y). \tag{1.56}$$

As a unique entropy identified by the principles of thermodynamics, stability, and axiomatic foundations, it is widely used for featuring complex systems, especially systems with long-range correlation.

1.3.2 Von Neumann Information

The Shannon information may be regarded as quantities defined over a given random variable. In quantum settings, Boltzmann, Gibbs, and von Neumann have proposed similar definitions von Neumann (1927) before Shannon information. In fact, consider a general state ρ on Hilbert space $\mathbb{H}$. The spectra decomposition is given by as: $\rho = \sum_{i=1}^{n} \lambda_i |\psi_i\rangle\langle\psi_i|$, where $\lambda_i \geq 0$, $\{|\psi_i\rangle\}$ are orthnormal states. The von Neumann entropy von Neumann (1927) is defined as

$$S(\rho) = -\mathrm{Tr}(\rho \log \rho) = -\sum_{k=1}^{n} \lambda_k \log \lambda_k. \tag{1.57}$$

Here, one may regard the von Neumann entropy as the Shannon information by performing the projection measurement on a given state ρ under the orthogonal basis $\{|\psi_i\rangle\langle\psi_i|\}$. In this case, all the von Neumann entropy of ρ is operationally recovered by the Shannon information by classical statistics.

In what follows, we present some important features of quantum entropy. Most of the features can be proved by using the definition (1.57). For the further results, one may read the book Nielsen and Chuang (2000).

Theorem 1.3.2 *The von Neumann entropy satisfies the following properties.*

1. ***Purity***. *$S(\rho) = 0$ for $\rho \in \mathcal{D}(\mathbb{H})$ if and only if ρ is a pure state.*
2. ***Invariance***. *The entropy is unchanged under the unitary operation, i.e., $S(U\rho U^\dagger) = S(\rho)$ for any $U \in SU(\mathbb{H})$ and $\rho \in \mathcal{D}(\mathbb{H})$.*
3. ***Maximum***. *If $\rho \in \mathcal{D}(\mathbb{H})$ has d non-zero eigenvalues, then $S(\rho) \leq \log d$, where the equality holds if and only if all nonzero eigenvalues are equal (maximum randomness).*
4. ***Concavity***. *The von Neumann entropy is larger if one knows less about how the state was prepared. For $\lambda_i \geq 0$ and $\sum_i \lambda_i = 1$,*

$$S\left(\sum_i \lambda_i \rho_i\right) \geq \sum_i \lambda_i S(\rho_i), \tag{1.58}$$

   where $\rho_i \in \mathcal{D}(\mathbb{H})$.
5. *Suppose that one measures $\rho \in \mathcal{D}(\mathbb{H})$ under the basis $\{|a_y\rangle\langle a_y|\}$ and the outcome a_y occurs with the probability $p(a_y) = \langle a_y|\rho|a_y\rangle$. The Shannon entropy for the*

ensemble of measurements outcomes $Y = \{a_y, p(a_y)\}$ *satisfies*

$$H(Y) \geq S(\rho), \tag{1.59}$$

where the equality holds if A *and* ρ *commute.*

6. ***Subadditivity***. *For a bipartite system* AB *in the state* $\rho_{AB} \in \mathcal{D}(\mathbb{H}_A \otimes \mathbb{H}_B)$, *the joint entropy satisfies*

$$S(\rho_{AB}) \leq S(\rho_A) + S(\rho_B), \tag{1.60}$$

where the equality holds if $\rho_{AB} = \rho_A \otimes \rho_B$.

7. ***Strong subadditivity***. *For any tripartite state* $\rho_{ABC} \in \mathcal{D}(\mathbb{H}_A \otimes \mathbb{H}_B \otimes \mathbb{H}_C)$, *it satisfies*

$$S(\rho_{ABC}) + S(\rho_B) \leq S(\rho_{AB}) + S(\rho_{BC}). \tag{1.61}$$

This reduces to the subadditivity when B *is one-dimensional.*

8. ***Araki-Lieb inequality*** *Araki and Lieb (1970). For a bipartite state* $\rho_{AB} \in \mathcal{D}(\mathbb{H}_A \otimes \mathbb{H}_B)$ *it satisfies*

$$S(\rho_{AB}) \geq |S(\rho_A) - S(\rho_B)|. \tag{1.62}$$

Similar to Shannon mutual information we may define the quantum mutual information for a quantum communication model by the von Neumann entropy.

Definition 1.3.2 For a bipartite quantum system XY in the state ρ on Hilbert space $\mathbb{H}_A \otimes \mathbb{H}_B$, the quantum mutual information between X and Y is defined by

$$I(X;Y)_\rho = S(\rho_X) + S(\rho_Y) - S(\rho_{XY}). \tag{1.63}$$

For the consistency of classical settings, we sometimes make use of the notion as $S(X)_\rho$ for $S(\rho_X)$ in the following discussions. This information has also shown the common information of two systems X and Y. Another useful quantity is conditional mutual information for tripartite systems.

Definition 1.3.3 For a tripartite quantum system in the state ρ on Hilbert space $\mathbb{H}_X \otimes \mathbb{H}_Y \otimes \mathbb{H}_Z$, the quantum conditional mutual information between X and Z conditioned on Y is defined by

$$I(X;Y|Z)_\rho = S(\rho_{XY}) + S(\rho_{XZ}) - S(\rho_{XYZ}) - S(\rho_Y), \tag{1.64}$$

where ρ_{XY}, ρ_{XZ} and ρ_Y are reduced density matrices.

Instead of common information defined by $I(X;Y)_\rho$, one may consider the difference of two density matrices. One way is to extend the relative entropy to quantum scenarios.

Definition 1.3.4 Quantum relative entropy between two density matrices ρ_1 and ρ_2 is defined as

$$S(\rho_1||\rho_2) = \mathrm{Tr}(\rho_1 \log \rho_1 - \rho_1 \log \rho_2). \tag{1.65}$$

A simple fact is $S(\rho_1||\rho_2) = 0$ if and only if $\rho_1 = \rho_2$. The quantum relative entropy shows that different information may have arisen from two quantum states.

Another quantity used in quantum communication theory is quantum coherent information. Consider a simple communication model. Let a pure state go through the channel. One part has remained at the input, and the channel will produce a new state at the output. A natural problem is to describe how much remained information after the communication. This may be resolved by quantum coherent information.

Definition 1.3.5 The quantum coherent information is defined by

$$I(A\rangle B)_\rho = S(\rho_B) - S(\rho_{AB}), \tag{1.66}$$

where $S(\cdot)$ denotes the von Neumann entropy.

Here, ρ_B in applications denotes the output state while ρ_{AB} is the channel environment of the input state.

So far, we have introduced some quantities related to von Neumann entropy. A directive extension can be followed for other quantum entropies. In fact, from Rényi entropy or Tsallis entropy, the quantum Rényi entropy or quantum Tsallis entropy are defined as

$$R_\alpha(\rho) = \frac{1}{1-\alpha}\mathrm{Tr}(\ln \rho^\alpha), \tag{1.67}$$

$$T_q(\rho) = \frac{1}{q-1}\mathrm{Tr}(1-\rho^q). \tag{1.68}$$

Both entropies will converge to von Neumann entropy. Similar to classical settings, these generalizations show different features from von Neumann entropy. For example, the quantum Tsallis entropy with $q > 1$ satisfies the subadditivity as

$$T_q(\rho \otimes \sigma) \leq T_q(\rho) + T_q(\sigma), \tag{1.69}$$

while for $0 < q < 1$ it satisfies

$$T_q(\rho \otimes \sigma) \geq T_q(\rho) + T_q(\sigma). \tag{1.70}$$

1.3.3 Entanglement Measure

Entanglement as an important feature is useful for distinguishing quantum theory from classical physics. For a long time, this was only stated qualitatively as either entangled or not for a given joint system. A natural problem is to determine different degrees of entanglement. This question plays an essential role in quantum information theory. Surprisingly, the answer to this question is nontrivial. Different ways are proposed to characterize the degree of entanglement of a system. Informally, these approaches are generally named entanglement measures . These entanglement measures have required to satisfy a common set of operational properties that can capture the essential entangled features. These postulates for entanglement measures are presented here Plenio and Virmani (2024).

1. A bipartite entanglement measure $E(\rho)$ is a mapping (functional) from density matrices into positive real numbers: $E(\rho) : \rho \mapsto \mathbb{R}^+$. $E(\rho) = 0$ if the state ρ is separable.
2. E does not increase on average under LOCC i.e.

$$E(\rho) \geq \sum_i p_i E\left(\frac{A_i \rho A_i^\dagger}{\mathrm{Tr}(A_i \rho A_i^\dagger)}\right) \tag{1.71}$$

   where A_i are the Kraus operators that describe the LOCC operation. The probability of obtaining outcome i is given by $p_i = \mathrm{Tr}(A_i \rho A_i^\dagger)$.
3. For pure state $|\phi\rangle$ on Hilbert space $\mathbb{H}_A \otimes \mathbb{H}_B$ the measure reduces to the entropy of entanglement, i.e., $E(|\phi\rangle) = S(\rho_A) = S(\rho_B)$ with reduced density matrices ρ_A and ρ_B.
4. E is convex, i.e., $E(\sum_i p_i \rho_i) \leq \sum_i p_i E(\rho_i)$.

If the first two conditions are satisfied it is named Entanglement Monotone. Sometimes, one may require more strong condition of additivity $E(\rho^{\otimes n}) = nE(\rho)$ or even full additivity $E(\rho \otimes \sigma) = E(\rho) + E(\sigma)$. There are lots of entanglement measures and most of them are based on one quantum entropy. One example is through von Neumann entropy, named as the entanglement of formation (EOF) Wootters (1998) . For a pure state $|\phi\rangle_{AB}$ on Hilbert space $\mathbb{H}_A \otimes \mathbb{H}_B$ it is defined as:

$$E(|\phi\rangle_{AB}) = S(\rho_A) = -\mathrm{Tr}(\rho_A \log_2 \rho_A), \tag{1.72}$$

where ρ_A denotes the reduced density matrix of the subsystem A, and $S(\rho_A)$ denotes the von Neumann entropy of the system A. The von Neumann entropy satisfies the additivity. This measure may be extended for a bipartite mixed state ρ_{AB} by using externalization as

$$E(\rho_{AB}) = \inf_{\{p_i, |\phi_i\rangle\}} \sum_i p_i E(|\phi_i\rangle_{AB}), \tag{1.73}$$

where the infimum is over all pure-state decompositions of $\rho_{AB} = \sum_i p_i |\phi_i\rangle_{AB}\langle\phi_i|$ with a probability distribution $\{p_i\}$, and some pure states $|\phi_i\rangle_{AB}$ on $\mathbb{H}_A \otimes \mathbb{H}_B$.

The second entanglement measure is based on the Rényi entropy. The so-called Rényi-α entanglement measure $\mathcal{R}_\alpha$ is defined by using the Rényi-α entropy as Kim and Sanders (2010):

$$\mathcal{R}_\alpha(|\phi\rangle_{AB}) = R_\alpha(\rho_A) = \frac{1}{1-\alpha} \log_2 \mathrm{Tr}\rho_A^q, \tag{1.74}$$

where $R_\alpha(\rho_A)$ denotes the Rényi-α entropy of the system A. The third one is from Tsallis entropy. For a bipartite pure state $|\phi\rangle_{AB}$, the Tsallis entropy entanglement measure is defined as

$$\mathcal{T}^{(q)}(|\phi\rangle_{AB}) = T_q(\rho_A) = \frac{1}{q-1}(1 - \mathrm{Tr}\rho_A^q), \tag{1.75}$$

where $T_q(\rho_A)$ denotes the Tsallis-q entropy of the system A, $q > 0$ and $q \neq 1$.

Instead, another entanglement measure Kim and Sanders (2011) is based on a two-parameter entropy, Unified (q, s)-entropy Rathie and Taneja (1991)

$$U_{q,s}(\rho) = \frac{1}{(1-q)s}((\mathrm{Tr}\rho^q)^s - 1) \tag{1.76}$$

where $q, s \geq 0$ and $q \neq 1$, $s \neq 0$. This entropy includes Rényi entropy and Tsallis entropy as special cases. For $q \to 1$, it converges to the von Neumann entropy Nielsen and Chuang (2000). For $q \geq 1$ and $s \geq 0$, the quantum Unified (q, s)-entropy satisfies the subadditivity property

$$U_{q,s}(\rho \otimes \sigma) \leq U_{q,s}(\rho) + U_{q,s}(\sigma), \tag{1.77}$$

while for $0 < q < 1$ and $s > 0$ it satisfies

$$U_{q,s}(\rho \otimes \sigma) \geq U_{q,s}(\rho) + U_{q,s}(\sigma). \tag{1.78}$$

The Unified (q, s) entanglement measure is defined for a bipartite pure state $|\phi\rangle_{AB}$ by

$$\mathcal{U}^{(q,s)}(|\phi\rangle_{AB}) = U_{q,s}(\rho_A) \tag{1.79}$$

for each $q, s \geq 0$.

So far, we have presented some entanglement measures based on quantum entropy. There are other definitions. One is the concurrence Hill and Wootters (1997) based on the linear entropy and can be defined as

$$C(|\phi\rangle) = \sqrt{2 - 2\mathrm{Tr}\rho_A^2}, \tag{1.80}$$

where ρ_A denotes the reduced density matrix of a given bipartite pure state $|\phi\rangle_{AB}$. The second one is entanglement cost which is used for quantifying the nonseparability Bennett et al. (1999). Especially, the nonseparability can be measured by using the amount of entanglement that is required by local measurement on separated subsystems under LOCC. This is related to another concept of distillable entanglement under LOCC, for more details see review Horodecki et al. (2009).

Different from the bipartite entanglement measure, another generalization is to characterize the genuine multipartite entanglement beyond biseparable states. This has attracted great interest in many-body entanglement. Similar to entanglement measure for bipartite systems, a multipartite entanglement measure $E(\rho)$ may satisfy some common postulates Ma et al. (2011) as

1. $E(\rho_{bs}) = 0$ for any biseparable state ρ_{bs};
2. $E(\rho_{gme}) > 0$ for any genuine multipartite entanglement ρ_{gme};
3. E is convex, i.e., $E(\sum_i p_i\rho_i) \leq \sum_i p_i E(\rho_i)$.
4. E does not increase on average under LOCC i.e.

$$E(\rho) \geq \sum_i p_i E\left(\frac{A_i\rho A_i^\dagger}{\mathrm{Tr}(A_i\rho A_i^\dagger)}\right) \tag{1.81}$$

   where A_i are the Kraus operators that describe the LOCC operation. The probability of obtaining outcome i is given by $p_i = \mathrm{Tr}(A_i\rho A_i^\dagger)$.

There are other conditions such as additivity may be considered in applications. Note that for any genuine multipartite entanglement, it is also bipartite entanglement for a given bipartition. In this sense, any proper bipartite entanglement measure can be extended as a genuine multipartite entanglement measure. One example is to define multipartite concurrence for an n-partite pure state $|\Phi\rangle$ on Hilbert space $\otimes_{i=1}^n \mathbb{H}_{a_i}$ as

$$C_n(|\Phi\rangle) = \min_{\Gamma,\overline{\Gamma}} \sqrt{2 - 2\mathrm{Tr}\rho_\Gamma^2}, \tag{1.82}$$

where Γ and $\overline{\Gamma} = \{a_1, \cdots, a_n\} - \Gamma$ denotes all possible bipartitions of $\{a_1, \cdots, a_n\}$, and ρ_Γ denotes the reduced density matrix of particles in Γ. Take $n = 3$ as an example. For a tripartite pure state $|\Phi\rangle \in \otimes_{i=1}^3 \mathbb{H}_{a_i}$, there are three bipartitions $\Gamma = \{1\}, \{2\}, \{3\}$ up to the symmetry. The multipartite concurrence requires an evaluation of the bipartite concurrence for all three bipartitions. The multipartite concurrence can be further extended for mixed states by a convex roof construction

$$C_n(\rho) = \inf_{\{p_i, |\phi_i\rangle\}} \sum_i p_i C(|\Phi_i\rangle), \tag{1.83}$$

where the infimum is taken over all possible pure-state decompositions of $\rho = \sum_i p_i |\Phi_i\rangle\langle\Phi_i|$ with a probability distribution $\{p_i\}$, and some pure states $|\Phi_i\rangle$ on

$\otimes_{i=1}^n \mathbb{H}_{a_i}$. In general, we define multipartite entanglement measure a given bipartite entanglement measure E as

$$E_n(|\Phi\rangle) = \min_{\Gamma} E_{\Gamma}(|\Phi\rangle), \tag{1.84}$$

where $E_{\Gamma}(|\Phi\rangle)$ denotes bipartite entanglement measure E of $|\Phi\rangle$ for a given bipartition inspired by Γ. E may be any bipartite entanglement measure stated above. This provides a general method for featuring genuine multipartite entanglement. In general, a genuine multipartite entanglement measure may be represented by a function of bipartite entanglement measures as

$$E_n(|\Phi\rangle) = F(E_{\Gamma}(|\Phi\rangle), \forall \Gamma), \tag{1.85}$$

where F is a function depending on all the bipartite entanglement measures $E_{\Gamma}(|\Phi\rangle)$, $\forall \Gamma$. The main problem is to find a special form of F in applications. As the computation of any proper entanglement is in an NP-hard problem Gurvits (2003), it is interesting to explore simple and computable functions. Another method is to quantify entanglement by using reliable bounds, especially for mixed states.

1.4 Network

A general network is a collection of objects connected to allow data sharing. Our consideration in this book is more restricted to quantum scenarios, i.e., a collection of quantum objects including quantum computers, quantum devices, or other quantum and classical devices connected to allow classical or quantum data sharing. Our goal is to transmit quantum or classical data over quantum devices. With this consideration, we only introduce some fundamental notions for describing networks. Most of the results may be found in various textbooks Williamson (2019).

For most topics in a given network one has to represent a network with two types of information:

- the network topology, that is, the network's node and arc structure;
- data such as costs, capacities, and supplies/demands associated with the network's nodes and arcs.

Usually, the way for storing the network's topology may provide a natural way for storing the associated node and arc information. This arises an interesting problem of network representation.

In classical networks, the graph representation can reveal its connectivity in the best reasonable way. It consists of nodes and edges where nodes denote the objects and edges represent their connection relationships. It may be regarded as an abstract visual representation or a diagram that indicates the connection of systems by nodes.

In the communication view a graph may define the flow of information with the following different classes:

- Directed edges: Directed edges represent a directed and defined flow through the nodes in a hierarchy form.
- Undirected edges: Undirected edges represent an undirected or undefined flow through the nodes in an unorganized form.
- Weighted edges: Weighted edges represent the transmission capacity of flow over the connections.

Shannon theory provides the mathematical foundations of point-to-point communication including the Shannon channel capacity for featuring the maximal information transmission ability in a given channel. This can arise a natural problem of network scenarios involving multiple senders and receivers simultaneously. Unfortunately, this problem is generally NP-hard even in the classical communication theory. So, it is interesting to explore special models of network communication. One is to investigate the transmission ability of any pair. This provides a reasonable way for characterizing the transmission ability of a given network.

Definition 1.4.1 A network $\mathcal{N} = (G, c, s, t)$ consists of a directed graph $G = (V, E)$, with set of vertices V and set of edges E, a function $c : E \to \mathbb{R}_+ := \{x \in \mathbb{R} \mid x \geqslant 0\}$, associating each edge $e \in E$ with a non-negative number $c(e)$, called its capacity, and two distinguished vertices, s and t, called the source and the sink, respectively.

A flow network may be visualized as a watering system with edges being pipes and vertices being junctions. Fluid enters the pipe from the node s and exits at another t. At any time, the flux cannot exceed the pipe capacity. Meanwhile, the total flux is conserved at any junction. This idea may be represented by the mathematical notion of flow.

Definition 1.4.2 A flow $f : E \to \mathbb{R}_+$ in a network $\mathcal{N}$ is an assignment of non-negative numbers to the edges, subject to the following conditions.

1. For every edge e, the flow is upper bounded by the capacity, namely $f(e) \leqslant c(e)$;
2. The total flow entering in each node v except for the source and the sink equals to its output, namely $\sum_{u\in V,(u,v)\in E} f(u, v) = \sum_{u\in V,(v,u)\in E} f(v, u)$.

The value f_s denotes the total flow from the source, namely

$$f_s := \sum_{v\in V,(s,v)\in E} f(s, v). \tag{1.86}$$

The maximum of f_s over all possible flows is called the *max-flow* of the network, and is denoted as

$$\text{max-flow}(\{\mathcal{N}\}) := \max_f f_s. \tag{1.87}$$

Intuitively, f_s is the maximal flow entering into a network. This can be upper bound by evaluating all the flows from the source to the sink. To address this problem it requires to find all the possible paths from the source to the sink. This may be resolved by its graph representation.

For a given network, consider its graph representation as $\mathcal{G} = (V, E)$, where V denotes the set of all vertexes and E denotes the set of all connections of nodes. A cut of the graph $\mathcal{G} = (V, E)$ associated with the network is a cross-section of the edges that separate the source and the sink.

Definition 1.4.3 A cut of $\mathcal{G} = (V, E)$ is a partition of all the vertices into two disjoint subsets $C_s, C_t \subset V$, with $s \in C_s$ and $t \in C_t$.

Denote the cut of $\mathcal{G} = (V, E)$ as $C = (C_s, C_t)$. An edge $(u, v) \in E$ is called a cut edge if $u \in C_s$ and $v \in C_t$. Since all the flow from the source has to pass through the cut C, the total capacity of the edges associated to cut edges provides an upper bound to the transmitted flow. Denote the capacity of the cut (C_s, C_t) as $c(C_s, C_t)$ which contains the sum of the capacities of the cut edges, namely

$$c(C_s, C_t) := \sum_{u \in C_s v \in C_b (u,v) \in E} c(u, v). \tag{1.88}$$

It follows an upper bound as

$$\text{max-flow } (\mathcal{N}) \leqslant c(C_s, C_t) \tag{1.89}$$

for each cut (C_s, C_t). The optimal bound is by choosing the cut with minimum capacity. The minimal capacity over all possible cuts is named as the *min-cut*, and is denoted by

$$\text{min-cut}(\mathcal{G}) := \min_{(C_s, C_t)} c(C_s, C_t). \tag{1.90}$$

A classical result shows the maximum flow from any pair of two nodes is *exactly* equal to the capacity of the minimal cross-section of the edges, i.e., the min-cut of its graph Pierce (2019).

Theorem 1.4.1 *For a given network $\mathcal{N}$, its max-flow is equivalent to the min-cut of graph $\mathcal{G}$ associated with $\mathcal{N}$, that is,*

$$\textit{max-flow}(\mathcal{N}) = \textit{min-cut}(\mathcal{G}_N). \tag{1.91}$$

An example of a flow network is shown in Fig. 1.1. The second special case is a multicast task including one sender and multiple receivers. This general communication requires a special transmission strategy such as network coding to realize the achievable multicast capacity Medard and Sprintson (2011). Another one is multiple

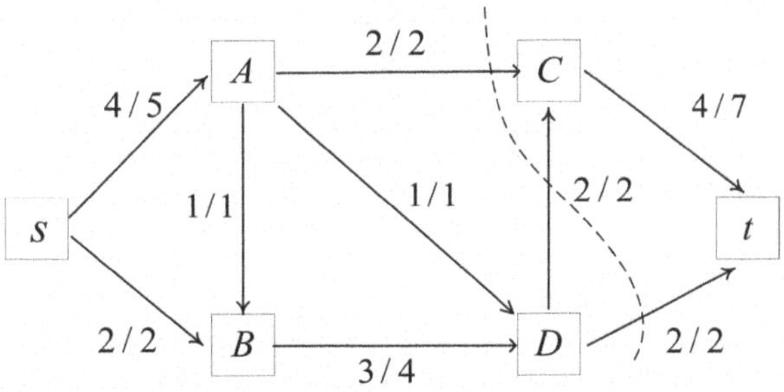

Fig. 1.1 A communication network. The numbers on each edge indicate the flow $f(e)$ and the capacity $c(e)$ of the edge, in the form $f(e)/c(e)$. The dashed line indicates a cut (C_s, C_t), with vertices to its left belonging to C_s, vertices to its right belonging to C_t

access channels including multiple senders and one receiver Zigangirov (2004). The quantum extension of these communications has recently attracted great interest. We will discuss it in the following chapters.

1.5 Conclusion

In this chapter, we have introduced some fundamental notions and definitions of quantum information. With the great development of quantum science, many interesting results cannot be included in this book. For further results on quantum foundations of computations, one may see books Nakahara and Ohmi (2008), Nielsen and Chuang (2000). For recent results on quantum information theory and quantum information processing, one may see books Wilde (2013), Hayashi (2017), Bergou et al. (2021).

As for entanglement theory, we present some well-known entanglement measures of bipartite systems. This topic and related multipartite entanglement have attracted great attention. For further reading see book Brody (2020) or the following chapter for multipartite entanglement. While communication over classical channels has been widely discussed in textbooks Cover and Thomas (2012), Pierce (2019), Medard and Sprintson (2011), Zigangirov (2004), we introduce simple definitions of information, i.e., entropy and its quantum extensions in this chapter. Further results will be discussed in the following chapters.

References

A. Acín, A. Andrianov, L. Costa, E. Jané, J.I. Latorre, R. Tarrach, Generalized Schmidt decomposition and classification of three-quantum-bit states. Phys. Rev. Lett. **85**, 1560 (2000)

H. Araki, E. Lieb, Entropy inequalities. Commun. Math. Phys. **18**, 160–170 (1970)

H. Barnum, C.M. Caves, C.A. Fuchs, R. Jozsa, B. Schumacher, Noncommuting mixed states cannot be broadcast. Phys. Rev. Lett. **76**, 2818 (1996)

C.H. Bennett, D.P. DiVincenzo, C.A. Fuchs, T. Mor, E. Rains, P.W. Shor, J.A. Smolin, W.K. Wootters, Quantum nonlocality without entanglement. Phys. Rev. A **59**, 1070 (1999)

J.A. Bergou, M. Hillery, M. Saffman, *Quantum Information Processing: Theory and Implementation*, 2nd edn. (Springer, 2021)
J. Brody, *Quantum Entanglement* (MIT Press, Essential Knowledge Series, 2020)
P.J. Coles, M. Berta, M. Tomamichel, S. Wehner, Entropic uncertainty relations and their applications. Rev. Mod. Phys. **89**, 015002 (2017)
T.M. Cover, J.A. Thomas, *Elements of Information Theory*, 2nd edn. (Wiley, 2012)
P.A.M. Dirac, *Principles of Quantum Mechanics*, 4th edn. (Clarendon Press, 1981)
W. Dür, G. Vidal, J.I. Cirac, Three qubits can be entangled in two inequivalent ways. Phys. Rev. A **62**, 062314 (2000)
A. Einstein, B. Podolsky, N. Rosen, Can quantum-mechanical description of physical reality be considered complete? Phys. Rev. **47**, 777–780 (1935)
H. Fan, Y.-N. Wang, Jing Li, J.-D. Yue, H.-D. Shi, Y.-L. Zhang, L.-Z. Mu, Quantum cloning machines and the applications. Phys. Reps. **544**, 241–322 (2014)
D.M. Greenberger, M.A. Horne, A. Zeilinger, in *Bell's Theorem, Quantum Theory and Conceptions of the Universe*. ed. by M. Kafatos (Kluwer, Dordrecht, 1989), pp.69–72
L. Gurvits, Classical deterministic complexity of Edmonds' problem and quantum entanglement, in *Proc. of the 35th ACM symp. on Theory of Comp.* (New York, ACM Press, 2003), pp. 10–19
M. Hayashi, *Quantum Information Theory* (Springer, 2017)
S.A. Hill, W.K. Wootters, Entanglement of a pair of quantum bits. Phys. Rev. Lett. **78**, 5022 (1997)
R. Horodecki, P. Horodecki, M. Horodecki, K. Horodecki, Quantum entanglement. Rev. Mod. Phys. **81**, 865 (2009)
R. Jozsa, Fidelity for mixed quantum states. J. Mod. Opt. **41**, 2315 (1994)
E.H. Kennard, The uncertainty relation for joint measurement of position and momentum. Z. Phys. **44**, 326 (1927)
J.S. Kim, B.C. Sanders, Monogamy of multi-qubit entanglement using Rényi entropy. J. Phys. A: Math. Theor. **43**, 445305 (2010)
J.S. Kim, B.C. Sanders, Unified entropy, entanglement measures and monogamy of multi-party entanglement. J. Phys. A: Math. Theor. **44**, 295303 (2011)
H.R. Li, M.X. Luo, H. Lai, Generalized quantum no-go theorems of pure states. Quantum Inf. Proc. **17**, 168 (2018)
M.X. Luo, H.R. Li, H. Lai, X. Wang, Unified quantum no-go theorems and transforming of quantum states in a restricted set. Quantum Inf. Proc. **16**, 297 (2017)
Z.-H. Ma, Z.-H. Chen, J.-L. Chen, C. Spengler, A. Gabriel, M. Huber, Measure of genuine multipartite entanglement with computable lower bounds. Phys. Rev. A **83**, 062325 (2011)
M. Medard, A. Sprintson, *Network Coding: Fundamentals and Applications* (Academic Press, 2011)
N.D. Mermin, Extreme quantum entanglement in a superposition of macroscopically distinct states. Phys. Rev. Lett. **65**, 1838 (1990)
M. Nakahara, T. Ohmi, *Quantum Computing* (From Linear Algebra to Physical Realizations (CRC Press, Boca Raton, 2008)
M.A. Nielsen, I.L. Chuang, *Quantum Computation and Quantum Information* (Cambridge University Press, New York, 2000)
M. Oszmaniec, A. Grudka, M. Horodecki, A. Wójcik, Creating a superposition of unknown quantum states. Phys. Rev. Lett. **116**, 110403 (2016)
A.K. Pati, General impossible operations in quantum information. Phys. Rev. A **66**, 062319 (2002)
A.K. Pati, B.C. Sanders, No-partial erasure of quantum information. Phys. Lett. A **359**, 31–36 (2006)
A. Peres, *Quantum Theory: Concepts and Methods* (Springer, 2006)
J.R. Pierce, *Information theory: Essentials of Mathematical Methods in Science and Engineering* (Wiley, 2019)
M.B. Plenio, S. Virmani, An introduction to entanglement measures. arXiv:quant-ph/0504163
P.N. Rathie, I.J. Taneja, Unified (r, s)-entropy and its bivariate measures. Inf. Sci. **54**, 23–39 (1991)
H.P. Robertson, The uncertainty principle. Phys. Rev. **34**, 163 (1929)

C.E. Shannon, A mathematical theory of communication. Bell Syst. Tech. J. **27**(379–423), 623–656 (1948)

G. Strang, *Introduction to Linear Algebra*, 6th edn. (MIT, 2023)

G. Svetlichny, Distinguishing three-body from two-body nonseparability by a Bell-type inequality. Phys. Rev. D **35**, 3066 (1987)

C. Tsallis, *Nonextensive Statistical Mechanics-Approaching a Complex World*, 1st edn. (Springer, New York, NY, USA, 2009)

J. von Neumann, Thermodynamik quantummechanischer Gesamheiten. Gott. Nach. **1**, 273–291 (1927)

M.M. Wilde, *Quantum Information Theory* (Cambridge University Press, 2013)

D.P. Williamson, *Network Flow Algorithms* (Cambridge University Press, 2019)

W.K. Wootters, Entanglement of formation of an arbitrary state of two qubits. Phys. Rev. Lett. **80**, 2245 (1998)

W.K. Wootters, W.H. Zurek, A single quantum cannot be cloned. Nature **299**, 802 (1982)

K.S. Zigangirov, *Theory of Code Division Multiple Access Communication* (Institute of Electrical and Electronics Engineers, Inc., 2004)

Chapter 2
Quantum Internet

A quantum internet, as one of the most important next-generation networks, has attracted great interest from many countries around the world. Building a quantum internet will provide competitive advantages in related industries and may open up a new world of innovations. Quantum internet is fast, secure, and scalable. The goal of this chapter is to present a comprehensive framework for a quantum internet.

2.1 Introduction

The classical internet (cInternet) consists of end-nodes connected by routers, and uses packet transmission to transmit voice, video, or data. The OSI and TCP/IP models provide a standardized approach to designing complex communication networks by separating functions into layers with specific protocols. The qInternet extends the cInternet to interconnect quantum networks for transmitting qubits and distributing entangled states. While classical protocols can't be used in quantum communications due to the special laws of quantum mechanics that govern the behavior of qubits, entanglement provides stronger correlations among legal participants. The qInternet is composed of both classical and quantum nodes, and hybrid networks can complete complex services using different network parts. Quantum computers can also provide powerful computational capabilities for optimizing computation tasks in networks.

The cInternet is comprised of various end-nodes connected by network routers through access links like WiFi and DSLs. Information is transmitted across the cInternet through packet transmission, and most devices are clustered into autonomous systems in a small world. The OSI and TCP/IP models provide a standardized approach to designing complex communication networks by separating functions into layers with specific protocols, shown in Fig. 2.1. Both models are based on the separation of concerns, also known as divide and conquer, which groups all involved functionalities into different stacks of layers. Each layer provides specific communication functionalities with specialized protocols. This allows for the

M.-X. Luo, *Quantum Networks*,
https://doi.org/10.1007/978-981-97-6226-2_2

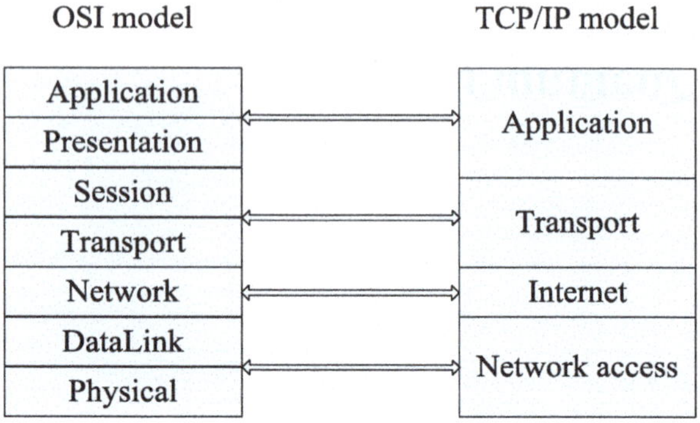

Fig. 2.1 OSI model versus TCP/IP model. OSI reference model (left) consists of seven layers, whereas TCP/IP reference model (right) consists of four layers. OSI model as a unified reference model is aim to provide a precise guide for network design. TCP/IP model is to interconnect different, heterogeneous networks

easy and efficient design of complex communication networks, including worldwide networks, irrespective of optimization over the past five decades.

The qInternet extends the cInternet to interconnect quantum networks for transmitting qubits and distributing entangled states Kimble (2008). In this model, quantum end-nodes are part of the qInternet and may be connected by quantum repeaters/routers. Classical fiber optics provides a potential connection way in the qInternet while optical lasers might be used for transmission in free space. The main reason is photonic connections are easy to code information and can be transmitted over long distances. qInternet will transmit qubits according to quantum principles. However, classical protocols cannot be used in quantum communications due to the special laws of quantum mechanics that govern the behavior of qubits. Entanglement provides stronger correlations among legal participants, allowing them to complete specific tasks without leaking information to potential adversaries.

One potential approach to developing the qInternet involves replacing or extending classical protocols with quantum counterparts without modifying the overall protocol stack. However, this is impossible due to the special laws of quantum mechanics that govern the behavior of qubits. Unlike classical communications, which assume that information can be read and duplicated faithfully at every point in the network, the quantum measurement postulate and the no-cloning theorem make it impossible to read and copy qubits without altering them. This presents significant challenges in designing quantum networks that can interface with the existing internet. Another challenge arises from quantum multipartite correlations, specifically quantum entanglement, which can redefine network communication.

The qInternet is composed of both classical and quantum nodes, and hybrid networks can complete complex services using different network parts. Quantum computers can also provide powerful computational capabilities for optimizing computation tasks in networks. The qInternet generally enhances rather than replaces the cInternet, introducing new communication paradigms into the existing Internet.

2.2 QInternet Structure

2.2.1 *Terms and Acronyms List*

In this subsection, we present some basic terms and acronyms that will be used in the following description of QInternet. Some notions such as qubit, entanglement, and quantum operations are defined in Chap. 1.

- *cInternet*-A classical Internet is used to transmit packets with bits between nodes. Some applications in the classical Internet may be enhanced by the quantum Internet. Here, we only consider the functionalities provided by classical Internet but not specific protocols (e.g., TCP, IP, UPD) being used.
- *LOCC*-The local operations assisted by classical communication (LOCC) has been widely used for quantum information processing. Here, the classical information in QInternet are easily transmitted by using classical in Internet; (2) The quantum operations can be generally performed on local systems; (3) The local quantum operations may be adapted in each round conditional on previous information.
- *EPR-Pairs*-As the EPR state in the form of $(|00\rangle + |11\rangle)/\sqrt{2}$ provides an easy way to build any entangled quantum networks Einstein et al. (1935). It can be regarded as important quantum resources for QInternet.
- *GHZ-Pairs*-A special type of maximally entangled multi-qubit quantum state. One example is $(|000\rangle + |111\rangle)/\sqrt{2}$ with three qubits Greenberger et al. (1996). Different from Bell-pairs, the GHZ-pairs are fundamental resources for multipartite quantum networks because they can generate special correlations for multiple participants. They are useful for building multipartite quantum channels in large-scale quantum networks.
- *Teleportation*-As a special quantum technique for remotely preparing quantum information it is based on entangled quantum channels assisted by LOCC Bennett et al. (1993). Suppose that two parties, A and B, share a Bell pair. A can perform a local projection on their share and an unknown qubit state, under the Bell basis, and send out the measurement outcomes. B can then recover the same state using local operations that are conditional on the measurement outcomes. This means that A can transfer an unknown qubit to B without physically sending it.
- *Entanglement Swapping*-Entanglement swapping is at the core of general quantum networks for completing large-scale tasks Zukowski et al. (1993). It can be used to create new entanglement between two parties who have not previously shared any entanglement. The process is based on the quantum teleportation protocol. As an example, consider a tripartite network consisting of three nodes A, B, and C. Each pair (A, B) or (B, C) shares an EPR pair. The node B can perform local projections on their shares to help A and C build a new Bell pair. This procedure is completed with the assistance of local operations and classical communication. Further analysis can extend this process to more general networks.
- *Fast Byzantine Agreement*-A Byzantine Agreement is an interactive protocol to reach a common agreement with a certain minimum number of nodes. It is of important protocol for classical secure applications with multiple players. This

can be extended for quantum methods Ben-Or and Hassidim (2005), Taherkhani et al. (2018).

- *NISQ*-While building large-scale quantum computers is difficult in a short time, it is reasonable to consider potential applications with noisy intermediate-scale quantum (NISQ) Preskill (2018). This kind of computer may have obvious features. The size might be 50 to hundreds of qubits. The other is noise.
- *Packet*-A specific format of the group contains some classical bits and qubits.
- *Prepare-and-Measure*-As some applications of quantum may be completed by only performing local measurements. This implies a simple setting of preparing the state and then measuring the states for quantum nodes in QInternet. One example is quantum key distribution Bennett and Brassard (2014).
- *Quantum Computer (QC)*-In QInternet most operations will be quantum, this means most nodes should have abilities of storage, processing, and preparation.
- *Quantum End-node*-A quantum end-node which may be a client, server, or peer-to-peer has users' applications and interfaces on the Internet. A quantum end-node is a quantum device that can generate, transmit, receive, and process quantum states or classical packets.
- *Quantum Key Distribution (QKD)* -A point-to-point protocol can generate a common random key for two remote parties. This can be realized in a device-independent manner, i.e., the security is proved based only on the statistical distribution of measurement outcomes Bennett and Brassard (2014), Ekert (1991), Lo and Chau (1999), Lo et al. (2005), Wang (2005), Lucamarini et al. (2018).
- *Entangled Quantum Network*-A new type of network consists of entangled states as quantum channels. This network is useful for transmitting quantum states between nodes. The entangled quantum network in general takes use of entangled channels assisted by cInternet.
- *QInternet*-The QInternet here is a new hybrid Internet by combining with cInternet and Quantum Entangled Networks. The QInternet can provide either classical applications or new quantum applications.

A quantum internet has the potential to connect global quantum devices like the classical internet. However, one of the biggest challenges is finding a reasonable physical realization for quantum computers, which has become the main focus of various countries and companies. The second is quantum applications beyond classical computers such as remote quantum information preparing in long-distance Illiano et al. (2022), Wehner et al. (2018), Castelvecchi (2018), Pirker and Dür (2019), Alshowkan et al. (2021) or QKD Alia et al. (2022). In this book, we will pay more attention to the networking infrastructure and protocols that may be applicable in a large-scale quantum network. Another important consideration in infrastructure is exploring hybrid networks of classical and quantum devices Townsend (1997), Alia, et al. (2022), Geng (2021).

Our goal in this chapter is to introduce a general networking infrastructure and simplified packet-switched quantum network for both problems.

2.2.2 QInternet Protocol

In what follows, we propose a QInternet protocol inspired by both the cInternet and quantum information. As a hybrid network, the QInternet should be compatible with the cInternet, meaning that classical tasks with packets can be easily implemented. This motivates the definition of network functionalities similar to those of the cInternet. However, quantum states exhibit special properties such as unfaithful cloning, short-time decoherence, noise, and entanglement, which impose temporal constraints on quantum information beyond those of the cInternet. Another difference between the QInternet and cInternet is how information is accessed. In the cInternet, classical bits can be freely measured, read, generated, and sent by router nodes. However, this cannot be faithfully implemented for unknown quantum states due to the quantum measurement postulate, which presents a great challenge for quantum routers in a quantum network due to the no-cloning theorem. To solve this problem, we assume that the network nodes may generate and distribute entangled states freely, i.e., quantum channels.

In general, classical bits and qudits (d-dimensional quantum states) may be applied as a hybrid state (*hstate* for short, with the first symbol h denoting hybrid), which is then combined into a packet routed through the network. Both classical and quantum information in the packet can flow through the network to reach the destination without interacting with other systems. This demonstrates a local assumption of information transmission.

Note that some functionalities in classical networks require local operations (routing protocols) that act on packets even without state information. This means that bits are *stateless*, and any node can store and read them easily without any additional assumptions. However, an unknown quantum state is generally *stateful* due to temporal constraints (*T-constraints*) and requires additional assumptions of decoherence and measurement for storing and reading. The first problem can be experimentally resolved by improving the physical systems, fast operations, and transmissions. These experimental issues will not be included in the following discussions. Instead, one may use logical methods such as decoherence-free quantum codes. When combined with error protection, this can be regarded as a fault-tolerant task similar to error correction in classical networks. It does not need to be considered in-network functionality design.

When designing network functionalities, it is important to consider entanglement. Entanglement is a correlated state among multiple states and is generally considered in a correlating manner, meaning it may be shared by different network nodes in a non-local scope. Otherwise, we may regard it as a high-dimensional quantum state. This means that any local operations may have an instantaneous effect on other correlated systems. For information transmission, we may assume that a single network node owns local information independent of others' local information. However, for entanglement distribution or quantum channel building, different network nodes may be involved. In this case, we should distinguish the roles of entanglement. One role is for the quantum channel, such as quantum teleportation. The other role is as correlated states where the transmitted particles are correlated with others. An example of this is quantum computation, which is a special application for quantum networks.

To sum up we introduce the following basic principles for designing a logical quantum Internet with specified network functionalities:

- **Universality**: A universal quantum network should be able to accommodate almost all types of applications, including classical communication, quantum communication for transmitting quantum states (such as entanglement swapping), quantum cloud computation, and quantum secure applications. This differs from previous definitions that only focused on entanglement distribution Wehner et al. (2018), Pirker and Dür (2019), Alshowkan et al. (2021). Essentially, a potential quantum internet should function like the classical internet by serving all users. The corresponding quantum networking protocols and layers should therefore satisfy this universality.
- **Coexistence**: Both classical and quantum information can be transmitted over the same physical channels, such as optical networks or free space links. In this case, it is reasonable to find a medium for deploying both classical and quantum signals. This allows the coexistence of both kinds of signals on the same network infrastructure, including routing protocols, layers, and standards, although they may differ in some details. The main difference between quantum and classical information, from our perspective, lies in their logic encoding. We believe that the quantum internet differs from the classical internet for certain applications due to the principles of quantum mechanics.
- **Scalability**: The hybrid network model should be scalable in network sizes, which requires networking protocols that can adapt to the addition of new networks with varied services. The classical internet achieves this property through the integration of various software and hardware devices based on standardized approaches. We believe that a similar approach can also be applied to quantum networks.

The principles of universality, coexistence, and scalability are regarded as fundamental for the development of future hybrid quantum-classical internet. The first attempt is proposed for quantum communication protocol Wang et al. (2018) and a quantum internet protocol Meter et al. (2008), Meter and Matsuo (2019). Here, EPR pairs are used to divide the quantum networking protocol into five layers based on entanglement purification Bennett et al. (1996), Gyongyosi and Imre (2020), Erhard et al. (2020), Riera-Sábat et al. (2021), Yan et al. (2023) and swapping. This protocol was designed as the first generation of quantum repeaters and was later improved by introducing a network architecture with quantum recursive (QRNA) Meter et al. (2008). The functionalities and protocols belong to the lower layers of the ISO model, namely the Physical and Link layers. The physical layer is been designed for building entanglement channels using a Midpoint Heralding Protocol (MHP), as well as for time synchronization. The link layer is for specifying different quantum operations performed by the physical layer. It includes a Hardware Abstraction Layer (HAL) which is for transferring the information (eg., the commands and outcomes) between the adjacent stacks. The network layer is used as an entanglement manager which builds entanglement for any two potential nodes. The last is the transport layer which can transfer the quantum states using teleportation for the application layer.

The quantum network model is recently extended to multipartite entangled states Pirker and Dür (2019). This model allows three phases for a quantum network: *dynamic* for distributing entanglement, *static* for sharing entanglement, and *adaptive* for manipulating entanglement. They then proposed a four-layer model, including the *physical*, *connectivity*, *link*, and *network* layers. Differently from Meter's model, the link layer can also transmit the quantum information qubits without error correction or entanglement distillation. The network layer is for building entangled inter-network by network devices of quantum routers. This proposal is updated as a *genuine quantum network* model Miguel-Ramiro et al. (2021) by coupling the target state with a quantum control register Gyongyosi and Imre (2019).

Our consideration is inspired by the OSI model versus TCP/IP model for hybrid networks with both quantum and classical devices. Figure 2.2 provides a quantum OSI model versus a quantum TCP/IP model.

As a reference model, the quantum model demonstrates the basic frameworks of information communication on a hybrid quantum network. It provides a visual design of each layer, starting from the physical media to the application layer of a network. The quantum model is shown as follows.

- Layer 1. **The physical layer**
  Transporting quantum information by specific interfaces in this layer. It specifies a way to complete physical connections in the hybrid network and represent classical bits or quantum states by physical signals, such as electrical, optical, or radio waves. This layer also builds physical transmission channels, such as electrical, optical, radio waves, or entangled states, between different nodes. The Midpoint Heralding Protocol (MHP) in the Link layer above determines this process. Both layers work together to complete the Channel Generation Protocol

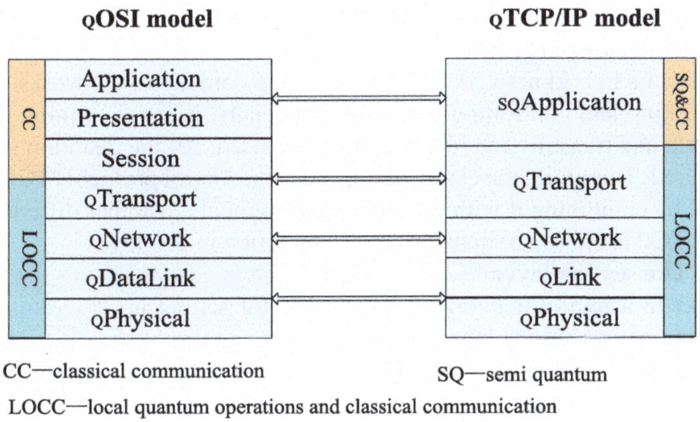

Fig. 2.2 Schematic quantum OSI model versus quantum TCP/IP model

(CGP), which receives requirements from the higher layer specified by several parameters, including remote node ID, information of the channel, whether it is a classical or quantum channel (entangled state, number of entangled pairs, minimum fidelity, request type, measurement basis, etc.). If an entangled channel is required from the layer above, the Physical layer uses entanglement distribution, local entanglement purification, and classical communication to build a faithful entangled channel. In this layer, an additional classical communication channel is combined with the above processes if it is necessary. The entanglement may be multipartite.

- Layer 2. **The data-link layer**
  This layer in a hybrid network is responsible for handling incoming and outgoing data through a physical link, as well as addressing transmission errors that may occur. It consists of three sublayers:

  - The information representation control (IRC) sublayer is for determining the operations in the following layers in terms of quantum states or classical bits.
  - The logical link control (LLC) sublayer is for logical link operations, such as multiplexing, flow control, acknowledgment, and notifying upper layers of any transmit/receive (TX/RX) errors in terms of different information representations.
  - The media access control (MAC) sublayer will make use of MAC addresses in the sending and receiving hardware for tracking transmitted frames. In this sublayer, each frame will be reorganized and marked for identifying the starting and ending such that the frame can be transmitted over the physical medium.

- Layer 3. **The network layer**
  The main function of this layer is to forward data between different subnetworks. This can be completed by packaging network addresses and routes with the transmitted data, and then sending it to the transport layer assisted by classical address protocol. This layer handles both classical and quantum information packets by combining them with the classical IP address protocol.
- Layer 4. **The transport layer**
  This layer in a hybrid network is for transferring data across networks. It allows error-checking and controlling data flow. Especially, it can determine the transmitted data and transmission rate which will be used. All the quantum states will be measured. Therefore, the classical transmission control protocol (TCP) may be extended by combining it with quantum measurements. Another difference from classical TCP is the use of quantum error correction in this layer.
- Layer 5. **The session layer**
  This layer in a hybrid network is responsible for setting up, coordinating, and terminating conversations between applications, similar to the classical session layer. It includes authentication and reconnection after an interruption. The main difference may be in the waiting strategies used, as in quantum information, a more rapid reply may be required due to the shorter decoherence time.

- Layer 6. **The presentation layer**
  This layer is a simple quantum extension of the classical presentation layer, for formatting data for potential tasks in the application layer. It is generally dependent on the semantics requirements of the specific application. It also includes classical cryptographic specifications. Quantum encryption and decryption, however, are regarded as special services that will be completed in the data link layer.
- Layer 7. **The application layer**
  This layer is a simple quantum extension of the classical application layer. It allows classical users to read the received messages, transfer the classical files, or complete special network tasks, while quantum users can operate states or perform quantum operations for specific goals.

The quantum TCP/IP model is inspired by Wehner et al. (2018). Note that quantum state transmission can be realized using quantum teleportation, which requires both quantum and classical channels. For quantum state-related tasks, a simplified model with four layers can be considered, as shown in Fig. 2.2. The main details are outlined below.

- **Physical layer**
  This layer is for building node-to-node quantum entanglement channels between different nodes in the network. This is achieved through the use of an midpoint heralding protocol (MHP) in the link sublayer. As a quantum layer, its main function is to enable the distribution and management of quantum information between nodes.
- **Link layer**
  This layer manages the entanglement in a hybrid network. The upper layer is the Link layer for transferring commands between the physical and other layers. Both the Physical and Link layers work together to realize a node-to-node quantum protocol (nQP) for distributing maximally entangled bipartite states. This layer requires a classical channel for each pair of nodes to complete faithful entanglement distribution. Additionally, local entanglement purification or quantum state operations are allowed in this layer.
- **Network layer**
  This layer produces long-distance or multipartite entanglement through generalized entanglement swapping. A general fact about entanglement distribution is that any multipartite entangled quantum network can be obtained by using quantum networks consisting of maximally entangled bipartite states, such as EPR states, assisted by LOCC. So, one may regard preparing a multipartite network as a network task by using all maximally entangled bipartite states.
- **Transport layer**
  This layer transmits quantum states by using teleportation according to the requests of *application layer*. This is completed assisted by LOCC.
- **Application layer**
  This layer is designed for specific applications of different users, including classical users with additional quantum measurement devices. It is responsible for

completing classical encryption and decryption operations in this layer. Additionally, it enables users to perform quantum operations and measurements for specific tasks related to their applications.

2.2.3 Packet Switching

Classical communication on networks has changed from the initial stage of circuit switching into packet switching, providing advantages such as time synchronization and low-latency transmissions Roberts (1978). They may bring network traffic problems, which were resolved by the introduction of packet switching. In the early stages of quantum networks, limited user numbers made an equivalent design similar to packet switching costly. Currently, the largest QKD network for distributing random keys without trusted parties serves only a few to tens of users. A more recent work Chen, et al. (2021) reports 46 nodes with switches and trusted nodes at a rate of 49.5 kb/s. It is based on a circuit-switched way without packet logic.

For future quantum networks with many users and high communication rates, packet switching is believed to be explored, similar to classical networks. Optical communication makes use of laser pulses for encoding information and transfers it over optical fibers, even in free space. The control information and payload of the packet can be converted into electrical signals, allowing for measurement of the optical signal and computer-readable formats. The node can robustly store the payload.

Another way is to keep the payload information in the optical domain Keiser (2021). This can be realized by using switches or wavelength selective switches, or add-drop multiplexers Saleh and Simmons (2012). But this method faces challenges related to quantum storage and error correction. There are two methods to address these challenges. One is a special packet design. For example, in photon encoding, all control information is represented by optical signals under a computational basis while the quantum information may be superposed. The quantum information can be stored locally in a cavity using photon-atom swapping and then read without measurement through another photon-atom swapping process. However, because of the principles of quantum no-cloning and quantum no-broadcasting, the quantum router cannot copy incoming states for broadcast-related tasks. Error correction is also not allowed for the quantum router since they do not know the transmitted information, similar to classical communication over VPN channels. The second method is to develop optical memories Arnold et al. (2021), Azuma et al. (2015). The signal amplification provides an efficient way for overcoming attenuation effects. However, as unknown states cannot be copied Wootters and Zurek (1982), this forbids amplifying weak quantum signals. Therefore, low-loss optical hardware must be constructed to improve network performance and reliability.

Building large-scale quantum networks presents challenges in technology and network models for transporting quantum states, but an abstract packet-switching model for quantum networks has potential applications in quantum networking. Developing such a model requires overcoming technical difficulties related to quantum measurement postulates, decoherence of quantum states, and establishing stable storage devices with long-decoherence times. Additionally, quantum no-cloning and quantum no-broadcasting principles and error correction must be taken into consideration when designing quantum routers capable of copying incoming states.

2.2.4 Hybrid Frame Structure

Switching the packet over the quantum network requires a hybrid frame to represent the transmitted data. Especially, there are a classical header and trailer in each payload as:

$$\boxed{\text{Classical header}|\text{quantum payload}|\text{classical trailer}} \tag{2.1}$$

The classical header contains special data for completing various tasks such as routing or error-correcting. A control unit can generate a classical header and trailer, and the quantum payload. The classical signals are represented by the computational basis $|0\rangle, |1\rangle, \ldots$.

The classical transmitter encodes the classical information of the header and trailer, while the quantum transmitters generate quantum signals. Both signals may use one physical medium, such as photons, or two physical media that can be multiplexed in a predetermined way before being sent over the fiber. There are two multiplexing methods: time division multiplexing with different time slots and wavelength division multiplexing with different wavelengths (Fig. 2.3).

In quantum networks, the classical header and trailer may be changed to adapt to noisy channels. One way is by using the link-layer discovery protocol frame (LLDP) Committee, et al. (2009). This may follow a new form of quantum ethernet discovery (QEP) as follows.

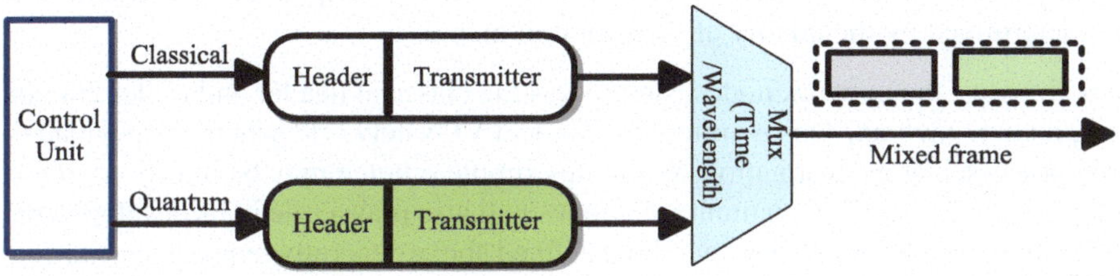

Fig. 2.3 Hybrid frame generation. The classical header and quantum payload can be generated from different photon sources and multiplexed into a hybrid data frame using a MUX in terms of time or wavelength division multiplexing

- The frame header contains the source and destination information and control codes.
- There is a data classification part called date-value (DV), which is used to specify whether the data being transmitted is classical or quantum.
- The data-header (DH) contains details about the quantum payload, including storing time, noise, cut-off time, and the quantum error correction protocol.
- The frame structure for the trailer may be similar to that of its LLDP counterpart.

Future studies should be performed for specifying detailed structures for the header and trailers based on the improvement of quantum techniques.

2.2.5 Quantum Packet Switching

In this subsection, we discuss packet switching based on the hybrid frame. In classical networks, forwarding an incoming packet to its next hop is straightforward due to the ability to amplify, copy, and error-correct bits. However, in quantum settings, these operations need clarification due to certain no-go theorems. To address this, we propose some simple switching schemes inspired by classical schemes used over optical channels.

(1) For small-scale quantum networks, we propose a simple way to avoid the need for quantum memory by using burst switching. A routing strategy is established before the payload arrives for a given number of hops from source to destination. If the required information processing can be completed, then the payload can be forwarded. Otherwise, it will be discarded and a retransmission will be attempted with more time. This approach is a hybrid between packet switching and circuit switching and does not require quantum memories Cho (2016), Abobeih (2018).
(2) For general quantum networks, we propose a simple way to incorporate quantum memory. The network nodes are equipped with quantum memory capabilities and a reasonable guard time, allowing them to store quantum payloads. When a packet arrives at a network node, the payload will be stored in quantum memory until the next node is reachable. The fidelity of quantum information is determined by the quality of the quantum memory.

In both cases, the network nodes can process the classical header and trailer but not the quantum payload. This is similar to classical VPN networks where the payload is only processed at its destination. While this simple scheme may be improved in the future, advancements in quantum techniques will also play a role in its development.

Realizing the hybrid frame switching defined above generally requires specialized network hardware at the nodes. One key component may be a quantum reconfigurable optical add drop multiplexer (qROADM). As a quantum switch, qROADM is capable of routing quantum states through a large optical network Wang (2019). To ensure faithful transmission, an amplification unit, and insertion loss reduction unit may be included. While this may be useful for near-term applications in circuit-switched networks such as QKD, it may not be suitable for packet-switched networks.

For a general quantum network, it is more convenient to use quantum units such as payload storage and dynamic switching instead of specialized hardware. This approach opens up new possibilities for realizing quantum hardware for hybrid frames. One way to achieve this is by adding limited functionality to existing components, while another is by designing more complex components.

In the first case, fundamental features can be presented by ignoring error correction, signal conversion, or other optimizations. A decoder of a multiplexer is used to read the header information, while the quantum payload is stored in memory and forwarded to the next station. This simple scheme requires transmission protocols that are designed with limited quantum memory.

In the second case, the design is inspired by classical ROADM technology, such as qROADM, but incorporates more complex components including well-known information processing units such as error correction and signal conversion. This design is intended for multiple users with multiple access channels. The decoder is used to read the classical header, while the memory is used to store quantum payloads. The node can take advantage of optical amplification, network traffic control, or other processing units to meet special requirements for hybrid packet routing.

Building a fully functional qROADM may require fundamental improvements in quantum technologies such as quantum memory and quantum repeaters. For some simple applications with few users, one may trade off near-term technologies for functional network hardware. However, for a general quantum Internet, entanglement generation, error correction, and quantum memory techniques are necessary.

2.3 Applications

The quantum Internet is expected to have various applications beyond those of the classical Internet. While the potential applications are still being explored, the field of quantum information processing is rapidly developing. In this section, we present some examples as an introduction to the quantum Internet.

Many applications can be grouped by their usage. This provides a simple classification of applications on the quantum Internet into the following categories:

- Quantum cryptography—The applications of quantum Internet are based on quantum cryptographic techniques such as quantum key distribution, quantum encryption, quantum commitment, and quantum authentication.
- Quantum computation—This provides a general kind of computing task based on quantum computers (e.g., distributed quantum computing).
- Quantum sensors—The quantum Internet can be applied for distributed sensing, long-baseline telescopes, or clock synchronization.

2.3.1 Quantum Key Distribution

Key distribution is crucial for modern cryptography applications, requiring a trusted party to distribute random keys to remote parties over the Internet. This challenge is addressed in quantum information processing by QKD Bennett and Brassard (2014), which provides provable security. One well-known model is device-independent, which uses only measurement statistics to detect potential adversaries by violating special Bell inequalities. This will be discussed in the following chapter.

For quantum networks with QKD or QKD networks, one may integrate the present model with QKD techniques. All classical processing on measurement outcomes may be added as an additional step for each node. In this case, the packet-switching protocol remains unchanged. Another simplification is using point-to-point QKD, which does not require quantum repeaters or quantum memory even for small-size quantum users. However, in some network traffic scenarios, packet switching may be more convenient.

After QKD is completed, all users can use classical cryptosystems such as AES or a one-time-pad, depending on their security requirements. One example is secure backbone network communication in financial systems.

A more advanced application is to build quantum financial systems that include quantum coins, quantum money, and even quantum banks for offline or online financial applications. Another extension is quantum conferencing, which will be discussed in the following chapter.

2.3.2 Secure Multiparty Computation

Secure multiparty computation problem accounts for active and malicious behavior Lamport et al. (1982). It generally requires more computational and communication resources than its counterpart for fail-stop faults in a distributed system. Its goal is to agree on a unique command under the condition that less than one-third of parties are internal adversaries. As one example, a Byzantine agreement is defined on broadcast networks. Solutions to the Byzantine problem rely on specific assumptions and generally include the following models:

- *Communication model*: One assumption for solving the Byzantine agreement problem is the communication model used in the distributed system. This includes synchronous and asynchronous models. All messages are sent in well-defined communication rounds in the synchronous model. There is no common clock for defining communication rounds in the asynchronous model.
- *Adversary*: The adversary can be static or dynamic in the protocol. The static adversary chooses its actions before the protocol begins, while the adaptive adversary can change its behavior during the protocol.
- *Computational model*: Another assumption is the computational power of the adversary, which can be either computationally bounded or unbounded. With

bounded computational power, the adversary can only access information specifically allowed within the distributed system. There may be private channels that the adversary cannot access, preventing them from obtaining all information generated and communicated during the protocol. With unbounded computational power, the adversary can obtain full access to all internal states.

The Byzantine problem can be resolved using quantum-based methods Ben-Or and Hassidim (2005) that reduce the number of communication rounds required compared to classical negotiation methods. It has assumed full-duplex ideal quantum channels and many tolerating faulty players for adaptive and computationally unbounded adversaries. It has been extended for use in quantum repeater networks Taherkhani et al. (2018), which reduces the depth of the circuit and the number of qubits. These types of agreements can be useful for improving consensus protocols in distributed quantum algorithms.

2.3.3 Distributed Quantum Computation

In previous distributed quantum algorithms , a task was completed by remote quantum nodes assisted by quantum communication, also known as a quantum repeater network (QRN) Kimble (2008), Meter (2014), Meter and Touch (2013), Hucul et al. (2015), Ritter (2012), Caleffi and Cacciapuoti (2020), Meter and Touch (2013), Munro et al. (2015). QRNs provide an infrastructure for distributed nodes using entanglement swapping, quantum teleportation, and quantum error correction. Here, one can make use of entangled states or high-dimensional states. There are several primitives for purifying the noisy entanglement Dür et al. (1999), correcting error Jiang et al. (2009), Fowler et al. (2010), and realizing quasi-asynchronous Munro et al. (2010).

The main goal of this book is to discuss general theoretical methods for distributed quantum computation. This model consists of a collection of small-scale quantum computers that are connected to complete a large task cooperatively. It relies on both general quantum entangled networks and classical communication networks Gyongyosi and Imre (2020).

Interestingly, there are nonlocal tasks that do not require classical communication, which is called nonlocal games. These tasks rely specifically on shared quantum entanglement and provide evidence of quantum entangled networks beyond classical networks. We will present some nonlocal games with quantum advantages over classical network resources.

Another application is quantum cloud computing, similar to classical cloud computing, which provides special delegation services for remote clients. The so-called quantum delegation computation relies on a special quantum computation model, the measurement-based quantum computation or one-way quantum computer. This is a special hybrid model with privacy preservation, allowing classical clients to delegate their tasks to quantum computers without leaking task information Fitzsimons (2017).

Distributed quantum computation is expected to be one of the most important applications for quantum networks, though quantum computing still faces many challenges in its early stages.

2.3.4 Quantum Sensing

Classical sensing or metrology techniques are very important in precision measurement and sensors, and they have recently been greatly improved by using quantum techniques. One possible way to improve sensing capabilities is by leveraging quantum entanglement, superposition, interference, and squeezing properties to build quantum sensors. These devices can offer high sensitivity and resolution, making them useful for a variety of applications. One important application is network clock synchronization such as atomic clocks which provide a precise clock signal Komar et al. (2014). This has practical implications for a wide range of fields including finance, telecommunications, and navigation.

Another promising application for quantum sensors is nanoscale sensing of physical magnitudes Hill (2019). Quantum sensors can be used to detect extremely small forces, such as those generated by single molecules or individual atoms, with high sensitivity and resolution. This has great potential for biological and chemical sensing, as well as for materials science and engineering. For example, quantum sensors could be used to study the behavior of proteins and other biomolecules in real time, offering insights into fundamental processes that drive life at the molecular level.

In addition to clock synchronization and nanoscale sensing, quantum sensors have other potential applications, including quantum radar, navigation, or UAV, which are particularly important in national defense. These applications rely on a network of entangled quantum sensors, even global networks with quantum satellites Proctor et al. (2018), Ge et al. (2018). By leveraging the unique properties of quantum systems, these sensors could provide unprecedented capabilities for detecting and tracking targets, navigating in challenging environments, and conducting remote sensing missions. As quantum technologies continue to advance, we will likely see even more applications emerge for quantum sensors in a variety of fields.

2.4 QInternet Application Scenarios

The QInternet can facilitate different applications and deployment setups. This section outlines some crucial scenarios that showcase the advantages of the QInternet. An application scenario usually consists of different interactions between nodes and users in a specific environment.

2.4.1 Communication Setup

Imagine an application scenario with two agents, A and B, who require secure communication. They should first share a random bit series securely. This is known as a communication setup for ensuring security. This can be realized if both A and B are basic quantum nodes or quantum computers. The potential application showcases how the quantum Internet can enhance the security of the classical Internet.

To establish the communication setup, it is crucial to protect against classical or quantum computing attacks. QKD offers a solution and enables the secure establishment of a secret key between two quantum nodes, utilizing two-way classical authentication and a one-way quantum channel. It is important to safeguard the QKD system against physical attacks such as injecting, phase-remapping, decoy state, and photon number splitting. So far, it remains one of the most well-known applications. A recent standard document ETSI ETSI GR QKD V2.1.1 (2018) has specified interfaces for users and devices.

In prepare-and-measure protocols like BB84 Bennett and Brassard (2014), the source uses qubit to encode classical bits, i.e., generating two random qubit strings according to the randomly chosen basis of $\{|0\rangle, |1\rangle\}$ or $\{|+\rangle, |-\rangle\}$. Here, the classical bit 0 is encoded by qubits as $|0\rangle$ or $|+\rangle$, for 1, she generates qubits $|1\rangle$ or $|-\rangle$. These qubits are sent from the source to the destination node over a quantum channel. The receiver measures each qubit in one of the two bases chosen randomly and identifies the corrected measurement outcomes assisted by the information of the prepared basis. Afterward, a post-processing procedure including parameter estimation, error correction, and privacy amplification is conducted over authenticated classical channels to generate the final secret key. If the error exceeds a threshold, the protocol is aborted; otherwise, they proceed to error correction. The entropic uncertainty relation property of quantum mechanics ensures both parties can generate information-theoretical secure keys Coronas (2011), Tang et al. (2019). The post-processing stage requires classical authentication Kiktenko et al. (2020).

Another realization is based on entanglement Treiber (2009), Hong et al. (2016), Hong et al. (2017), Hong et al. (2017), Hong and Luo (2018), Hong et al. (2018), Hong et al. (2018), that work differently than the prepare-and-measure steps described earlier. Typically, entanglement-based schemes involve the preparation of entangled states externally to the source and destination nodes. These schemes are not typically categorized as "prepare-and-measure" Wehner et al. (2018). However, there are other entanglement-based schemes where entanglement is generated within the source node, which may be regarded as "prepare-and-measure". Enhanced QKD protocols Bennett and Brassard (2014) have been developed to address loopholes caused by imperfect measurement devices. For example, measurement-device-independent QKD Hong and Luo (2018), Erkılıç et al. (2023), Zhang, et. al. (2019), Mizutani et al. (2015), Takeoka et al. (2014), Pirandola et al. (2017), Lucamarini et al. (2018), Yin and Fu (2019) offers solutions to mitigate these attacks. These enhanced QKD protocols can work differently than the steps of the BB84 protocol Bennett and Brassard (2014).

QKD Networks (QKDN) can be regarded as a special quantum Internet for large-scale applications. A QKDN is composed of three layers: application layer, network layer, and link layer. Between the source and the destination node, there may be one or multiple trusted QKD relays Zhang et al. (2018) within the QKDN. These relays facilitate the connection between the nodes. A QKDN may depend on entanglement distribution and quantum repeaters/routers. The practical vulnerabilities will be addressed with Measurement Device-Independent QKD (MDIQKD) protocols.

CV-QKD protocols, which are compatible with classical coherent detection schemes in high bandwidth communication systems Grosshans and Grangier (2002), has gained interest recently. These protocols utilize readily available telecommunications equipment and offer the potential for secure key distribution over limited distances. However, the absence of quantum repeaters for continuous variable systems restricts their application to short-distance communications or trusted relay-based QKD networks. The establishment of a quantum Internet necessitates entanglement managing, repeaters/routers, or trusted relays in large-scale deployments.

2.4.2 Privacy Preservation

In secure computation with privacy preservation, a client may delegate the computation task to a remote server while ensuring the data is undisclosed. Note that the remote node may be untrusted. The terminal node may be any classical node, quantum node, or quantum computer. While the terminal node can utilize cloud computing to offload the data and avoid local computing overhead, this may raise privacy concerns.

To address these privacy concerns, there are many protocols Fitzsimons (2017) which have been designed for delegated quantum computing or Blind Quantum Computation (BQC) . This kind of quantum computation can achieve secure delegated computation while simultaneously preserving privacy. As an example of this scenario, the data stays local and is not sent to the remote computation node, which could compromise the privacy of the data.

BQC as a client/server quantum model enables the terminal nodes to delegate computation tasks to the server without sending original qubits. Instead, the client sends transformed qubits to the server, which performs the computation and sends temporary result qubits back to the client. The terminal client transforms the receiver qubits into final qubits. There are two types of BQC models: circuit-based and measurement-based. Circuit-based BQC Childs (2005), which uses quantum circuits for qubit transformation, while the server performs a sequence of quantum logic gates. Measurement-based BQC, such as Universal BQC in Broadbent et al. (2009), leverages entangled states and repeated quantum teleportation to perform quantum computation.

A classical client can interact with the quantum server using classical information communication Huang (2017) where two entangled servers are designed for performing the computation. Another is to design verifiable BQC Fitzsimons (2017).

2.4.3 Distributed Quantum Computing

There are two models for completing quantum computing in a distributive manner.

In one case, quantum mechanics is utilized to enhance classical distributed computing problems such as the election. By leveraging entangled quantum states, the measurement of these states at each party can improve the election process without the requirement of classical communications among the distributed parties Pal et al. (2003). The entanglement will be pre-shared for the remote parties. LOCC operations allow for completing the distributive task.

Another way involves the distribution of quantum computing functions among physically separate quantum computers. Here, a quantum task is divided and allocated among different quantum computers that are physically isolated from each other. To transmit quantum states, pre-shared entanglement can be employed, similar to quantum teleportation. One example is to implement a remote CNOT gate Gottesman and Chuang (1999), Eisert et al. (2000). This model involves connecting NISQ computers Preskill (2018). This scenario allows for higher computation power before the development of quantum computers. Distributed quantum computing can enable complex problems, such as chemical interactions or secure multi-party quantum computation (MPQC) Caleffi et al. (2018), Cacciapuoti et al. (2020), Cacciapuoti et al. (2020), Crépeau et al. (2002).

To transmit sensitive data from one NISQ computer to another, teleportation can be employed Cuomo et al. (2020), Meter et al. (2006), Meter et al. (2008). The process involves locally preparing qubits and entanglement distribution. It allows a secure transfer assisted by LOCC and entanglement detection. Another way is from a one-time pad assisted by QKD.

Distributed quantum computing is a process of performing quantum computations by sharing and processing information among multiple remote quantum processors or nodes. This approach can overcome the limitations of current quantum technologies, such as limited qubit connectivity and high error rates, by leveraging the strengths of distributed systems, such as fault tolerance, scalability, and redundancy. Distributed quantum computing has the potential to revolutionize fields such as cryptography, optimization, and simulation by solving problems that are intractable for classical computers. However, this field is still in its infancy, and many technical and conceptual challenges need to be addressed before it can be effectively implemented at scale.

2.4.4 Clock Synchronization

Clock synchronization in quantum networks can be achieved by exploiting quantum entanglement. By creating entangled qubits and distributing them to different locations, it is possible to synchronize clocks across long distances André et al. (2004).

One approach to clock synchronization using quantum networks is through the use of entangled photon pairs Giovannetti et al. (2004). These entangled photons can be generated and distributed to different locations. By measuring the arrival times of the entangled photons at each location, it is possible to synchronize the clocks. The entangled photons act as reference clocks, and any deviations in the arrival times can be used to adjust the local clocks and achieve synchronization.

Another approach to clock synchronization is through the use of quantum teleportation. Quantum teleportation allows the transfer of quantum states from one location to another, without physically moving the particles. By teleporting the state of a local clock to a remote location, it is possible to achieve synchronization between the clocks at the two locations.

Clock synchronization using quantum networks offers several advantages over traditional methods. Firstly, it allows for precise synchronization over long distances, overcoming the limitations of the finite speed of light and uncertainties introduced by the communication medium. Secondly, quantum networks provide a secure communication channel, ensuring that any eavesdropping attempts are immediately detected. Finally, quantum networks have the potential to be scaled up to accommodate increasing data demands and the addition of more telescopes.

2.4.5 Long-Baseline Telescopes

Long-baseline telescopes play a crucial role in modern astronomy, enabling scientists to observe and study celestial objects with unprecedented precision. However, these telescopes face limitations in terms of data transfer rates and synchronization due to the vast distances between the telescopes. Quantum networks, with their ability to transmit information securely and efficiently over long distances, offer a promising solution to these challenges.

Long-baseline telescopes are designed to overcome the limitations of traditional telescopes, such as atmospheric distortion and diffraction. By combining the signals received from multiple telescopes spread across vast distances, astronomers can achieve higher angular resolution and capture finer details of celestial objects such as the Very Long Baseline Array (VLBA) Giovannetti et al. (2004).

The main challenges faced by long-baseline Telescopes are as follows. The first is data transfer rates. Long-baseline telescopes generate enormous amounts of data, which need to be transmitted and processed in real time. The current data transfer rates are limited by the available communication infrastructure, leading to potential data loss or delays. The other is a synchronization of the signals received from different telescopes which is essential for interferometry, where the signals are combined to create a high-resolution image Giovannetti et al. (2004). Achieving accurate synchronization over long distances is challenging due to the finite speed of light and the uncertainties introduced by the communication medium.

Quantum networks may provide potential solutions for Long-Baseline Telescopes. Quantum networks can significantly enhance the data transfer rates of long-baseline telescopes. By encoding the astronomical data into qubits and transmitting them through quantum channels, the data transfer rates can be significantly increased, overcoming the limitations of classical communication infrastructure. Additionally, quantum clock synchronization may achieve precise synchronization between the signals received by different telescopes is critical for interferometry. Quantum networks offer the potential for accurate clock synchronization by exploiting quantum entanglement Albert and Uhlmann (2018). The entangled qubits can be used as reference clocks, allowing for synchronized measurements across vast distances. Developing and maintaining quantum network infrastructure can be costly and requires expertise in both quantum physics and network engineering.

Long-baseline telescopes, with their ability to capture high-resolution images of celestial objects, have revolutionized our understanding of the universe. The integration of quantum networks with these telescopes holds immense potential to overcome the limitations of data transfer rates and synchronization. By leveraging the power of quantum communication and clock synchronization, long-baseline telescopes can enhance their capabilities and pave the way for groundbreaking astronomical discoveries. However, significant technological challenges and investments are required to realize the full potential of quantum networks in the field of astronomy O'Brien and Furusawa (2009), Pirandola et al. (2019).

2.4.6 *Quantum Sensing*

Quantum sensing is a rapidly advancing field that explores the use of quantum systems to achieve highly precise and sensitive measurements Degen et al. (2017). Traditional sensing techniques face limitations in terms of sensitivity and accuracy, especially when dealing with small signals or noise. Quantum sensing, on the other hand, takes advantage of quantum properties such as superposition and entanglement to enhance measurement capabilities. The use of quantum networks in quantum sensing offers even greater potential by enabling distributed sensing, where multiple sensors are interconnected and share information through quantum entanglement.

Quantum sensing using quantum networks involves the use of entangled qubits as sensors. These entangled qubits can be distributed to different locations and used to measure physical quantities such as magnetic fields, electric fields, or temperature Degen et al. (2017). By exploiting quantum entanglement, the measurements from multiple sensors can be combined to achieve higher precision and sensitivity compared to traditional sensing techniques.

One approach to quantum sensing using quantum networks is through the use of quantum metrology Gilmore (2021), Giovannetti et al. (2011), Pezzé et al. (2018), Casacio (2021). Quantum metrology aims to exploit quantum properties to achieve measurements beyond the limitations imposed by classical physics. By creating entangled qubits and distributing them to different locations, it is possible to perform measurements with higher precision and sensitivity. The entangled qubits act as sensors that can detect small changes in the physical quantity being measured.

Another approach to quantum sensing is through the use of quantum-enhanced sensors. Quantum-enhanced sensors utilize quantum states to enhance the sensitivity and accuracy of measurements. By preparing the sensor in a superposition of different states, it is possible to achieve higher sensitivity to small changes in the physical quantity being measured. Additionally, quantum entanglement can be used to enhance the precision of measurements by combining information from multiple sensors.

Quantum networks enable distributed sensing, where multiple sensors are interconnected and share information through quantum entanglement. This distributed sensing approach offers several advantages over traditional sensing methods. Firstly, it allows for measurements to be performed in multiple locations simultaneously, providing a more comprehensive understanding of the physical quantity being measured. Secondly, distributed sensing can enhance the overall sensitivity and accuracy of measurements by combining information from multiple sensors. Lastly, quantum networks provide a secure communication channel, ensuring that any eavesdropping attempts are immediately detected. However, several challenges need to be addressed to realize the full potential of quantum sensing using quantum networks Pirandola et al. (2019). While there are challenges to overcome, the development of quantum sensing using quantum networks holds great promise for a wide range of applications, including healthcare, environmental monitoring, and fundamental research Taylor and Bowen (2016).

2.5 General Requirements

2.5.1 Background

Quantum technologies are undergoing rapid development, and predicting their future milestones is a challenging task. As discussed in Grumbling and Horowitz (2019), current quantum computers based on NISQ technology can achieve up to hundreds of qubits. The error rates of basic gates have been decreasing over the years. Nevertheless, it is difficult for processing multiple qubits. So, while the potential of quantum computing is promising, further advances in hardware and software are needed to improve the performance and reliability of quantum computers Miguel-Ramiro et al. (2021).

On the network level, there are six different stages for developing quantum Internet Wehner et al. (2018): trusted repeater, prepare-measure, entanglement distribution, Quantum memory, fault-tolerant few qubits, and quantum computing networks.

The roadmap towards achieving the full-blown qInternet, as described in Wehner et al. (2018), involves several stages of development. Each stage requires new techniques for developing specific functionality, applications, and characteristics. The first stage establishes a trusted repeater, while the final is for quantum computing networks with a larger number of qubits. Table 2.1 illustrates application scenarios

Table 2.1 Schematically application scenarios in different quantum internet stages Wehner et al. (2018)

Procedures	Applications	Features
1	Setup using QKD	Trusted nodes
2	Communication with QKD	PM capability
3	Communication with EQKD	ED
4	BQC	Quantum memory
5	Clock synchronization	Fault tolerance
6	DQC	More qubits

for the qInternet mapped to each stage. For instance, a secure communication setup can be supported in stages 1, 2, or 3 using different quantum key distribution (QKD) solutions.

In the first stage, basic QKD is used to support secure communication, but trusted nodes are required to provide end-to-end security. This stage primarily requires trusted nodes to establish secure communication. In the second stage, users can prepare, receive, and measure quantum states. Third the entanglement distribution can enable long-distance QKD with end-to-end security.

Stage 4 can store and manipulate entangled states in memories. This enables sophisticated applications such as delegated computing, election, and secret sharing. In stage five, repeaters may correct errors, making fault-tolerant quantum computations possible. This stage allows for running distributed quantum computing and quantum sensor applications over a smaller number of qubits. The last is for realizing distributed quantum computing, and for developing even more advanced applications and technologies.

2.5.2 Requirements

From a networking perspective, there are several general and functional requirements for the qInternet, based on the identified application scenarios. Firstly, methods for facilitating efficient interaction between quantum applications and entangled qubits are necessary to enable designated entangled qubits to be distributed to any other node in the network. Specific operations, such as entanglement swapping and distillation, need to be performed on entangled qubits to trigger their distribution. Quantum end nodes, repeaters/routers, and computers may all be involved in this process.

Secondly, quantum repeaters/routers should be used to complete high-fidelity entanglement connections between two nodes. This requires generating an entangled pair on each hop of the path, assisted by entanglement swapping.

Thirdly, quantum nodes can transfer classical information to aid the transmission of individual qubits with repeaters/receivers. Since qubits do not have any associated packet overhead, extra information is required for routing Gyongyosi and Imre (2018), Chakraborty et al. (2019), Pant et al. (2019), Hahn et al. (2019), Li et al. (2021), Patil et al. (2022), identification, and other purposes. For efficiency reasons, in an anticipated quantum network beyond point-to-point communication, it is preferable that many parties can communicate simultaneously over the underlying infrastructure; however, bottlenecks in the network may cause delays.

Finally, managing and controlling the QInternet is necessary. Quantum nodes may have resources such as quantum memory, channels, qubits, and established quantum connections. It can be used to monitor status or identify potential issues. New management information models may need to be developed for the QInternet.

To sum up, several challenges need to be addressed to realize the full potential of quantum networks. Firstly, building a robust and reliable quantum network infrastructure over long distances is a significant technological challenge. Quantum systems are inherently prone to errors due to environmental noise and decoherence, and implementing error correction techniques is crucial for maintaining the integrity of the measurements. Additionally, developing and maintaining quantum network infrastructure can be costly and requires expertise in both quantum physics and network engineering.

2.6 Conclusion

Integrating classical and quantum networks where both the control and data plane co-exist is a challenging task that requires incremental steps. Although achieving full integration of classical and quantum hardware at the physical layer is unlikely, we foresee future networks merging the hardware into a single device capable of supporting both data types and intelligently distinguishing the processing path.

This chapter has discussed the potential infrastructure of quantum Internet including QInternet structure Illiano et al. (2022), Wehner et al. (2018), Pirker and Dür (2019), Alshowkan et al. (2021), Townsend (1997), Alia, et al. (2022), Geng (2021), Packet switching for optical quantum networks and quantum packet switching Keiser (2021), Arnold et al. (2021), and some applications related to QKD, Byzantine negotiation, distributed quantum computation, and quantum sensing. The journey towards completely integrated classical-quantum networks will require overcoming various challenges, including amplifying quantum signals, reducing the geographical distance between quantum switches, addressing cross-talk between bright classical signals and weak quantum signals, and accommodating state-of-the-art quantum technologies that require careful environmental control.

Despite these challenges, we anticipate that future hybrid networks will provide significant benefits for applications such as distributed quantum computing, secure communication, and quantum sensing. Furthermore, advances in quantum repeater and error correction technology, and novel fiber technologies may accelerate the development of the necessary hardware and infrastructure for such networks.

References

M.H. Abobeih et al., One-second coherence for a single electron spin coupled to a multi-qubit nuclear-spin environment. Nat. Commun. **9**, 2552 (2018)

C. Albert, A. Uhlmann, *Quantum Information Theory* (Cambridge University Press, 2018)

O. Alia, et al., DV-QKD coexistence with 1.6 Tbps classical channels over hollow core fibre (2022) arXiv:2203.14621

O. Alia, R.S. Tessinari, E. Hugues-Salas, G.T. Kanellos, R. Nejabati, D. Simeonidou, Dynamic DV-QKD networking in trusted-node-free software-defined optical networks. J. Lightwave Tech. **40**, 5816–5824 (2022)

M. Alshowkan, N. Rao, J. Chapman, B. Williams, P. Evans, R. Pooser, J. Lukens, N. Peters, Lessons Learned on the Interface between Quantum and Conventional Networking, in *Driving Scientific and Engineering Discoveries Through the Integration of Experiment, Big Data, and Modeling and Simulation*, vol. 1512. (SMC, Communications in Computer and Information Science, Springer, 2021)

A. André, A. Sørensen, M. Lukin, Stability of atomic clocks based on entangled atoms. Phys. Rev. Lett. **92**, 230801 (2004)

N.T. Arnold, M. Victora, M.E. Goggin, P.G. Kwiat, 1.5-THz-Bandwidth digital photonic quantum memory, in *Frontiers in Optics* (Washington, DC United States, 2021), pp. FW1E-2

K. Azuma, K. Tamaki, H.-K. Lo, All-photonic quantum repeaters. Nat. Commun. **6**, 6787 (2015)

C.H. Bennett, G. Brassard, Quantum cryptography: Public key distribution and coin tossing, in *Proceedings of IEEE International Conference on Computers, Systems, and Signal Processing* (Bangalore, 2014), pp. 175–179

C.H. Bennett, G. Brassard, C. Crépeau, R. Jozsa, A. Peres, W.K. Wootters, Teleporting an unknown quantum state via dual classical and Einstein-Podolsky-Rosen channels. Phys. Rev. Lett. **70**, 1895 (1993)

C.H. Bennett, G. Brassard, S. Popescu, B. Schumacher, J.A. Smolin, W.K. Wootters, Purification of noisy entanglement and faithful teleportation via noisy channels. Phys. Rev. Lett. **76**, 722 (1996)

M. Ben-Or, A. Hassidim, Fast quantum byzantine agreement, *Proceedings of the Thirty-seventh Annual ACM Symposium on Theory of Computing* (2005), pp. 481–485

A. Broadbent, J. Fitzsimons, E. Kashefi, Universal blind quantum computation, in *Proceedings of the 50th Annual IEEE Symposium on Foundations of Computer Science* (FOCS, 2009), pp. 517–526

A.S. Cacciapuoti, M. Caleffi, F. Tafuri, F.S. Cataliotti, S. Gherardini, G. Bianchi, Quantum internet: networking challenges in distributed quantum computing. IEEE Network **34**, 137–143 (2020)

A.S. Cacciapuoti, M. Caleffi, R. Van Meter, L. Hanzo, When entanglement meets classical communications: quantum teleportation for the quantum internet. IEEE Trans. Comm. **68**, 3808–3833 (2020)

M. Caleffi, A.S. Cacciapuoti, G. Bianchi, Quantum internet: From communication to distributed computing, in *NANOCOM '18: Proceedings of the 5th ACM International Conference on Nanoscale Computing and Communication* (2018)

M. Caleffi, A.S. Cacciapuoti, Quantum switch for the quantum internet: Noiseless communications through noisy channels. IEEE J. Sel. Area Comm. **38**, 575–588 (2020)

C.A. Casacio et al., Quantum-enhanced nonlinear microscopy. Nature **594**, 201 (2021)

D. Castelvecchi, The quantum Internet has arrived (and it hasn't). Nature **554**, 289–292 (2018)

K. Chakraborty, F. Rozpedeky, A. Dahlbergz, S. Wehner, Distributed routing in a quantum internet (2019). arXiv:1907.11630v1

T.-Y. Chen, et al., Implementation of a 46-node quantum metropolitan area network. npj Quantum Inf. **7**, 134 (2021)
A.M. Childs, Secure assisted quantum computation. Quantum Inf. Comput. **5**, 456–466 (2005)
Y.-W. Cho et al., Highly efficient optical quantum memory with long coherence time in cold atoms. Optica **3**, 100 (2016)
L.S. Committee, et al., IEEE standard for local and metropolitan area networks, station and media access control connectivity discovery. Standard IEEE 802.1ab (2009)
D.E. Coronas, Information reconciliation for quantum key distribution. Quantum Inf. Comput. **11**, 0226–0238 (2011)
C. Crépeau, D. Gottesman, A. Smith, Secure multi-party quantum computation, in *34th Symposium on Theory of Computing (STOC)* (2002), pp. 643–652
D. Cuomo, M. Caleffi, A.S. Cacciapuoti, Towards a distributed quantum computing ecosystem. IET Quantum Comm. **1**, 3–8 (2020)
C.L. Degen, F. Reinhard, P. Cappellaro, Quantum sensing. Rev. Mod. Phys. **89**, 035002 (2017)
W. Dür, H.-J. Briegel, J.I. Cirac, P. Zoller, Quantum repeaters based on entanglement purification. Phys. Rev. A **59**, 169 (1999)
A. Einstein, B. Podolsky, N. Rosen, Can quantum-mechanical description of physical reality be considered complete? Phys. Rev. **47**, 777 (1935)
J. Eisert, K. Jacobs, P. Papadopoulos, M.B. Plenio, Optimal local implementation of nonlocal quantum gates. Phys. Rev. A **101**, 032332 (2000)
A.K. Ekert, Quantum cryptography based on Bell's theorem. Phys. Rev. Lett. **67**, 661–663 (1991)
M. Erhard, M. Krenn, A. Zeilinger, Advances in high-dimensional quantum entanglement. Nat. Rev. Phys. **2**, 365–381 (2020)
ETSI GR QKD V2.1.1, Quantum Key Distribution (QKD); Components and Internal Interfaces (2018)
J.F. Fitzsimons, Private quantum computation: an introduction to blind quantum computing and related protocols. *npj Quantum Inf.***3**, 23 (2017)
A.G. Fowler, D.S. Wang, C.D. Hill, T.D. Ladd, R. Van Meter, L.C. Hollenberg, Surface code quantum communication. Phys. Rev. Lett. **104**, 180503 (2010)
W. Ge, K. Jacobs, Z. Eldredge, A. Gorshkov, M. Foss-Feig, Distributed quantum metrology with linear networks and separable inputs. Phys. Rev. Lett. **121**, 043604 (2018)
J.-Q. Geng et al., Coexistence of quantum key distribution and optical transport network based on standard single-mode fiber at high launch power. Opt. Lett. **46**, 2573 (2021)
K.A. Gilmore et al., Quantum-enhanced sensing of displacements and electric fields with two-dimensional trapped-ion crystals. Science **373**, 673 (2021)
V. Giovannetti, S. Lloyd, L. Maccone, Quantum-enhanced measurements: beating the standard quantum limit. Science **306**, 1330 (2004)
V. Giovannetti, S. Lloyd, L. Maccone, Advances in quantum metrology. Nat. Photon. **5**, 222 (2011)
D. Gottesman, I. Chuang, Demonstrating the viability of universal quantum computation using teleportation and single-qubit operations. Nature **402**, 390–393 (1999)
D.M. Greenberger, M.A. Horne, A. Zeilinger, in *Bell's Theorem, Quantum Theory, and Conceptions of the Universe*, ed. by M. Kafatos (Kluwer, Dordrecht, 1989, 1996), pp. 69–72
F. Grosshans, P. Grangier, Continuous variable quantum cryptography using coherent states. Phys. Rev. Lett. **88**, 057902 (2002)
E. Grumbling, M. Horowitz, *Quantum Computing: Progress and Prospects* (The National Academies Press, 2019)
L. Gyongyosi, S. Imre, Decentralized base-graph routing for the quantum internet. Phys. Rev. A **98**, 022310 (2018)
L. Gyongyosi, S. Imre, Entanglement access control for the quantum Internet. Quantum Inf. Process. **18**, 107 (2019)
L. Gyongyosi, S. Imre, Entanglement concentration service for the quantum Internet. Quantum Inf. Process. **19**, 221 (2020)

L. Gyongyosi, S. Imre, Resource prioritization and balancing for the quantum internet. Sci. Rep. **10**, 22390 (2020)

F. Hahn, A. Pappa, J. Eisert, Quantum network routing and local complementation. npj Quantum Inf. **5**, 76 (2019)

R.M. Hill et al., A tool for functional brain imaging with lifespan compliance. Nat. Commun. **10**, 4785 (2019)

L. Hong, M.-X. Luo, High-rate and high-capacity measurement-device-independent quantum key distribution with Fibonacci matrix coding in Free space. Sci. China Inf. Sci. **61**, 062501 (2018)

L. Hong, M.-A. Orgun, J. Pieprzyk, L. Jing, M.-X. Luo, F. Xiao, High-capacity quantum key distribution using Chebyshev-map values corresponding to Lucas numbers coding. Quantum Inf Proc. **15**, 4663–4679 (2016)

L. Hong, M.-X. Luo, J. Pieprzyk, J. Zhang, L. Pan, S. Li, M. Orgun, Fast and simple high-capacity quantum cryptography with error detection. Sci. Reps. **7**, 46302 (2017)

L. Hong, M.-X. Luo, C. Zhan, J. Pieprzykc, M.A. Orgune, An improved coding method of quantum key distribution protocols based on Fibonacci-valued OAM entangled states. Phys. Lett A **381**, 2922–2926 (2017)

L. Hong, M.-X. Luo, M.A. Orgun, J. Pieprzyk, F. Xiao, C. Zhan, Tunable multi-party high-capacity quantum key distribution based on m-generalized Fibonacci sequences using golden coding. Quantum Inf. Proc. **17**, 246 (2018)

L. Hong, M.-X. Luo, Y. Xu, J. Pieprzyk, J. Zhang, L. Pan, M.A. Orgun, Round-robin-differential-phase-shift quantum key distribution based on wavelength division multiplexing. Laser Phys. Lett. **15**, 115201 (2018)

H. Huang et al., Experimental blind quantum computing for a classical client. Phys. Rev. Lett. **119**, 050503 (2017)

D. Hucul, I.V. Inlek, G. Vittorini, C. Crocker, S. Debnath, S.M. Clark, C. Monroe, Modular entanglement of atomic qubits using photons and phonons. Nat. Phys. **11**, 37–42 (2015)

J. Illiano, M. Caleffi, A. Manzalini, A.S. Cacciapuoti, Quantum internet protocol stack: a comprehensive survey. Computer Networks **213**, 109092 (2022)

L. Jiang, J.M. Taylor, K. Nemoto, W.J. Munro, R. Van Meter, M.D. Lukin, Quantum repeater with encoding. Phys. Rev. A **79**, 032325 (2009)

G. Keiser, *Fiber Optic Communications* (Springer, Singapore, 2021)

E.O. Kiktenko, A.O. Malyshev, M.A. Gavreev, A.A. Bozhedarov, N.O. Pozhar, M.N. Anufriev, A.K. Fedorov, Lightweight authentication for quantum key distribution. IEEE Trans. Inf. Theory **66**, 6354–6368 (2020)

H.J. Kimble, The quantum internet. Nature **453**, 1023 (2008)

P. Komar, E.M. Kessler, M. Bishof, L. Jiang, A.S. Srensen, J. Ye, M.D. Lukin, A quantum network of clocks. Nat. Phys. **10**, 582–587 (2014)

W. Kozlowski, S. Wehner, R.V. Meter, B. Rijsman, A.S. Cacciapuoti, M. Caleffi, Architectural Principles for a Quantum Internet, Work in Progress, Internet-Draft, draft-irtf-qirg-principles-07 (2021)

L. Lamport, R. Shostak, M. Pease, The Byzantine generals problem. ACM Trans. Progr. Lang. Sys. **4**, 382–401 (1982)

C. Li, T. Li, Y.-X. Liu, P. Cappellaro, Effective routing design for remote entanglement generation on quantum networks. npj Quantum Inf. **7**, 10 (2021)

H.-K. Lo, H.F. Chau, Unconditional security of Quantum Key Distribution over arbitrarily long distances. Science **283**, 2050–2056 (1999)

H.-K. Lo, X. Ma, K. Chen, Decoy state quantum key distribution. Phys. Rev. Lett. **94**, 230504 (2005)

M. Lucamarini, Z.L. Yuan, J.F. Dynes, A.J. Shields, Overcoming the rate-distance limit of quantum key distribution without quantum repeaters. Nature **557**, 400 (2018)

M. Lucamarini, Z.L. Yuan, J.F. Dynes, A.J. Shields, Overcoming the rate-distance limit of quantum key distribution without quantum repeaters. Nature **557**, 400–403 (2018)

R.V. Meter, *Quantum Networking* (John Wiley and Sons Ltd, 2014)

R.V. Meter, T. Matsuo, Connection Setup in a Quantum Network, Work in Progress, Internet-Draft, draft-van-meter-qirg-quantum-connection-setup-01 (2019)
R.V. Meter, J. Touch, Designing quantum repeater networks. IEEE Commun. Mag. **51**, 64–71 (2013)
R.V. Meter, J. Touch, Designing quantum repeater networks. IEEE Commun. Maga. **51**, 64–71 (2013)
R.V. Meter, K. Nemoto, W.J. Munro, K.M. Itoh, Distributed arithmetic on a quantum multicomputer. ACM SIGARCH Comput. Arch. **34**, 354–365 (2006)
R.V. Meter, T.D. Ladd, W.J. Munro, K. Nemoto, System design for a long-line quantum repeater. IEEE/ACM Trans. Networking **17**, 1002–1013 (2008)
R.V. Meter, W.J. Munro, K. Nemoto, *Architecture of a quantum multicomputer implementing Shor's algorithm, in 3rd Workshop on Theory of Quantum Computation, Communication, and Cryptography* (Japan, Tokyo, TQC, 2008), pp.105–114
J. Miguel-Ramiro, A. Pirker, W. Dür, Genuine quantum networks with superposed tasks and addressing. npj Quantum Inf. **7**, 135 (2021)
A. Mizutani, M. Curty, C.C.W. Lim, N. Imoto, K. Tamaki, Finite-key security analysis of quantum key distribution with imperfect light sources. New J. Phys. **17**, 093011 (2015)
W.J. Munro, K.A. Harrison, A.M. Stephens, S.J. Devitt, K. Nemoto, From quantum multiplexing to high-performance quantum networking. Nat. Photon. **4**, 792–796 (2010)
W.J. Munro, K. Azuma, K. Tamaki, K. Nemoto, Routing on quantum repeater networks. IEEE J. Sel. Top. Quant. **21**, 78 (2015)
Ö. Erkılıç, L. Conlon, B. Shajilal, S. Kish, S. Tserkis, Y.-S. Kim, P.K. Lam, S.M. Assad, Surpassing the repeaterless bound with a photon-number encoded measurement-device-independent quantum key distribution protocol. npj Quantum Inf **9**, 29 (2023)
J.L. O'Brien, A. Furusawa, Quantum information processing and communication. Science **318**, 1567–1570 (2009)
S.P. Pal, S.K. Singh, S. Kumar, Multi-partite quantum entanglement versus randomization: fair and unbiased leader election in networks (2003) arXiv:quant-ph/0306195v1
M. Pant, H. Krovi, D. Towsley, L. Tassiulas, L. Jiang, P. Basu, D. Englund, S. Guha, Routing entanglement in the quantum Internet. npj Quantum Inf. **5**, 25 (2019)
A. Patil, M. Pant, D. Englund, D. Towsley, S. Guha, Entanglement generation in a quantum network at distance-independent rate. npj Quantum Inf. **8**, 51 (2022)
L. Pezzé, A. Smerzi, M.K. Oberthaler, R. Schmied, P. Treutlein, Quantum metrology with nonclassical states of atomic ensembles. Rev. Mod. Phys. **90**, 035005 (2018)
S. Pirandola, R. Laurenza, C. Ottaviani, L. Banchi, Fundamental limits of repeaterless quantum communications. Nat. Commun. **8**, 15043 (2017)
S. Pirandola, S.L. Braunstein, S. Lloyd, Quantum internet: A vision for the road ahead. Nat. Photon. **12**, 724–726 (2019)
A. Pirker, W. Dür, A quantum network stack and protocols for reliable entanglement-based networks. New J. Phys. **21**, 033003 (2019)
J. Preskill, Quantum computing in the NISQ era and beyond. Quantum **2**, 79 (2018)
T.J. Proctor, A. Paul, Knott, A. Jacob, Dunningham, Multiparameter estimation in networked quantum sensors. Phys. Rev. Lett. **120**, 080501 (2018)
F. Riera-Sábat, P. Sekatski, A. Pirker, W. Dür, Entanglement-assisted entanglement purification. Phys. Rev. Lett. **127**, 040502 (2021)
S. Ritter et al., An elementary quantum network of single atoms in optical cavities. Nature **484**, 7393 (2012)
L.G. Roberts, Principles and lessons in packet communications. Proc. IEEE **66**, 1320–1329 (1978)
A.A. Saleh, J.M. Simmons, All-optical networking-evolution, benefits, challenges, and future vision. Proc. IEEE. **100**, 1105 (2012)
M.A. Taherkhani, K. Navi, R. Van Meter, Resource-aware system architecture model for implementation of quantum aided byzantine agreement on quantum repeater networks. Quantum Sci. Technol. **3**, 014011 (2018)

M. Takeoka, S. Guha, M.M. Wilde, Fundamental rate-loss tradeoff for optical quantum key distribution. Nat. Commun. **5**, 5235 (2014)

B.Y. Tang, B. Liu, Y.P. Zhai, C.Q. Wu, W.R. Yu, High-speed and large-scale privacy amplification scheme for Quantum Key Distribution. Sci. Reps. **9**, 15733 (2019)

M.A. Taylor, W.P. Bowen, Quantum metrology and its application in biology. Phys. Rep. **615**, 1 (2016)

P.D. Townsend, Simultaneous quantum cryptographic key distribution and conventional data transmission over installed fibre using wavelength-division multiplexing. Electronics Lett. **33**, 188 (1997)

A. Treiber et al., A fully automated entanglement-based quantum cyptography system for telecom fiber networks. New J. Phys. **11**, 045013 (2009)

X.-B. Wang, Beating the photon-number-splitting attack in practical quantum cryptography. Phys. Rev. Lett. **94**, 230503 (2005)

R. Wang et al., End-to-end quantum secured inter-domain 5G service orchestration over dynamically switched flex-grid optical networks enabled by a q-ROADM. J. Light. Techn. **38**, 139 (2019)

F. Wang, M.X. Luo, G. Xu, X.-B. Chen, Y.X. Yang, Photonic quantum network transmission assisted by the weak cross-Kerr nonlinearity. Sci. China Phys. Mech. Astr. **61**, 060312 (2018)

S. Wehner, D. Elkouss, R. Hanson, Quantum internet: A vision for the road ahead. Science **362**, 9288 (2018)

W.K. Wootters, W.H. Zurek, A single quantum cannot be cloned. Nature **299**, 802 (1982)

P.-S. Yan, L. Zhou, W. Zhong, Y.-B. Sheng, Advances in quantum entanglement purification. Sci. China Phys. Mech. Ast. **66**, 250301 (2023)

H.-L. Yin, Y. Fu, Measurement-device-independent twin-field quantum key distribution. Sci. Rep. **9**, 3045 (2019)

P. Zhang, et. al., Integrated relay server for measurement-device-independent Quantum Key Distribution (2019) arxiv:1912.09642

Q. Zhang, F. Xu, Y.-A. Chen, C.-Z. Peng, J.-W. Pan, Large scale quantum key distribution: challenges and solutions. Opt. Exp. **26**, 24260 (2018)

M. Zukowski, A. Zeilinger, M.A. Horne, A.K. Ekert, Event-ready-detectors Bell experiment via entanglement swapping. Phys. Rev. Lett. **71**, 4287–4290 (1993)

Chapter 3
Quantum Network Communication

In this chapter, we introduce some main ideas of quantum communication in quantum networks. For general network communication with multiple senders and receivers, it is generally NP-hard to find all the possible communication capacities. Instead, each communication protocol may be divided into three sub-protocols: one is point-to-point communication of any two pair (point-to-point communication); the second is from multiple senders to one sender (multiple-access communication); the last is from one sender to multiple receivers (multicast communication). With this communication strategy, we present quantum communication for these three scenarios.

3.1 Quantum Point-To-Point Communication

We introduce some main ideas of quantum communication in quantum point-to-point communication . We will provide a unified framework of quantum communication in different settings.

3.1.1 Introduction

All the considerations in this chapter include the following four capacities:

- The ordinary classical capacity C, is defined as the maximum asymptotic rate at which classical bits can be transmitted reliably through the channel, with the help of a quantum encoder and decoder.
- The ordinary quantum capacity Q, which is the maximum asymptotic rate at which qubits can be transmitted under similar circumstances.

M.-X. Luo, *Quantum Networks*,
https://doi.org/10.1007/978-981-97-6226-2_3

- The classically assisted quantum capacity Q_2, which is the maximum asymptotic rate of reliable qubit transmission with the help of unlimited use of a 2-way classical side channel between sender and receiver.
- The entanglement-assisted classical capacity C_E, which is the maximum asymptotic rate of reliable bit transmission with the help of unlimited prior entanglement between the sender and receiver.

In most cases, only the upper and lower bounds on these capacities are known, not the capacities themselves.

Aside from the constraints $Q \leq C \leq C_E$, and $Q \leq Q_2$, which are obvious consequences of the definitions, the four capacities appear to vary rather independently. It is conjectured that $Q_2 \leq C$, but this has not been proved to date. Except in special cases, it is not possible, without knowing the parameters of a channel, to infer any one of its four capacities from the other three.

With Wiesner's superdense coding , the classical capacity of a noiseless quantum channel is doubled by prior entanglement, i.e., $C_E = 2C$ for any noiseless quantum channel. This result persists even for some noisy channels where the quantum capacities Q and Q_2 both vanish. The main ingredient is that the channel itself can be simulated by LOCC between the sender and receiver. In general, for any quantum channel, we have $C \leq C_E \leq C_{fc}$, where C_{fc} denotes the forward classical communication cost, i.e., the forward classical capacity needed, in conjunction with other resources, to simulate the quantum channel. The comparison of the capacities of several simple channels is shown King (2002).

3.1.2 Unified Model

In this subsection, we provide a unified model for describing three different communication models. Here, we consider only unit-preserving quantum evolutions operators as

$$\mathcal{F}(\rho) = \sum_{\mu} A_{\mu}^{\dagger} \rho A_{\mu}, \tag{3.1}$$

where $\{A_\mu\}$ is a set of operators satisfying $\sum_\mu A_\mu A_\mu^\dagger = \sum_\mu A_\mu^\dagger A_\mu = \mathbb{1}$ with the identity operator $\mathbb{1}$.

The problem of coding is generated by practical problems, which are connected with compressing of a message or data transferring in a channel without noise.

Definition 3.1.1 Define the coding $\mathcal{E}$ for a given state $\rho = \sum_i p_i |\psi_i\rangle\langle\psi_i|$ as

$$|\psi_i\rangle\langle\psi_i| \mapsto \mathcal{E}(|\psi_i\rangle\langle\psi_i|) = \gamma_i, \tag{3.2}$$

$$\rho \mapsto \rho_1 = \sum_i p_i \gamma_i. \tag{3.3}$$

The decoding $\mathcal{D}$ on a given state ρ_1 is defined as

$$\gamma_i \mapsto \mathcal{D}(\gamma_i) = \omega_i, \tag{3.4}$$

$$\rho_1 \mapsto \rho_2 = \sum_i p_i \omega_i, \tag{3.5}$$

where $\dim \rho = \dim \rho_2 \geq \dim \rho_1$.

The faithful transmission means for specific $\mathcal{E}$ and $\mathcal{D}$ with the minimal dimension of ρ_1 for a fixed state ρ such that the fidelity of ρ and ρ_2 closes to 1. In particular, $\mathcal{E}$ and $\mathcal{D}$ may be linear combinations of unitary operations, measurements, partial traces, and others.

Definition 3.1.2 The rate of coding for quantum source is defined as

$$R = \log \dim \rho_1. \tag{3.6}$$

The quantity of coding rate characterizes the degree of compressing for ρ.

We describe the coding problem for a quantum noisy channel, defined by mapping $\mathcal{N}$. Similar to Shannon's coding theory, the meaning of source coding is to compress the redundancy. In noisy channel coding, one obtains faithful transmitting of density matrix ρ by introducing the redundancy. In general, the quantum state over a noisy channel can be described by the following input-output system as

$$\rho \xrightarrow{\text{Encoding}} \mathcal{E}(\rho) \xrightarrow{\text{Transmitting}} \mathcal{N} \circ \mathcal{E}(\rho) \xrightarrow{\text{Decoding}} \mathcal{D} \circ \mathcal{N} \circ \mathcal{E}(\rho) = \rho_2. \tag{3.7}$$

Here, $\dim \rho = \dim \rho_2 \leq \dim \mathcal{E}(\rho)$. The goal is to find $\mathcal{E}$ and $\mathcal{D}$ for a fixed ρ with the fidelity F_e between ρ and ρ_2 being close to 1.

Definition 3.1.3 A classical-quantum (CQ) channel $\mathcal{N}_{\mathcal{X}\to B}$ describes a physical setup in which the sender can remotely prepare any of conditional density matrices $\{\rho_x^B\}_{x\in\mathcal{X}}$ in the receiver.

Definition 3.1.4 A *quantum channel* $\mathcal{N}_{A\to B}$ means a CPTP map from density matrices on A to those on B.

Definition 3.1.5 Given a channel $\mathcal{N}_{A\to B}$, there is an *isometry* channel $\mathcal{U}_{A\to BE}$ which maps the state in A into an extended state in BE where E denotes the unobservable environment, i.e., $\mathcal{N}_{A\to B} = \mathrm{Tr}_E \mathcal{U}_{A\to BE}$. Here, $\mathcal{U}_{A\to BE}$ is named as an *isometric extension* of $\mathcal{N}_{A\to B}$.

Denote $\mathcal{E}$ as an encoding protocol by Alice, in which she receives an m-qubit state $\psi \in \Gamma_m$, and produces n possibly entangled inputs to the channel $\mathcal{N}$. Let $\mathcal{D}$ be a decoding protocol by Bob, in which he receives n (possibly entangled) channel outputs and produces an m-qubit output. Denote $F_e(\rho, \mathcal{E}, \mathcal{D}, \mathcal{N})$ as the fidelity of this output relative to the input state ψ, i.e., the probability that the output state passes a test determining whether it equals to the input.

Denote C_X as a generalized capacity from the sender Alice to receiver Bob. We present a general capacity of a quantum channel as follows.

Definition 3.1.6 The capacities of a quantum channel $\mathcal{N}$ is defined within a common framework as

$$C_X(\mathcal{N}) = \lim_{\epsilon \to 0} \lim_{n \to \infty} \sup \left\{ \frac{m}{n} : \exists \mathcal{E}, \mathcal{D}, \forall \rho \in \Gamma_m, F_e(\rho, \mathcal{E}, \mathcal{D}, \mathcal{N}) > 1 - \epsilon \right\}. \quad (3.8)$$

The capacities depend on the specification of Γ_m, $\mathcal{E}$ and $\mathcal{D}$. The classical capacities C and C_E are defined by restricting the input state ψ to a standard orthogonal set of states, without loss of generality the "Boolean" states labeled by bit strings $\Gamma_m = \{|0\rangle, |1\rangle\}^{\otimes m}$.

The classical capacity of a classical discrete memoryless channel is also given by an expression of the same form, with ψ restricted to Boolean values; the encoder $\mathcal{E}$, decoder $\mathcal{D}$, and channel $\mathcal{N}$ all being restricted to be classical stochastic maps. The fidelity F_e is defined as the probability that the (Boolean) output is equal to the input state ψ. We will sometimes indicate these restrictions implicitly by using upper case italic letters (e.g. N_c) for classical stochastic maps, and lower case italic letters (e.g. x) for classical discrete data.

Definition 3.1.7 The classical capacity is defined by

$$\begin{aligned} C(N_c) = \lim_{\epsilon \to 0} \lim_{n \to \infty} \sup\{\frac{m}{n} &: \exists \mathcal{E}, \mathcal{D}, \forall x \in \{0, 1\}^m, \\ &F_e(x, \mathcal{E}, \mathcal{D}, N_c) > 1 - \epsilon\}. \end{aligned} \quad (3.9)$$

A classical stochastic map, or classical channel, may be defined in quantum terms. Specially, it is completely dephasing in the Boolean basis both with regard to its inputs and its outputs.

Definition 3.1.8 A completely dephasing channel makes a von Neumann measurement in the Boolean basis and resends the result of the measurement.

Dephasing only the inputs or only the outputs is insufficient to abolish all quantum properties of a quantum channel $\mathcal{N}$.

Definition 3.1.9 A channel is classical if and only if it can be represented as a composition of the completely dephasing channel $\mathcal{D}_e$ on the input Hilbert space, a general quantum channel $\mathcal{N}$, and the completely dephasing channel $\mathcal{D}'_e$ on the output Hilbert space, i.e.,

$$N_c = \mathcal{D}'_e \circ \mathcal{N} \circ \mathcal{D}_e. \quad (3.10)$$

For the quantum capacities Q and Q_2, it follows that $\Gamma_m = \mathbb{H}_2^{\otimes m}$. For the capacities Q and C, the protocols between Alice and Bob are the CPTP maps from $\mathbb{H}_2^{\otimes m}$ to the input space of $\mathcal{N}^{\otimes n}$, and from the output space of $\mathcal{N}^{\otimes n}$ back to $\mathbb{H}_2^{\otimes m}$, respectively.

For capacities C_E and Q_2, the protocols are more complicated. In the first case, suppose that both parties share lots of ebits (maximally entangled pairs of qubits). And then, they use a 2-way classical channel. The definition of Q_2 includes interactive protocols, in which the classical communication may be implemented in rounds.

The notion of capacity may be further generalized to define a capacity of one channel $\mathcal{N}$ to simulate another channel $\mathcal{M}$. This implies a new form of unified capacity as

$$C_X(\mathcal{N} \to \mathcal{M}) = \lim_{\epsilon \to 0} \lim_{n \to \infty} \sup \big\{ \tfrac{m}{n} : \exists \mathcal{E}, \mathcal{D}, \forall \psi \in \mathbb{H}_M^*, \\ F_e(\mathcal{M}^{\otimes m}(\psi), \mathcal{E}, \mathcal{D}, \mathcal{N}) > 1 - \epsilon \big\}. \tag{3.11}$$

where $\mathcal{E}$ and $\mathcal{D}$ are respectively Alice's and Bob's protocols which together enable Alice to receive an input $\psi \in \mathbb{H}_M^{\otimes m}$ and, making n forward use of the simulating channel $\mathcal{N}$, allow Bob to produce some output state, and $F_e(\mathcal{M}^{\otimes m}(\psi), \mathcal{E}, \mathcal{D}, \mathcal{N})$ is the fidelity of this output state with respect to the state that would have been generated by sending the input ψ through $\mathcal{M}^{\otimes m}$.

These definitions of capacity are all asymptotic, depending on the properties of $\mathcal{N}^n$ in the limit $n \to \infty$. However, several of the capacities are given by, or closely related to, non-asymptotic expressions involving input and output entropies for a single use of the channel. As an example, a quantum system A, initially in mixed state ρ, is sent through the channel, emerging in a mixed state $\mathcal{N}(\rho)$. It is useful to think of the initially mixed state as being part of an entangled pure state $|\Phi\rangle_p^{AR}$, where R is some reference system that is never operated upon physically. Similarly, the channel can be thought of as a unitary interaction U between the quantum system A and some environment subsystem E, which is initially supplied in a standard pure state $|0\rangle_E$, and leaves the interaction in a mixed state $\mathcal{E}_e(\rho)_E$. Thus $\mathcal{N}$ and $\mathcal{E}_e$ are completely positive maps relating the final states of the channel output and environment, respectively, to the initial state of the channel input, when the initial state of the environment is fixed. For this kind of communication, quantum coherent information will be used.

3.1.3 Quantum Coding Theorems

For generality, we formulate the coding theorems in terms of blocks with length n. Denote $\rho^n = \rho \otimes \cdots \otimes \rho$. The direct theorems of information theory have asymptotic characters and are valid only for $n \gg 1$. Here, $\mathcal{E}^n$, $\mathcal{D}^n$, and $\mathcal{N}^n$ act on ρ^n. If $\mathcal{E}^n = \mathcal{E} \otimes \cdots \otimes \mathcal{E}$, then we have memoryless channel (source). We work with any n and with channel (source) with any memory.

Theorem 3.1.1 (Quantum source coding theorem) *If $R < S(\rho)$ then the fidelity F_e between the input states ρ^n and the decoded states ρ_2^n is not close to 1 for any n, encoding and decoding mappings.*

Proof We present the skeptical idea of the proof. There exists some δ such that $F_e \leq \delta \leq 1$. We obtain that

$$nR \geq S(\rho_1^n) \geq I(\rho^n; \mathcal{E}^n(\rho^n)), \tag{3.12}$$

where I denotes the quantum mutual information. From the assumption of transformations $\mathcal{E}^n$, it follows that

$$nR \geq I(\rho^n; \mathcal{D}^n \circ \mathcal{E}^n(\rho^n)) = S(\rho_2^n) - S(\rho^n \mapsto \rho_2^n). \tag{3.13}$$

With the help of H-theorem Czekaj (2011), we obtain that

$$nR \geq nS(\rho) - S(\rho^n \mapsto \rho_2^n). \tag{3.14}$$

If $R = S(\rho) - \delta$, with $\delta > 0$ the quantum Fano inequality implies that

$$\delta \leq \frac{1}{n}((1 - F_e)\log(d^{2n} - 1) + h(F_e)), \tag{3.15}$$

where $h(F_e)$ denotes the binary Shannon entropy. For any n, the fidelity F_e is not close to 1. If $n \to \infty$, we obtain that

$$\delta \leq 2(1 - F_e)\log d. \tag{3.16}$$

The capacity C_s of a quantum source is defined as follows.

- If $R \leq C_s$, then there exist some $\mathcal{D}^n$ and $\mathcal{E}^n$ with F_e being close to 1.
- If $R \geq C_s$, then for any $\mathcal{D}^n$ and $\mathcal{E}^n$, the fidelity F_e is not close to 1.

It is well-known that there exist $\mathcal{D}^n$ and $\mathcal{E}^n$ with F_e being close to 1, and $\log\dim(\rho) = nS(\rho)$ for large n Schumacher (1995). The present inequality shows that $C_s \geq S(\rho)$ for sufficiently general coding and decoding schemes. □

We now go to the converse theorem for the noisy quantum channel. In channel coding, we introduce the redundancy in the step $\rho \mapsto \rho^n$ (in source coding problem we directly work with ρ^n). After this, $\mathcal{E}^n$ becomes a unitary transformation. The concrete form of this transformation depends on the concrete coding schemes. The decoding transformation $\mathcal{D}^n$ of $\mathcal{E}^n$ may be nonunitary because it requires determining an unknown state. In other words, in channel coding $\mathcal{E}^n$ is a unitary transformation, and $\mathcal{D}^n$ is a general but unit-preserving.

We introduce the quantity C_s as

$$C_s = \max_{\{p_i\}} \frac{1}{n} I(\rho^{(n)}; \mathcal{W}^n(\rho^{(n)})), \tag{3.17}$$

where $\mathcal{W}^n$ is a quantum channel. We define the rate of noisy channel coding as

$$R_c = \max_{\{p_i\}} S(\rho). \tag{3.18}$$

If all the states $|\psi_i\rangle$ satisfy $\langle\psi_i|\psi_j\rangle = \delta_{ij}$, we obtain a classical-like input ensemble of random variables. Here, the entropy $S(\rho^n)$ is maximized by the uniform distribution of p_i. This yields to the classical expression for R_c as

$$R_c = \log d. \tag{3.19}$$

With the channel coding, we can prove the following theorem.

Theorem 3.1.2 (Quantum converse source coding theorem) *If $R_c \geq \tilde{C}$, then the fidelity F_e between the states ρ^n and ρ_2^n is not close to 1 for any n, encoding mapping $\mathcal{E}^n$ and decoding mapping $\mathcal{D}^n$, i.e., there exists some δ such that $F_e \leq \delta < 1$.*

In what follows, we introduce three kinds of von Neumann entropies. One is the input entropy $S(\rho_A) = -\operatorname{Tr}\rho_A \log \rho_A$. The second is the output entropy $S(\mathcal{N}(\rho)_A)$. The third is the entropy exchange that is given by

$$S((\mathcal{N} \otimes \mathbb{1})(\Phi_\rho)_{AR}) = S(\mathcal{E}(\rho)_E). \tag{3.20}$$

The left side represents the entropy of the joint state of the subsystem A transmitted through the channel, and the reference system R. The density operator $(\mathcal{N} \otimes \mathbb{1})(\Phi_\rho)$ is the quantum analog of a joint input-output probability distribution. Without the reference system, a joint input-output mixed state may be wrong because the input and output are not present at the same time. The entropy exchange is equal to the final entropy of the environment $S(\mathcal{E}(\rho))$, because the tripartite system ARE remains in a pure state, which make two complementary subsystems E and AR always isospectral. For the relations between these entropies and quantum channels, see refs. Schumacher and Nielsen (1996), Holevo and Werner (2001).

By the Shannon coding theorem Cover (1972), the capacity of a classical channel N_c is the maximum of the input-output mutual information over input distributions. The quantum generalization of mutual information for a bipartite mixed state ρ_{AB}, which reduces to classical mutual information when ρ_{AB} is diagonal in a product basis of the two subsystems, is given by

$$I_{\rho_{AB}^c}(A; B) = S(\rho_A^c) + S(\rho_B^c) - S(\rho_{AB}^c), \tag{3.21}$$

where ρ_A^c and ρ_B^c are reduced density matrices defined by $\rho_A^c = \operatorname{Tr}_B \rho_{AB}^c$ and $\rho_B^c = \operatorname{Tr}_A \rho_{AB}^c$.

The capacity of a classical channel can be expressed as

$$C(N_c) = \max_{\rho^c \in \Delta} \{S(\rho^c) + S(N_c(\rho^c)) - S((N_c \otimes \mathbb{1})(\Phi_{\rho^c}))\}, \tag{3.22}$$

where Δ denotes the set consisting of density operators on the channel's input Hilbert space that are diagonal in the Boolean basis. The third term (entropy exchange), for a classical channel N_c, is the joint Shannon entropy of the classically correlated Boolean input and output because the von Neumann entropies reduce to Shannon entropies when evaluated in the Schmidt basis of Φ_{ρ^c}, with respect to which all states are diagonal.

The restriction to classical inputs $\rho \in \Delta$ can be removed, because any nondiagonal elements in ρ would only reduce the first term, while leaving the other two terms unchanged. A generalization to quantum channels $\mathcal{N}$ of a classical channel's maximal input-output mutual information is given by

$$I(A; B) = \max_{\rho \in \mathbb{H}_{\text{in}}} \{S(\rho) + S(\mathcal{N}(\rho)) - S((\mathcal{N} \otimes \mathbb{1})\Phi_\rho)\}. \tag{3.23}$$

This is equal to the classical capacity when $\mathcal{N}$ is classical.

Consider the following protocol for the classical information transmission through a quantum channel $\mathcal{N}$ from Alice to Bob. Systems A and B of the same dimension are in an entangled (pure) state ρ_{AB}. Alice does some encoding $i \rightarrow \mathcal{E}_A^i$ depending on a classical signal i with probabilities p_i and sends the shared state through the channel $\mathcal{N}$ to Bob. Thus Bob obtains the states $(\mathcal{N} \otimes \mathbb{1}_B)(\gamma_{AB}^i)$, where $\gamma_{AB}^i = (\mathcal{E}_A^i \otimes \mathbb{1}_B)(\rho_{AB})$ with probabilities p_i. Bob is then to extract the maximal classical information by measuring these states. Similar to the dense coding, but instead of the ideal channel, they use a noisy channel $\mathcal{N}$. We now explore the classical capacity of this protocol, which is called entanglement-assisted classical capacity C_E of the channel $\mathcal{N}$.

Definition 3.1.10 The *one-shot entanglement-assisted classical capacity* is given by

$$\begin{aligned} C_{ea}^{(1)}(\mathcal{N}) = & \max_{\{p_i\}, \mathcal{E}_A^i, \rho_{AB}} \{S(\sum_i p_i (\mathcal{N} \otimes \mathbb{1}_B)(\gamma_{AB}^i)) \\ & - \sum_i p_i S((\mathcal{N} \otimes \mathbb{1}_B)(\gamma_{AB}^i))\}. \end{aligned} \tag{3.24}$$

Using n times of the channel, assume Bob can perform collective measurement. This implies that

$$C_{ea}^{(n)}(\mathcal{N}) = C_{ea}^{(1)}(\mathcal{N}^n). \tag{3.25}$$

The full entanglement-assisted classical capacity is then given by

$$C_{ea}(\mathcal{N}) = \lim_{n \to \infty} \frac{1}{n} C_{ea}^{(1)}(\mathcal{N}^n). \tag{3.26}$$

Theorem 3.1.3 *The entanglement-assisted classical capacity is given by*

$$C_{ea}(\mathcal{N}) = \max_{\rho_A} I(\rho_A; \mathcal{N}), \tag{3.27}$$

where $I(\rho_A; \mathcal{N})$ *is the quantum mutual information defined by* $I(\rho_A; \mathcal{N}) = S(\rho_A) + S(\mathcal{N}(\rho_A))$ -$S(\rho_A, \mathcal{N})$ *with the entropy exchange* $S(\rho_A, \mathcal{N})$.

The Theorem 3.1.3 provides the classical capacity in entanglement-assisted quantum communication models Bennett et al. (2002). It may be achieved asymptotically by combining the superdense coding with Schumacher-coding of the states ρ^n for large n. The capacity is evaluated by using Holevo's formula Holevo (1998). Further results of point-to-point communication, see book Wilde (2013).

3.2 Quantum Multiple-Access Communication

Multi-access communication is important network models that allow several senders to communicate with the same receiver via shared channels. The channel of this communication is called multiple access channels (MAC). The network channel in applications may be a single fiber connecting multiple nodes, or a portion of the wireless spectrum. We present quantum communication over MACs in this section.

3.2.1 Classical Capacity

We consider the multi-sender communication with s spatially separated senders $\vec{\mathsf{A}} := (\mathsf{A}_1, \cdots, \mathsf{A}_s)$ and one receiver B. The receiver B obtains some output data b that depends on the collection of messages $\vec{a} = (a_1, \cdots, a_s)$ chosen by the senders. Ideally, $\vec{a}$ would be a perfect copy of all the s messages. However, in practice, some physical limitations prevent perfect communication. In such scenarios, the communication is described by the transition probabilities $p(b|\vec{a})$. The distributions $\{p(b|\vec{a})\}$ collectively represent a MAC, where the probabilities $p(b|\vec{a})$ are determined by the particular physical system used in applications.

For the special case of $(\mathcal{X} \times \mathcal{Y}, p(z|x, y), \mathcal{Z})$ with two senders and one receiver, the capacity involves finding the region of achievable transmission rates R_1 and R_2 for two senders. The classical capacity region $C_{\mathrm{MAC}}(R_1, R_2)$ of a MAC was found Ahlswede (1971), Liao (1972) as

$$C_{\mathrm{MAC}}(R_1, R_2) := \mathrm{Cl}\left\{ (R_1, R_1) \,\middle|\, \begin{array}{rl} R_1 & \leq I(X; Z|Y); \\ R_2 & \leq I(Y; Z|X); \\ R_1 + R_2 & \leq \ I(XY; Z) \end{array} \right\} \tag{3.28}$$

for some product distribution $P(x, y) = p(x)p(y)$, where Cl denotes the closure of the convex hull of all possible rates (R_1, R_2). The joint distribution of XYZ is given by $p(x)p(y)p(z|x, y)$. The conditional mutual information is defined as $I(X; Z|Y) = I(X; YZ) - I(X; Y)$ in terms of Shannon entropy. We consider a similar problem of quantum multiple access channels as follows.

This is the simplest situation of multi-user communication, as shown in Fig. 3.1, where all senders transmit general quantum states, i.e., quantum multiple access channels (QMAC) . For the special case of transmitting classical states or classical messages, the sender A_i uses a (finite) alphabet X_i with a priori probability distribution p_i Winter (2001). The receiver has partial access the output system. Their task is to reconstruct the input messages. One possible way is by using block coding and coding/decoding schemes.

Definition 3.2.1 The QMAC is represented by a mapping defined by

$$\mathcal{W} : \vec{\mathcal{X}} \triangleq \mathcal{X}_1 \times \cdots \times \mathcal{X}_s \to \mathcal{D}(\mathbb{H}) \tag{3.29}$$

from the input alphabets into the set $\mathcal{S}(\mathbb{H})$ of states of the (finite dimensional Hilbert space $\mathbb{H}$), which maps the input messages $\vec{x} = (x_1, \cdots, x_s)$ to the output state $W_{\vec{x}}$.

If all the output states $W_{\vec{x}}$ commute with each other (hence have a common diagonalization) the channel is called quasi-classical, and classical if $\mathcal{D}$ is a commutative algebra.

For fixed a priori distributions, define the channel state as

$$\gamma = \sum_{\forall i, x_i \in X_i} p(x_1) \cdots p(x_s) |\vec{x}\rangle\langle\vec{x}| \otimes W_{\vec{x}} \tag{3.30}$$

on $\mathcal{X}_1 \otimes \cdots \otimes \mathcal{X}_s \otimes \mathcal{D}$. This serves as a quantum analog of the joint distribution of the random variables in the classical case. This can be interpreted as the joint state of the system after channel usage, where the senders kept a record of their letters.

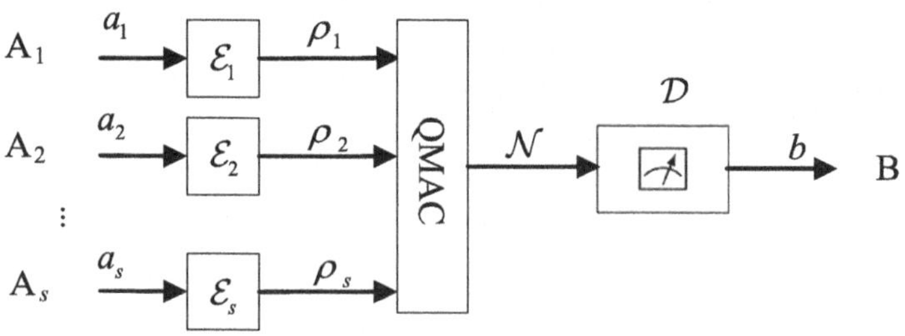

Fig. 3.1 A quantum MAC system with s senders $\mathsf{A}_1, \cdots, \mathsf{A}_s$ who want to transmit the state $\rho_1, \cdots, \rho_s$ to B, respectively. Here, $\mathcal{E}_i$ denotes the quantum coding mapping by A_i. $\mathcal{D}$ denotes the quantum decoding mapping by B on all the received states. $\mathcal{N}$ denotes the quantum channels

We will use block coding on the discrete memoryless channel generated by $\mathcal{W}$: for sequences $\vec{x}_1^n, \cdots, \vec{x}_s^n$ with $\vec{x}_i^n = (x_{i1}, \cdots, x_{in}) \in X_i^n$. Specially, the n-block channel $\mathcal{W}^n : X_1^n \times \cdots \times X_s^n \to \mathfrak{S}(\mathcal{D}^{\otimes n})$ is defined by the output state as

$$W_{\vec{x}_1^n}^n \triangleq W_{\vec{x}_1} \otimes W_{\vec{x}_2} \otimes \cdots \otimes W_{\vec{x}_n}. \tag{3.31}$$

We introduce some notation to describe the channel by a subset of senders, while the others enter only stochastically. For a subset $J \subset [s] = \{1, \cdots, s\}$, denote $\mathcal{P}_J = \otimes_{i\in J}\mathcal{P}_i$, i.e., $P_J(x_i|i \in J) = \prod_{i\in J} P_i(x_i)$, and $X_J = \prod_{i\in X_i}$. Similarly, denote $\mathcal{X}_J = \otimes_{i\in J}\mathcal{X}_i$. If $J = [s]$, denote $\vec{\mathcal{P}} = \otimes_{i\in J}\mathcal{P}_i$, i.e., $\vec{P}(x_i|i \in J) = \prod_{i\in J} P_i(x_i)$, and $\vec{X} = \prod_{i\in X_i}$. Similarly, denote $\vec{\mathfrak{X}} = \otimes_{i\in J}\mathfrak{X}_i$.

Define the reduced channel $\mathcal{P}_{J^c} \circ \mathcal{P} : \mathcal{X}_J \to \mathfrak{S}(\mathfrak{D})$ by

$$(P_{J^c}W)(x_i|i \in J) \triangleq \sum_{\forall i\in J^c:x_i\in X_i} P_{J^c}(x_i|i \in J^c)W_{\vec{x}}, \tag{3.32}$$

where J^c denotes the complement subset $[s] \setminus J$ of J in $[s]$. The transmission takes use of n-block codes:

Definition 3.2.2 (n-block-code Winter (2001)) An n-block-code is a collection $(\vec{f}, \vec{D})$:= $(f_1, \cdots, f_s, D_1, \cdots, D_r)$ of mappings $f_i : \mathcal{M}_i \to X_i^n$, where $\mathcal{M}_i$ is the set of messages of the sender i. The decoding mapping is given by $D_j \subset \mathfrak{D}_j^{\otimes n}$ and indexed by $\vec{\mathcal{M}} := \mathcal{M}_1 \times \cdots \times \mathcal{M}_s$, i.e.,

$$D_j = \{D_{jt} \in \mathfrak{D}_j : t \in \vec{\mathcal{M}}\}, \tag{3.33}$$

such that $D_{jt} \geq 0$ and $\sum_t D_{jt} = \mathbb{1}$.

In general, we consider all the r error probabilities of the code, i.e., the average probability that the receiver cannot decode correctly the transmitted messages. By using the uniform distribution of the input messages, we obtain that the error probability of decoding the received states as

$$P_{e_j}(\vec{f}, D_j) = 1 - \frac{1}{\prod_{i=1}^s |\mathcal{M}_i|} \sum_{\forall i:m_i\in\mathcal{M}_i} \mathrm{Tr}(W_{f(m_1)\cdots f(m_s)}^n D_{j,\vec{m}}), \tag{3.34}$$

where $(\vec{f}, \vec{D})$ is regarded as an (n, ϵ)-code if $P_{e_j}(\vec{f}, D_j) \leq \epsilon$ for any j.

The rates of an n-block-code are given by $R_i = \frac{1}{n}\log|\mathcal{M}_i|$. A tuple $(R_1, \cdots, R_s)$ is achievable, if for any $\epsilon, \delta > 0$ there is an (n, ϵ)-code with the i-th realized rate no less than $R_i - \delta$ for any large enough n. All the achievable tuples are named as the capacity region of the given channel, which is generally closed. There are some simple facts about the block-code.

- The capacity region is convex. The main reason is from a time-sharing principle for two block codes. In fact, for any two m-block and n-block codes with rate

tuples $(R_1, \cdots, R_s)$ and $(R'_1, \cdots, R'_s)$, respectively, the concatenated codewords lead to an $(m+n)$-block code. By using the tensored decoding observables, the error probability is no more than the summation of error probabilities for two block codes. The rate is given by $(mR_i + nR'_i)/(m+n)$. The concatenating operation may be iteratively used for multiple block codes.
- It requires a rate tuple but not a single rate for characterizing the channel. Similar to the interference in classical multi-user channels, the other senders' actions in quantum networks may be regarded as noise for specific senders.

With these definitions for the channel state γ, we obtain that

$$\begin{aligned} I(X_J; \mathfrak{D}|\mathfrak{X}_{J^c}) &= I(X_J; \mathfrak{D}\mathfrak{X}_{J^c}) \\ &= S(P_J W|\mathcal{P}_{J^c}) - S(W|\vec{P}), \end{aligned} \tag{3.35}$$

For a channel $\mathcal{N} : \mathcal{D} \to \mathfrak{S}(\mathfrak{Z})$ and a probability distribution p on $\mathcal{A}$, let

$$S(\mathcal{N}|p) = \sum_{a\in\mathcal{A}} p(a) S(X_a), \tag{3.36}$$

with the von Neumann entropy S. Similar to the subadditivity of classical channels, we can easily prove the following lemma for quantum channels.

Lemma 3.2.1 *Let $\mathcal{N}_k : X_k \to \mathfrak{S}(\mathfrak{Z}_k)(k = 1, 2, \cdots, s)$ be s channels, and p a probability distribution on $X_1 \times \cdots \times X_s$. By forming the channel state*

$$\gamma = \sum_{x_k\in X_k : k=1,\cdots,s} p(x_1, \cdots, x_s) \otimes_{i=1}^{s} |x_i\rangle\langle x_i| \otimes \mathcal{N}_{i,x_i}, \tag{3.37}$$

we obtain the following subadditivity relation as

$$I(\mathcal{X}_1 \otimes \cdots \otimes \mathcal{X}_s; \mathfrak{Z}_1 \otimes \cdots \otimes \mathfrak{Z}_s) \le \sum_{i=1}^{s} I(\mathcal{X}_i; \mathfrak{Z}_i). \tag{3.38}$$

Lemma 3.2.2 (Quantum Fano inequality) *Winter (2001) Let $\mathfrak{X}, \mathfrak{Y}$ be commuting algebras, and $\mathfrak{X}$ be commutative. For a state ρ on $\mathfrak{X}\mathfrak{Y}$, consider POVMs $X \subset \mathfrak{X}$ and $Y \subset \mathfrak{Y}$ running over the same index set. Then the probability of the event $X \neq Y$, i.e., $P_e := \Pr[X \neq Y] = 1 - \sum_j \mathrm{Tr}(\rho X_j Y_j)$ satisfies*

$$S(\mathfrak{X}|\mathfrak{Y}) \le 1 + P_e \log \mathrm{Tr} \mathbb{1}_{\mathcal{X}}. \tag{3.39}$$

Theorem 3.2.1 (Quantum Coding Theorem) *For a priori distribution P_i on the $\mathcal{X}_i$, all the nonnegative rates $R_1, \cdots, R_s$ satisfy for $\forall J \subset [s]$ that*

$$\sum_{i\in J} R_i \le I(\mathfrak{X}_J; D|\mathfrak{X}_{J^c}). \tag{3.40}$$

For any $\epsilon, \delta > 0$ *and sufficiently large* n*, there are* (n, ϵ)*-codes with rates* $\frac{1}{n}\log|\mathcal{M}_i| \geq R_i - \delta$.

Proof The main idea is to make use of the time-sharing principle for the upper extremal points of the involved region. From the symmetry, assume for $i = 1, \ldots, s$ we obtain that

$$\begin{aligned} R_i &= I(\mathfrak{X}_i \wedge \mathfrak{Y}\mathfrak{X}_1 \cdots \mathfrak{X}_{i-1}) \\ &= S(\mathfrak{Y}|\mathfrak{X}_1 \cdots \mathfrak{X}_{i-1}) - S(\mathfrak{Y}|\mathfrak{X}_1 \cdots \mathfrak{X}_i) \\ &= S(P_{j>i-1}W|P_{j\leq i-1}) - S(P_{j>i}W|P_{j\leq i}). \end{aligned} \tag{3.41}$$

The main idea is to decode m_1 from the first sender using the incoming signal. And then, the receiver can decode m_2 from the second sender using m_1 and the new incoming signal. This procedure may be iteratively to decode m_s from the s-th sender using all previous messages $m_1, \ldots, m_{s-1}$ and the incoming signal.

Let $\epsilon, \delta > 0$. Consider s families of codewords $\mathcal{C}_i \subset \mathcal{X}_i^n$ of size $L_i = |\mathcal{C}_i| = \lceil 2^{n(R_i-\delta)} \rceil$, drawn independently from $\mathcal{X}_i^n$ according to a priori distribution $P_i^{\otimes n}$. For a given i, define a quantum channel for $x_i^n \in \mathcal{X}_i^n$ as

$$\mathcal{E}^{(i)} : x_i^n \mapsto \rho_{x_i^n}^{(i)} = \frac{1}{\prod_{j\neq i} L_j} \sum_{\forall j\neq i:\ c_j^n \in \mathcal{C}_j} \otimes_{j<i} c_j^n \otimes W^n_{c_1^n \ldots x_i^n \ldots c_s^n}. \tag{3.42}$$

Denote these states as $\rho^{(i)}$. The channel belongs to the channel state γ, which is reduced to $(\mathfrak{X}_1 \cdots \mathfrak{X}_{i-1}\mathfrak{Y})^{\otimes n}$.

The average of $\rho_{x_i^n}^{(i)}$ over $\times_{j\neq i}\mathcal{C}_j$ is a product state given by

$$\begin{aligned} \langle \rho_{x_i^n}^{(i)} \rangle_{\mathcal{C}_i:i\neq j} &= \sum_{\forall j\neq i:\ x_j^n \in \mathcal{X}_j^n} P_{\{\neq i\}}(x_j^n | j \neq i) \otimes_{j<i} x_j^n \otimes W^n_{x_1^n \ldots x_s^n} \\ &= \sum_{\forall j<i:\ x_j^n \in \mathcal{X}_j^n} P_{\{<i\}}(x_j^n | j < i) \otimes_{j<i} x_j^n \otimes (P_{\{>i\}}W)^n_{x_1^n \ldots x_i^n} \\ &= V^n_{x_i^n}, \end{aligned} \tag{3.43}$$

where V_{x_i} is defined by $V_{x_i} = \sum_{\forall j<i:\ x_j \in \mathcal{X}_j} P_{j<i}(x_j | j < i) \otimes_{j<i} x_j \otimes (P_{j>i}W)_{x_1 \ldots x_i}$.

The decoding observable Winter (2001) for the channel V and set $\mathcal{C}_i$ of codewords can be constructed with an average error probability

$$P_{e;V^n}(\mathcal{C}_i, D_i) = 1 - \frac{1}{|\mathcal{C}_i|} \sum_{c_i^n \in \mathcal{C}_i} \mathrm{Tr}(V^n_{c_i^n} D_{i,c_i^n}) < \frac{1}{s}\epsilon \tag{3.44}$$

for large enough n. In the decoding, we assure that $D_{i,c_i^n} \in (\mathfrak{X}_1 \cdots \mathfrak{X}_{i-1}\mathfrak{Y})^{\otimes n}$ for all $c_i^n \in \mathcal{C}_i$. For any D_i, $V^n_{c_i^n}$ are density operators on some Hilbert space $\mathbb{H}$ such that

$$V_{c_i^n}^n \in (\mathfrak{X}_1 \cdots \mathfrak{X}_{i-1}\mathfrak{Y})^{\otimes n} \subset \mathfrak{L}(\mathbb{H}). \tag{3.45}$$

Consider to embed this subset by using E which satisfies

$$\mathrm{Tr}(V_{c_i^n}^n D_{i,c_i^n}) = \mathrm{Tr}(E(V_{c_i^n}^n) D_{i,c_i^n}) = \mathrm{Tr}(V_{c_i^n}^n E^*(D_{i,c_i^n})), \tag{3.46}$$

where E^* is the adjoint map of E. The decoding error probability is then unchanged by replacing D_{i,c_i^n} with $E^*(D_{i,c_i^n})$.

Consider the decoder D_i with the signal ρ. From Eq. (3.44), the average error probability is given by

$$\begin{aligned} P_{e_i} &= \bar{e}_{\rho^{(i)}}(\mathcal{C}_i, D_i) \\ &= 1 - \frac{1}{L_i} \sum_{c_i^n \in \mathcal{C}_i} \mathrm{Tr}(\rho_{c_i^n} D_{i,c_i^n}) \\ &= P_{e;V^n}(\mathcal{C}_i, D_i). \end{aligned} \tag{3.47}$$

This further implies that

$$\langle \sum_{i=1}^{s} Pe_i \rangle_{\mathcal{C}_1 \cdots \mathcal{C}_s} \leq \epsilon. \tag{3.48}$$

Now, consider the operations Δ_i of D_i. For each i, we obtain that Winter (2001):

$$\frac{1}{\prod_{i=1}^{s} L_i} \sum_{\forall j: c_j^n \in \mathcal{C}_j} \| \otimes_{j \leq i} c_j^n \otimes W_{c_1^n \ldots c_s^n}^n - \Delta_i(\otimes_{j<i} c_j^n \otimes W_{c_1^n \ldots c_s^n}^n) \|_1 \leq 3\sqrt{\epsilon}. \tag{3.49}$$

The decoder process will use the concatenation of $\Delta = \Delta_s \circ \cdots \circ \Delta_1$. This implies that

$$\frac{1}{\prod_{i=1}^{s} L_i} \sum_{\forall j: c_j^n \in \mathcal{C}_j} \| \otimes_{j=1}^{s} c_j^n \otimes W_{c_1^n \ldots c_s^n}^n - \Delta(W_{c_1^n \ldots c_s^n}^n) \|_1 \leq 3s\sqrt{\epsilon}. \tag{3.50}$$

By tracing out $\mathfrak{Y}^{\otimes n}$, we obtain that $\widetilde{\Delta} = \mathrm{Tr}_{\mathfrak{Y}^{\otimes n}} \Delta$, which satisfies that

$$\frac{1}{\prod_{i=1}^{s} L_i} \sum_{\forall j: c_j^n \in \mathcal{C}_j} \| \otimes_{j=1}^{s} c_j^n - \widetilde{\Delta}(W_{c_1^n \ldots c_s^n}^n) \|_1 \leq 3s\sqrt{\epsilon}. \tag{3.51}$$

This means that $\widetilde{\Delta}$ is a quantum operation with classical outcomes in $(c_1^n, \ldots, c_s^n) \in \mathcal{C}_1 \times \cdots \times \mathfrak{C}_s$ from $W_{c_1^n \ldots c_s^n}^n$. □

Theorem 3.2.2 (Weak converse) *The capacity region* C_{QMAC} *of QMAC for the input distribution* q_u *and channel states* γ_u *is given by*

$$C_{\mathrm{QMAC}} \subseteq \mathrm{Cl}\left\{(R_1, \cdots, R_s) \middle| \begin{array}{c} \sum_{i\in J} R_i \leq \sum_u q_u I_{\gamma_u}(\mathfrak{X}_J; \mathfrak{D}_j|\mathfrak{X}_{J^c}), \\ \forall J \subset [s], j \in [r] \end{array} \right\}.$$

Proof Let $(\vec{f}, \vec{D})$ be any (n, ϵ)-code with the rate tuple $(R_1, \cdots, R_s)$. Consider a uniform distribution on the codewords. The channel state γ on $(\mathfrak{X}_1 \otimes \cdots \otimes \mathfrak{X}_s \otimes \mathfrak{D})^n$ is given by

$$\gamma = \frac{1}{\prod_{i=1}^s |\mathcal{M}_i|} \sum_{\forall i: m_i \in \mathcal{M}_i} \otimes_i f_i(m_i) \otimes W^n_{f_1(m_1)\cdots f_s(m_s)}. \tag{3.52}$$

Denote γ_i as the restriction to the i-th copy. Let $j \in [r]$ and $J \subset [s]$. From Quantum Fano inequality Wilde (2013), we obtain that

$$S(\mathfrak{X}^n_J|\mathfrak{D}^n_j\mathfrak{X}^n_{J^c}) \leq 1 + \epsilon \cdot nR_J, \tag{3.53}$$

where $S(\mathfrak{X}^n_J|\mathfrak{D}^n_j\mathfrak{X}^n_{J^c})$ is defined by

$$\begin{aligned} S(\mathfrak{X}^n_J|\mathfrak{D}^n_j\mathfrak{X}^n_{J^c}) &= S(\mathfrak{X}^n_J) - I(\mathfrak{X}^n_J; \mathfrak{D}^n_j\mathfrak{X}^n_{J^c}) \\ &= nR_J - I(\mathfrak{X}^n_J; \mathfrak{D}^n_j\mathfrak{X}^n_{J^c}). \end{aligned} \tag{3.54}$$

From Lemma 3.2.1, it implies that

$$\begin{aligned} (1-\epsilon)R_J &\leq \frac{1}{n} + \frac{1}{n} I_\gamma(\mathfrak{X}^n_J; \mathfrak{D}^n_j\mathfrak{X}^n_{J^c}) \\ &\leq \frac{1}{n} + \frac{1}{n}\sum_{u=1}^n I_{\gamma_u}(\mathfrak{X}_J; \mathfrak{D}_j\mathfrak{X}_{J^c}). \end{aligned} \tag{3.55}$$

This has completed the proof. □

We combine the coding theorem and the converse theorem to obtain the capacity region as follows.

Theorem 3.2.3 *The capacity region* C_{QMAC} *of QMAC is given by*

$$C_{\mathrm{QMAC}} = \mathrm{Cl}\left\{(R_1, \cdots, R_s) \middle| \sum_{i\in J} R_i \leq I(\mathfrak{X}_J; \mathfrak{D}|\mathfrak{X}_{J^c}), \forall J \subset [s] \right\} \tag{3.56}$$

for some input distribution q_u *and corresponding channel state* γ_u.

3.2.2 *Private Classical Capacity*

For simplicity, the encoding for classical MAC may be reduced to a point-to-point manner. One way is from successive decoding Chou and Yener (2018) or rate-splitting Rimoldi and Urbanke (1996), Grant et al. (2001). Both strategies work well without privacy constraints Winter (2001) because the involved capacity region is a polymatroid. Unfortunately, in the presence of privacy constraints, the decoding task is challenging, even for classical two transmitters Chou and Yener (2017). The main reason is that the secure capacity region is unknown to be a polymatroid. This subsection devotes to private classical communication over QMAC. For an arbitrary number of transmitters, we may derive a regularized expression of the capacity region. This can be regarded as a quantum version of the classical multiple access wiretap channel Tekin and Yener (2008).

3.2.2.1 Rate-Splitting

Consider s independent senders $\mathsf{A}_1, \cdots, \mathsf{A}_s$ and one receiver B, and one eavesdropper E, as shown in Fig. 3.2. The quantum channel $\mathcal{C}$ is modelled as a CPTP mapping from Hilbert spaces $\otimes_{i=1}^{s}\mathbb{H}_{A_i} \otimes \mathbb{H}_B$ to $\mathbb{H}_C \otimes \mathbb{H}_E$. The sender A_i obtains a message $m_{a_i} \in [M_{A_i}]$ from independent distributions P_{X_i}, and encodes m_{a_i} into a density matrix $\sigma_{m_{a_i}}^{A_i} \in \mathbb{H}_{A_i}$. The input state of the channel $\mathcal{C}$ is defined by the joint state of $\otimes_{i=1}^{s}\sigma_{m_{a_i}}^{A_i}$ while the output state is defined by $\rho_{m_{a_1}\cdots m_{a_s}}^{CE}$. Here, the output corresponding to each input tuple $(x_1, \cdots, x_s) \in \mathcal{X}_1 \times \cdots \times \mathcal{X}_s$ is a quantum state shared by one receiver and the eavesdropper Eve. The situation is modeled by the control state as

$$\rho^{\vec{X}CE} := \sum_{\vec{x}\in\vec{\mathcal{X}}} \prod_i P_{X_i}(x_i)|x_1\rangle\langle x_1|_{X_1} \otimes \cdots \otimes |x_s\rangle\langle x_s|_{X_s} \otimes \rho_{\vec{x}}^{CE}, \tag{3.57}$$

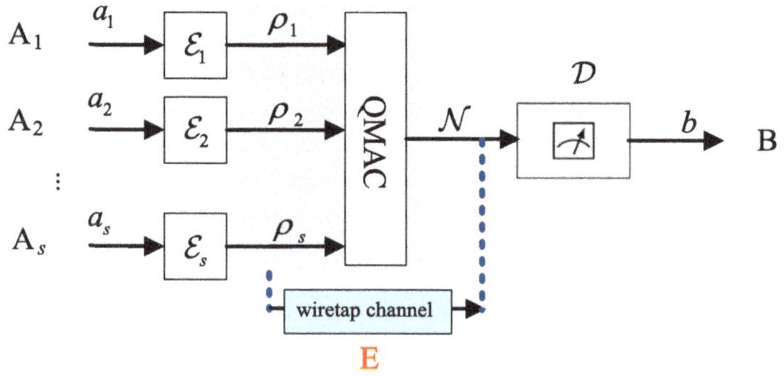

Fig. 3.2 A quantum multiple access wiretap channel with s senders $\mathsf{A}_1, \cdots, \mathsf{A}_s$ who want to transmit the state $\rho_1, \cdots, \rho_s$ to B, respectively. Here, E denotes eavesdropper Eve who wants to get information about the transmitted states

where $\vec{x} = x_1 \cdots x_s$, similar definitions hold for $\vec{X}$ and $\vec{\mathcal{X}}$.

The goal for all senders is to send messages to the receiver such that Eve does not gain any information about the sent message tuple, yet the receiver can decode all messages with high probability.

Let $0 < \epsilon, \delta < 1$. A faithful communication protocol should satisfy the following properties:

- **Correctness**-For all messages $(m_{a_1}, \cdots, m_{a_s}) \in [M_{A_1}] \times \cdots \times [M_{A_s}]$, the decoding error probability satisfies

$$\Pr[(m'_{a_1}, \cdots, m'_{a_s}) \neq (m_{a_1}, \cdots, m_{a_s})] \leq \epsilon \tag{3.58}$$

where $(m'_{a_1}, \cdots, m'_{a_s})$ are decoded messages by the receiver.
- **Secrecy**-There is a state σ^E such that, for all tuples $(m_{a_1}, \cdots, m_{a_s})$

$$\|\rho^E_{f_1(m_{a_1}), \cdots, f_s(m_{a_s})} - \sigma^E\|_1 \leq \delta. \tag{3.59}$$

where $\rho^E_{m_{a_1}, \cdots, m_{a_s}}$ is Eve's system depending on the sent messages $m_{a_1}, \cdots, m_{a_s}$ and encoding mappings $f_i : [M_{A_i}] \to \mathcal{X}_i$.

Definition 3.2.3 The tuple of $(R_{A_1}, \cdots, R_{A_s})$ is an achievable rate for private classical communication over the channel $\mathcal{C}$ with an error ϵ and leakage δ when $[M_{A_1}] \times \cdots \times [M_{A_s}] = [2^{R_{A_1}}] \times \cdots \times [2^{R_{A_s}}]$.

Similar definition may be extended to multiple senders.

Denote the bipartite smooth maximal mutual information $I^\delta_{\max}(X; Y)$ as

$$I^\delta_{\max}(X; Y) = \inf_{\|\rho' - \rho\| \leq \epsilon} D^\delta_{\max}(\rho_{XY} \| \rho_X \otimes \rho_Y), \tag{3.60}$$

where the maximal relative entropy D of ρ^{XY} with respect to the product state $\rho_X \otimes \rho_Y$ is defined by

$$D_{\max}(\rho_{XY} \| \rho_X \otimes \rho_Y) = \inf\{\lambda | \rho \leq 2^\lambda \rho_X \otimes \rho_Y\}. \tag{3.61}$$

To show the achievable communication region, we need the following distributed covering theorem.

Theorem 3.2.4 *Given the control state $\rho^{\vec{X}CE}$, $\delta > 0$ and $0 < \epsilon' < \delta$, let $\{x_1^1, \cdots, x_1^{K_1}\}, \cdots, \{x_s^1, \cdots, x_s^{K_s}\}$ be independent and identically distributed samples from the probability distributions $P_{X_1}, \cdots, P_{X_s}$, respectively. We obtain that*

$$\mathbb{E}_{\vec{x}_1, \cdots, \vec{x}_n} \left\| \frac{1}{\prod_{i=1}^s K_i} \sum_{j=1}^s \sum_{i_j \in [K_j]} \rho^E_{x_1^{i_1}, \cdots, x_s^{i_s}} - \rho^E \right\| \leq c\delta^{1/8} \tag{3.62}$$

with some constant $c > 0$ *if the rates* $K_1, \cdots, K_s$ *satisfy*

$$\log K_1 \ge I_{\max}^{\delta-\epsilon'}(X_1; E)_\rho + \log \frac{3}{\epsilon'^3} - \frac{1}{4}\log\frac{1}{\delta},$$
$$\log K_i \ge I_{\max}^{\delta-\epsilon'}(X_i; EX_1 \cdots X_{i-1})_\rho + \log \frac{3}{\epsilon'^3} - \frac{1}{4}\log\frac{1}{\delta} + O(1), \forall i \le s.$$

The special case of $s = 2$ is proved in Chakraborty et al. (2021). Combined with a rate splitting scheme Grant et al. (2001), we can prove the result for $s \ge 2$.

Proof Suppose that there is a cq state given by

$$\rho^{X_1\cdots X_j E} := \sum_{x_1\in\mathcal{X}_1,\cdots,x_j\in\mathcal{X}_j} \prod_{i=1}^{j} p_{X_i}(x_i)|x_{[j]}\rangle\langle x_{[j]}|\rho^E_{x_1\cdots x_j}, \tag{3.63}$$

where $|x_{[j]}\rangle = \otimes_{i=1}^{j}|x_i\rangle^{X_i}$. Define the quantity for $j \ge 2$ as

$$\begin{aligned}\lambda_j &:= \tilde{I}^{\epsilon}_{max}(X_j; EX_1\cdots X_{j-1})_\rho \\ &:= \inf_{\|\rho'_j-\rho_j\|_1\le\epsilon} D_{\max}(\rho'^{X_1\cdots X_j E}\|\rho^{X_j}\otimes\rho'^{X_1\cdots X_{j-1}E}).\end{aligned} \tag{3.64}$$

Since the optimizer ρ^* satisfies

$$\begin{aligned}&\rho^*_j \le 2^{\tilde{I}^{\epsilon}_{max}(X_j;EX_1\cdots X_{j-1})}\rho^{X_j}\otimes\rho^{*X_1\cdots X_{j-1}E},\\ &\|\rho^*_j-\rho_j\|_1 \le \epsilon,\end{aligned} \tag{3.65}$$

we may assume $\rho^*_j = \tilde{\rho}^{X_1\cdots X_j E}$ to be a multipartite cq state and

$$\tilde{\rho}_j := \sum_{x_1,\cdots,x_j} \tilde{P}_{X_1\cdots X_j}(x_1,\cdots,x_j)\otimes_{i=1}^{j}|x_i\rangle\langle x_i|^{X_i}\otimes\tilde{\rho}^E_{x_1\cdots x_j} \tag{3.66}$$

for $j \ge 2$. This implies that

$$\|\tilde{P}_{X_1\cdots X_j} - P_{X_1}\cdots P_{X_j}\|_1 \le \epsilon \tag{3.67}$$

for $j \ge 2$.

From Eq. (3.67) it follows that

$$\|\tilde{P}_{X_i} - P_{X_i}\|_1 \le \epsilon, \tag{3.68}$$

which implies that

$$\|\tilde{P}_{X_1\cdots X_j} - \tilde{P}_{X_1}\cdots\tilde{P}_{X_{j-1}}P_{X_j}\|_1 \le 2\epsilon. \tag{3.69}$$

This can further be written as

$$\mathbb{E}_{\tilde{P}_{X_1}\cdots\tilde{P}_{X_{j-1}}P_{X_j}}\left[\frac{\tilde{P}_{X_1\cdots X_j}(X_1,\cdots,X_j)}{\tilde{P}_{X_1}(X_1)\cdots\tilde{P}_{X_{j-1}}(X_{j-1})P_{X_j}(X_j)}-1\right]\leq 2\epsilon \tag{3.70}$$

From the Markov inequality Wilde (2013), it follows that

$$\Pr_{\tilde{P}_{X_1}\cdots\tilde{P}_{X_{j-1}}P_{X_j}}\left[\frac{\tilde{P}_{X_1\cdots X_j}(X_1,\cdots,X_j)}{\prod_{i=1}^{j-1}\tilde{P}_{X_i}(X_i)}-1\geq\sqrt{\epsilon}\right]\leq 2\sqrt{\epsilon} \tag{3.71}$$

for $j\geq 2$.

Now, by using the definition of $\tilde{\rho}$ and the classical systems $X_1,\cdots,X_j$, we obtain for all $(x_1,\cdots,X_j)$ that

$$\tilde{P}_{X_1\cdots X_j}(x_1,\cdots,X_j)\,\tilde{\rho}^E_{x_1,\cdots,x_j}\leq 2^{\lambda_j}\prod_{i=1}^{j-1}\tilde{P}_{X_i}(X_i)P_{X_j}(x_j)\tilde{\rho}^E_{x_1\cdots x_{j-1}}, \tag{3.72}$$

where $\tilde{\rho}^E_{x_1\cdots x_{j-1}}$ is defined by $\tilde{\rho}^E_{x_1\cdots x_{j-1}}:=\sum_{x_j}\tilde{P}_{X_j}(x_j|x_1\cdots x_{j-1})\tilde{\rho}^E_{x_1,\cdots,x_j}$ for $j=2,\cdots,s$.

Combining with Eq. (3.71) it implies that

$$\tilde{\rho}^E_{x_1,\cdots,x_j}\leq 2^{\lambda}\frac{1}{1-\sqrt{\epsilon}}\tilde{\rho}^E_{x_1\cdots x_{j-1}} \tag{3.73}$$

for $j\geq 2$.

Define the function $\mathbf{1}_{x_1,\cdots,x_j}$ as the indicator, which is 1 when Eq. (3.73) holds. Moreover, define $\mathbf{1}_{x_1\cdots x_{j-1}}:=\sum_{x_j}P_{X_j}(x_j)\mathbf{1}_{x_1\cdots x_j}$ with $j=2,\cdots,s$. Intuitively, $\mathbf{1}_{x_1\cdots x_{j-1}}$ is the probability that, for a fixed $x_1,\cdots,x_{j-1}$, the tuples $(x_1,\cdots,x_j)$ satisfy Eq. (3.73) over x_j. We know that

$$\sum_{x_1,\cdots,x_{j-1}}\tilde{P}_{X_1}(x_1)\cdots\tilde{P}_{X_{j-1}}(x_{j-1})\mathbf{1}_{x_1\cdots x_{j-1}}\geq 1-c\sqrt{\epsilon} \tag{3.74}$$

for some constant c, and $j=2,\cdots,s$. From the Markov inequality Wilde (2013) it implies that

$$\Pr_{\tilde{P}_{X_1}\cdots\tilde{P}_{X_{j-1}}}[\{x|\mathbf{1}_{x_1\cdots x_{j-1}}\geq 1-\epsilon^{1/4}\}]\geq 1-c\epsilon^{1/4} \tag{3.75}$$

for $j=2,\cdots,s$. We define

$$\mathrm{NI}_{X_1\cdots X_{j-1}}:=\{x_1,\cdots,x_{j-1}|\mathbf{1}_{x_1\cdots x_{j-1}}\geq 1-\epsilon^{1/4}\} \tag{3.76}$$

for $j = 2, \cdots, s$. For any $x_1, \cdots, x_{j-1} \in \mathrm{NI}_{X_1 \cdots X_{j-1}}$, the probability over x_j that $(x_1, \cdots, x_j)$ satisfies Eq. (3.73) is at least $1 - \epsilon^{1/4}$. The probability that a random $x_1, \cdots, x_{j-1}$ is picked from $\mathrm{NI}_{X_1 \cdots X_{j-1}}$ is at least $1 - c\epsilon^{1/4}$ under the distribution $\tilde{P}_{X_1} \cdots \tilde{P}_{X_{j-1}}$.

Define

$$\mathrm{NIC}_{X_1 \cdots X_{j-1}} := \mathrm{NI}_{X_1 \cdots X_{j-1}} \cap \{x \,|\, \|P_{X_j|X_1 \cdots X_{j-1}} - P_{X_j}\|_1 \le \sqrt{\epsilon}\} \tag{3.77}$$

for $j = 2, \cdots, s$. From Eq. (3.67) we obtain that

$$\mathbb{E}_{\tilde{P}_{X_1}, \cdots, \tilde{P}_{X_{j-1}}}[\|\tilde{P}_{X_j|X_1 \cdots X_{j-1}} - P_{X_j}\|_1] \le 2\epsilon \tag{3.78}$$

for $j = 2, \cdots, s$. Combining with the Markov inequality Wilde (2013) it follows that

$$\Pr_{\tilde{P}_{X_1}, \cdots, \tilde{P}_{X_{j-1}}}[\{x_1, \cdots, x_{j-1} | \|P_{X_j|X_1 \cdots X_{j-1}} - P_{X_j}\|_1 \le \sqrt{\epsilon}\}] \ge 1 - 2\sqrt{\epsilon}, \tag{3.79}$$

which further implies that

$$\Pr_{\tilde{P}_{X_1}, \cdots, \tilde{P}_{X_{j-1}}}[\mathrm{NIC}_{X_1 \cdots X_{j-1}}] \ge 1 - c^2 \epsilon^{1/4}. \tag{3.80}$$

Since $\|\tilde{P}_{X_i} - P_{X_i}\|_1 \le \epsilon$, this implies that

$$\Pr_{\tilde{P}_{X_1}, \cdots, \tilde{P}_{X_{j-1}}}[\mathrm{NIC}_{X_1 \cdots X_{j-1}}] \ge 1 - (c^2 + 1)\epsilon^{1/4} \tag{3.81}$$

Moreover, consider the state

$$\rho'_j := \sum_{x_1, \cdots, x_j} \prod_{i=1}^{j} P_{X_i}(x_i) |x_j\rangle\langle x_j| \otimes \tilde{\rho}^E_{x_1, \cdots, x_j}. \tag{3.82}$$

It follows that

$$\begin{aligned} \|\rho'_j - \rho_j\|_1 &\le \|\rho'_j - \tilde{\rho}_j\|_1 + \|\tilde{\rho}_j - \rho_j\|_1 \\ &\le 3\epsilon. \end{aligned} \tag{3.83}$$

This implies that

$$\begin{aligned} &\Pr_{P_{X_1}, \cdots, P_{X_{j-1}}}[\| \textstyle\sum_{x_n} P_{X_n}(x_n) |x_j\rangle\langle x_j|^{X_j} \otimes \tilde{\rho}^E_{x_1, \cdots, x_j} \\ &- \textstyle\sum_{x_j} P_{X_j}(x_j) |x_j\rangle\langle x_j|^{X_j} \otimes \rho^E_{x_1, \cdots, x_j} \|_1 \ge \sqrt{\epsilon}] \le 3\sqrt{\epsilon}. \end{aligned} \tag{3.84}$$

Let the event is the last probability by $B_{X_1 \cdots X_{j-1}}$. Define $\mathrm{GD}_{X_1 \cdots X_{j-1}} := \mathrm{NIC}_{X_1 \cdots X_{j-1}} \cap B_{X_1 \cdots X_{j-1}}$. This implies that

$$\Pr_{P_{X_1},\cdots,P_{X_{j-1}}}[\mathrm{GD}_{X_1\cdots X_{j-1}}] \geq 1 - 2(c^2+1)\epsilon^{1/4} \tag{3.85}$$

for $j = 2, \cdots, s$. Fix an $x_1, \cdots, x_{j-1} \in \mathrm{GD}_{X_1\cdots X_{j-1}}$, we obtain that

$$\begin{aligned} &\|\tilde{P}_{X_j|x_1\cdots x_{j-1}} - P_{X_j}\|_1 \leq \sqrt{\epsilon} \\ \text{or } &\mathbb{E}_{P_{X_j}}|\frac{\tilde{P}_{X_j|x_1\cdots x_{j-1}}(X_j)}{P_{X_j}(X_j)} - 1| \leq \sqrt{\epsilon}. \end{aligned} \tag{3.86}$$

By the Markov inequality Wilde (2013) we obtain that

$$\Pr_{P_{X_j}}\left[|\frac{\tilde{P}_{X_j|x_1\cdots x_{j-1}}(X_j)}{P_{X_j}(X_j)} - 1| \geq \epsilon^{1/4}\right] \leq \epsilon^{1/4} \tag{3.87}$$

with $j = 2, \cdots, s$.

Define the set

$$\begin{aligned} \mathrm{GD}_{X_j|x_1\cdots x_{j-1}} := \{x_j|(x_1, \cdots, x_j) \text{ s.t. } Eq.(3.73), \\ |\frac{\tilde{P}_{X_j|x_1\cdots x_{j-1}}(X_j)}{P_{X_j}(X_j)} - 1| \leq \epsilon^{1/4}\} \end{aligned} \tag{3.88}$$

for $j = 2, \cdots, s$. Under the distribution P_{X_j}, it follows that

$$\Pr_{P_{X_j}}[\mathrm{GD}_{X_j|x_1\cdots x_{j-1}}] \geq 1 - 2\epsilon^{1/4} \tag{3.89}$$

Define the subdistribution $\bar{P}_{X_j}$ as

$$\bar{P}_{X_j|x_1\cdots x_{j-1}} = \begin{cases} \tilde{P}_{X_j|x_1\cdots x_{j-1}}, & x_j \in \mathrm{GD}_{X_j|x_1\cdots x_{j-1}}; \\ 0, & \text{Otherwise.} \end{cases} \tag{3.90}$$

It holds that

$$\begin{aligned} \bar{\rho}^{X_jE}_{x_1\cdots x_{j-1}} &:= \sum_{x_j} \bar{P}_{X_j|x_1\cdots x_{j-1}}(x_j|x_1\cdots x_{j-1})x_j^{X_j} \otimes \tilde{\rho}^E_{x_1,\cdots,x_j} \\ &\leq 2^{\lambda_j}\frac{1+\epsilon^{1/4}}{1-\sqrt{\epsilon}}\rho^{X_j} \otimes \tilde{\rho}^E_{x_1\cdots x_{j-1}}. \end{aligned} \tag{3.91}$$

Moreover, we obtain that

$$\begin{aligned} \bar{\rho}^E_{x_1\cdots x_{j-1}} &\leq \sum_{x_n} \tilde{P}_{X_j|x_1\cdots x_{j-1}}(x_j|x_1\cdots x_{j-1})\tilde{\rho}^E_{x_1,\cdots,x_j} \\ &= \tilde{\rho}^E_{x_1\cdots x_{j-1}}. \end{aligned} \tag{3.92}$$

Using the convex split lemma Chakraborty et al. (2021) we obtain for some K that

$$\left\| \frac{1}{K}\sum_{i=1}^{K} \bar{\rho}^{Y_i E}_{x_1\cdots x_{j-1}} \otimes_{j\neq i} \rho^{Y_j} - \tilde{\rho}^{E}_{x_1\cdots x_{j-1}} \otimes_{i=1}^{K} \rho^{Y_i} \right\|_1 \leq 2^{\lambda/2}\epsilon^{1/4}, \tag{3.93}$$

where X_j is split into $Y_1, \cdots, Y_K$.

Next, we estimate the distance between two states $\rho^{X_j E}_{x_1\cdots x_{j-1}}$ and $\bar{\rho}^{X_j E}_{x_1\cdots x_{j-1}}$. By using the triangle inequality, it follows that

$$\begin{aligned} \|\rho_j - \bar{\rho}_j\|_1 \leq & \sum_{x_j : P_{X_j|x_1\cdots x_{j-1}}(x_j|x_1\cdots x_{j-1})=0} \tilde{P}_{X_j|x_1\cdots x_{j-1}}(x_j) \\ & + \|\tilde{P}_{X_j|x_1\cdots x_{j-1}} - P_{X_j}\|_1 \\ & + \sum_{x_j} P_{X_j}(x_j)\|\tilde{\rho}^{E}_{x_1,\cdots,x_j} - \rho^{E}_{x_1,\cdots,x_j}\|_1 . \end{aligned} \tag{3.94}$$

For the first term, the summation is over those x_j's which do not belong to $\mathrm{GD}_{X_j|x_1\cdots x_{j-1}}$. Under the distribution P_{X_j}, this set has probability at most $2\epsilon^{1/4}$. By the definition of $\mathrm{GD}_{X_1\cdots X_{j-1}}$, the first term is at most $O(c^2)\epsilon^{1/4}$ for some constant c. The second and third terms can be bounded similarly from $\mathrm{GD}_{X_1\cdots X_{j-1}}$ by $\sqrt{\epsilon}$. Thus, we obtain that

$$\|\rho^{X_j E}_{x_1\cdots x_{j-1}} - \bar{\rho}^{X_j E}_{x_1\cdots x_{j-1}}\|_1 \leq O(c^2)\epsilon^{1/4} \tag{3.95}$$

Moreover, from the inequality of $\|\tilde{\rho}^{E}_{x_1\cdots x_{j-1}} - \rho^{E}_{x_1\cdots x_{j-1}}\|_1 \leq 2\sqrt{\epsilon}$, by using a convex split lemma, we obtain that

$$\mathbb{E}_{x_j^1,\cdots,x_j^K \sim P_{X_j}} \left\| \frac{1}{\prod_{t=1}^{j-1} K_t} \sum_{i_1=1}^{K_1} \cdots \sum_{i_{j-1}=1}^{K_{j-1}} \rho^{E}_{x_1,\cdots,x_{j-1},x_j^i} - \rho^{E}_{x_1\cdots x_{j-1}} \right\|_1 \leq O(c^2)\epsilon^{1/4} \tag{3.96}$$

for some $K_1, \cdots, K_{j-1}$.

In what follows, we define the successive cancellation step. Define $\epsilon_0 := O(c^2)\epsilon^{1/4}$. Suppose that we sample K' independently from the distribution $P_{X_1\cdots X_{j-1}}$. From the Markov inequality Wilde (2013), it follows

$$\Pr\left[\sum_{i_1=1}^{K_1} \cdots \sum_{i_{j-1}=1}^{K_{j-1}} I_{X_1^i\cdots X_{j-1}^i \notin \mathrm{GD}_{X_1\cdots X_{j-1}}} \geq \sqrt{\epsilon_0}\cdot \prod_{t=1}^{j-1} K_t \right] \leq \sqrt{\epsilon_0} \tag{3.97}$$

Suppose that $x_1^1\cdots x_1^{K'}, x_2^1\cdots x_2^{K'}, \cdots, x_{s-1}^1\cdots x_{s-1}^{K'}$ are $n-1$ sequences which have at most $\sqrt{\epsilon_0}K'$ samples from $\mathrm{GD}^c_{X_1\cdots X_{s-1}}$. For these fixed sequences, we get

$$
\begin{aligned}
&\mathbb{E}_{x_s^1,\cdots,x_n^K\sim P_{X_s}}\left\|\frac{1}{\prod_{t=1}^{s}K_t}\sum_{i_1=1}^{K_1}\cdots\sum_{i_s=1}^{K_s}\rho^E_{x_1^{i_1}\cdots x_s^{i_s}}-\rho^E_s\right\|_1\\
\leq&\mathbb{E}_{x_s^1,\cdots,x_s^K\sim P_{X_s}}\left[\|\frac{1}{\prod_{t=1}^{s}K_t}\sum_{i_1=1}^{K_1}\cdots\sum_{i_s=1}^{K_s}\rho^E_{x_1^{i_1}\cdots x_s^{i_s}}\right.\\
&\left.-\frac{1}{\prod_{t=1}^{s-1}K_t}\sum_{i_1=1}^{K_1}\cdots\sum_{i_{s-1}=1}^{K_{s-1}}\rho^E_{x_1^{i_1}\cdots x_{s-1}^{i_{s-1}}}\|_1\right]\\
&+\left\|\frac{1}{\prod_{t=1}^{s-1}K_t}\sum_{i_1=1}^{K_1}\cdots\sum_{i_{s-1}=1}^{K_{j-1}}\rho^E_{x_1^{i_1}\cdots x_{s-1}^{i_{s-1}}}-\rho^E_s\right\|_1\\
\leq&O(c^2)\epsilon^{1/8}
\end{aligned}
\tag{3.98}
$$

for large numbers of $K_1, \cdots, K_s$. This has completed the proof. □

To overcome simultaneous smoothing conjecture for the entire inverted pentagonal region, one may prove a sequential covering lemma for a corner point of the region. In fact, if the sender A_i randomises over a block of size K_i with $\log K_i > I^{\delta}_{\max}(X_i; X_1 \cdots X_{i-1}E)$ and A_{i+1} randomises over a block of size $\log K_{i+1} > I^{\delta}_{\max}(X_{i+1}; X_1 \cdots X_i E)$, then indeed the inequality (3.62) holds, albeit with a worse dependence in δ. A similar statement holds for other corner points. Moreover, to construct a decoder, there are two kinds of decoders. One is the successive cancellation. The other is simultaneous decoding. Both may arise in different achievable regions. For the special case of $s = 2$, the successive cancellation can be regarded as a deformed version of the other Chakraborty et al. (2021).

For the special case of $s = 2$, denote the achievable region as $\mathcal{S}_\theta$, for some fixed $\theta \in [0, 1]$. Consider the block sizes $(K_1 + K_3, K_2)$. Define

$$
\mathcal{T}_\theta = \{(\log K, \log K')|K \geq K_1K_3, K' \geq K_2\}. \tag{3.99}
$$

The Theorem 3.2.4 implies the following inner bound on the private region over the QMAC.

Theorem 3.2.5 *Chakraborty et al. (2021) Given a QMAC, the control state (3.57) and a split* $(P^\theta_U, P^\theta_V, f)$ *of the distribution* P_X*, for some* $\theta \in [0, 1]$*, the rate pairs in* $\mathcal{T}_\theta$ *are achievable for private transmission of messages on the channel* $(\cup_{\theta\in[0,1]}(\mathcal{S}_\theta - \mathcal{T}_\theta))^+$ *with a decoding error at most* $49\sqrt{\epsilon}$ *and privacy leakage at most* $40\delta^{1/8}$*, where* $\epsilon, \delta > 0$ *and* $0 < \epsilon < \delta$*. The operation* $(A - B)^+$ *denotes by* $\{\max(a - b, 0)|a \in A, b \in B\}$.

Consider a quantum state with s classical inputs defined as

$$
\rho^{\vec{x}C} := \sum_{\vec{x}}\prod_{i=1}^{s} P_{X_i}(x_i)|x_1\rangle\langle x_1|_{X_1} \otimes \cdots \otimes |x_s\rangle\langle x_s|_{X_s}\rho^C_{\vec{x}}. \tag{3.100}
$$

Let $J \subset [s]$. For some $\epsilon > 0$ and an integer m, the following results hold

$$\lim_{\epsilon\to 0}\lim_{m\to\infty}\frac{1}{m}I_H^{\epsilon}(X_J^m; C^m X_{J^c}^m)_{\rho^{\otimes m}} = I(X_J; CX_{J^c})_\rho, \tag{3.101}$$

$$\lim_{\epsilon\to 0}\lim_{n\to\infty}\frac{1}{m}I_{\max}^{\epsilon}(X_J^m; C^m X_{J^c}^m)_{\rho^{\otimes nm}} = I(X_J; CX_{J^c})_\rho, \tag{3.102}$$

where X_J is defined by $X_J = \prod_{j\in J} X_j$. The one-shot technique is useful for getting the expected private capacity region of CQMAC in the limit of asymptotically many channel uses.

Theorem 3.2.6 *Given a s-sender quantum access channel, if asymptotically many channel uses are allowed, we obtain the achievable private rate region C_{pQMAC} as*

$$C_{\text{pQMAC}} = \text{Cl}\left\{(R_{A_1}^{\text{p}}, \cdots, R_{A_s}^{\text{p}}) \,\middle|\, \begin{array}{l} R_{A_i}^{\text{p}} < I(X_i; \overline{X}_i C) - I(\overline{X}_i; E), i \in [s]; \\ \sum_{i=1}^{s} R_{A_i}^{\text{p}} < I(\vec{X}; C) - I(\vec{X}; E) \end{array}\right\}, \tag{3.103}$$

where $\overline{X}_i$ is defined by $\overline{X}_i = X_1 \cdots X_{i-1}X_{i+1}\cdots X_s$.

3.2.2.2 Random Binning Techniques

Instead of using successive decoding or rate-splitting, we present another point-to-point coding technique in this subsection. The main idea is from techniques Csiszar (1996), which is sufficient to build good codes for classical point-to-point .

Consider a QMAC $\mathcal{N}_{\vec{A}'\to B} : \otimes_{\ell\in[s]}\mathcal{S}(\mathbb{H}_{A'_\ell}) \to \mathcal{S}(\mathbb{H}_B)$ with s senders $\mathsf{A}_1, \cdots, \mathsf{A}_s$, where $\vec{A}' = A'_1 \cdots A'_s$. Let $\mathcal{U}^{\mathcal{N}}_{\vec{A}'\to BE}$ be an isometric extension of the channel $\mathcal{N}_{\vec{A}'\to B}$ such that the complementary channel to the environment $\mathcal{N}_{\vec{A}'\to E}$ satisfies $\mathcal{N}^c_{\vec{A}'\to E}(\rho) = \text{Tr}_B[\mathcal{U}^{\mathcal{N}}_{\vec{A}'\to BE}]$ for $\rho \in \otimes_{\ell\in[s]}\mathcal{S}(\mathbb{H}_{A'_\ell})$.

Definition 3.2.4 (Private Quantum Multiple Access Code) For the channel $\mathcal{N}_{\vec{A}'\to B}$, a classical private quantum multiple access code (cpQMAC) $(t, (2^{tR_i})_{i\in[s]})$ is defined as tuples of $(\mathcal{M}_1, \cdots, \mathcal{M}_s, \mathcal{E}_1, \cdots, \mathcal{E}_s, (D_{\vec{m}})_{\vec{m}\in\vec{\mathcal{M}}})$, i.e.,

- $\mathcal{M}_i = [2^{tR_i}]$ are message sets, $i \in [s]$;
- $\mathcal{E}_i : \mathcal{M}_i \to \mathcal{S}(\mathcal{H}_{A'_i})$ are encoding mappings, $i \in [s]$;
- $(D_{\vec{m}})_{\vec{m}\in\vec{\mathcal{M}}}$ is a set of POVM for decoding, $\vec{\mathcal{M}} = \times_{i\in[s]}\mathcal{M}_i$.

The operation procedure is as follows. Sender A_i, $i \in [s]$ chooses a message $m_i \in M_i$, generates the state $\rho_{A'_i}^{m_i} = \mathcal{E}_i(m_i)$ and send into the channel $\mathcal{N}_{\vec{A}'^t\to B^t}$. The output state is given by

$$\omega_{B^t}^{\vec{m}} \triangleq \mathcal{N}_{\vec{A}'^t\to B^t}(\rho_{\vec{A}'^t}^{\vec{m}}), \tag{3.104}$$

where $\rho^{\vec{m}}_{\vec{A}'^t} = \otimes_{i\in[s]}\rho^{m_i}_{A_i'^t}$ and $\vec{m} = m_1 \cdots m_s$. The decoding POVM is used to detect the sent messages. The complementary channel output is denoted by $\omega^{\vec{m}}_{E^t} = \mathcal{N}^c_{\vec{A}'^t\to E^t}(\rho^{\vec{m}}_{\vec{A}'^t})$.

Definition 3.2.5 A rate-tuple $(R_1, \cdots, R_s)$ is achievable if there exists a sequence of cpQMACs $(t, (2^{tR_i})_{i\in[s]})$ such that for some classical states (ρ_{E^t})

$$\text{Reliability}: \lim_{t\to\infty} \max_{\vec{m}\in\vec{M}} \mathrm{Tr}[(\mathbb{1} - D_{\vec{m}})\omega^{\vec{m}}_{B^t}] = 0, \tag{3.105}$$

$$\text{Indistinguishability}: \lim_{t\to\infty} \max_{\vec{m}\in\vec{M}} \|\omega^{\vec{m}}_{B^t} - \rho_{E^t}\|_1 = 0. \tag{3.106}$$

The classical private capacity region C_{PMAC} of a quantum multiple-access channel (QMAC) $\mathcal{N}_{\vec{A}'\to B}$ is defined as the closure of achievable rate-tuples $(R_1, \cdots, R_s)$.

Theorem 3.2.7 *The private capacity region of a QMAC $\mathcal{N}_{\vec{A}'\to B}$ is given by*

$$C_{\text{pQMAC}} = \text{Cl}\left\{ (R^{\text{p}}_{A_1}, \cdots, R^{\text{p}}_{A_s}) \middle| \begin{array}{l} \cup_{n=1}^{\infty} \frac{1}{n}\mathcal{R}(\mathcal{N}^n); \\ \sum_{i\in J} R_i \le [I(X_J; B|X_{J^c})_\rho \\ \quad -I(X_J; E)_\rho]^+, \forall J \subseteq [s] \end{array} \right\} \tag{3.107}$$

for some classical-quantum states as

$$\rho_{\vec{X}\vec{A}'} = \otimes_{i\in[n]}(\sum_{x_i} p_{X_i}(x_i)|x_i\rangle\langle x_i| \otimes \rho^{x_i}_{A_i'}), \tag{3.108}$$

and $\rho_{\vec{X}BE} = \mathcal{U}^{\mathcal{N}}_{\vec{A}'\to BE}(\rho_{\vec{X}\vec{A}'})$ with $\mathcal{U}^{\mathcal{N}}_{\vec{A}'\to BE}$ as an isometric extension of $\mathcal{N}_{\vec{A}'\to B}$.

Proof Consider a classical-quantum multiple-access wiretap channel (cqMAWC), i.e., a mapping $W : \mathcal{X}_{[s]} \to \mathcal{S}_=(\mathbb{H}_B \otimes \mathbb{H}_E)$, which maps $x_{[s]} \in \mathcal{X}_{[s]}$ to $\bar{\rho}^{x_{[s]}}_{BE} \in \mathcal{S}_=(\mathbb{H}_B \otimes \mathbb{H}_E)$. For any probability distribution $p_{X_{[s]}} \triangleq \prod_{l\in[s]} p_{X_l}$, it is sufficient to prove the following region is achievable, i.e.,

$$\mathcal{R}(W, p_{X_{[s]}}) \triangleq \{(R_{l\in[s]}) : R_{\mathcal{S}} \le [I(X_{\mathcal{S}}; B|X_{\mathcal{S}^c})_\rho - I(X_{\mathcal{S}}; E)_\rho]^+, \forall \mathcal{S} \subseteq [s]\}, \tag{3.109}$$

where $\rho_{X_{[s]}BE}$ is defined by $\rho_{X_{[s]}BE} = \sum_{x_{[s]}} p_{X_{[s]}}(x_{[s]})|x_{[s]}\rangle\langle x_{[s]}| \otimes \bar{\rho}^{x_{[s]}}_{BE}$.

The main idea is to use distributed source coding and randomness extraction. The proof is consisting of three steps.

Firstly, we define a stochastic channel that can simulate the inversion of multiple hash functions. It can approximate the joint distribution of both inputs and outputs of hash functions Renes and Renner (2011), Chou and Bloch (2016). In fact, consider $X^t_{[s]}$ distributed according to some arbitrary product distribution $p_{X^s_{[n]}} \triangleq \prod_{l\in[s]} p_{X^t_l}$, and s universal hash functions $F_{[s]}$ uniformly distributed over $\mathcal{F}_{[s]}$. Denote $(tR^{\text{U}}_l)_{l\in[s]}$ as the output lengths of the hash functions. Define $\widetilde{W}_{[s]}$ be the channel with the conditional probability distribution $p_{X^t_{[s]}|F_{[s]}(X^t_{[s]})F_{[s]}} \triangleq \prod_{l\in[s]} p_{X^t_l|F_l(X^t_l)F_l}$. $\widetilde{W}_l$ is the channel with $p_{X_l^t|F_l(X_l^t)F_l}$, $l \in [s]$. For $l \in [s]$, let U_l^s be uniformly distributed on $\mathcal{U}_l^t \triangleq [2^{tR_l^{\mathrm{U}}}]$. Define

$$\widetilde{p}_{X_{[s]}^t U_{[s]}^t F_{[s]}} := p_{X_{[s]}^t|F_{[s]}(X_{[s]}^t)F_{[s]}} p_{U_{[s]}^t} p_{F_{[s]}}, \tag{3.110}$$

where $p_{U_{[s]}^t}$ is the uniform distribution on $\mathcal{U}_{[s]}^t$. $\widetilde{p}_{X_{[s]}^t U_{[s]}^t F_{[s]}}$ denotes the joint probability distribution of the input $(U_{[s]}^t, F_{[s]})$ and output $\widetilde{X}_{[s]}^t \triangleq \widetilde{W}_{[s]}(U_{[s]}^t, F_{[s]})$ of the channel $\widetilde{W}_{[s]}$.

Secondly, we construct a QMAC code for jointly uniform input distributions (in the absence of any privacy constraint) for the channel $W \circ \widetilde{W}_{[s]}$. Let $m \in \mathbb{N}$. For the classical-quantum product state $\widetilde{\rho}_{U_{[s]}^t B^t}^{\otimes m}$, there is a $(2^{mtR_l^{\mathrm{DC}}})_{l\in[s]}$ distributed code Chou (2022, Lemma 4), where $\widetilde{\rho}_{U_{[s]}^t B^t}$ is defined by

$$\widetilde{\rho}_{U_{[s]}^t B^t} := \frac{1}{|\mathcal{U}_{[s]}^t|} \sum_{u_{[s]}^t \in \mathcal{U}_{[s]}^t} |u_{[s]}^t\rangle\langle u_{[s]}^t| \otimes \bar{\rho}_{B^t}^{\widetilde{W}_{[s]}(u_{[s]}^t)} \tag{3.111}$$

and $(tR_l^{\mathrm{DC}})_{l\in[s]} \in \mathcal{C}(\widetilde{\rho}_{U_{[s]}^t B^t})$. Moreover, there are s encoders e_l, $l \in [s]$, and one decoder d Chou (2022, Lemma 5) such that $R_l = R_l^{\mathrm{U}} - R_l^{\mathrm{DC}}$ as $t \to \infty$, and

$$\lim_{k\to\infty} \mathbb{P}[d(\bar{\rho}_{B^{kt}}^{\widetilde{W}_{[s]}^{\otimes k}(e_{[s]}(M_{[s]}^k))}) \neq M_{[s]}^k] = 0, \tag{3.112}$$

where $e_{[s]}(M_{[s]}^k) = (e_l(M_l^k))_{l\in[s]}$.

The last step is to define encoders for the wiretap channel as

$$\mathsf{E}_l : M_l^k \mapsto \widetilde{W}_l^{\otimes k}(e_l(M_l^k)), l \in [s], \tag{3.113}$$

and the decoder as

$$\mathsf{D} : \bar{\rho}_{B^{kt}}^{\phi_{\mathcal{L}}(M_{[s]}^k)} \mapsto d(\bar{\rho}_{B^{kt}}^{\mathsf{E}_{[s]}(M_{[s]}^k)}), \tag{3.114}$$

where $\phi_{[s]}(M_{[s]}^k) = (\mathsf{E}_l(M_l^k))_{l\in[s]}$. This implies the average reliability as

$$\mathbb{P}[\psi(\bar{\rho}_{B^{kt}}^{\mathsf{E}_{[n]}(M_{[s]}^k)}) \neq M_{[s]}^k] = \mathbb{P}\left[d(\bar{\rho}_{B^{kt}}^{\widetilde{W}_{[s]}^{\otimes k}(e_{[s]}(M_{[s]}^k))}) \neq M_{[s]}^k\right] \xrightarrow{m\to\infty} 0. \tag{3.115}$$

The average message indistinguishability can be proved by using the random choice of the encoders Chou (2022). The converse of theorem is similar to the point-to-point channels Wilde (2013). It is sufficient to exchange private randomness but not private messages between the transmitters and the legitimate receiver Chou (2022). □

Definition 3.2.6 A degradable QMAC $\mathcal{N}_{\vec{A}'\to B}$ means that there exists a channel $\mathcal{M}_{B\to E}$ such that $\mathcal{M}_{B\to E}\circ\mathcal{N}_{\vec{A}'\to B}=\mathcal{N}_{\vec{A}'\to E}$.

Define $C^{\mathrm{sum}}_{\mathrm{PMAC}}$ as the supremum of all achievable sum-rates in $C_{\mathrm{PMAC}}(\mathcal{N})$. This implies an asymptotical private region as

$$C^{\mathrm{sum}}_{\mathrm{PMAC}}(\mathcal{N}) = \lim_{t\to\infty}\frac{1}{t}P^{\mathrm{sum}}_{\mathrm{MAC}}(\mathcal{N}^{\otimes t}) \tag{3.116}$$

with $P^{\mathrm{sum}}_{\mathrm{MAC}}(\mathcal{N})=\max_{\rho}[I(\vec{X};B)_{\rho}-I(\vec{X};E)_{\rho}]^{+}$, where the maximization is defined in Theorem 3.2.7. This implies the best achievable sum rates in the classical private capacity region.

Theorem 3.2.8 *For a degradable QMAC $\mathcal{N}_{\vec{A}'\to B}$ we have*

$$C^{\mathrm{sum}}_{\mathrm{PMAC}}(\mathcal{N}) = P^{\mathrm{sum}}_{\mathrm{MAC}}(\mathcal{N}). \tag{3.117}$$

Definition 3.2.7 Consider a QMAC $\mathcal{N}_{\vec{A}'\to B}$. Quantum sum rates is defined by

$$Q^{\mathrm{sum}}_{\mathrm{MAC}}(\mathcal{N}) = \max_{\varphi_{\vec{A}\vec{A}'}}\{S(B)_{\rho}-S(\vec{A}B)_{\rho}\}, \tag{3.118}$$

where the maximization is over states $\varphi_{\vec{A}\vec{A}'}=\otimes_{i\in[n]}\varphi_{A_iA_i'}$ with a pure state $\varphi_{A_iA_i'}$, and $\rho_{\vec{A}B}=\mathcal{N}_{\vec{A}'\to B}(\varphi_{\vec{A}\vec{A}'})$.

It is easy to show the additivity for the quantum coherent information on degradable channels, i.e., for two degradable QMACs $\mathcal{N}$ and $\mathcal{M}$, we obtain that

$$Q^{\mathrm{sum}}_{\mathrm{MAC}}(\mathcal{N}\otimes\mathcal{M}) = Q^{\mathrm{sum}}_{\mathrm{MAC}}(\mathcal{N})+Q^{\mathrm{sum}}_{\mathrm{MAC}}(\mathcal{M}). \tag{3.119}$$

This further implies the consistency of private sum rates and quantum sum rates.

Theorem 3.2.9 *For a degradable QMAC $\mathcal{N}_{\vec{A}'\to B}$, we have*

$$C^{\mathrm{sum}}_{\mathrm{PMAC}}(\mathcal{N}) = Q^{\mathrm{sum}}_{\mathrm{MAC}}(\mathcal{N}). \tag{3.120}$$

The random binning idea depends on quantum source coding. Instead, another capacity-achieving approach can explore the reliability and privacy constraints by quantum channel coding and universal hashing Hayashi (2015). The quantum channel coding seems natural because the wiretap channel is a generalization of a channel coding problem.

3.2.3 Entanglement-Assisted Classical Capacity

In this subsection, we consider the entanglement-assisted classical capacity over QMACs. As shown in Fig. 3.3, consider a QMAC with s senders, each one sends

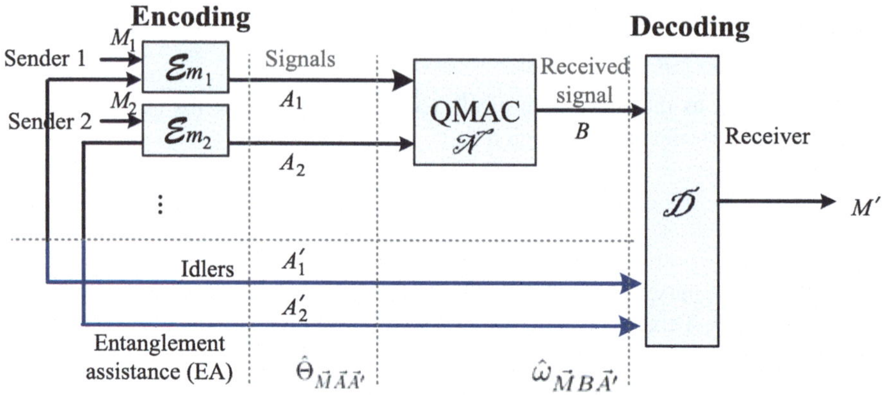

Fig. 3.3 Schematic EA-QMAC communication protocol. The s senders apply independent quantum encoding operations $\mathcal{E}_{m_i}$ on the input state σ_{m_i} dependent of the message m_i, $i = 1, \cdots, s$. Denote the entire message as $\vec{m} = m_1 \cdots m_s$. The encoded state is $\hat{\theta}^m$. The senders' encoded systems $\vec{A} = A_1 \cdots A_s$ are passed through the QMAC $\mathcal{N}$, leading to the output system B. The receiver applies the decoding operation $\mathcal{D}$ on the joint state $\hat{\varrho}^m$ of B and the pre-shared systems $\vec{A}' = A_1' \cdots A_s'$

a message $m_i (1 \leq i \leq s)$ sampled from a message space M_i. The overall messages $\vec{m} = m_1 \cdots m_s$ is sampled from the message space $\vec{M} = \otimes_{i=1}^{s} M_i$. To send each message m_i, the i-th sender A_i performs a quantum mapping $\mathcal{E}_{m_i}$ to produce a quantum system A_i. Following Hsieh et al. (2008), we introduce EA, namely the receiver has a local system A_i' (idler) pre-shared as the $\Phi_{\vec{A}'}$ with the i-th sender.

We introduce some results of the two-party entanglement-assisted direct coding theorem Fanizza et al. (2020), Bennett and Wiesner (1992). Especially, suppose that two parties Alice and Bob are connected by n copies of quantum channel $\mathcal{N}$: $A' = A_1' \cdots A_n' \to B = B_1 \cdots B_n$. Alice controls its input system A' and Bob has access to its output B. They also share n copies of some pure bipartite entanglement $\varphi_{A'B'}$. Alice and Bob use these resources to communicate similar to the superdense coding Lloyd (1997). Based on her message Alice implements a quantum operation on the shared entanglement and sends the outcomes through the quantum channel. Bob implements decoding on the joint system of the channel output and his shared entanglement. We define an (t, R, ρ, ϵ) entanglement-assisted code for this two-party quantum communication as follows.

Definition 3.2.8 An (t, R, ρ, ϵ) EA code consists of:

- a set of unitary encoding maps $\{\mathcal{E}_i, i \in [2^{tR}]\}$ acting on A'^k by Alice;
- a decoding measurement $\{\mathcal{D}_i, i \in [2^{tR}]\}$ acting on $B^t B'^t$ by Bob, such that for all $i \in [2^{tR}]$

(i) $\mathrm{Tr}[(\mathcal{D}_i \circ ((\mathcal{N}^t \circ \mathcal{E}_i) \otimes \mathbb{1}))(\varphi^t)] \geq 1 - \epsilon$;
(ii) the encoded density matrix satisfies $\mathcal{E}_i(\rho^t) = \rho^t$;
(iii) $\mathrm{Tr}[(\mathcal{D} \otimes \mathbb{1}_{E'}) \circ ((\mathcal{N}^t \circ \mathcal{E}_i) \otimes \mathbb{1}) - (\mathcal{N}^t \otimes \mathbb{1}))(\varphi^t)] \leq \epsilon$.

With this code, one can prove the entanglement-assisted direct coding theorem which implies the Holevo-Schumacher-Westmoreland Theorem. Another proof is from the method of conditionally typical subspaces.

Theorem 3.2.10 *Holevo (2001), Schumacher (1995), Hsieh et al. (2005) (Holevo-Schumacher-Westmoreland) Define* $R = I(A; B)_\theta$ *and* $\theta_{AB} = (\mathbb{1} \otimes \mathcal{N})(\varphi_{AA'})$. *For any* $\epsilon, \delta > 0$ *and large n, there exists an* $(n, R - \delta, \rho, \epsilon)$ *EA code.*

The special case of two-sender QMAC was given in Hsieh et al. (2008). Although the superadditivity in the capacity region Czekaj and Horodecki (2009), Czekaj (2011) and EA advantage have been proved in a QMAC Leditzky et al. (2020), it is unclear how much advantage the entanglement can provide for a QMAC Hsieh et al. (2005), Shi et al. (2021).

In a QMAC, multiple senders individually communicate with a single receiver. Besides a transmitter that sends encoded messages, each sender may have access to the entanglement pre-shared with the receiver. The receiver decodes all messages by performing joint measurements on all received states and the stored EA.

Definition 3.2.9 A quantum measurement $\{\mathcal{D}_m, m \in [n]\}$ is a set of cp (completely positive) maps $\mathcal{D}_m$,

$$\mathcal{D}_m : \rho \to \sum_k D_{km} \rho D_{km}^\dagger. \tag{3.121}$$

The sum of the cp maps $\mathcal{D} = \sum_{m \in [s]} \mathcal{D}_m$ is trace preserving, and $\sum_{km} D_{km}^\dagger D_{km} = \mathbb{1}$.

The measurement has one quantum input and two outputs, classical and quantum. The probability of classical outcome m and corresponding quantum output $\mathcal{D}_m(\rho)/\mathrm{Tr}\mathcal{D}_m(\rho)$ is $\mathrm{Tr}\mathcal{D}_m(\rho)$. Ignoring the classical output reduces the measurement to the quantum map $\mathcal{D}$. Ignoring the quantum output reduces the measurement to the set of POVMs (positive operator valued measure) $\{\Gamma_m\}$ with $\Gamma_m = \sum_k D_{km}^\dagger D_{km}$.

We now turn to the scenario with two senders, A_1 and A_2, and one receiver, B. They are connected by a large number n of uses of QMAC $\mathcal{M} : A_1' A_2' \to B$. A_1 and A_2 control the input systems A_1' and A_2', respectively. B can access to the output B. Each sender shares a large number of entanglements with the receiver, for simplicity, it may be in arbitrary pure entanglement $\gamma_{A_1 B_{A_1}}$ and $\gamma'_{A_2 B_{A_2}}$. A_1 has the system A_1 and A_2 has A_2. The system $B_{A_1} B_{A_2}$ belongs to Bob. Based on her message A_1 implements encoding locally, likewise for A_2. The encoded states are then sent into the quantum channel. The receiver decodes the messages by performing decoding on joint states of all the received states and the shared entanglement. This may be formalized as the following information processing task.

Definition 3.2.10 An (n, R_1, R_2, ϵ) EA code is given by:

- two sets of encoding CTPT maps $\{\mathcal{E}_i, i \in [2^{nR_1}]\}$ taking A_1 to $A_1'^n = A_{1,1}' \cdots A_{1,n}'$ for A_1, and $\{\mathcal{E}_i', i \in [2^{nR_2}]\}$ taking A_2 to $A_2'^n = A_{2,1}' \cdots A_{2,n}'$ for A_2.

- a decoding POVM measurement $\{\mathcal{D}_{k,i}, k \in [2^{nR_1}], i \in [2^{nR_2}]\}$ acting on $B_{A_1} B_{A_2}$ for B such that

$$\mathrm{Tr}\left[[\mathcal{D}_{k,i} \circ ((\mathcal{M}^n \circ (\mathcal{E}_k \otimes \mathcal{E}'_i) \otimes \mathbb{1}_{B_{A_1} B_{A_2}})](\gamma \otimes \gamma')\right] \geq 1 - \epsilon \tag{3.122}$$

The tuple of (R_1, R_2) is achievable rate if for any $\epsilon > 0$, $\delta > 0$ and large n, there is an $(n, R_1 - \delta, R_2 - \delta, \epsilon)$ EA code. The EA capacity region $C_E(M)$ is the closure of the set of all the achievable rates. Our goal here is to find $C_E(M)$.

Consider a two quantum access channel $\mathcal{N} : A'_1 A'_2 \to B$. For some states $\rho_{A'_1}$ and $\rho'_{A'_2}$, define $\theta_{A_1 A_2 B} = (\mathbb{1}^{A_1 A_2} \otimes \mathcal{N})(\varphi_{A_1 A'_1} \otimes \varphi'_{A_2 A'_2})$, where $\varphi_{A_1 A'_1}$ and $\varphi'_{A_2 A'_2}$ are purifications of $\rho_{A'_1}$ and $\rho'_{A'_2}$, respectively.

Define the channel $\mathcal{N}_1 : A'_1 \to B$ by

$$\mathcal{N}_1 : \omega \mapsto \mathcal{N}(\omega \otimes \rho'). \tag{3.123}$$

Here, $\mathcal{N}_1$ is an effective channel from A_1 to B when A_2's input to $\mathcal{N}^n$ is ρ'^n. Define $\hat{\mathcal{N}}_1 : A'_1 \to B_{A_1} B$ by

$$\hat{\mathcal{N}}_1 : \omega \mapsto (\mathbb{1} \otimes \mathcal{N})(\omega \otimes \varphi'). \tag{3.124}$$

Here, $\hat{\mathcal{N}}_1$ is an extension of $\mathcal{N}_1$.

Define the channel $\mathcal{N}_2 : A'_1 \to B$ by

$$\mathcal{N}_2 : \omega \mapsto (\mathbb{1} \otimes \mathcal{N})(\varphi \otimes \omega), \tag{3.125}$$

where $\mathcal{N}_2^n$ is an effective channel from A_2 to B when A_1 inputs A'_1 of the purification $\gamma_{A'_1 B_{A_1}}$ without encoding.

For a given $\epsilon > 0$, $\delta > 0$, define two quantities

$$\begin{aligned} R_1 &= I(A_1; B)_\theta - \delta, \\ R_2 &= I(A_2; B|A_1)_\theta - \delta. \end{aligned} \tag{3.126}$$

Definition 3.2.11 Define the two-dimensional capacity region C_E, by the set of pairs of nonnegative rates (R_1, R_2) satisfying

$$C_E := \left\{ \begin{array}{c} R_1 \leq I(A_1; B|A_2)_\theta; \\ R_2 \leq I(A_2; B|A_1)_\theta; \\ R_1 + R_2 \leq I(A_1 A_2; B)_\theta. \end{array} \right\} \tag{3.127}$$

Define $\hat{C}_E(\mathcal{N})$ as the union of the $C_E(\mathcal{N}, \rho, \rho')$ regions taken over all states ρ, ρ', i.e.,

$$\hat{C}_E(\mathcal{N}) = \bigcup_{\rho, \rho'} C_E(\mathcal{N}, \rho, \rho'). \tag{3.128}$$

From each pair of A_i and B we can obtain EA code. This implies that for a large n there is an (n, R_1, ρ, ϵ)-EA code $(\mathcal{E}, \mathcal{D})$ for $\mathcal{N}_1$ and an $(n, R_2, \rho', \epsilon)$-EA code $(\mathcal{E}', \mathcal{D}')$ for $\mathcal{N}_2$ such that for all $k \in [2^{nR_1}]$, $i \in [2^{nR_2}]$,

- $\mathrm{Tr}[(\mathcal{D}_k \circ ((\mathcal{N}_1^n \circ \mathcal{E}_k) \otimes \mathbb{1}_{B_{A_1}}))(\gamma)] \geq 1 - \epsilon$;
- $\|[(\mathcal{D} \otimes \mathbb{1}) \circ ((\hat{\mathcal{N}}_1^n \circ \mathcal{E}_k) \otimes \mathbb{1}_{B_{A_1}}) - (\hat{\mathcal{N}}_1^n \otimes \mathbb{1}_{B_{A_1}})](\gamma)\| \leq \epsilon$;
- $\mathrm{Tr}[[\mathcal{D}_i' \circ ((\mathcal{N}_2^n \circ \mathcal{E}_i') \otimes \mathbb{1}_{B_{A_1}})](\gamma')] \geq 1 - \epsilon$;
- The encoded density operator satisfies $\mathcal{E}_i'(\rho'^n) = \rho'^n$.

We now define a code for QMAC $\mathcal{N}$. A_1 and A_2 encode the input messages according to encoding mappings $\mathcal{E}_k$ and $\mathcal{E}_i'$, respectively. Define an measurement $\Delta = \{\Delta_{k,i}\}$ on $B_{A_1} B_{A_2}$ by

$$\Delta_{k,i} = \mathcal{D}_i' \circ (\mathcal{D}_k \otimes \mathbb{1}_{B_{A_2}}). \tag{3.129}$$

The B's decoding POVM $\{\mathcal{D}_{k,i}\}$ is the restriction of $\{\mathcal{D}_{k,i}\}$. It is easy to prove Hsieh et al. (2005) that the success probability of decoding A_1's message m_k satisfies

$$\mathrm{Tr}[((\mathcal{D}_{m_k} \otimes \mathbb{1}_{B_{A_2}}) \circ (\mathcal{N}^n \circ (\mathcal{E}_{m_k} \otimes \mathcal{E}_i') \otimes \mathbb{1}_{B_{A_1} B_{A_2}}))(\gamma \otimes \gamma')] \geq 1 - \epsilon. \tag{3.130}$$

This implies the following coding theorem.

Theorem 3.2.11 *For a quantum two-access channel $\mathcal{N} : A_1' A_2' \to B$, the EA capacity region C_E is given by the regularized expression*

$$C_E = \mathrm{Cl}\{(R_1, R_2) | \cup_{n=1}^{\infty} \frac{1}{n} \hat{C}_E(\mathcal{N}^{\otimes n})\}, \tag{3.131}$$

where Cl indicates the closure operation.

This theorem provides an additional single-letter upper bound on the sum-rate as

$$R_1 + R_2 \leq \max_{\rho, \rho'} I(A_1 A_2; B)_\theta. \tag{3.132}$$

Take the case $|A_1'| = |A_2'| = d \geq 2$ as an example. The collective phase-flip channel Bennett et al. (2002) $\mathcal{N}_p : A_1' A_2' \to B$ is defined as

$$\mathcal{N}_p(\rho) = \sum_{k=0}^{d-1} (\hat{Z}_k \otimes \hat{Z}_k) \rho (\hat{Z}_k \otimes \hat{Z}_k)^\dagger, \tag{3.133}$$

where $\hat{Z}_k = \sum_j e^{i2\pi jk/d} |j\rangle\langle j|$ is the generalized Pauli phase operator. Define the corresponding state θ given by

$$\theta_{ABC} = (\mathbb{1}_{AB} \otimes \mathcal{N}_p)(\Phi_{A_1 A_1'} \otimes \Phi_{A_2 A_2'}), \tag{3.134}$$

where $\Phi = \frac{1}{\sqrt{d}} \sum_i |i\rangle_{A_1} |i\rangle_{A_2}$ is the maximal entanglement. With a forward evaluation, we obtain $I(A_1; B|A_2) = I(A_2; B|A_1) = 2 \log d$ and $I(A_1 A_2; B) = 4 \log d - H(p)$. The capacity region for the multiple access phase-flip channel $\mathcal{N}_p$ assisted by entanglement is given by

$$\mathcal{C}_E := \left\{ (R_1, R_2) \middle| \begin{array}{l} R_1 \leq 2 \log d; \\ R_2 \leq 2 \log d; \\ R_1 + R_2 \leq 4 \log d - H(p). \end{array} \right\} \tag{3.135}$$

We turn to the case of s senders. Suppose a simple case that each pair of sender and the receiver share one entanglement $\hat{\theta}_{A_i A_i'}$. The total entangled system is defined as $\hat{\theta}_{\vec{A}\vec{A}'} = \otimes_{i=1}^{s} \hat{\theta}_{A_i A_i'}$, where $\vec{A} = A_1 \cdots A_s$ and $\vec{A}' = A_1' \cdots A_s'$.

All the senders implement the encoding locally. The encoded system is defined as the composite system $\vec{A}$, which contains all systems $A_1 \cdots A_s$. These systems are then input the QMAC $\mathcal{N}_{\vec{A} \to B}$. The receiver decodes the messages jointly with the output system B and the EA. For convenience, define the output state without the encoding as

$$\hat{\rho}_{B\vec{A}'} = [\mathcal{N}_{\vec{A} \to B} \otimes \mathbb{1}](\hat{\theta}_{\vec{A}\vec{A}'}), \tag{3.136}$$

where $\mathbb{1}$ is the identity channel. The goal is to maximize the transmission rate R_i between the i-th sender and the receiver, $i = 1, \cdots, s$.

The rates $(R_1, \cdots, R_s)$, in general, have non-trivial tradeoffs with each other. In the simplest case without $E_{\vec{A}}$, the capacity region is well-established in Winter (2001).

To show the rate region, we divide the senders into two subsets $J \subset [n]$ and the complementary set J^c. Any s-fold quantity is then written as a composition of two parts, e.g., message space $\vec{M} = M_J M_{J^c}$, with $M_J = \otimes_{i \in J} M_i$, $M_{J^c} = \otimes_{i \in J^c} M_i$. Similarly, denote the message as $\vec{m} = m_J m_{J^c}$.

Definition 3.2.12 For a s-sender QMAC $\mathcal{N}$, the EA classical capacity region is given by

$$\mathcal{C}_E(\mathcal{N}) = \mathrm{Cl}\{(R_1, \cdots, R_s) | \cup_{\ell=1}^{\infty} \frac{1}{\ell} \mathcal{C}_E^{(1)}(\mathcal{N}^{\otimes \ell})\}, \tag{3.137}$$

where $\mathcal{C}_E^{(1)}(\mathcal{N})$ is the one-shot capacity region defined by $\mathcal{C}_E^{(1)}(\mathcal{N}) = \mathrm{Cv}(\cup_{\hat{\theta}} \tilde{\mathcal{C}}_E(\mathcal{N}, \hat{\theta}))$ and $\tilde{\mathcal{C}}_E(\mathcal{N}, \hat{\theta}))$ is the one-shot capacity region depending on the state $\hat{\theta}$.

Inspired by Theorems 3.2.1 and 3.2.2, we obtain the following one-shot and one-encoding rate regions.

Theorem 3.2.12 *The one-shot and one-encoding rate region $\tilde{\mathcal{C}}_E(\mathcal{N}, \hat{\theta})$ for the $2s$-partite pure product state $\hat{\theta}_{\vec{A}\vec{A}'}$ is given by*

$$\tilde{C}_E(\mathcal{N}, \hat{\theta}) = \left\{ (R_1, \cdots, R_s) | \sum_{k\in J} R_k \leq I(A'_J; B|A'_{J^c})_{\hat{\rho}}, \forall J \right\}, \quad (3.138)$$

where the conditional quantum mutual information is evaluated over the output state $\hat{\rho}_{B\vec{A}'} = [\mathcal{N}_{\vec{A}\to B} \otimes \mathbb{1}](\hat{\theta}_{\vec{A}\vec{A}'})$.

The EA capacity theorem 3.2.12 quantifies the trade-off between the communication rates of senders. The capacity is dependent on conditional mutual information similar to the classical scenarios. Compared with the communication scenario without EA Yen and Shapiro (2005), entanglement provides great advantages.

Take an optical or microwave communication scenario as an example, where the QMAC is a bosonic thermal-loss. Denote the input modes as $\hat{a}_{A_1} \cdots \hat{a}_{A_s}$ from s senders. In the QMAC $\mathcal{N}$, all the input modes pass through a beam splitter (BS) to generate a mixing mode

$$\hat{a}_{A_{mix}} = \sum_{i=1}^{s} \sqrt{\eta_i} \hat{a}_{A_i}, \quad (3.139)$$

while other ports of BS may be discarded. Here, $\{\eta_i\}$ are non-negative and normalized weights. The mixed mode is then going through a bosonic thermal-loss channel $\mathcal{L}^{\tau;N_B}$ represented by

$$\hat{a}_B = \sqrt{\tau}\hat{a}_{A_{mix}} + \sqrt{1-\tau}\hat{a}_E, \quad (3.140)$$

where $\hat{a}_E$ denotes the environment mode with a mean photon number $\langle \hat{a}_E^\dagger \hat{a}_E \rangle = N_B/(1-\tau)$. Similar to the communication scenario without EA, the outer bounds for the EA classical capacity are given by

$$R_k \leq C_E(N_{S,K}; \mathcal{L}^{\tau;N_B}), 1 \leq k \leq s, \quad (3.141)$$

$$\sum_{k=1}^{s} R_k \leq C_E(\sum_{k=1}^{s} N_{S,K}; \mathcal{L}^{\tau;N_B}), \quad (3.142)$$

by reducing it to the single-sender EA classical capacity, where $C_E(N_{S,K}; \mathcal{L}^{\tau;N_B})$ is given by

$$C_E(N_{S,K}; \mathcal{L}^{\tau;N_B}) = S(N_S) + S(N'_S) - S(A_+) - S(A_-) \quad (3.143)$$

with $A_\pm = (\sqrt{(N'_S + N_S + 1)^2 - 4\tau N_S(N_S + 1)} - 1 \pm (N'_S - N_S)/2)$ and $N'_S = \tau N_S + N_B$.

3.3 Quantum Broadcast Communication

Different from the quantum communication over QMAC, we consider quantum broadcast communication in this subsection. Here, one sender will send messages to different receivers, as shown in Fig. 3.4.

In classical communication, a simple broadcast channel is modeled as a discrete memoryless channel with one sender, Alice, and two receivers, Bob and Charlie. This kind of channel was initiated by Cover Cover (1972). The main goal is to superimpose information for larger achievable rates than those by timesharing protocols. Denote R as the common information rate for two senders, R_Y and R_Z are personal information rates for two senders. The triple tuple of (R, R_Y, R_Z) satisfies

$$\begin{cases} R_Y \leq I(X; Y|T), \\ R + R_Z \leq I(T; Z), \end{cases}$$

over a degraded broadcast channel Bergmans (1973), Gallager (1980), where $|T| \leq \min\{|X|, |Y|, |Z|\}$.

Wyner reconsidered this problem Wyner (1975) with wiretap channels. Due to an operational correspondence between the privacy and quantum coherence Devetak (2005), the private information rate R_Y for Bob, as well as public common information rate R can be ensured over a broadcast channel if and only if there exists probability functions $p(t, v)p(x|v)$ such that Csiszar and Körner (1978):

$$\begin{cases} 0 \leq R_Y \leq I(V; Y|T) - I(V; Z|T), \\ 0 \leq R \leq \min\{I(T; Y), I(T; Z)\}. \end{cases}$$

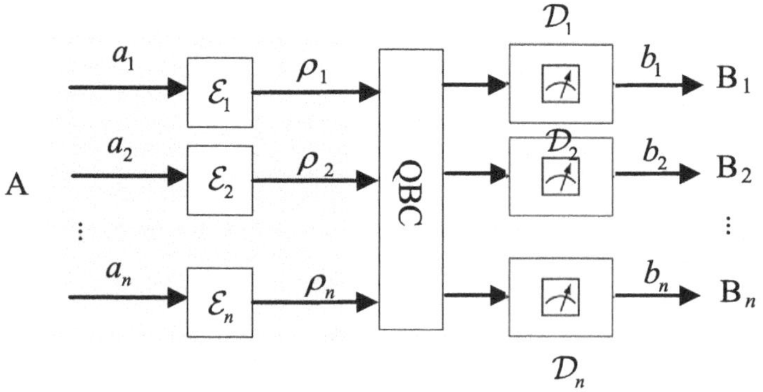

Fig. 3.4 Schematic broadcast communication over a quantum broadcast channel. The sender A encodes its message a_i on its share and sends the signal ρ_i to the receiver B_i, $i = 1, \cdots, n$. Each receiver decodes the messages by measuring the received states

In this section, we consider quantum generalizations of this problem Yard et al. (2011). Consider a quantum broadcast communication. One sender A sends messages to a group of receivers $\mathsf{B}_1, \cdots, \mathsf{B}_n$. A general assumption is to send different messages to different receivers.

Definition 3.3.1 A quantum broadcast channel (QBC) $\mathcal{N}_{A\to B_1\cdots B_n}$ consists of a single input state $\mathbb{H}_A$ and s output states on Hilbert space $\mathbb{H}_{B_1}\otimes\cdots\otimes\mathbb{H}_{B_n}$.

The channel from the sender A to the receiver B_i is defined by partial trace of $\mathcal{N}_{A\to B_1\cdots B_n}$ as $\mathcal{N}_{A\to B_i} := \mathrm{Tr}_{B_j, j\neq i}\mathcal{N}_{A\to B_1\cdots B_n}$.

3.3.1 Classical Public and Private Capacities

For the special case of $n=2$, a QBC has a Kraus representation as

$$\mathcal{N}_{A\to B_1B_2}(\rho_A) = \sum_j D_j\rho_A D_j^\dagger \tag{3.144}$$

for some set of Kraus operators D_j satisfying $\sum_j D_j^\dagger D_j = \mathbb{1}$.

The classical broadcast channel as a special case uses random variables X (input) and Y_1, Y_2 (outputs), where Kraus operators are $D_{x,y_1,y_2} = \sqrt{P_{Y_1Y_2|X}(y_1,y_2|x)}|y_1y_2\rangle\langle x|$ for some probability kernel $P_{Y_1Y_2|X}$ and classical orthogonal bases $\{|x_i\rangle\}$ and $\{|y_1,y_2\rangle\}$. For a given input $x\in X$, the output state of a classical broadcast channel is given by

$$\mathcal{N}^{C1}_{X\to Y_1Y_2}(|x\rangle\langle x|) = \sum_{(y_1,y_2)\in\mathcal{Y}_1\times\mathcal{Y}_2} P_{Y_1Y_2|X}(y_1,y_2|x)|y_1,y_2\rangle\langle y_1,y_2|. \tag{3.145}$$

Definition 3.3.2 A QBC $\mathcal{N}_{A\to B_1\cdots B_n}$ is chained degraded if there are *degrading channels* $\mathcal{N}_{B_i\to B_{i+1}}$ from B_i to B_{i+1} such that $\mathcal{N}_{B_i\to B_{i+1}}\circ\mathcal{N}_{A\to B_i} = \mathcal{N}_{A\to B_{i+1}}$, $i=2,\cdots,n$.

This is named as *stochastically degraded* in the classical scenarios, which means $X, Y_1, \cdots, Y_n$ can form a Markov chain $X\to Y_1\to\cdots\to X_n$.

Definition 3.3.3 A classical-quantum (CQ) channel $\mathcal{N}_{\mathcal{X}\to B}$ means that the sender can remotely prepare any one of conditional states $\{\rho_x^B\}_{x\in\mathcal{X}}$ for the receiver.

A classical-quantum broadcast (CQB) channel $\mathcal{N}_{\mathcal{X}\to B_1\cdots B_n}$ from A to B_1, ..., B_n, means A prepares any one of n-partite conditional states $\{\rho_x^{B_1\cdots B_n}\}_{x\in\mathcal{X}}$ for n receivers.

We consider degraded messages for the CQ channel. A sequence $x_1x_2\cdots x_n$, with each x_i belonging to some set $\mathcal{X}$ will be denoted by $x_{[n]}$. For a CQB channel $\mathcal{N}_{\mathcal{X}\to B_1\cdots B_n}$, suppose that A sends a private message to B_i, $i\in J\subset[n]$ while simultaneously sending common message to all receivers. If $\mathcal{N}$ has conditional states $\rho_x^{B_1\cdots B_n}$, we define an (R, R_J, t, ϵ) code with one party in J as follows. Here, $R_J = (R_i, j\in J)$. Similar notations will be used in what follows.

Definition 3.3.4 An (R, R_p, t, ϵ) code for the channel $\mathcal{N}_{\mathcal{X}\to B_1\cdots B_n}$ and $J \subset [n]$ consists of an encoding $\{x_{[t]}(m, k) \in \mathcal{X}^n\}$ with $(m, k) \in 2^{tR} \times 2^{tR_p}$, a POVM $\{M_m^{(i)}\}$ on B_i^t $(i \in [n])$and a POVM $\{M_{mk}\}$ on B_n^s, which satisfy

$$\mathrm{Tr}[\rho_{x_{[t]}(m,k)}(\otimes_{j\in[n]} M_m^{(i)} \otimes M_{mk})] \geq 1-\epsilon \tag{3.146}$$

for any $(m, k) \in 2^{tR} \times 2^{tR_p}$.

A pair (R, R_p) is achievable rate if there is a sequence of (R, R_p, t, ϵ_t) codes with $\epsilon_t \to 0$. The classical private capacity region $\mathcal{C}$ of $\mathcal{N}$ is defined as the closure of all achievable rate triples, i.e.,

$$\mathcal{C} := \mathrm{Cl}\{(R, R_p)|\exists(R, R_p, t, \epsilon_s) \text{ codes}, \epsilon_t \to 0\} \tag{3.147}$$

with the closure operation Cl.

Let ρ and M be uniformly chosen $d \times d$ matrices such that ρ is a density matrix and $0 \leq M \leq 1$, which satisfy $\mathbb{E}[\mathrm{Tr}\rho M] \geq 1-\epsilon$. This implies an average gentle measurement given by Winter (2001):

$$\mathbb{E}[|\sqrt{M}\rho\sqrt{M} - \rho|_1] \leq 3\sqrt{\epsilon}. \tag{3.148}$$

We state an average error version of the Holevo-Schumacher-Westmoreland Theorem for CQ codes with codewords chosen independent and identically distributed (i.i.d.) according to a product distribution.

Lemma 3.3.1 (Holevo-Schumacher-Westmoreland Random Coding Theorem) *Yard et al. (2011). For a given CQ state $\varrho^{XQ} = \oplus_x p(x)\rho_x^Q$ and R with $0 \leq R < I(X; Q)_\varrho$, for any $\epsilon > 0$, there is large t so that if 2^{tR} codewords $\mathcal{C} = \{X^t(m)\}$ are chosen i.i.d. according to the product distribution $p(x_{[t]}) = \prod_{i=1}^t p(x_i)$ corresponding to the input $\rho_{x_{[t]}} = \otimes_i \rho_{x_i}$, there is a POVM $\{M_m\}$ on Q^t with an average error probability less than ϵ, in the sense that*

$$\mathbb{E}_{\mathcal{C}}[2^{-tR}\sum_{m=1}^{2^{tR}} \mathrm{Tr}\rho_{X^t(m)} M_m] \geq 1-\epsilon. \tag{3.149}$$

Lemma 3.3.2 Yard et al. (2008, Lemma 1) *Given density matrices ρ, σ and an operator Λ with $0 \leq \Lambda \leq \mathbb{1}$ we obtain that*

$$\mathrm{Tr}\Lambda\sigma \geq \mathrm{Tr}\Lambda\rho - \|\rho - \sigma\|_1. \tag{3.150}$$

Consider a CQ channel $W_{\mathcal{X}\to B_{[n]}C}$, where A sends some common messages to all receivers $\mathsf{B}_1, \cdots, \mathsf{B}_n$ and C, while she also sends some personal information to C. Sender B_i has the system B_i while receiver C has the system C. With the stated lemmas, we prove the following result for achievable rates.

Theorem 3.3.1 *Given a CQ channel $W_{\mathcal{X}\to B_{[n]}C}$ with conditional states $\{\rho_x^{B_{[n]}C}\}$, a triple of (R, R_B, R_C) is achievable if*

$$\begin{cases} R_C \leq I(X; C|T), \\ R + nR_B \leq \sum_{i\in[n]} I(T; B_i), \\ R + nR_B + R_C \leq I(X; C), \end{cases}$$

for some $p(t, x)$ with $|\mathcal{T}| \leq \min\{|\mathcal{X}|, |B_i|^2 + |C|^2 - 1, i = 1, \cdots, n\}$ and the state

$$\sigma^{TXB_{[n]}C} = \oplus_{t,x} p(t, x)\rho_x^{B_{[n]}C}. \tag{3.151}$$

Proof From the assumption of broadcast, all the receivers may have the same decoding strategy. From a similar sub-channel for each pair of A and B_i, assume that the achievability for B_i's have the same rate R_B. Hence, it only needs to show the achievability for $(R + nR_B, nR_B, 0)$, where all B_i ignores common messages intended only for C. Note that

$$I(X; C) = I(TX; C) = I(T; C) + I(X; C|T). \tag{3.152}$$

The region (3.151) with $R_B = 0$ is equivalent to

$$\begin{cases} R_C \leq I(T; C), \\ R \leq \min\{I(T; B_i), I(T; C), i \in [n]\}. \end{cases} \tag{3.153}$$

In what follows, we prove the achievability with $R_B = 0$. In fact, let $\mathcal{N}_{\mathcal{X}\to B_{[n]}C}$ be a CQ broadcast channel with $\rho_x^{B_{[n]}C}$ and $p(t, x)$ be arbitrary nonnegative functions. These define a joint CQ state

$$\sigma^{TXB_{[n]}C} = \oplus_{t,x} p(t, x)\rho_x^{B_{[n]}C} \equiv \oplus_t p(t)\sigma_t^{XB_{[n]}C}. \tag{3.154}$$

The corresponding conditional distribution $p(x|t)$ implies conditional states for a new CQ channel $\mathcal{V}_{\mathcal{T}\to B_{[n]}C}$ as

$$\psi_t^{B_{[n]}C} = \sum_x p(x|t)\rho_x^{B_{[n]}C} = \mathrm{Tr}_X \sigma_t^{XB_{[n]}C} \tag{3.155}$$

as a backed up version of $\mathcal{N}$. These states are rewritten into

$$\sigma^{TB_{[n]}C} = \mathrm{Tr}_X \sigma^{TXB_{[n]}C} = \oplus_t p(t)\psi_t^{B_{[n]}C}. \tag{3.156}$$

For any $\epsilon, \delta > 0$ and large n, we show that for the rates R_B and R satisfying $I(X; C|T)_\sigma - (1 + |\mathcal{X}|)\delta \leq R_C < I(X; C|T)_\sigma$ and $0 \leq R_C = \min\{I(T; B_i)_\sigma, I(T; C)_\sigma, i \in [n]\} - \delta$, there is an $(R, R_C, s, 16\sqrt{\epsilon})$ code for $\mathcal{N}_{\mathcal{X}\to B_{[n]}C}$.

From Yard et al. (2011, Proposition 2), there is an (R, t, ϵ) code $\{u^t(m), \Lambda_m, \Lambda'_m\}_{m \in 2^{tR}}$ for the channel $\mathcal{V}_{\mathcal{T} \to B_{[n]}C}$ with codewords of type $\hat{p}$ satisfying $|\hat{p} - p|_1 \leq \delta$. For each m we obtain that

$$\mathrm{Tr}[(\Lambda_m \otimes \Lambda'_m)\psi_m^{B^t_{[n]}C^t}] \geq 1 - \epsilon, \tag{3.157}$$

where $\psi_m = \otimes_i \psi_{u_i(m)}$.

For each k, define $u_k = s\hat{p}(k), \epsilon_k = \epsilon\hat{p}(k), \delta_k = \delta\hat{p}(k)$ and $R_k = \frac{1}{n}I(X; B_{[n]})_{\sigma_k} - \delta_k \leq |\mathcal{X}|$. It follows from the Lemma 3.3.1 that for each k, there is an (R_k, t_k, ϵ_k) random code $\{X^{t_k}(l_k|k), \Lambda_l^{(k)}\}_{l_k \in 2^{tR_k}}$ for the channel $\mathcal{N}_{\mathcal{X} \to C}$ to C that satisfies

$$\mathbb{E}[2^{-tR_k} \sum_{l_k=1}^{2^{tR_k}} \mathrm{Tr}[\rho_{l_k}^{C^k} \Lambda_{l_k}^{(k)}] \geq 1 - \epsilon_k, \tag{3.158}$$

where the expectation is over the randomness in the HSW codes. Here, we have used $C^k \equiv C^{t_k}$ and taken $\rho_{l_k}^{C^k} = \otimes_{i=1}^{t_k} \rho_{X_i(l_k|k)}^{C^i}$. Each $X_i(l_k|k)$ is chosen independently according to $p(x|k)$, so that $\mathbb{E}\rho_{l_k} = \psi_k^{\otimes t_k}$. By the symmetry of the random code construction, Eq. (3.158) may imply

$$\mathbb{E}[\mathrm{Tr}\rho_1^{C^k} \Lambda_1^{(k)}] \geq 1 - \epsilon_k. \tag{3.159}$$

Since $I(X; C|T)_\sigma = \sum_k p(k) I(X; C)_{\sigma_k}$, it follows that

$$\begin{aligned} R_C &= \sum_k \hat{p}(t) R_k \\ &\geq \sum_k p(k) R_k - \|\hat{p} - p\|_1 |\mathcal{X}| \\ &\geq I(X; C|T) - (|\mathcal{X}| + 1)\delta. \end{aligned}$$

Moreover, from $R_C < I(X; C|T)$, C can recover personal message $l \in 2^{tR_C}$ with messages $\{l_k \in 2^{tR_k}\}_k$. A's (random) encoding is defined by

$$X^t(m, l) = \mathsf{g}(m)(X^{t_1}(l_1|1) X^{t_2}(l_2|2) \cdots X^{t_d}(l_d|d)). \tag{3.160}$$

with $\rho_{ml}^{B^t_{[n]}C^t} = \rho_{X^t(ml)}^{B^t_{[n]}C^t}$. Here, $d = |\mathcal{T}|$, $k^t(1) = 1^{t_1}2^{t_1} \cdots d^{t_d}$ for identifying permutations $\{\mathsf{g}(m) \colon \mathcal{T}^t \to \mathcal{T}^t\}$ with $k^t(m) = \mathsf{g}(m)(k^t(1))$. For each l, we have $\mathbb{E}\rho_{ml} = \tau_m$.

Receiver C firstly measures $\{\Lambda_m\}$ while all B_i's measures $\{\Lambda'_m\}$, and declare respective results to be the common message M. C permutes the systems C^t according to $\mathsf{g}^{-1}(m)$ and obtains a state approximate to $\rho_{1m}^{C^t}$. For each k, he/she measures each block of l_k outputs with $\{\Lambda_{l_k}^{(k)}\}$ to get personal message $(l_1, \cdots, l_s) = l$. This procedure is defined by POVMs $\{\Lambda_{ml}\}$, where $\Lambda_{ml} = \sqrt{\Lambda_m}\Lambda_{l|m}\sqrt{\Lambda_m}$ and $\Lambda_{l|m} = \mathsf{g}(m)(\otimes_k \Lambda_{l_k}^{(k)})$. Denote

$$P_{ml} = \mathrm{Tr}[(\Lambda_{ml} \otimes \Lambda'_m)\rho_{ml}^{B^t_{[n]}C^t}],$$
$$\tilde{\rho}_{ml} = (\sqrt{\Lambda_m} \otimes \sqrt{\Lambda'_m})\rho_{ml}^{B^t_{[n]}C^t}(\sqrt{\Lambda_m} \otimes \sqrt{\Lambda'_m}).$$

This implies that

$$\begin{aligned}\mathbb{E}P_{ml} &= \mathbb{E}[\mathrm{Tr}\tilde{\rho}_{ml}^{B^t_{[n]}C^t}\Lambda_{l|m}] \\ &\geq \mathbb{E}[\mathrm{Tr}\rho_{ml}^{C^t}\Lambda_{l|m}] - \mathbb{E}\|\tilde{\rho}_{ml}^{B^t_{[n]}C^t} - \rho_{ml}^{B^t_{[n]}C^t}\|_1 \\ &\geq 1 - 4\sqrt{\epsilon}.\end{aligned}$$

from Lemma 3.3.2 and the inequality (3.148). By de-randomizing the common randomness, it follows that

$$2^{-t(R_C+R)}\sum_{m=1}^{2^{tR}}\sum_{l=1}^{2^{tR_B}} P_{ml} \geq 1 - 4\sqrt{\epsilon}.$$

Combined with the Markov inequality, it completes the proof. □

The optimality of this Theorem can be proved for a degraded channel.

Theorem 3.3.2 *Suppose that the conditional density matrices* $\{\rho_x^{B_{[n]}C}\}$ *of a CQ channel* $\mathcal{N}_{\mathcal{X}\to B_{[n]}C}$ *satisfy that* $\rho_x^C = \mathcal{M}(\rho_x^{B_{[n]}})$ *for a degraded channel* $\mathcal{M}_{B_1\to\cdots B_n\to C}$. *It follows that*

$$\begin{cases} nR_B \leq \sum_{i\in[n]} I(X; B_i|T)_\sigma, \\ R \leq I(T; C)_\sigma, \end{cases} \tag{3.161}$$

for some $\sigma^{TXB_{[n]}C} = \oplus_x p(x)\rho_x^T \otimes \rho_x^{B_{[n]}C}$ *with* $p(x)$ *and auxiliary density matrices* ρ_x^T. *If* $\rho_x^{B_i}$ *commute, then the same region with* $\sigma^{TXB_{[n]}C}$ *of the form (3.151) with* $|\mathcal{T}| \leq \min\{|\mathcal{X}|, |B_i|^2, i \in [n]\}$ *is consistent with Theroem 3.3.1.*

Proof Similar to Theorem 3.3.1, we may assume that the achievabilities for B_i's have the same R_B. Moreover, assume (R, R_B, R_C) is achievable. $(0, R_B, R + R_C)$ is also achievable. Let $\{x^t(k, m)\}$, $\{\Lambda_k\}$ and $\{\Lambda'_m\}$ comprise a $(0, R_B, R + R_C, t, \epsilon_t)$ code. Here all B_i's messages are k, and C's massage is m. Denote

$$\Pi_{km}^{KMX^t} = |m\rangle\langle m| \otimes |k\rangle\langle k|_{[n]} \otimes |x^t(k, m)\rangle\langle x^t(k, m)|. \tag{3.162}$$

Define the state induced by choosing the messages K and M uniformly at random as

$$\omega^{KMX^tB^t_{[n]}C^t} = 2^{-t(R+nR_B+R_C)}\sum_{k=1}^{2^{tnR_B}}\sum_{m=1}^{2^{t(R+R_C)}} \Pi_{km} \otimes \rho_{km}^{B^t_{[n]}C^t}.$$

Let $\Omega^{KM\hat{K}_{[n]}\hat{M}}$ be the joint state after the decoding if B_i stores the decoded messages in $\hat{K}_i$ and C stores in $\hat{M}$. For some $\epsilon_t', \epsilon_t'', \epsilon_t''' \to 0$, we obtain that

$$\begin{aligned} t(R+R_C) &= H(M) \\ &\leq I(M;\hat{M})_\Omega + t\epsilon_t' \\ &\leq I(M;C^n)_\omega + t\epsilon_t', \end{aligned} \tag{3.163}$$

from the Fano inequality and the Holevo bound Wilde (2013). From the chained degraded channel, we obtain that

$$\begin{aligned} tnR_B &\leq t\sum_{i\in[n]} R_{B_i} \\ &= \sum_{i\in[n]} H(K_i) \\ &\leq \sum_{i\in[n]} I(K_i;B_i^t M)_\omega + s\epsilon_t''' \\ &\leq \sum_{i\in[n]} I(X^n;B_i^t|M)_\omega + s\epsilon_t''', \end{aligned} \tag{3.164}$$

from the Fano inequality, the Holevo bound, data processing, and the Markov chain Wilde (2013).

The next step is to single-letterize the bounds from the classical proof Gallager (1980), Devetak and Winter (2005), Yard et al. (2011). □

3.3.2 Classical and Quantum Capacities

Consider a scenario in which the sender sends quantum information to C at rate Q, while sending the classical message at rate R for all receivers $\mathsf{B}_1, \cdots, \mathsf{B}_n$ and C. The sender prepares one of many states $\{|\Upsilon_m\rangle^{AA''}\}$ which are entangled between a system A and the inputs of parallel identical broadcast channels. Receiver C performs quantum measurement $\mathcal{D}^{C^t\to\hat{A}M_C}$ for recovering the classical message, as well as holding the entangled part $\hat{A}$. Each B_i performs a measurement $\mathcal{D}_i^{B_i^t\to M_{B_i}}$ to recover the common classical message. These components may be used to define an (R, Q, t, ϵ) CQ entanglement generation code .

Definition 3.3.5 An (R, Q, t, ϵ) classical-quantum entanglement generation (CQEG) code for the broadcast channel $\mathcal{N}_{A'\to BC}$ consisting of states $\{|\Upsilon_m\rangle^{AA''}\}$, measurements $\mathcal{D}_{C^t\to\hat{A}M_C}$, and $\mathcal{D}^i_{B_i^t\to M_{B_i}}$, $i=1,\cdots,n$, which satisfy

$$F(|m\rangle^{M_{B_{[n]}}M_C}|\Phi_Q\rangle^{A\hat{A}}, (\mathcal{D}\otimes_{i\in[n]}\mathcal{D}^i)\circ\mathcal{N}^{\otimes n}(\Upsilon_m^{AA''})) \geq 1-\epsilon \tag{3.165}$$

for each m, where the fidelity F is defined as:

$$F(\rho,\sigma) = \mathrm{Tr}\sqrt{\rho^{\frac{1}{2}}\sigma\rho^{\frac{1}{2}}}, \tag{3.166}$$

$|m\rangle^{M_{B_{[n]}}M_C} \equiv |m\rangle^{M_{B_{[n]}}}|m\rangle^{M_C}$ and $|\Phi_Q\rangle^{A\hat{A}}$ is an EPR state $|\Phi_Q\rangle^{A\hat{A}} = \frac{1}{\sqrt{2^{tQ}}}\sum_{a=1}^{2^{tQ}}$ $|a\rangle^A|a\rangle^{\hat{A}}$ with some fixed rate Q.

A pair of (R, Q) is called an achievable CQ rate for entanglement generation if there is a sequence of (R, Q, t, ϵ_t) CQEG codes with $\epsilon_t \to 0$. Its CQEG capacity region $\mathcal{C}_Q(\mathcal{N})$ is defined as the closure of all achievable CQ rates, i.e.,

$$\mathcal{C}_Q := \mathrm{Cl}\{(R, Q)|\exists(R, Q, t, \epsilon_t) \text{ CQEG codes}, \epsilon_t \to 0\}. \tag{3.167}$$

Lemma 3.3.3 (LSD Random Coding Theorem) *Yard et al. (2011). For a given channel $\mathcal{N} : A' \to B$, a density matrix $\rho_{A'}$, and a number R with $0 \le R < I_c(\rho; \mathcal{N})$, for any $\epsilon > 0$, there is large t so that there are random $(2^{tR}, t, \epsilon)$ CQEG codes $(p_\beta, |\Gamma_\beta\rangle_{AA'^t}, \mathcal{D}^\beta_{B^t\to\hat{A}})$ with average density operator $\varrho_{A'^t} = \sum_\beta p_\beta \mathrm{Tr}_A\Gamma_\beta$ satisfying $|\varrho - \rho^{\otimes t}|_1 \le \epsilon$. Moreover, each code satisfies that $F(|\Phi\rangle_{A\hat{A}}, \mathcal{D}_\beta \circ \mathcal{N}^{\otimes t}(|\Gamma_\beta\rangle_{AA'^t})) \ge 1 - \epsilon$ for any random β.*

The following theorem describes $\mathcal{C}_Q(\mathcal{N})$ of any broadcast channels as a regularized union of rectangles.

Theorem 3.3.3 *Let $\mathcal{N}_{A'\to B_{[n]}C}$ be arbitrary broadcast channel. Then $\mathcal{C}_Q$ consists of nonnegative CQ rates (Q, R) given by*

$$\mathcal{C}_Q \subseteq \left\{(Q, R)\,\middle|\, \begin{array}{l} Q \le I(A\rangle CT)_\sigma, \\ R < \min\{I(T; B_i)_\sigma, I(T; C)_\sigma, i \in [s]\} \end{array}\right\} \tag{3.168}$$

for some state $\sigma_{TAB_{[n]}C} = \oplus_k p(k)\mathcal{N}(\phi_k^{AA'})$. Moreover, it is sufficient to take $|T| \le \min\{|A'|^2, |B_i|^2 + |C|^2 - 1, i \in [n]\}$.

Proof This proof is inspired by Yard et al. (2011) for a special case of $n = 1$. Let $\mathcal{N}_{A'\to B_{[n]}C}$ be a broadcast channel. Given bipartite pure states $\{p(k), |\phi_k\rangle^{A''A'}\}$, for any $\epsilon, \delta > 0$ and large t, we show that there is an $(R, Q, t, 12\sqrt{\epsilon})$ CQEG code

$$\{|\Upsilon_m\rangle^{AA''}, \mathcal{D}_{C^t\to M_C\hat{A}}, \mathcal{D}^i_{B_i^t\to M_{B_i}}\}_{m\in 2^{tR}, i\in[n]}, \tag{3.169}$$

if $0 \le Q = I(A''\rangle CT)_\sigma - \delta$ and $0 \le R = \min\{I(T; B_i)_\sigma, I(T; C)_\sigma, i \in [n]\} - \delta$, where $\sigma^{TAB_nC} = \oplus_k p(k)\mathcal{N}(|\phi_k\rangle^{A''A'})$. This can be completed by defining two POVMs: $\{\Lambda_m\}_{m\in 2^{tR}}$ on C^t and $\{\Lambda'_m\}_{m\in 2^{tR}}$ on B_i^t, and maps $\mathcal{D}^m_{C^t\to\hat{A}}$ satisfying the following results: $\{\mathcal{D}^m(\sqrt{\Lambda_m}(\cdot)\sqrt{\Lambda_m})\}_{m\in 2^{tR}} \subset \mathcal{D}$ and $\mathcal{D}^i = \{\Lambda'_m\}_{m\in 2^{tR}}$.

Let $\rho_k^{A'} = \mathrm{Tr}_{A''}(\phi_k)$ and $\tau_k^{B_{[n]}C} = \mathcal{N}(\rho_k)$. Define a CQ channel as $\mathcal{V}_{T\to B_{[n]}C}$ with conditional states $\tau_k^{B_{[n]}C}$. For a large t, there is an (R, t, ϵ) code given by

$\{k^t(m), \Lambda_m, \Lambda'_m\}_{m\in 2^{tR}}$ for $\mathcal{V}$ with type $\hat{p}$ satisfying $\|\hat{p}-p\|_1 \le \epsilon$. For each m, denote $\rho_m^{A''} := \otimes_i \rho_{k_i(m)}^{A'}$ we get

$$\mathrm{Tr}(\Lambda_m \otimes \Lambda'_m)\mathcal{N}^{\otimes t}(\rho_m^{A''}) \ge 1-\epsilon. \tag{3.170}$$

From Theorem 3.3.1, we define $t_k = t\hat{p}(k)$, $\epsilon_k = \epsilon\hat{p}(k)$ and $\delta_k = \delta\hat{p}(k)$. Assume that for $|\mathcal{T}| = d$, the first codeword is $k^t(1) = 1^{t_1}2^{t_2}\cdots d^{t_d}$ so that there are permutations $\mathsf{g}(m)$ of $\mathcal{T}^t$ satisfying $k^t(m) = \mathsf{g}(m)(k^t(1))$.

For each $k \in \mathcal{T}$, set $Q_k = I_c(\tau_k;\mathcal{N}) - \delta_k$. There exists a (Q_k, t_k, ϵ_k) random CQEG code Yard et al. (2011): $\{|\Upsilon_k\rangle^{A_k A''_k}, \mathcal{D}_k^{C^{t_k}\to\hat{A}_k}\}$ whose average code density operator $\varrho_k^{A''_k} = \mathbb{E}[\mathrm{Tr}_{A_k}\Upsilon_k]$ satisfies

$$\|\varrho_k^{A''_k} - \rho_k^{\otimes t_k}\|_1 \le \epsilon_k, \tag{3.171}$$

$$\xi_k^{A_k B_{[n]}^{t_k} C^{t_k}} \equiv \mathcal{N}^{\otimes t}(\Upsilon_k), \tag{3.172}$$

which is created by the k-th random quantum code with rate Q_k entanglement shared by A and C with a high fidelity as

$$F(|\Phi_{Q_k}\rangle^{A_k\hat{A}_k}, \mathrm{Tr}_{B^{t_k}}\mathcal{D}_k(\xi_k)) \ge 1-\epsilon_k. \tag{3.173}$$

Denote $A := \bigotimes_k A_k$ and

$$|\Upsilon_1\rangle^{AA''} = \otimes_k |\Upsilon_k\rangle^{A_k A''_k},$$
$$|\Upsilon_m\rangle^{AA''} = (1^A \otimes \mathsf{g}(m))|\Upsilon_1\rangle^{AA''}.$$

Define the average code for the new code as $\varrho_m^{A''} = \mathbb{E}[\mathrm{Tr}_A \Upsilon_m]$. We obtain that

$$\|\varrho_m - \rho_m\|_1 = \|\otimes_k \varrho_k^{A''_k} - \otimes_k \rho_k^{\otimes t_k}\|_1 \le \epsilon. \tag{3.174}$$

The sender A prepares the state $|\Upsilon_m\rangle^{AA''}$ to transmit common message m. The decoder is similar to Theorem 3.3.1. All receivers perform their respective measurements to recover the classical message. And then C permutes her output systems and implements quantum decoder $\mathcal{D} \equiv \otimes_k \mathcal{D}_k$.

The joint state after the sender A sends the encoding state through the channel is given by

$$\vartheta_m^{AB_{[n]}^t C^t} = \mathcal{N}^{\otimes t}(\Upsilon_m^{AA''}). \tag{3.175}$$

In this case, the state corresponding to the first message is given by $\vartheta_1^{AB_{[n]}^t C^t} = \otimes_k \xi_k^{A_k B^{t_k} C^{t_k}}$. After the decoder $\mathcal{D}^{C^s\to\hat{A}}$ being performed on ϑ_1, the resultant shared by the sender A and the receiver C is approximately maximally entangled, i.e.,

$$F(|\Phi_Q\rangle^{A\hat{A}}, \mathrm{Tr}_{B^t_{[n]}}\mathcal{D}(\vartheta_1)) = \prod_k F(|\Phi_{Q_k}\rangle^{A_k\hat{A}_t}, \mathrm{Tr}_{B^{t_k}_{[n]}}\mathcal{D}_k(\xi_k))$$
$$\geq 1-\epsilon \tag{3.176}$$

from the inequality (3.173).

Define the normalized states as

$$\tilde{\vartheta}_m^{AB^t_{[s]}C^t} = (\sqrt{\Lambda_m}\otimes\sqrt{\Lambda'_m})\vartheta_m^{AB^t_{[n]}C^t}(\sqrt{\Lambda_m}\otimes\sqrt{\Lambda'_m}). \tag{3.177}$$

From Lemma 3.3.2, the expect probability is bounded as

$$\begin{aligned}\mathbb{E}[\mathrm{Tr}\tilde{\vartheta}_m^{AB^t_{[n]}C^t}] &= \mathrm{Tr}(\Lambda_m\otimes\Lambda'_m)\cdot\mathbb{E}[\mathrm{Tr}_A\vartheta_m^{AB^t_{[n]}C^t}]\\ &\geq \mathrm{Tr}(\Lambda_m\otimes\Lambda'_m)\mathcal{N}^{\otimes t}(\rho^{A''}) - \|\varrho^{A''}-\rho^{A''}\|_1\\ &\geq 1-2\epsilon,\end{aligned} \tag{3.178}$$

from the inequalities (3.170) and (3.174). The expected fidelity F_m is given by

$$\begin{aligned}\mathbb{E}[F_m] &= \mathbb{E}[F_1]\\ &\equiv \mathbb{E}[F(|\Phi_Q\rangle^{A\hat{A}}, \mathrm{Tr}_{B^t_{[n]}}\mathcal{D}(\tilde{\vartheta}_1^{AB^t_{[n]}C^t}))]\\ &\geq 1-\mathbb{E}[\|\Phi_Q - \mathrm{Tr}_{B^t_{[n]}}\mathcal{D}(\tilde{\vartheta}_1)\|_1]\\ &\geq 1-6\sqrt{\epsilon}\end{aligned} \tag{3.179}$$

from the inequalities (3.148), (3.176) and (3.178). After de-randomizing, we obtain

$$\mathbb{E}[2^{-tR}\sum_m F_m] \geq 1-6\sqrt{\epsilon}. \tag{3.180}$$

By ignoring the worst codewords, from the Markov inequality we obtain a code with

$$F_m \geq 1-12\sqrt{\epsilon} \tag{3.181}$$

for each m. This has completed the encoding theorem.

The proof of the converse theorem is similar to the special case of $n=1$ Yard et al. (2011). It is mainly based on the Fano inequality, the Holevo bound, and following continuity lemmas. □

Lemma 3.3.4 Yard et al. (2011, Lemma 3) *If* $\|\rho^{AB}-\sigma^{AB}\|_1\leq\delta$ *for some* $0\leq\delta\leq 1/e$, *then the following inequalities hold:*

$$|S(A|B)_\rho - S(A|B)_\sigma| \leq 2S(\delta) + 4\delta\log|AB|,$$
$$|I(A;B)_\rho - I(A;B)_\sigma| \leq 3S(\delta) + 6\delta\log|AB|.$$

Based one Theorem 3.3.3 and Lemma 3.3.4 we may extend for a single-letter characterization of $\mathcal{C}_Q$, when the receiver C holds part of a generalized dephasing channel from the sender A to all the other receivers B_i's.

Theorem 3.3.4 *Let $\mathcal{N}_{A'\to B_{[n]}C}$ be an isometric extension of the channel given by $\mathcal{U} = \sum_x |x\rangle_C |\psi_x\rangle_{B_{[n]}E} \langle x|_{A'}$ so that $\mathcal{N}_{A'\to C}$ is a chained dephasing channel. Then $\mathcal{C}_Q(\mathcal{N})$ is given by*

$$\mathcal{C}_Q = \mathrm{Cl}\left\{(R, Q) \middle| \begin{array}{l} Q \leq S(X|T) - \min\{S(B_i E|T), i \in [n]\}; \\ R \leq \min\{I(T; B_i), i \in [n]\} \end{array}\right\} \tag{3.182}$$

for some state $\omega_{TXCE} = \oplus_{t,x} p(t,x)\psi_x^{B_{[n]}E}$ with $|\mathcal{T}| \leq |\mathcal{X}|$.

3.3.3 Entanglement-Assisted Classical and Quantum Capacities

Consider a quantum transmission for the broadcast channel $\mathcal{N}$. Here, the sender A attempts to share a large bipartite entangled state with the receivers $\mathsf{B}_1, \cdots, \mathsf{B}_n$ while aims to generate a large GHZ state with the receiver B_i's and C. The sender prepares a state $|\Upsilon\rangle_{ADA'^s}$ which is entangled with the inputs of a large number s of instances of $\mathcal{N}_{A'\to B_{[n]}C}$. All the receivers implement local decoders $\mathcal{D}^i_{B_i^s\to \hat{A}D_{B_i}}$ and $\mathcal{D}_{C^s\to D_C}$. These may consist a (Q, Q_C, s, ϵ) EG code for $\mathcal{N}$ if they generate a rate Q_C EPR state $|\Phi_{Q_C}\rangle_{A\hat{A}}$ and a rate Q GHZ state

$$|\Gamma_Q\rangle_{DD_{B_{[n]}}D_C} = \frac{1}{\sqrt{2^s Q}} \sum_{m=1}^{2^s Q} |m\cdots m\rangle_{DD_{B_{[n]}}D_C} \tag{3.183}$$

with high fidelity as

$$F(|\Phi_{Q_C}\rangle_{A\hat{A}} |\Gamma_Q\rangle_{DD_{B_{[n]}}D_C}, (\mathcal{D}^1 \circ \mathcal{D}^2) \circ \mathcal{N}^{\otimes n}(\Upsilon_{DAA'^s})) \geq 1 - \epsilon. \tag{3.184}$$

Achievable rates and the capacity region $\mathcal{Q}(\mathcal{N})$ are defined similarly to classical scenarios.

We now give a multi-letter formula for $\mathcal{Q}(\mathcal{N})$ in the case, where $\mathcal{N}$ is an isometry.

Theorem 3.3.5 *Let $\mathcal{N}_{A'\to B_{[n]}C}$ be an isometric broadcast channel. The capacity $\mathcal{Q}$ consisting of nonnegative quantum rates (Q_C, Q) is given by*

$$\mathcal{Q} = \mathrm{Cl}\left\{(Q_C, Q) \middle| \begin{array}{l} Q_C \leq I(A\rangle CT)_\sigma; \\ Q < \min\{I(T; C)_\sigma, I(T; B_i^k)_\sigma, i \in [n]\} \end{array}\right\} \tag{3.185}$$

The achievable region regularizes to give the entire capacity region as well.

The main idea of the proof is as follows. Let $\mathcal{N}^{A'\to B_{[n]}C}$ be an isometric broadcast channel. Denote $\mathcal{N}_C = \mathrm{Tr}_{B_{[n]}}\mathcal{N}$. For any bipartite state ensemble $\{p(t), |\phi_t\rangle^{A''A''}\}_{m\in 2^{tr}}$ and any $\epsilon > 0$, for large t, there is a $(Q, Q_B, t, 6\sqrt{\epsilon})$ CQEG code $\{|\Upsilon_m\rangle, \mathcal{D}_m, \{\Lambda_m\}, \{\Lambda'_m\}\}$ for $\mathcal{N}^{A'\to B_{[n]}C}$, if

$$\begin{cases} \mathcal{Q}_C < I(A''\rangle CT)_\sigma; \\ \mathcal{Q} < \min\{I(T;C)_\sigma, I(T;B_i)_\sigma, i \in [n]\}. \end{cases} \tag{3.186}$$

Moreover, it is easy to construct coherent local measurements $\mathcal{V}^i_{B_i^t\to B_i^t D_{B_i}}$ and $\mathcal{W}_{C^t\to C^t D_C}$ satisfying for each m that

$$\langle m|^{D_C D_{B_{[n]}}} \langle \Upsilon'_m|(\otimes_{i\in[n]}\mathcal{V}^i \otimes \mathcal{W})|\Upsilon'_m\rangle \geq 1-\epsilon. \tag{3.187}$$

Define the sender's encoding as

$$|\Upsilon\rangle^{GAA''} = \frac{1}{\sqrt{2^{tQ}}}\sum_m |m\rangle^D |\Upsilon_m\rangle^{AA''}, \tag{3.188}$$

which can be used for encoding with high fidelity. A similar idea may be useful as an isometric extension of a chained dephasing channel.

Theorem 3.3.6 *Given a quantum broadcast channel $\mathcal{N}_{A'\to B_{[n]}C}$, if it is an isometric extension of a chained dephasing channel to C with*

$$\mathcal{N} = \sum_x |x\rangle^C |\psi_x\rangle^{B_{[n]}} \langle x|^{A'}. \tag{3.189}$$

Then $\mathcal{Q}$ consists of nonnegative quantum rates $(\mathcal{Q}_C, \mathcal{Q})$ given by

$$\mathcal{Q} = \mathrm{Cl}\left\{ (\mathcal{Q}_C, \mathcal{Q}) \,\middle|\, \begin{array}{l} \mathcal{Q}_C \leq S(X|T)_\omega - \min\{S(B_i|T)_\omega, i \in [n]\}; \\ \mathcal{Q} \leq \min\{I(T;B_i)_\omega, i \in [n]\}, \end{array} \right\} \tag{3.190}$$

where $\omega^{TXB_{[n]}} = \oplus_{x,t} p(t,x)\psi_x^{B_{[n]}}$ and $|T| \leq |\mathcal{X}|$.

These results may provide improvements over standard entanglement distillation Devetak and Winter (2005) without accomplished state merging. A more detailed analysis may be used to design state merging protocols with the negative cost. Another extension is for free communication of all receivers.

3.3.4 Public, Private and Confidential Messages

Consider a quantum broadcast communication model. A quantum broadcast channel $\mathcal{N}_{A\to B_{[n]}C}$ with isometric extension $\mathcal{V}_{\mathcal{N}}^{A\to B_{[n]}CF}$ consists of one sender A with A system and $n + 1$ receivers: n primary receivers ($\mathsf{B}_1, \cdots, \mathsf{B}_n$) with $B_{[n]}$ and another receiver C with system C. The sender attempts to send three messages simultaneously: a common message M_0 for all receivers, a private message M_1 for C without secrecy requirement, and a confidential message M_s for C which should not be leaked to all B_i's. The special case of $n = 1$ is considered in Salek et al. (2020).

The sender encodes each message (M_0, M_1, M_s, M_r) into a quantum state ρ_A, where M_r denotes the randomness used by the sender for encryption. Upon receiving $B_{[n]}$ and C systems, the receiver C recovers $\hat{M}_0$, $\hat{M}_1$ and $\hat{M}_s$, while each B_i recovers the common message $\hat{M}_0$. A tradeoff among these rates will be made to ensure reliability and security.

Definition 3.3.6 A (R_0, R_1, R_s) one-shot code $\mathcal{C}$ for the QBC $\mathcal{N}_{A\to B_{[n]}C}$ consists of:

- three message sets $[2^{R_0}]$, $[2^{R_1}]$ and $[2^{R_s}]$;
- a source of local randomness $[2^{R_r}]$;
- an encoding operator $\mathcal{E} : M_0 \times M_1 \times M_s \times M_r \to \rho_A$ to get a code state ρ_A from each (M_0, M_1, M_s, M_r);
- a decoding POVM $\mathcal{D}_C : C \to M_0 \times M_1 \times M_s$ for estimating a triple $(\hat{M}_0, \hat{M}_1, \hat{M}_s) \in [2^{R_0}] \times [2^{R_1}] \times [2^{R_s}]$ from each received state ρ_C;
- a decoding POVM $\mathcal{D}_{B_i} : B_i \to M_0$ for estimating $\tilde{M}_0^i \in [2^{R_0}]$ from each state ρ_{B_i}, $i = 1, \cdots, n$.

The (R_0, R_1, R_s) one-shot code and the source of randomness M_r are known by all parties. The realizations of M_r are accessible by the sender. All the message triple (M_0, M_1, M_r) is uniformly distributed on $\in [2^{R_0}] \times [2^{R_1}] \times [2^{R_s}]$ so that $R_0 = H(M_0)$, $R_1 = H(M_1)$ and $R_s = H(M_s)$ with Shannon entropy function $H(\cdot)$.

We define the security of reliable performance and the ignorance as:

- **Reliability**–The reliability of the code $\mathcal{C}$ is featured by its average error probability as

$$P_e = \Pr\{(\hat{M}_0, \hat{M}_1, \hat{M}_s) \neq (M_0, M_1, M_s) \text{ or } \tilde{M}_0^i \neq M_0, i \in [n]\}; \quad (3.191)$$

- **Ignorance**–The indication of $\mathsf{B}_{[n]}$'s ignorance about the confidential message is characterized by

$$\forall m_0 : P_{secrecy}(m_0) = \max\left\{\frac{1}{2^{R_s}}\sum_{m_s}\frac{1}{2}\|\rho^{B_i}_{m_0,m_s} - \sigma^{B_i}_{m_0}\|_1, i \in [n]\right\}. \quad (3.192)$$

The secrecy indicates $\mathsf{B}_{[n]}$'s ignorance about m_s conditional on the correct decoded message m_0.

Definition 3.3.7 A rate quadruple (R_0, R_1, R_s, R_r) is ϵ-achievable if there is a one-shot code $\mathcal{C}$ satisfying

$$P_e \leq \epsilon, \tag{3.193}$$

$$P_{sec}(m_s) \leq \epsilon, \forall m_s. \tag{3.194}$$

The ϵ-achievable rate region $\mathcal{R}^{\epsilon}(\mathcal{N})$ is the closure of all ϵ-achievable rates (R_0, R_1, R_s, R_r).

Definition 3.3.8 For any two states ρ and σ on the same Hilbert space $\mathbb{H}$, the max-relative entropy is defined by $D_{\max}(\rho\|\sigma) = \inf\{\lambda \in \mathbb{R} : \rho' \leq 2^{\lambda}\sigma\}$

With forward evaluation, the max-relative entropy is monotonically non-increasing with CPTP maps $\mathcal{E} : \mathbb{H} \times \mathbb{H}$ as

$$D_{\max}(\mathcal{E}(\rho)\|\mathcal{E}(\sigma)) \leq D_{\max}(\rho\|\sigma). \tag{3.195}$$

With these properties, we may define the smooth max-relative entropy as follows.

Definition 3.3.9 For a parameter $\epsilon \in (0, 1)$, the smooth max-relative entropy is defined as

$$D^{\epsilon}_{\max}(\rho\|\sigma)_{\rho} = \min_{\rho' \in \mathbb{B}_{\epsilon}(\rho)} D_{\max}(\rho'\|\sigma), \tag{3.196}$$

where $\mathbb{B}_{\epsilon}(\rho)$ denotes an ϵ-ball around the state ρ.

The smooth max-relative entropy may be used to feature the mutual information of two states. For a bipartite state ρ_{AC} and a parameter $\epsilon \in (0, 1)$, define a mutual information quantity Salek et al. (2020) as

$$\begin{aligned} \tilde{I}^{\epsilon}_{\max}(A; C)_{\rho} &:= \inf_{\|\rho'_{AC} - \rho\| \leq \epsilon} D_{\max}(\rho'_{AC}\|\rho'_A \otimes \rho_C) \\ &\geq D_{\max}(A; C)_{\rho} \end{aligned} \tag{3.197}$$

with $\mathrm{Tr}_C \rho'_{XAC} = \sum_x p'(x)|x\rangle\langle x|_X \otimes \rho'^{A}_x$.

Let $\rho_{XAC} := \sum_x p(x)|x\rangle\langle x|_X \otimes \rho^{AC}_x$ and $\rho_{A-X-C} := \sum_x p(x)|x\rangle\langle x|_X \otimes \rho^A_x \otimes \rho^C_x$ be quantum states with classical X and $\epsilon \in (0, 1)$. Define

$$D^{\epsilon}_{\max}(A; C|X)_{\rho} := D^{\epsilon}_{\max}(\rho_{XAC}\|\rho_{A-X-C})_{\rho}, \tag{3.198}$$

$$\tilde{I}^{\epsilon}_{\max}(A; C|X)_{\rho} := \min_{\rho' \in \mathbb{B}(\rho)} (\rho'^{XAC}\| \sum_x p'(x)|x\rangle\langle x|^X \rho'^{A}_x \otimes \rho^C_x). \tag{3.199}$$

It follows that Salek et al. (2020):

$$I(A; C|X)_{\rho} \leq D^{\epsilon}_{\max}(A; C|X)_{\rho}$$

$$\leq \tilde{I}^{\epsilon}(A:C|X)_{\rho}$$
$$\leq D_{\max}^{\epsilon/2}(A;C|X)_{\rho} + \log(\frac{2}{2-\sqrt{4-\epsilon^2}}+1) \tag{3.200}$$

for $\epsilon \leq 1/\sqrt{2}$.

Our goal is then to find encoding and decoding for all parties. Consider a QBC $\mathcal{N}_{A\to B_{[n]}C}$. Let the random variables U, V, and X be distributed according to a distribution $P(u,v,x)$ which can be factorized as $P(u,v,x) = p(u,v)p(x|v)$. Define a CQ state as

$$\rho_{UVXA} = \sum_{u,v,x} p(u,v,x)|u\rangle\langle u|_U \otimes |v\rangle\langle v|_V \otimes |x\rangle\langle x|_X \otimes \rho_x^A. \tag{3.201}$$

Consider the following encoding-decoding procedure.

Codebook generation

Fix a probability function $P(u,v,x) = p(u,v)p(x|v)$, all the parties share 2^{R_0} copies of the classical state given by

$$\rho_{U_A U_{B_{[n]}} U_C} := \sum_u p(u)|u\rangle\langle u|_{U_A} \otimes |u_{[n]}\rangle\langle u_{[n]}|_{U_{B_{[n]}}} \otimes |u\rangle\langle u|_{U_C}, \tag{3.202}$$

where the sender A possesses U_A systems, the receiver B_i has U_{B_i} systems, and C has U_C systems. The shared state is used to construct the first layer of code. The parties A and C are assumed to share $2^{R_s+R_1}$ copies of the state $\rho_{V_A V_C} = \sum_v p(v)|vv\rangle\langle vv|_{V_A V_C}$. The set $[2^{R_s+R_1}]$ is partitioned into 2^{R_s} equal size bins. This will constitute the second layer of the code. Finally, for each and everyone of the states $\rho_{V_A V_{B_{[n]}} V_C}$, the parties will share 2^{R_r} copies of the state $\rho_{X_A X_{B_{[n]}} X_C} := \sum_x p(x)|xx_{[n]}x\rangle\langle xx_{[n]}x|_{X_A X_{B_{[n]}} X_C}$, as mentioned below for the i-th state $\sum_{u,v} p(u,v)|vvv\rangle\langle vvv|_{V_A^i V_{B_{[n]}}^i V_C^i} \otimes (\rho_{X_A X_{B_n} X_C})^{\otimes 2^{R_r}}$. These states construct the third layer and are assumed to be available for all parties.

Encoding

To send a message triple (m_0, m_1, m_s), the encoder firstly chooses a message $m_r \in [2^{R_r}]$. In the first layer, the encoder finds the m_0-th state $\rho_{U_A^{m_0}}$. And then, looks for the m_s-th substate inside, and selects the state associated to the private message m_1. Finally, the encoder picks the m_r-th state $\rho_{X_{m_r}^A}$ among those which are combined with the states found in the preceding step. These classical states are then passed through a modulator to get a quantum code state ρ_x^A.

Decoding

The receiver C performs a two-phase decoding strategy. Specially, she recovers the common message in the first phase, and confidential and private messages in the subsequent phase. For transmitting the m_0-th common message, her state is given by

$$\rho_{U_1} \otimes \cdots \otimes \rho_{U_{m_0} C} \otimes \cdots \otimes \rho_{U_{2^{R_0}}}. \tag{3.203}$$

Here, C find the position in which the received system C is tied to her U system. This arises a 2^{R_0}-ary hypothesis testing problem which will be solved by using position-based decoding as follows. Define POVM $\{T^{UC}, \mathbb{1} - T^{UC}\}$ for discriminating $\{\rho^{UC}, \rho^U \otimes \rho^C\}$. The correct deciding probability of T^{UC} is no less than $1 - (\epsilon' - \delta_1)$. Reciver C uses the following square-root measurement to recover the common message:

$$\Omega_{m_0} := (\sum_{m_0'=1}^{2^{R_0}} M_{m_0'})^{-\frac{1}{2}} M_{m_0} (\sum_{m_0'=1}^{2^{R_0}} M_{m_0'})^{-\frac{1}{2}}, \tag{3.204}$$

where $M_{m_0} := \mathbb{1}_r \otimes T^{U_{m_0}C}$. We obtain $\mathrm{Tr}[M_{m_0}(\rho^{U_1} \otimes \cdots \otimes \rho^{U_{m_0}C} \otimes \cdots \otimes \rho^{U_{R_0}})] = \mathrm{Tr}[M_{m_0}\rho^{U_{m_0}C}]$ and for any $m_0' \neq m_0$, and $\mathrm{Tr}[M_{m_0}(\rho^{U_1} \otimes \cdots \otimes \rho^{U_{m_0'}C} \otimes \cdots \otimes \rho^{U_{R_0}})] = \mathrm{Tr}[M_{m_0}(\rho^{U_{m_0}} \otimes \rho^C)]$. This implies for $m_0 = 1$ that

$$\begin{aligned}\Pr(\hat{M}_0 \neq 1 | M_0 = 1) &= \mathrm{Tr}[(\mathbb{1} - \Omega_1)(\rho^{U_1B} \otimes \cdots \otimes \rho^{U_{R_0}})] \\ &\leq O(\epsilon' - \delta_1) + \alpha 2^{R_0 - I_S^{\epsilon'-\delta_1}(U;C)_\rho}\end{aligned} \tag{3.205}$$

for some constant α. Hence, we obtain that

$$R_0 = I_S^{\epsilon'-\delta_1}(U; B)_\rho - O(\log_2 \frac{\epsilon'}{\delta_1^2}). \tag{3.206}$$

Moreover, for the common message m_0, the selection of (m_s, m_1) will induce a state as $\rho_{V_{(m_0,1,1)}} \otimes \cdots \otimes \rho_{V_{(m_0,m_s,m_1)}C} \otimes \cdots \otimes \rho_{V_{(m_0,R_s,R_1)}}$. Receiver C implements the second position-based POVM to solve $2^{R_s+R_1}$-ary hypothesis testing problem by using the square-root measurement POVM as

$$\Gamma_{m_s,m_1} := (\sum_{m_s'=1}^{2^{R_s}} \sum_{m_1'=1}^{2^{R_1}} M'_{m_s',m_1'})^{-\frac{1}{2}} M'_{m_s,m_1} (\sum_{m_s'=1}^{2^{R_s}} \sum_{m_1'=1}^{2^{R_0}} M'_{m_s',m_1'})^{-\frac{1}{2}}, \tag{3.207}$$

where $M'_{m_s,m_1} = \mathbb{1}_r \otimes L^{UV_{m_s,m_1}C}$ with identity operator $\mathbb{1}_r$ and a binary operator $L^{UV_{m_s,m_1}C}$ for discriminating $\rho^{UVC} = \sum_u p(u)|u\rangle\langle u|^U \otimes \rho_u^{VC}$ and $\rho^{V-U-C} = \sum_u p(u)|u\rangle\langle u|^U \otimes \rho_u^V \otimes \rho_u^C$. By using the symmetry of the random code construction Salek et al. (2020), the average error probability is given by

$$R_s + R_1 = I_S^{\epsilon'-\delta_3}(V; B|U)_\rho - O(\log_2 \frac{\epsilon'}{\delta_3^2}). \tag{3.208}$$

Meanwhile, each receiver B_i implements the position-based decoding POVM to recover the common message. Its state is given by $\rho_{U_1} \otimes \cdots \otimes \rho_{U_{m_0}B_i} \otimes \cdots \otimes \rho_{U_{R_0}}$. Because of the same decoding strategy for the common message, all the receivers can recover the same number of messages. This means that the achievabilities for common messages have the same R_0 approximately.

Combined with recent results Salek et al. (2020), the present encoding-decoding strategy implies the following achievable region.

Theorem 3.3.7 *For small positive constants ϵ', ϵ'', δ_1, δ_2, δ_3, η with $0 < \delta_i < \epsilon'$ and $0 < \eta < \epsilon''2$, we obtain that*

$$\cup \mathcal{C}^{(in)}(\rho) \subseteq \mathcal{C}^{\epsilon}(\mathcal{N}), \tag{3.209}$$

where $\epsilon = \max\{\sqrt[4]{\epsilon}, \sqrt[4]{\epsilon''}\}$, the union is over all $\rho_{UVXB_{[n]}C}$, and the achievable rate region $\mathcal{C}^{(in)}(\rho)$ consists of quadruples (R_0, R_1, R_s, R_r) satisfying

$$\begin{cases} R_0 \le \min\{I_S^{\epsilon'-\delta_1}(U;C)_\rho - O_{log}(\epsilon,\delta_1), \\ \quad I_S^{\epsilon'-\delta_2}(U;B_i)_\rho - O_{log}(\epsilon,\delta_2), i \in [n]\}, \\ R_0 + R_1 + R_s \le I_S^{\epsilon'-\delta_3}(V;C|U)_\rho \\ \quad +\min\{I_S^{\epsilon'-\delta_1}(U;B)_\rho - O_{log}(\epsilon',\delta_1), \\ \quad I_S^{\epsilon'-\delta_2}(U;B_i)_\rho - O_{log}(\epsilon',\delta_2), i \in [n]\} - O_{log}(\epsilon',\delta_3), \\ R_s \le I_S^{\epsilon'-\delta_3}(V;C|U)_\rho - \{I_{\max}^{\epsilon''}(V;B_i|U)_\rho, i \in [n]\} \\ \quad -O_{log}(\epsilon',\delta_3) + O_{log}(\eta), \\ R_1 + R_r \ge \min\{I_{\max}^{\epsilon''}(V;B_i|U)_\rho + \tilde{I}_{\max}^{\epsilon''}(X;B_i|V)_\rho \\ \quad -O_{log}(\eta), i \in [n]\}, \\ R_r \ge \min\{I_{\max}^{\epsilon''}(X;B_i|V)_\rho - O_{log}(\eta), i \in [n]\} \end{cases} \tag{3.210}$$

and $O_{log}(\epsilon',\delta_i) = O(\log_2(\epsilon'/\delta_i^2))$.

Its converse region is given as follows.

Theorem 3.3.8 *For a small nonnegative ϵ, we obtain that*

$$\mathcal{C}^{\epsilon}(\mathcal{N}) \subseteq \mathcal{C}^{(co)}(\rho), \tag{3.211}$$

where the union is over all classical states $\rho_{UVXB_{[n]}C}$ arising from the channel, and the achievable rate region $\mathcal{C}^{(co)}(\rho)$ consist of quadruples (R_0, R_1, R_s, R_r) satisfying

$$\begin{cases} R_0 \le \min\{I_S^{\epsilon}(U;C)_\rho, I_S^{\epsilon}(U;B_i)_\rho, i \in [n]\}, \\ R_0 + R_1 + R_s \le I_S^{\epsilon}(V;C|U)_\rho \\ \quad +\min\{I_S^{\epsilon}(U;C)_\rho, I_S^{\epsilon}(U;B_i)_\rho, i \in [n]\}, \\ R_s \le I_S^{\epsilon}(V;C|U)_\rho - \min\{D_{\max}^{\sqrt{2\epsilon}}(V;B_i|U)_\rho, i \in [n]\}, \\ R_1 + R_r \ge \min\{D_{\max}^{\sqrt{2\epsilon}}(V;B_i|U)_\rho + D_{\max}^{\sqrt{2\epsilon}}(X;B_i|V)_\rho, i \in [n]\}, \\ R_r \ge \min\{D_{\max}^{\sqrt{2\epsilon}}(X;B_i|V)_\rho, i \in [n]\}. \end{cases} \tag{3.212}$$

So far, we have studied the achievable rate region for one-shot quantum channel and non-zero error and secrecy parameters. In the asymptotic regime, the achievable bounds, the error, and the secrecy parameters may be assumed to be vanishing asymptotically. Formally, the rate region in the asymptotic regime is defined by

$$\mathcal{C}_\infty(\mathcal{N}) := \lim_{\epsilon\to 0}\lim_{\ell\to\infty}\frac{1}{\ell}\mathcal{R}^\epsilon(\mathcal{N}^{\otimes \ell}), \tag{3.213}$$

where $\mathcal{N}^{\otimes \ell}$ indicates the ℓ independent channel $\mathcal{N}$. Similar to Theorem 3.3.7, we can prove the following result for the asymptotic regime.

Theorem 3.3.9 *The asymptotic rate region $\mathcal{C}_\infty(\mathcal{N})$ of the quantum broadcast channel $\mathcal{N}_{A\to BC}$ is given as follows:*

$$\mathcal{C}_\infty(\mathcal{N}) = \lim_{\epsilon\to 0}\lim_{\ell\to\infty}\frac{1}{\ell}\mathcal{C}_\infty^{(1)}(\mathcal{N}^{\otimes \ell}), \tag{3.214}$$

where $\mathcal{C}_\infty^{(1)}(\mathcal{N}) := \cup_{\rho_{UVXB_{[n]}C}}\mathcal{C}_\infty^{(2)}(\mathcal{N})$, in which $\mathcal{C}_\infty^{(2)}(\mathcal{N})$ consists of (R_0, R_1, R_s, R_r) satisfying

$$\begin{cases} R_0 \le \min\{I(U;C)_\rho, I(U;B_i)_\rho, i\in[n]\} \\ R_0+R_1+R_s \le I(V;C|U)_\rho + \min\{I(U;B_i)_\rho, I(U;C)_\rho, i\in[n]\} \\ R_s \le I(V;C|U)_\rho - \min\{I(V;B_i|U)_\rho, i\in[n]\} \\ R_1+R_r \ge \min\{I(V;B_i|U)_\rho + I(X;B_i|V)_\rho, i\in[n]\} \\ R_r \ge \min\{I(X;B_i|V)_\rho, i\in[n]\} \end{cases} \tag{3.215}$$

and $\rho_{UVXB_{[n]}C}$ is the classical state from the channel $\mathcal{N}$.

Theorem 3.3.7 shows the tradeoff among the common, private, and confidential messages in the one-shot QBC. The achievability is based on a generalized convex-split lemma Salek et al. (2020). One can further prove Theorem 3.3.9 for an asymptotic i.i.d setting. For the special case of $n = 1$, Theorems 3.3.7 and 3.3.9 are proved in Salek et al. (2020). If without the confidential consideration, Theorem 3.3.9 implies a capacity region for the common and private message Yard et al. (2011). It further reduces to the rate-limited randomness in broadcast channels Watanabe and Oohama (2015).

3.3.5 Cooperating Decoders

In broadcast communication over a QBC, there are lots of receivers. They may cooperate to decode the received states Pereg et al. (2021). The first form of cooperation addressed is a classical conference, where one receiver can send classical messages to the other. Another cooperation setting involves a quantum conference, where one receiver teleports a state to the other. When the first receiver only helps the transmission to the other, it reduces to the quantum relay channel. The quantum conference is intimately related to quantum repeaters, as the sender, and both receivers can be viewed as the transmitter, the repeater, and the destination receiver, respectively. Our goal is to show the achievable capacity for this model.

The first form of cooperation between the decoders involves conferences. We consider a broadcast channel, where one group receiver B_J with $J \subset [n]$ can transmit information to the other receivers B_{J_c} with $J_c = [n] - J$ using a classical conference link of capacity $\mathcal{C}_{12}$. Here, each receiver in J or J_c is independent, as shown in Fig. 3.5.

Definition 3.3.10 A $(2^{tR_0}, 2^{tR_1}, t)$ classical code for the QBC $\mathcal{N}_{A\to B_J B_{J_c}}$ with degraded message sets and conference consists of the following:

- three index sets $[2^{tR_0}]$, $[2^{tR_1}]$, and $[2^{t\mathrm{C}_{12}}]$, corresponding to the common message for all receivers, the private message of all receivers in B_J, and the conference message, respectively;
- an encoding map $\mathcal{E}_{M_0M_1\to A}$, where M_0 and M_1 are classical registers that store the common and private messages, respectively;
- a decoding POVM $\{\Lambda_{B_i^t}^{m_0,m_1,g}\}$ for the receiver B_i, $i \in J$, where the measurement outcome is a triplet of indices $(m_0, m_1, g) \in [2^{tR_0}] \times [2^{tR_1}] \times [2^{\mathcal{C}_{12}}]$;
- a collection of decoding POVMs $\{\Gamma_{B_j^t|g}^{m_0}\}$, $g \in [2^{\epsilon\mathcal{C}_{12}}]$, for the receiver B_j, $j \in J_c$, where the measurement outcome is an index in $[2^{tR_0}]$.

We denote the code by $(\mathcal{E}, \Lambda, \Gamma)$.

The communication scheme is shown in Fig. 3.5. The sender has the system A^t, and the receivers B_J and B_{J_c} have the systems B_J^t and B_J^n, respectively. The sender chooses a common message $m_0 \in [2^{tR_0}]$ and a private message $m_1 \in [2^{tR_1}]$ for B_J, and stores them in the classical registers M_0 and M_1, respectively. She encodes the messages by applying the encoding map $\mathcal{E}_{M_0M_1\to A^t}$ which results in an input state

$$\rho_{A^t}^{m_0,m_1} = \mathcal{E}_{M_0M_1\to A^t}(m_0, m_1) \tag{3.216}$$

and transmits the systems A^t over t channel uses of $\mathcal{N}_{A\to B_J B_{J_c}}$. Hence, the output state is

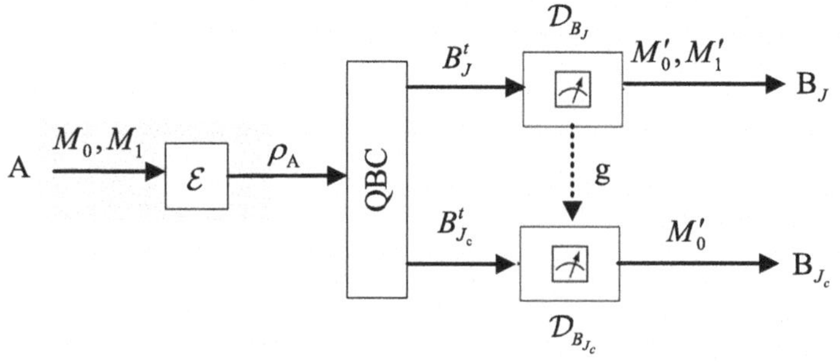

Fig. 3.5 Classical coding for a QBC $\mathcal{N}_{A\to B_J B_{J_c}}$ with conference and degraded message sets. There is a conference link between the decoders which allows each receiver B_i with $i \in J$ to send a conference message to B_j with $j \in J_c$ at the same rate $\mathcal{C}_{12}$

$$\rho_{B_J^t B_{J_c}^t}^{m_0,m_1} = \mathcal{N}_{A\to B_J B_{J_c}}^{\otimes t}(\rho_{A^t}^{m_0,m_1}). \tag{3.217}$$

The receiver B_i with $i \in J$ receives the channel output systems B_i^t and applies the POVM $\{\Lambda_{B_i^t}^{m_0,m_1,g}\}$. And then, he obtains an estimate of the message pair $(m_0', m_1') \in [2^{tR_0}] \times [2^{tR_1}]$ and a conference message $g \in [2^{t\mathcal{C}_{12}}]$. Next, B_i sends the conference message g to all receivers B_j with $j \in J_c$ vias classical broadcast channel. Here, all the receivers B_i with $i \in J$ can get the common message by using independent local measurements from the discussions in the subsection above. In this case, B_j will receive a common message from all receivers B_i with $i \in J$. Now, he chooses a POVM $\{\Gamma_{B_j^t|g}^{m_0}\}$ to perform on the channel output systems B_j^t, producing an estimate of the common message $m_0' \in [2^{tR_0}]$ as the measurement outcome.

Note that each receiver is obtained the estimate by local measurement. In this case, we can regard this scheme as quantum broadcast communication with independent receivers where the receivers may be regarded as local information. In this case, the conditional probability of error of the code, given that the message pair (m_0, m_1) was sent, is given by

$$P_{e|m_0,m_1}^{(t)}(\mathcal{E}, \Lambda, \Gamma) = 1 - \sum_{g=1}^{2^{t\mathcal{C}_{12}}} \mathrm{Tr}[(\Lambda_{B_J^t}^{m_0,m_1,g} \otimes \Lambda_{B_{J_c}^t|g}^{m_0})\rho_{B_J^t B_{J_c}^t}^{m_0,m_1}]. \tag{3.218}$$

Definition 3.3.11 A $(2^{tR_0}, 2^{tR_1}, t, \epsilon)$ classical code satisfies $P_{e|m_0,m_1}^{(t)}(\mathcal{E}, \Lambda, \Gamma) \le \epsilon$ for all $(m_0, m_1) \in [2^{tR_0}] \times [1 : 2^{tR_1}]$.

Definition 3.3.12 A rate pair (R_0, R_1) is called achievable with conference if for every $\epsilon > 0$ and sufficiently large n, there exists a $(2^{nR_0}, 2^{nR_1}, n, \epsilon)$ code.

The classical capacity region $\mathcal{R}_{\mathrm{Cl}}(\mathcal{N})$ is defined as the set of achievable pairs (R_0, R_1) with conference. Define the rate region

$$\mathcal{R}_{\mathrm{Cl}}(\mathcal{N}) \triangleq \cup \left\{ (R_0, R_1) \left| \begin{array}{l} R_0 \le I(X_0; B_2)_\rho + \mathcal{C}_{12}; \\ R_1 \le I(X_1; B_1|X_0); \\ R_0 + R_1 \le I(X_0, X_1; B_1), \end{array} \right. \right\} \tag{3.219}$$

where the union is over the set of all distributions $P_{X_0,X_1}(x_0, x_1)$ and state collection $\{\theta_A^{x_0,x_1}\}$, with

$$\begin{aligned} \rho_{X_0X_1B} = &\sum_{x_0\in\mathcal{X}_0} \sum_{x_1\in\mathcal{X}_1} p_{X_0,X_1}(x_0, x_1)|x_0\rangle\langle x_0| \\ &\otimes|x_1\rangle\langle x_1| \otimes \mathcal{N}_{A\to B_1B_2}(\theta_A^{x_0,x_1}). \end{aligned} \tag{3.220}$$

Before stating the capacity theorem, we give the following lemma which estimates the cardinality bounds for the auxiliary random variables X_0 and X_1. In principle, one can use those cardinality bounds to evaluate the region $\mathcal{R}_{\mathrm{Cl}}(\mathcal{N})$ numerically.

Lemma 3.3.5 *The union in (3.219) is exhausted by auxiliary random variables* X_0 *and* X_1 *of cardinality* $|\mathcal{X}_0| \leq |\mathbb{H}_A|^2 + 2$ *and* $|\mathcal{X}_1| \leq (|\mathbb{H}_A|^2 + 2)|\mathbb{H}_A|^2 + 1$.

Inspired by the method Pereg et al. (2021) for two receivers, we can prove the classical capacity region as follows.

Theorem 3.3.10 *The classical capacity region of the QBC* $\mathcal{N}_{A\to B_J B_{J_c}}$ *with conference and degraded messages is given by*

$$\mathcal{R}_{Cl} = \lim_{k\to\infty} \bigcup_{k=1}^{\infty} \mathcal{R}_{Cl}(\mathcal{N}^{\otimes k}). \tag{3.221}$$

Note that in the special case of a conference link with zero capacity, i.e., $\mathcal{C}_{12} = 0$. It recovers the result on the QBC without conference. The communication scheme may be further extended for sending different messages for receivers B_{J_c}.

Next, we consider the case where the messages are quantum. Given entanglement between the decoders, the classical conference link can be used to transfer *quantum* information from the receiver B_J to B_{J_c} using the teleportation protocol. This is thus equivalent to a conference link with quantum capacity $\mathcal{C}_{Q,12} = \frac{1}{2}\mathcal{C}_{12}$. Due to the no-cloning theorem, the encoder cannot transmit a quantum message to all receivers. We consider different private messages in what follows.

Definition 3.3.13 A $(2^{tQ_1}, 2^{tQ_2}, t)$ quantum code for quantum broadcast channel $\mathcal{N}_{A\to B_J B_{J_c}}$ with independent messages and conference consists of the following:

- a quantum message state $\rho_{M_1 M_2}$, where M_1 and M_2 are quantum systems that store the private messages of B_J and B_{J_c}, respectively. The dimension of each system is given by $|\mathcal{H}_{M_k}| = 2^{tQ_k}$ for $k = 1, 2$;
- an encoding map $\mathcal{E}_{M_1 M_2 \to A^t}$;
- a decoding map $\mathcal{D}_{B_i^t \to M_1' G}$ for B_i with $i \in J$, where G is a quantum register of dimension $2^{t\mathcal{C}_{Q,12}}$ that stores the conference message from B_J to B_{J_c};
- a decoding map $\mathcal{D}_{G' B_j^t \to M_0'}$ for B_j with $j \in J_c$.

We denote the code by $(\mathcal{E}, \mathcal{D}_J, \mathcal{D}_{J_c})$.

In this communication scheme, suppose that the sender has the systems M_1, M_2, and A^t, receiver B_i has the systems B_i^t, G, and M_1' for $i \in J$; and B_j with $j \in J_c$ has the systems B_j^t, G', and M_2'. The sender encodes the quantum state of the message systems M_1 and M_2 by applying the encoding map $\mathcal{E}_{M_1 M_2 \to A^t}$, which results in the input state

$$\rho_{A^t} = \mathcal{F}_{M_1 M_2 \to A^t}(\rho_{M_1 M_2}). \tag{3.222}$$

And then, she transmits the systems A^t over t channel uses of $\mathcal{N}_{A\to B_J B_{J_c}}$. The output state is given by

$$\rho_{B_J B_{J_c}} = \mathcal{N}^{\otimes t}_{A\to B_J B_{J_c}}(\rho_{A^t}). \tag{3.223}$$

All the receivers B_i with $i \in J$ receive the output systems B_i^t and applies the decoding map $\mathcal{D}_{B_i^t \to M_1' G}$. Their measurements will result in the state

$$\rho_{M_1' G B_{J_c}^t} = \circ_{i\in J} \mathcal{D}_{B_i^t \to M_1' G}(\rho_{B_J^t B_{J_c}^n}). \tag{3.224}$$

The reduced state of M_1' of B_i with $i \in J$ is the estimate of his private message system M_1. And then, B_i sends the conference message from G to G' using the noiseless conference link $\mathrm{id}_{G\to G'}$, hence $\rho_{M_1' G' B_{J_c}^t} = \mathrm{id}_{G\to G'}(\rho_{M_1' G B_{J_c}^t}) = \rho_{M_1' G B_{J_c}^t}$. All the receivers B_j with $j \in J_c$ receive the output systems B_j^t and the conference message in G', and applies the decoding map $\mathcal{D}_{G' B_j^t \to M_2'}$. The estimated state is then given by

$$\rho_{M_1' M_2'} = \circ_{j\in J_c} \mathcal{D}_{G' B_j^t \to M_2'}(\rho_{M_1' G B_j^t}) \tag{3.225}$$

and the estimation error is given by

$$P_e^{(t)}(\mathcal{E}, \mathcal{D}_J, \mathcal{D}_{J_c}, \rho_{M_1, M_2}) = \|\rho_{M_1 M_2} - \rho_{M_1' M_2'}\|_F. \tag{3.226}$$

Definition 3.3.14 Given a $(2^{tQ_1}, 2^{tQ_2}, t, \epsilon)$ quantum code, it satisfies the following inequality: $P_e^{(t)}(\mathcal{E}, \mathcal{D}_J, \mathcal{D}_{J_c}, \rho_{M_0 M_1}) \le \epsilon$ for all ρ_{M_0, M_1}.

A rate pair (Q_1, Q_2) is called achievable with conference if for every $\epsilon > 0$ and sufficiently large t, there exists a $(2^{tQ_1}, 2^{tQ_2}, t, \epsilon)$ code. The quantum capacity region $\mathcal{R}_Q(\mathcal{N})$ is defined as the set of achievable pairs (Q_1, Q_2) with conference. Inspired by the method Pereg et al. (2021) for two receivers, we can prove the quantum capacity region as follows.

Theorem 3.3.11 *An achievable rate pair* (Q_1, Q_2) *over QBC* $\mathcal{N}_{A\to B_J B_{J_c}}$ *with private messages and quantum conference satisfies the following inequalities*

$$\begin{cases} Q_1 \le \frac{1}{|J|} I(A_J\rangle B_J)_\rho; \\ Q_2 \le \frac{1}{|J_c|} I(A_{J_c}\rangle B_{J_c})_\rho + C_{Q,12}; \\ Q_1 + Q_2 \le \frac{1}{|J|} I(A_J\rangle B_J)_\rho + \frac{1}{|J_c|} I(A_{J_c}\rangle B_{J_c})_\rho, \end{cases} \tag{3.227}$$

for some input state $\rho_{A_J A_{J_c} A'}$.

The rate region in Theorem 3.3.11 depends on a greedy approach, where the conference link can increase the information rate of B_{J_c} by the expense of B_J. For the transmission of classical information, the optimal performance is achieved using superposition coding. Without conference, it is impossible for B_J to decode the message of B_{J_c} by the no-cloning theorem. Nevertheless, B_{J_c} can decode the state of M_2 only after B_J's local measurements.

The third form is to consider a QBC where all receivers share entanglement resources. Denote their entangled systems by S_{B_J} and $S_{B_{J_c}}$, respectively.

Definition 3.3.15 A $(2^{tR_0}, 2^{tR_1}, t)$ classical code for the quantum broadcast channel $\mathcal{N}_{A\to B_J B_{J_c}}$ with degraded message sets and entangled decoders consists of:

- two index sets $[2^{tR_0}]$ and $[2^{tR_1}]$, corresponding to the common message for both users and the private message of B_i with $i \in J$, respectively;
- an encoding map $\mathcal{E}_{M_0M_1\to A^t}$, where M_0 and M_1 are classical registers that store the common and private messages, respectively;
- a pure entangled state $|\Psi\rangle_{S_{B_J},S_{B_{J_c}}}$.
- two sets of decoding POVMs $\{\Lambda^{m_0,m_1}_{S_{B_i},B_i^t}\}$ for B_i with $i \in J$ and $\{\Gamma^{m_0}_{S_{B_j},B_j^t}\}$ for B_j with $j \in J_c$, where the measurement outcome $m_k \in [2^{tR_k}]$, for $k = 0, 1$.

In this case, suppose that the sender has the systems A^t, and the receivers B_J and B_{J_c} have the systems B_J^t, S_{B_J} and $B_{J_c}^t$, $S_{B_{J_c}}$, respectively. The sender chooses a common message $m_0 \in [2^{tR_0}]$ that is intended for both users and a private message $m_1 \in [2^{tR_1}]$ for B_J, and stores them in the classical registers M_0 and M_1, respectively. She encodes the messages by $\mathcal{E}_{M_0M_1\to A^n}$ to get $\rho_{A^t}^{m_0,m_1} = \mathcal{E}_{M_0M_1\to A^t}(m_0, m_1)$, and transmits the systems A^t over $\mathcal{N}^{\otimes t}_{A\to B_J B_{J_c}}$. The output state is given by

$$\rho^{m_0,m_1}_{B_J^t B_{J_c}^t} = \mathcal{N}^{\otimes t}_{A\to B_J B_{J_c}}(\rho_{A^t}^{m_0,m_1}). \tag{3.228}$$

All the receivers B_i with $i \in J$ receive the systems B_i^t, combines them with the entangled system S_{B_i}, and applies the POVM $\{\Lambda^{m_0,m_1}_{S_{B_i}B_i^t}\}$. They obtain estimated message pair $(m_0', m_1') \in [2^{tR_0}] \times [2^{tR_1}]$. Similarly, all the receivers B_j with $j \in J_c$ find a common message $\hat{m}_0 \in [2^{tR_0}]$ by $\{\Gamma^{m_0}_{S_{B_j}B_j^t}\}$ on the systems B_j^t and his entangled system S_{B_j}.

The conditional probability of error of the code, given that the message pair (m_0, m_1) was sent, is given by

$$P^{(t)}_{e;m_0,m_1}(\mathcal{E}, \Psi, \Lambda, \Gamma) = 1 - \mathrm{Tr}[(\Lambda^{m_0,m_1}_{S_{B_J}B_J^t} \otimes \Lambda^{m_0}_{S_{B_{J_c}}B_{J_c}^t})\rho^{m_0,m_1}_{B_J^t B_{J_c}^t}]. \tag{3.229}$$

An $(2^{tR_0}, 2^{tR_1}, t, \epsilon)$ classical code satisfies $Pe^{(t)}_{m_0,m_1}(\mathcal{E}, \Psi, \Lambda, \Gamma) \le \epsilon$ for all $(m_0, m_1) \in [2^{tR_0}] \times [2^{tR_1}]$.

A rate pair (R_0, R_1) is called achievable with entangled decoders if for every $\epsilon > 0$ and sufficiently large t, there exists a $(2^{tR_0}, 2^{tR_1}, t, \epsilon)$ code. The classical capacity region is defined as the set of achievable pairs (R_0, R_1) with entangled decoders.

Given a quantum conference link of capacity $\mathcal{C}_{Q,12}$, B_J can send conference messages to B_{J_c} at a rate $2 \cdot \mathcal{C}_{Q,12}$ using the superdense coding as follows. First, consider a QBC $\mathcal{N}_{EA\to B_J B_{J_c}}$ without a conference, classical capacity is unchanged with the entanglement resources. On the other hand, given a quantum conference link of capacity $\mathcal{C}_{Q,12}$, the classical capacity region with an entanglement between the decoders is given by the regularization of the region (3.219). The achievability follows by using the superdense coding to send classical conference messages from B_J to B_{J_c}. This concludes that entanglement between the decoders cannot enlarge the capacity region of the classical broadcast channel without a conference. A similar property holds for a broadcast channel with classical conferences.

3.4 Quantum Communication on Networks

This subsection devotes to the quantum capacity of general quantum networks, i.e., the quantum state transmission over a quantum network.

3.4.1 *Quantum Teleportation of Bipartite Channel*

Consider a point-to-point communication where one sender tries to send an unknown state to the receiver. One way is to send it through a physical link. Another is to send the information through a quantum-entangled channel. The so-called quantum teleportation is completed as follows Bennett et al. (1993). Both parties firstly share a two-qubit entangled state as

$$|\Psi\rangle_{23} = \frac{1}{\sqrt{2}}(|01\rangle + |10\rangle)_{23}. \tag{3.230}$$

Suppose that the sender's original particle is in the unknown state $|\phi\rangle$. These three particles may be of different kinds, e.g., one or more may be photons, the polarization degree of freedom having the same algebra as a spin.

Before the local measurement and classical communication, the shared entanglement contains no information about the transmitted state $|\phi\rangle$. The complete state of the three particles is given by

$$\begin{aligned} |\Omega\rangle_{123} &= \frac{a}{\sqrt{2}}(|0\rangle|0\rangle|1\rangle - |0\rangle|1\rangle|0\rangle) + \frac{b}{\sqrt{2}}(|1\rangle|0\rangle|1\rangle - |1\rangle|1\rangle|0\rangle) \\ &= \frac{1}{2}[|\Psi^-\rangle_{12}(-a|0\rangle - b|1\rangle) + |\Psi^+\rangle_{12}(-a|0\rangle + b|1\rangle) \\ &\quad +|\Phi^-\rangle_{12}(a|1\rangle + b|0\rangle) + |\Phi^+\rangle_{12}(a|1\rangle - b|0\rangle)], \end{aligned} \tag{3.231}$$

where the Bell basis is given by

$$|\Psi^\pm\rangle = \sqrt{\frac{1}{2}}(|01\rangle \pm |10\rangle), |\Phi^\pm\rangle = \sqrt{\frac{1}{2}}(|00\rangle \pm |11\rangle). \tag{3.232}$$

The sender performs a complete measurement on both particles 1 and 3. Regardless of the unknown state $|\phi\rangle_1$, the four outcomes have an equal probability 1/4. After the sender's measurement, the particle 3 is projected into one of the four pure states in Eq. (3.231). In each case, the receiver can reproduce a replica of the sender's state by local operations. Thus faithful teleportation can be achieved in all cases by the classical outcome of the sender's measurement. Our goal in this section is to extend to general quantum networks.

3.4.2 Entanglement Monogamy of Quantum Networks

In comparison to single entanglement, an n-partite quantum network consists of independent entangled states. The local unitary operations may entangle these independent states into a new entanglement in the global version. This generates high-dimensional quantum entangled states. We prove the monogamy relations for this high-dimensional entanglement based on von Neumann entropy and Rényi entropy, Tsallis entropy, and Unified entropy.

In Chap. 1, we have stated some bipartite entanglement measures, including the entanglement of formation (EOF), the Rényi-α entanglement measure $\mathcal{R}_\alpha$, the Tsallis-q entanglement measure $\mathcal{T}^{(q)}$, and Unified (q, s)-entropy. All of these are subadditive for some parameters, that is,

$$\mathcal{Q}(\rho \otimes \sigma) \leq \mathcal{Q}(\rho) + \mathcal{Q}(\sigma), \tag{3.233}$$

where $\mathcal{Q}$ can be any one of $\mathcal{E}, \mathcal{R}_\alpha, \mathcal{T}^{(q)}$ and $\mathcal{U}^{(q,s)}$. These quantities are useful to feature the entanglement of bipartite systems.

For multipartite systems, the entanglement is the impossibility of sharing entanglement freely. This provides a new feature of multipartite entanglement beyond any classical counterpart. The so-called monogamy of entanglement (MOE) implies that entanglement satisfies special constraints distributed on multipartite systems. The first result is Coffman, Kundu, and Wootters (CKW) Coffman et al. (2000) as

$$\mathcal{Q}_{A|BC}(\rho_{ABC}) \geq \mathcal{Q}_{A|B}(\rho_{AB}) + \mathcal{Q}_{A|C}(\rho_{AC}) \tag{3.234}$$

with specific bipartite entanglement measure $\mathcal{Q}$, where $\rho_{AB} = \mathrm{Tr}_C(\rho_{ABC})$ denotes the reduced state, and $A|BC$ denotes a bipartite split. The first example is related to the squared concurrence between bipartitions for tripartite states and N-qubit states.

Consider an n-partite quantum network $\mathcal{N}_q(\mathcal{A}, \sim)$, where $\mathcal{A}$ denotes parties $\mathsf{A}_1, \cdots, \mathsf{A}_n$, and ξ denotes entangled states including generalized EPR states and GHZ states. Assume that any two parties A_i and A_j share the states $\varrho_{ij}(\theta_1), \cdots, \varrho_{ij}(\theta_s)$, $\sigma_{ij}(\varphi_1), \cdots, \sigma_{ij}(\varphi_t)$, and $\delta_{ij}(\vartheta_1), \cdots, \delta_{ij}(\vartheta_k)$, where $\varrho(\theta)$ and $\sigma(\varphi)$ are density matrices of the following entangled states

$$|\phi(\theta)\rangle = \cos\theta|00\rangle + \sin\theta|11\rangle, \tag{3.235}$$

$$|\phi(\varphi)\rangle = \cos\varphi|0\rangle^{\otimes m} + \sin\varphi|1\rangle^{\otimes m} \tag{3.236}$$

with any integer $m \geq 3$, and $\delta(\vartheta) = \cos^2\vartheta|00\rangle\langle 00| + \sin^2\vartheta|11\rangle\langle 11|$ denotes the reduced density matrix of any two subsystems in a generalized GHZ state. Each party may entangle the local systems with local operations. The total system is regarded as an n-partite high-dimensional entanglement. The entanglement distribution for quantum networks satisfies a stronger monogamy inequality beyond the inequality (3.234).

Theorem 3.4.1 *The quantum network $\mathcal{N}_q(\mathcal{A}, \sim)$ satisfies the following inequality*

$$\mathcal{Q}_{A_i|\overline{A_i}}(\rho_{A_1\cdots A_n}) \geq \sum_{j=1,j\neq i}^{N} \mathcal{Q}_{A_i|A_j}(\rho_{A_iA_j}), \tag{3.237}$$

for EOF or Rényi-α entropy $\mathcal{Q}$. The equality holds if and only if all GHZ states are shared by two parties.

The proof of Theorem 3.4.1 is based on the additivity of quantum entropy, i.e., $\mathcal{Q}(\rho \otimes \sigma) = \mathcal{Q}(\rho) + \mathcal{Q}(\sigma)$. Theorem 3.4.1 indicates a mutually exclusive relation of quantum networks between any pair of parties A_i and A_j, which is beyond the single entanglement of GHZ states. The present result provides a generic monogamous relation for high-dimensional entanglement Yang and Luo (2021). Theorem 3.4.1 may provide a better-than single-entanglement performance in quantum tasks. One example is device-independent quantum key distribution Horodecki et al. (2009) where eavesdropper's attacks are limited by the monogamy of entanglement.

Consider a device-independent quantum key distribution model with entanglement ρ_{AB} Ekert (1991). An eavesdropper Eve is represented by E who may be correlated with ρ_{AB}. The total state is given by by ρ_{ABE} satisfying the equality of $\mathrm{Tr}_E(\rho_{ABE}) = \rho_{AB}$, as shown in Fig. 3.6a. Eve may control the source or quantum devices. The entanglement shared by Eve and Alice satisfies Zhu and Fei (2014):

$$\mathcal{E}_{\mathsf{A}|E}(\rho_{AE}) \leq (\mathcal{E}^{\sqrt{2}}_{\mathsf{A}|BC}(\rho_{ABE}) - \mathcal{E}^{\sqrt{2}}_{\mathsf{A}|B}(\rho_{AB}))^{\frac{1}{\sqrt{2}}}, \tag{3.238}$$

where $\mathcal{E}$ is the EOF measure. The eavesdropper may hold local systems correlated with ρ_{AB}, and fake entangled sources $\varphi_\beta\rangle$ and $|\varphi_\gamma\rangle$, as shown in Fig. 3.6b. This generates a new network that satisfies

$$\mathcal{E}_{\mathsf{A}|E}(\rho_{AE}) \leq \mathcal{E}_{\mathsf{A}|BE}(\rho_{ABE}) - \mathcal{E}_{\mathsf{A}|B}(\rho_{AB}). \tag{3.239}$$

Define the difference of two upper bounds (3.238) and (3.239) by

$$f(\mathcal{E}_{\mathsf{A}|BE}, \mathcal{E}_{\mathsf{A}|B}) = (\mathcal{E}^{\sqrt{2}}_{\mathsf{A}|BE} - \mathcal{E}^{\sqrt{2}}_{\mathsf{A}|B})^{\frac{1}{\sqrt{2}}} - (\mathcal{E}_{\mathsf{A}|BE} - \mathcal{E}_{\mathsf{A}|B}),$$

where $\mathcal{E}_{\mathsf{A}|BE}$ and $\mathcal{E}_{\mathsf{A}|B}$ satisfy $\mathcal{E}_{\mathsf{A}|BE} > \mathcal{E}_{\mathsf{A}|B}$. Note that the function $f(\mathcal{E}_{\mathsf{A}|BE}, \mathcal{E}_{\mathsf{A}|B})$ is always positive. This means that the inequality (3.239) gives a stronger constraint for information leakage.

Example 3.4.1 Consider a tripartite entangled W state given by

$$|W\rangle = \frac{1}{\sqrt{3}}(|100\rangle + |010\rangle + |001\rangle). \tag{3.240}$$

We obtain that

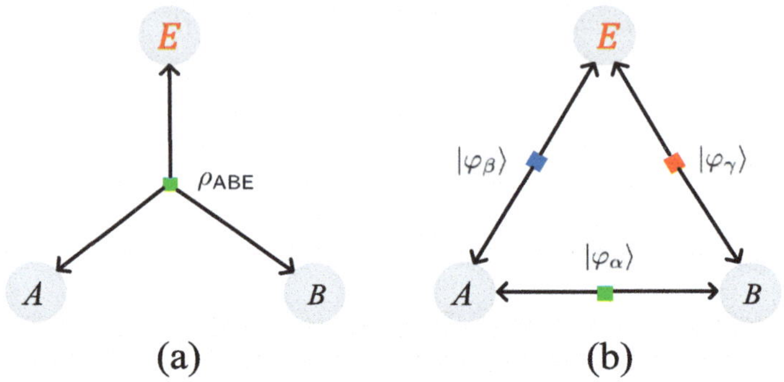

Fig. 3.6 (Color online). Device-independent QKD with an attacker E. **a** The standard attack model. **b** The new attack model is based on a triangle network

$$\begin{aligned}\mathcal{E}(|W\rangle_{A|B_1B_2}) &= H(\frac{1}{3}) \approx 0.918,\\ \mathcal{E}(\rho_{AB_i}) &= H(\frac{1+\sqrt{1-C^2(\rho_{AB_i})}}{2}) \approx 0.55\end{aligned} \tag{3.241}$$

for EOF $\mathcal{E}$, where $C(\rho_{AB_i}) = \frac{2}{3}$ $(i = 1, 2)$ is the concurrence Wootters (1998) of the two-qubit mixed state ρ_{AB_i}. From Eq. (3.241), we obtain that

$$\mathcal{E}(|W\rangle_{A|B_1B_2}) \leq \mathcal{E}(\rho_{AB_1}) + \mathcal{E}(\rho_{AB_2}), \tag{3.242}$$

that is, EOF $\mathcal{E}$ of W state is not monogamous. However, the monogamy relation of $\mathcal{E}$ holds for quantum networks consisting of GHZ states and W states. The monogamy of the quantum network reaches a good saturation in the case that the quantum network consists of $|W\rangle^{\otimes 5} \otimes |GHZ\rangle$. In fact, for the GHZ state, we obtain that

$$\mathcal{E}(|GHZ\rangle_{A|BC}) = 1, \mathcal{E}(\rho_{AB}) = \mathcal{E}(\rho_{AC}) = 0. \tag{3.243}$$

From Eqs. (3.241) and (3.243), we obtain the inequality

$$\mathcal{E}(|GHZ\rangle_{A|BC}) + n\mathcal{E}(|W\rangle_{A|B_1B_2}) \geq n\mathcal{E}(\rho_{AB_1}) + n\mathcal{E}(\rho_{AB_2}) \tag{3.244}$$

for $n \leq 5$. The inequality (3.244) is optimal in the case of $n = 5$, i.e.,

$$1 + 5H(\frac{1}{3}) \geq 5\mathcal{E}(\rho_{AB_1}) + 5\mathcal{E}(\rho_{AB_2}). \tag{3.245}$$

From the inequality (3.245), we can extend Theorem 3.4.1 to the network consisting of EPR states, GHZ states, and W states.

Theorem 3.4.2 *The quantum network $\mathcal{N}_{q_1}$ consisting of EPR states, GHZ states, and W states satisfies*

$$\mathcal{E}_{A_i|\overline{A_i}}(\rho_{A_1\cdots A_n}) \geq \sum_{j=1,j\neq i}^{n} \mathcal{E}_{A_i|A_j}(\rho_{A_iA_j}), \tag{3.246}$$

where $\mathcal{E}_{X|Y}$ represents EOF of a bipartite systems X and Y.

Consider an n-partite entangled quantum network $\mathcal{N}_{q_2}(\mathcal{A}, \sim)$, where any pair of A_i and A_j shares s_j pairs of arbitrary bipartite entangled pure states $|\varphi^1\rangle_{ij}, |\varphi^2\rangle_{ij}, \cdots, |\varphi^s\rangle_{ij}$. This network satisfies the following stronger monogamy relation.

Theorem 3.4.3 *The quantum network $\mathcal{N}_{q_2}$ satisfies*

$$\mathcal{Q}_{A_i|\overline{A_i}}(\rho_{A_1\cdots A_n}) = \sum_{j=1,j\neq i}^{n} \mathcal{Q}_{A_i|A_j}(\rho_{A_iA_j}), \tag{3.247}$$

where $\mathcal{Q}_{A_i|\overline{A_i}}$ and $\mathcal{Q}_{A_i|A_j}$'s are defined in Theorem 3.4.1.

Example 3.4.2 Consider a 3-star network consisting of four parties, where each pair of A_i and B shares one EPR state. This can be regarded as a 4 partite state given by

$$\begin{aligned}|\phi\rangle_{ABCD} = &\frac{1}{2\sqrt{2}}(|0000\rangle + |1001\rangle + |2010\rangle + |3011\rangle \\ &+|4100\rangle + |5101\rangle + |6110\rangle + |7111\rangle)_{ABCD}.\end{aligned} \tag{3.248}$$

The EOF of $|\phi\rangle_{A|BCD}$ is given by: $\mathcal{E}(|\phi\rangle_{A|BCD}) = 3$. For mixed state ρ_{AB}, by the Hughston-Jozsa-Wootters (HJW) Theorem Hughston et al. (1993), Vidal and Werner (2002) we have $\mathcal{E}(\rho_{AB}) = \mathcal{E}(\rho_{AC}) = \mathcal{E}(\rho_{AD}) = 1$.

Similarly, for the Rényi-α entanglement measure the entanglement distribution satisfies

$$\mathcal{R}_\alpha(|\psi\rangle_{A|BCD}) = \mathcal{R}_\alpha(\rho_{AB}) + \mathcal{R}_\alpha(\rho_{AC}) + \mathcal{R}_\alpha(\rho_{AD}). \tag{3.249}$$

Generally, it is interesting to show the equality (3.247) for the multipartite entangled pure state shared by two parties.

Theorem 3.4.4 *The entanglement distribution of $\mathcal{N}_{q_3}$ satisfies*

$$\mathcal{Q}_{A_i|\overline{A_i}}(\rho_{A_1\cdots A_n}) \geq \sum_{j=1,j\neq i}^{n} \mathcal{Q}_{A_i|A_j}(\rho_{A_iA_j}). \tag{3.250}$$

The entanglement distribution of $\mathcal{N}_{q_2}(\mathcal{A}, \xi)$ holds in terms of Tsallis-q entanglement and Unified (q, s)-entropy entanglement.

Theorem 3.4.5 *Assume that a quantum network $\mathcal{N}_{q_2}(\mathcal{A}, \xi)$ is connected by arbitrary multipartite entangled pure states.*

(i) For $0 < q < 1$, $\mathcal{N}_{q_2}$ *satisfies*

$$\mathcal{T}^{(q)}_{A_i|\overline{A_i}}(\rho_{A_1\cdots A_n}) \geq \sum_{j=1,j\neq i}^{n} \mathcal{T}^{(q)}_{A_i|A_j}(\rho_{A_iA_j}). \tag{3.251}$$

(ii) For $q > 1$, $\mathcal{N}_{q_2}$ *satisfies*

$$\mathcal{T}^{(q)}_{A_i|\overline{A_i}}(\rho_{A_1\cdots A_n}) \geq \sum_{j=1,j\neq i}^{n} \mathcal{T}^{(q)}_{A_i|A_j}(\rho_{A_iA_j}), \tag{3.252}$$

where $\mathcal{T}^{(q)}_{X|Y}$ *is Tsallis-q entanglement with respect to the bipartition X and Y.*

Theorem 3.4.6 *Assume that a quantum network* $\mathcal{N}_{q_2}(\mathcal{A}, \xi)$ *consists of arbitrary multipartite entangled pure states.*

(i) For $0 < q < 1$ *and* $s < 0$, *or* $q \geq 1$ *and* $s \geq 0$, $\mathcal{N}_{q_2}$ *satisfies*

$$\mathcal{U}^{(q,s)}_{A_i|\overline{A_i}}(\rho_{A_1\cdots A_n}) \geq \sum_{j=1,j\neq i}^{n} \mathcal{U}^{(q,s)}_{A_i|A_j}(\rho_{A_iA_j}). \tag{3.253}$$

(ii) For $q > 1$ *and* $s < 0$, *or* $0 < q < 1$ *and* $s > 0$, $\mathcal{N}_{q_2}$ *satisfies*

$$\mathcal{U}^{(q,s)}_{A_i|\overline{A_i}}(\rho_{A_1\cdots A_n}) \geq \sum_{j=1,j\neq i}^{n} \mathcal{U}^{(q,s)}_{A_i|A_j}(\rho_{A_iA_j}), \tag{3.254}$$

where $\mathcal{U}^{(q,s)}_{X|Y}$ *is Unified* (q, s)*-entropy entanglement with respect to the bipartition X and Y.*

The main idea is from the additivity of Tsallis-q entropy and the subadditivity of Unified (q, s)-entropy.

3.4.3 Quantum Max-Flow Min-Cut Theorem

Quantum networks provide new capabilities to generate and apply quantum entanglement. One fundamental problem is to determine the maximal rate achievable in the end-to-end transmission of states. Our goal here is to investigate the capacities of general quantum networks $\mathcal{N}_q$ consisting of EPR states and GHZ states. We present a one-shot scheme for characterizing the network capacity of a hybrid network. The new method has always polynomial complexity.

For any bipartite entanglement $|\varphi\rangle_{AB}$, it can be regarded as a quantum channel with the capacity Bennett et al. (1993):

$$\mathcal{C}_{AB} = S(\rho_A) = \mathcal{E}_{A|B}(|\varphi\rangle_{AB}), \tag{3.255}$$

where $S(\rho_A)$ is the von Neumann entropy of the party A. As a result, the total capacity of the parties A_i and A_j is given by

$$S(\rho_i^j) = \mathcal{E}_{A_i|A_j}(\rho_{A_iA_j}) = n_1 + n_2, \tag{3.256}$$

where n_1 and n_2 denote the respective number of EPR and GHZ states shared by two parties, $S(\rho_i^j)$ is the von Neumann entropy of ρ_i^j by tracing out the system of j. This equality allows us to evaluate the capacity of quantum networks via von Neumann entropy. Note that Rényi entropy is also useful for the channel capacity.

Similar to classical network theory Voloshin (2009). Consider an n-partite network $\mathcal{N}_q(\mathcal{A}, \Omega)$. Let $\mathbf{s}$ and $\mathbf{t}$ be the source and sink, respectively.

Definition 3.4.1 Define the *flow* $f_{i\to j}$ in $\mathcal{N}_q$ as the number of qubits reliably transmitted from A_i to A_j, that is,

$$f_{i\to j} \geq 0. \tag{3.257}$$

For the quantum channel consisting of EPR states and GHZ states, the capacity between A_i and A_j is given by Eq. (3.256), which shows the maximal number of qubits transmitted. For a given channel, the flow is bounded by the capacity, i.e.,

$$f_{i\to j} \leq \mathcal{C}(\xi_{ij}) = S(\rho_i^j), \tag{3.258}$$

where $H(\rho_i^j)$ has the formation of Eq. (3.256).

At any intermediate party, the input qubits will be transmitted through point-to-point communications. This means for any $i \in \mathcal{A} \setminus \{\mathbf{s}, \mathbf{t}\}$, we obtain that

$$\sum_{j\in\mathcal{A}} f_{i\to j} = 0. \tag{3.259}$$

Moreover, for the source $\mathbf{s}$ and the sink $\mathbf{t}$, we obtain that

$$\sum_{j\in\mathcal{A}} f_{\mathbf{s}\to j} = -\sum_{j\in\mathcal{A}} f_{\mathbf{t}\to j} = f, \tag{3.260}$$

where f is known as the value of the flow. Similar to classical networks, define a *flow* in quantum network as follows.

Definition 3.4.2 A flow of a quantum network $\mathcal{N}_q(\mathcal{A}, \Omega)$ from $\mathbf{s}$ to $\mathbf{t}$ is a function $f : \Omega \longrightarrow \mathbb{R}^+$ satisfying

(i) Capacity constraint: for any $\xi_{ij} \in \Omega$, $f_{i\to j}$ satisfies

$$f_{i\to j} \leq \mathcal{C}(\xi_{ij}). \tag{3.261}$$

(ii) Conservation of flow: we have

$$\sum_{\{i:\xi_{ij}\in\Omega\}} f_{i\to j} = \sum_{\{k:\xi_{jk}\in\Omega\}} f_{j\to k}, \tag{3.262}$$

for any party $j \in \mathcal{A}$, other than the sender (source) **s** and receiver (sink) **t**.

Herein, $f_{i\to j}$ denotes the number of qubits transmitted from the party A_i to A_j by the quantum channel ξ_{ij}. If $f_{i\to j} = \mathcal{C}(\xi_{ij})$, then the quantum channel ξ_{ij} is saturated. Otherwise, ξ_{ij} is unsaturated.

Definition 3.4.3 A maximum flow $MF(\mathcal{N})$ of $\mathcal{N}_q(\mathcal{A}, \Omega)$ is the maximum amount of flow f from **s** to **t**, that is,

$$f_{\max} = \max_{\xi_{st}}\{f\}, \tag{3.263}$$

over all possible channels ξ_{st} from **s** to **t**.

For a given network $\mathcal{N}_q(\mathcal{A}, \Omega)$, we first find the associated graph. Here, one vertex schematically denotes one party. One edge (i, j) denotes a generalized EPR state shared by two parties A_i and A_j. The generalized GHZ state shared by m parties is represented by a hyperedge with m vertices. Thus the quantum network $\mathcal{N}_q$ is represented by an undirected finite hypergraph $\mathcal{G} = (\mathcal{V}, E)$. With this correspondence, the weight of each edge (i, j) or hyperedge is the entanglement entropy of the generalized EPR state or GHZ state. We then adopt the standard definition of cut-set for graph Voloshin (2009). A *cut* T_{cut} of $\mathcal{G}$ is a bipartition $(\mathsf{S}_1, \mathsf{S}_2)$ of $\mathcal{V}$ such that $\mathbf{s} \in \mathsf{S}_1$ and $\mathbf{t} \in \mathsf{S}_2$. The cut-set $\widetilde{T}_{cut}$ of T_{cut} is the set of edges so that the removal of these edges disconnects the source **s** and the sink **t**.

Definition 3.4.4 For an edge cut set $\widetilde{T}_{cut}$ in terms of the bipartition S_1 and S_2 of all the vertexes, the capacity of an edge cut set T_{cut} is given by

$$\mathcal{C}(\mathsf{S}_1, \mathsf{S}_2) = \sum_{(i,j)\in\widetilde{T}_{cut}} \mathcal{C}(i, j), \tag{3.264}$$

where $\mathcal{C}(i, j)$ denotes the weight of edge (i, j).

The flow in a classical network obeys the celebrated Max-Flow Min-Cut Theorem Voloshin (2009). A similar result holds for the quantum settings.

Theorem 3.4.7 *For a given quantum network $\mathcal{N}_q$ associated with undirected graph $\mathcal{G}$, the maximal flow $f_{\max}$ is equal to the minimal cut of $\mathcal{G}$, that is,*

$$f_{\max} = \min_{T_{cut}} \mathcal{C}(\mathsf{S}_1, \mathsf{S}_2). \tag{3.265}$$

In what follows, we prove Theorem 3.4.7. Firstly, we resort to the following Lemma 3.4.1 for classical networks Elias et al. (1956), Ford and Fulkerson (1956).

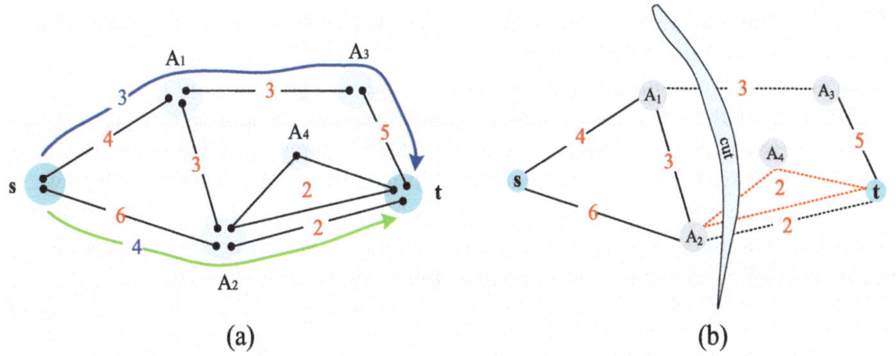

Fig. 3.7 (Color online) **a** Max-flow of 6-partite quantum network $\mathcal{N}_q$. **s** is source while **t** is sink. Three parties A_2, A_4, and **t** share two maximally entangled GHZ states, and the remaining parties share multiple maximally entangled EPR states. **b** Min-cut of the associated graph of $\mathcal{N}_q$. One edge denotes an EPR state while one hyperedge (red line) denotes a GHZ state

Lemma 3.4.1 *Let f be a flow on a network $\mathcal{N}$, and T_{cut} be a cut $T_{cut} = (\mathsf{S}_1, \mathsf{S}_2)$ with the capacity $\mathcal{C}(T_{cut})$. Then the flow satisfies*

$$f \leq \mathcal{C}(T_{cut}), \tag{3.266}$$

where the equality holds if and only if each edge (i, j) in the cut $T_{cut} = (\mathsf{S}_1, \mathsf{S}_2)$ is saturated.

Proof of Theorem 3.4.7. Consider a quantum network $\mathcal{N}_q(\mathcal{A}, \Omega)$ consisting of quantum channels defined as generalized EPR states and GHZ states. There exists a cut T_{cut} in $\widetilde{T}_{cut}$ of the associated graph $\mathcal{G}$ separating S_1 and S_2. Similar to Lemma 3.4.1, we obtain that

$$f(\mathsf{S}_1, \mathsf{S}_2) \leq \mathcal{C}(T_{cut}), \tag{3.267}$$

where the equality holds if and only if each channel ξ_{ij} in the cut $T_{cut} = (\mathsf{S}_1, \mathsf{S}_2)$ is saturated, and $f_{i\to j} = \mathcal{C}(\xi_{ij}) = \mathcal{C}(i, j)$. As shown in refs.Bennett et al. (1993), one can perfectly transmit one qubit using the quantum channel of generalized EPR state or generalized GHZ state in an asymptotic manner, i.e., many copies of quantum channels. Thereby, it is feasible to achieve $f_{i\to j} = \mathcal{C}(i, j)$ from the point of view of quantum communication. Note that the transmission of information in quantum communication is one-way, i.e., $f(\mathsf{S}_2, \mathsf{S}_1) = 0$. We obtain the following result.

Lemma 3.4.2 *If a flow f and cut $T_{cut} = (\mathsf{S}_1, \mathsf{S}_2)$ on a quantum network $\mathcal{N}_q(\mathcal{A}, \Omega)$ satisfy $f = \mathcal{C}(T_{cut})$, then f is a maximum flow and the cut T_{cut} is a minimum cut.*

Note that any flow f of the network $\mathcal{N}_q$ satisfies $f \leq \mathcal{C}(T_{cut})$. On the other hand, for any cut T'_{cut}, we have $\mathcal{C}(T'_{cut}) \geq f = \mathcal{C}(T_{cut})$. Thus f is the maximum flow and the cut T_{cut} is the minimum cut. We obtain Theorem 3.4.7 for the quantum network

$\mathcal{N}_q(\mathcal{A}, \Omega)$. The maximal flow of $\mathcal{N}_q(\mathcal{A}, \Omega)$ equals the minimal cut of the associated graph $\mathcal{G}$. This has completed the proof. □

Theorem 3.4.7 provides a general method similar to point-to-point quantum protocol Bennett et al. (1993) for general quantum networks. In applications, for a given quantum network $\mathcal{N}_q$ consisting of generalized EPR states and GHZ states, we may make use of extended Ford-Fulkerson Algorithm Cormen et al. (2001) as follows.

One example is shown in Fig. 3.7a by applying to an extended Ford-Fulkerson Algorithm. The maximum flow from **s** to **t** equals 7, which is consistent with the minimal cut of an associated undirected graph as shown in Fig. 3.7b.

3.4.4 Conclusion

Theorems 3.1.1 and 3.1.2 are proved in Schumacher (1995). Theorem 3.1.3 is proved in Holevo (2001). The Sect. 3.2.1 is based on results in Winter (2001). The regularization of a rate region $R(\mathcal{N})$ is defined as

$$R^{\infty}(\mathcal{N}) = \mathrm{Cl}\{\cup_{n=1}^{\infty} \frac{1}{n} R(\mathcal{N}^{\otimes n})\}. \tag{3.268}$$

This is meaningful even for the general case when an achievable region is not equal to the full capacity region.

The special case $n = 2$ of Theorems 3.2.4 and 3.2.6 is firstly proved in Chakraborty et al. (2021). A further introduction to quantum information theory may be Wilde (2013). The formation of $\lim_{s\to\infty} Q_{\mathrm{MAC}}^{sum}(\mathcal{N}^s)$ is a regularized expression for the largest achievable sum-rate for quantum communication over qMACs Yard et al. (2008). The Sect. 3.2.2.2 is based on the results in Chou (2022). The Theorem 3.2.11 is firstly proved in Hsieh et al. (2005). Theorem 3.2.12 is firstly proved in Shi et al. (2021).

References

R. Ahlswede, Multi-way communication channels, in *Proc. 2nd Int. Symp. Information Theory*. (Budapest, Hungary, Academia Kiado, 1971), pp. 23–52

C.H. Bennett, S.J. Wiesner, Communication via one-and two-particle operators on einstein-podolsky-rosen states. Phys. Rev. Lett. **69**, 2881 (1992)

C.H. Bennett, G. Brassard, C. Crépeau, R. Jozsa, A. Peres, W.K. Wootters, Teleporting an unknown quantum state via dual classical and Einstein-Podolsky-Rosen channels. Phys. Rev. Lett. **70**, 1895–1899 (1993)

C.H. Bennett, P.W. Shor, J.A. Smolin, A.V. Thapliyal, Entanglement-assisted capacity of a quantum channel and the reverse shannon theorem. IEEE Trans. Inf. Theory **48**, 2637 (2002)

P. Bergmans, Random coding theorem for broadcast channels with degraded components. IEEE Trans. Inf. Theory **19**, 197–207 (1973)

S. Chakraborty, A. Nema, and P. Sen, One-shot inner bounds for sending private classical information over a quantum MAC (2021). arXiv:2105.06100v1

R.A. Chou, Private classical communication over quantum multiple-access channels (2022). arXiv:2201.11899

R.A. Chou, A. Yener, The Gaussian multiple access wiretap channel when the eavesdropper can arbitrarily jam, in *IEEE International Symposium on Information Theory (ISIT)* (2017), pp. 1958–1962

R.A. Chou, M.R. Bloch, Polar coding for the broadcast channel with confidential messages: A random binning analogy. IEEE Trans. Inf. Theory **62**, 2410–2429 (2016)

R.A. Chou, A. Yener, Polar coding for the multiple access wiretap channel via ratesplitting and cooperative jamming. IEEE Trans. Inf. Theory **64**, 7903–7921 (2018)

V. Coffman, J. Kundu, W.K. Wootters, Distributed entanglement. Phys. Rev. A **61**, 052306 (2000)

T.H. Cormen, C.E. Leiserson, R.L. Rivest, C. Stein, *Introduction to Algorithms*, 2nd edn. (The MIT Press, 2001)

T. Cover, Broadcast channels. IEEE Trans. Inf. Theory **18**, 2–14 (1972)

I. Csiszar, Almost independence and secrecy capacity. Prob. Inf. Trans. **32**, 40–47 (1996)

I. Csiszar, J. Körner, Broadcast channels with confidential messages. IEEE Trans. Inf. Theory **24**, 339–348 (1978)

L. Czekaj, Subadditivity of the minimum output entropy and superactivation of the classical capacity of quantum multiple access channels. Phys. Rev. A **83**, 042304 (2011)

L. Czekaj, P. Horodecki, Purely quantum superadditivity of classical capacities of quantum multiple access channels. Phys. Rev. Lett. **102**, 110505 (2009)

I. Devetak, The private classical capacity and quantum capacity of a quantum channel. IEEE Trans. Inf. Theory **55**, 44–55 (2005)

I. Devetak, A. Winter, Distillation of secret key and entanglement from quantum states. Proc. Roy. Soc. Lond. Ser. A **461**, 207–235 (2005)

A.K. Ekert, Quantum cryptography based on Bell's theorem. Phys. Rev. Lett. **67**, 661–663 (1991)

P. Elias, A. Feinstein, C.E. Shannon, A note on the maximum flow through a network. IEEE Trans. Inf. Theory **2**, 117–119 (1956)

M. Fanizza, F. Kianvash, V. Giovannetti, Quantum flags and new bounds on the quantum capacity of the depolarizing channel. Phys. Rev. Lett. **125**, 020503 (2020)

L.R. Ford, D.R. Fulkerson, Maximal flow through a network. Cana. J. Math. **8**, 399–404 (1956)

R.G. Gallager, Capacity and coding for degraded broadcast channels. Probl. Pered. Inform. **16**, 17–25 (1980)

R.G. Gallager, Capacity and coding for degraded broadcast channels. Probl. Pered. Inform. **16**, 17–25 (1980)

A.J. Grant, B. Rimoldi, R.L. Urbanke, P.A. Whiting, Ratesplitting multiple access for discrete memoryless channels. IEEE Trans. Inf. Theory **47**, 873–890 (2001)

A. Grant, B. Rimoldi, R. Urbanke, P. Whiting, Rate-splitting multiple access for discrete memoryless channels. IEEE Trans. Inf. Theory **47**, 873–890 (2001)

M. Hayashi, Quantum wiretap channel with non-uniform random number and its exponent and equivocation rate of leaked information. IEEE Trans. Inf. Theory **61**, 5595–5622 (2015)

A.S. Holevo, On entanglement-assisted classical capacity (2001). arXiv:quant-ph/0106075v1

A.S. Holevo, The capacity of the quantum channel with general signal states. IEEE Trans. Inf. Theory **44**, 269–273 (1998)

A.S. Holevo, R.F. Werner, Evaluating capacities of bosonic Gaussian channels. Phys. Rev. A **63**, 032313 (2001)

R. Horodecki, P. Horodecki, M. Horodecki, K. Horodecki, Quantum entanglement. Rev. Mod. Phys. **81**, 865 (2009)

M.-H. Hsieh, I. Devetak, A. Winter, Entanglement-assisted capacity of quantum multiple access channels. IEEE Trans. Inf. Theory **54**, 1109 (2005)

M.-H. Hsieh, I. Devetak, A. Winter, Entanglement-assisted capacity of quantum multiple-access channels. IEEE Trans. Inf. Theory. **54**, 3078 (2008)

L.P. Hughston, R. Jozsa, W.K. Wootters, A complete classification of quantum ensembles having a given density matrix. Phys. Lett. A **183**, 1418 (1993)

C. King, Additivity for unital qubit channels. J. Math. Phys. **43**, 4641–4653 (2002)

F. Leditzky, M.A. Alhejji, J. Levin, G. Smith, Playing games with multiple access channels. Nat. Commun. **11**, 1 (2020)

H. Liao, Multiple access channels, Ph.D. dissertation (1972)

S. Lloyd, Capacity of the noisy quantum channel. Phys. Rev. A **55**, 1613 (1997)

U. Pereg, C. Deppe, H. Boche, Quantum broadcast channels with cooperating decoders: an information-theoretic perspective on quantum repeaters. J. Math. Phys. **62**, 062204 (2021)

J. Renes, R. Renner, Noisy channel coding via privacy amplification and information reconciliation. IEEE Trans. Inf. Theory **57**, 7377–7385 (2011)

B. Rimoldi, R. Urbanke, A rate-splitting approach to the Gaussian multiple-access channel. IEEE Trans. Inf. Theory **42**, 364–375 (1996)

F. Salek, M. Hsieh, J. Fonollosa, Single-serving quantum broadcast channel with common, individualized, and confidential messages. IEEE Trans. Inf. Theory **66**, 7752–7771 (2020)

B. Schumacher, Quantum coding. Phys. Rev. A **51**, 2738–2747 (1995)

B.W. Schumacher, M.A. Nielsen, Quantum data processing and error correction. Phys. Rev. A **54**, 2629 (1996)

H. Shi, M.-H. Hsieh, S. Guha, Z. Zhang, Q. Zhuang, Entanglement-assisted capacity regions and protocol designs for quantum multiple-access channels. Npj Quantum Inf. **7**, 74 (2021)

E. Tekin, A. Yener, The general Gaussian multiple-access and twoway wiretap channels: Achievable rates and cooperative jamming. IEEE Trans. Inf. Theory **54**, 2735–2751 (2008)

G. Vidal, R.F. Werner, Computable measure of entanglement. Phys. Rev. A **65**, 032314 (2002)

V.I. Voloshin, *Introduction to Graph and Hypergraph Theory* (Nova Science Publishers, 2009)

S. Watanabe, Y. Oohama, The optimal use of rate-limited randomness in broadcast channels with confidential messages. IEEE Trans. Inf. Theory **61**, 983–995 (2015)

M. M. Wilde, *Quantum Information Theory*. (Cambridge University Press, 2013)

A. Winter, The capacity of the quantum multiple-access channel. IEEE Trans. Inf. Theory **47**, 3059–3065 (2001)

W.K. Wootters, Entanglement of formation of an arbitrary state of two qubits. Phys. Rev. Lett. **80**, 2245 (1998)

A. Wyner, The wire-tap channel. Bell Syst. Tech. J. **54**, 1355–1387 (1975)

X. Yang, M.-X. Luo, Unified monogamy relation of entanglement measures. Quantum Inf. Proc. **20**, 108 (2021)

J. Yard, I. Devetak, P. Hayden, Capacity theorems for quantum multiple access channels, classical-quantum and quantum-quantum capacity regions. IEEE Trans. Inf. Theory **54**, 3091–3113 (2008)

J. Yard, P. Hayden, I. Devetak, Capacity theorems for quantum multiple-access channels: Classical-quantum and quantum-quantum capacity regions. IEEE Trans. Inf. Theory **54**, 3091–3113 (2008)

J. Yard, P. Hayden, I. Devetak, Quantum broadcast channels. IEEE Trans. Inf. Theory **57**, 7147–7162 (2011)

B.J. Yen, J.H. Shapiro, Multiple-access bosonic communications. Phys. Rev. A **72**, 062312 (2005)

X.N. Zhu, S.M. Fei, Entanglement monogamy relations of qubit systems. Phys. Rev. A **90**, 024304 (2014)

Chapter 4
QKD on Quantum Networks

In this chapter, we discuss quantum key distribution in quantum networks. Quantum key distribution, which is an application of quantum cryptography, aims to generate a random bit sequence with unconditional security guaranteed by quantum mechanics. The random key is distributed using quantum signals or qubits. Any attempt by an eavesdropper to recover useful information from the transmitted qubits will disrupt the quantum signal and their eavesdropping behavior can be detected by evaluating the transmission errors. This utilization of quantum principles helps to establish a random key in a secure manner, which can then be used to encrypt data to be transmitted.

A secure encryption scheme for transmitting data is the one-time pad as shown in Fig. 4.1. To encode her n-bit message, Alice adds it to a key of identical length and sends the resulting n-bit string to Bob. The decoding process is completed by subtracting the identical n-bit secret key. The key is used only once, and its length should be the same as the message. This type of encryption can provide unconditional security if the shared key is secure.

Alternatively, nonlinear encrypting operations combined with a random key can be used to encrypt data, such as DES, IDEA, and AES. Although these encryption systems cannot ensure unconditional security or even computational security, they do offer limited computational security. This means that an eavesdropper would need to spend many years to crack the system, which is still meaningful for most transmissions.

In both private key cryptographic systems, key security is a major requirement. In classical communication, trusted couriers or trusted parties are usually employed to distribute a random key to both parties through private communication links. However, this approach requires two basic conditions: a trusted party and a private channel, which are challenging to achieve in Internet communication. This problem can be resolved through the use of quantum key distribution (QKD).

M.-X. Luo, *Quantum Networks*,
https://doi.org/10.1007/978-981-97-6226-2_4

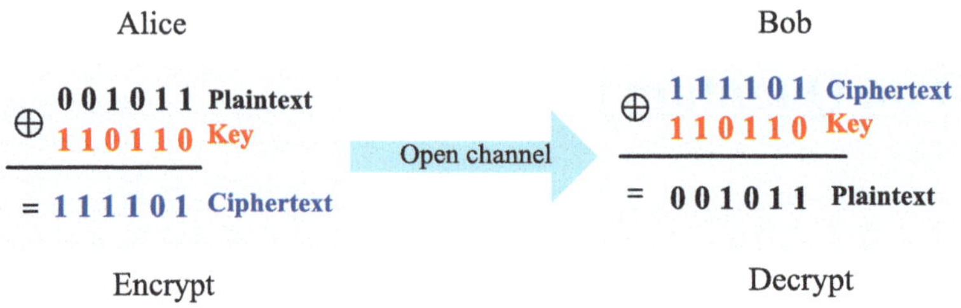

Fig. 4.1 Bipartite secure communication with a one-time pad

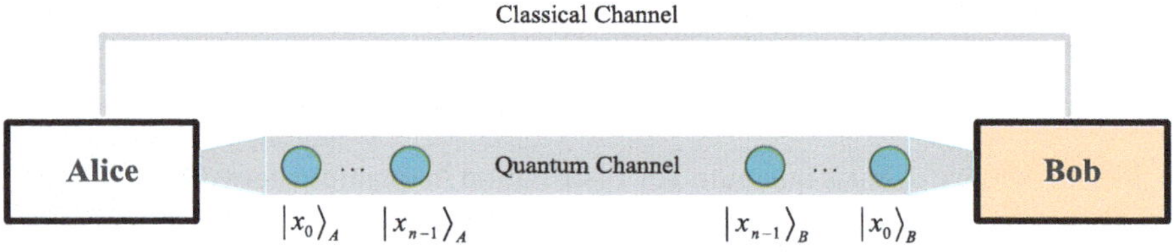

Fig. 4.2 Bipartite QKD. Alice and Bob are spatially separated. One party may create n pairs of qubits $|\phi\rangle$ and sends one of them to the other

4.1 Bipartite QKD

In this section, we provide a brief overview of bipartite quantum key distribution. The security of modern cryptography systems relies on the security of the key used. To establish a secure key, two users who do not share any secret information must use a reliable and secure channel for communication. It is difficult to detect interception by eavesdroppers in classical channels. But, this is not right for quantum channels. A quantum channel can distribute the random qubits without any "element of reality" associated with the transmitted qubits. The security is ensured by the completeness of quantum mechanics (Bennett and Brassard, 1984; Ekert, 1991).

4.1.1 BB84 Protocol

In QKD, the random bits are not generated at one site and then transported to the other side (Bennett and Brassard 1984). Instead, they are generated by both parties. The key is created by one party through the generation and transmission of encoded quantum states, and by the other party through measurements. The process involves an analysis of common information shared by both parties. The qubits are transmitted on a public quantum channel, as shown in Fig. 4.2. Photons can be used as qubits and transmitted on a fiber or through free space. The classical channel is used to evaluate shared information on qubit preparation, transmission, and measurement. One of the goals of using a classical channel is to detect potential eavesdroppers.

The BB84 protocol, also known as the Bennett-Brassard 1984 protocol (Bennett and Brassard 1984), is one of the earliest practical applications of quantum cryptography. It is based on the principles of quantum superposition and the uncertainty principle. In this protocol, Alice and Bob exchange qubits, which are quantum bits of information, using a quantum channel. Each qubit can be in one of four possible states, represented by two non-orthogonal bases: Pauli X and Z. Alice randomly chooses the basis in which to encode each qubit, and Bob randomly measures the qubits in one of the two bases. The protocol is comprised of four main steps: preparation, transmission, basis choice, and key distillation.

- **Preparation**-Alice prepares a stream of qubits that she sends to Bob. Each qubit can be prepared using any quantum system that exhibits two distinguishable states, such as photons or electrons. The qubits are in one of the four possible states, which can be represented by two non-orthogonal bases, Pauli X and Z. The X basis can be represented by the eigenstates of the Pauli matrix X, while the Z basis can be represented by the eigenstates of the Pauli matrix Z. Alice randomly chooses the basis in which to encode each qubit, and the basis can be represented by the binary variable $b = 0$ for the Z basis and $b = 1$ for the X basis. The qubits are then transmitted to Bob through a quantum channel.
- **Transmission**-Alice sends the qubits to Bob over a quantum channel, which must be a secure and reliable means of transmitting quantum information, such as a fiber optic cable or a line-of-sight laser link. During transmission, the qubits may be subjected to various types of noise and interference, which can cause errors in the received qubits. To minimize the errors, Alice and Bob can use various techniques, such as error correction codes and decoy states. Error correction codes can be used to detect and correct errors that occur during transmission, while decoy states can be used to estimate the error rate and detect eavesdropping attacks.
- **Basis Choice**-Bob chooses randomly one of the two bases, X or Z, in which to measure each qubit that he receives from Alice. The choice of basis is random and independent of Alice's choice of basis. If Bob chooses the same basis as Alice, he will measure the qubit correctly with a probability of 1. If he chooses a different basis from Alice, he will measure the qubit incorrectly with a probability of 1/2. The probability of Bob choosing the same basis as Alice is 1/2, and the probability of him choosing a different basis is 1/2.
- **Key Distillation**- After the transmission of qubits, two parties compare the bases in which each qubit was encoded and measured. If they used the same basis for a qubit, they keep the value of the qubit. If they used different bases, they discard the qubit. By comparing the values of the qubits that they kept, Alice and Bob can establish a shared secret key. The secret key can be used for secure communication between Alice and Bob, such as encrypting and decrypting messages. However, to ensure that the secret key is secure, Alice and Bob must perform key distillation. This is the process of removing any errors or eavesdropping that may have occurred during transmission. Key distillation can be performed using various techniques, such as privacy amplification and error correction codes. By performing key distillation,

Alice and Bob can ensure that the shared secret key is secure and can be used for cryptographic purposes without the risk of being compromised by an eavesdropper.

- **Privacy Amplification**-Privacy amplification is a process by which Alice and Bob can reduce the amount of information that an eavesdropper, Eve, may have obtained during transmission. Eve may have obtained some information about the qubits by intercepting and measuring them, and this information may be used to try to determine the secret key. To perform privacy amplification, Alice and Bob can use a hash function, which is a mathematical function that takes a large amount of data and produces a fixed-length output. The hash function can be used to reduce the amount of information that Eve may have obtained during transmission.
  Alice and Bob can agree on a random seed value, which is used as the input to the hash function. They then apply the hash function to the secret key and obtain a shorter and more secure key. The shorter key is used as the final secret key for secure communication between Alice and Bob.

Another way to distribute a random key is by bipartite entanglement (Ekert, 1991). Here, the quantum channel consists of a bipartite entangled source that generates pairs of two-qubit EPR states $\frac{1}{\sqrt{2}}(|00\rangle + |11\rangle)$. The particles are sent to two legitimate users. Each party performs random measurements on the shares along one of three directions $\vec{a}_i$ and $\vec{b}_j (i, j = 1, 2, 3)$ on Bloch sphere. One example is to take $\vec{a}_i$ and $\vec{b}_j$ lie in the x-y plane, and are represented by azimuthal angles: $\theta_1^a = 0$, $\theta_2^a = \frac{1}{4}\pi$, $\theta_1^3 = \frac{1}{2}\pi$, and $\theta_1^b = \frac{1}{4}\pi$, $\theta_2^b = \frac{1}{2}\pi$, $\theta_3^b = \frac{3}{4}\pi$. The superscripts a and b refer to Alice and Bob's devices, respectively. Each measurement follows two results, $+1$ and -1.

Denote $P_{\pm}(\vec{a}_i, \vec{b}_j)$ as the probability for measurement outcomes result ± 1 along $\vec{a}_i$ and $\vec{b}_j$. The correlation of the measurements is defined by

$$E(\vec{a}_i, \vec{b}_j) = P_{++}(\vec{a}_i, \vec{b}_j) + P_{--}(\vec{a}_i, \vec{b}_j) - P_{+-}(\vec{a}_i, \vec{b}_j) - P_{-+}(\vec{a}_i, \vec{b}_j). \quad (4.1)$$

According to the Born's rule, we rewrite the correlation as

$$E(\vec{a}_i, \vec{b}_j) = -\vec{a}_i \cdot \vec{b}_j. \quad (4.2)$$

For two pairs of measurement devices under the orientation ($\vec{a}_2$, $\vec{b}_1$ and $\vec{a}_3, \vec{b}_2$) quantum mechanics implies $E(\vec{a}_2, \vec{b}_1) = E(\vec{a}_3, \vec{b}_2) = -1$.

Now, we consider the protocol based on the CHSH inequality (Clauser et al., 1969), in which we define a CHSH correlator as

$$S = E(\vec{a}_1, \vec{b}_1) - E(\vec{a}_1, \vec{b}_3) + E(\vec{a}_3, \vec{b}_1) + E(\vec{a}_3, \vec{b}_3) \leq 2, \quad (4.3)$$

which holds for any bipartite separable state. For a bipartite entanglement, there are quantum correlations that violates the CHSH inequality as

$$S \leq -2\sqrt{2}. \quad (4.4)$$

After receiving all the qubits, Alice and Bob publicly announce the measurement bases used for each qubit. They can categorize the measurements into two subsets, based on whether they used the same or different measurement bases. For the subset where they used different measurement bases, they can publicly reveal the joint statistics $E(\vec{a}_i, \vec{b}_j)$ only. This enables them to perform a CHSH test to calculate the statistical value S. If the qubits were affected by noise or interference, the value of S may be no more than 2, indicating that the qubits should be disturbed. If the qubits do not disturbed, the value of S should be close to the upper bound given by Eq. (4.4). This ensures that Alice and Bob obtain consistent results for the subset where they used the same measurement bases. Finally, the measurement outcomes are converted into a bit string that can be used as a shared secret key if the CHSH test is successful.

The original scheme is based on a single photon of four indistinguishable states $\{|0\rangle, |1\rangle, |+\rangle, |-\rangle\}$ (Bennett and Brassard, 1984). This famous protocol makes use of the no-cloning principle because of the quantum superposition assumption (Wootters and Zurek, 1982). The rigid security is based on the entanglement realization (Ekert, 1991) or quantum error-correction code (Shor and Preskill, 2000; Lo and Chau, 1999; Gisin et al., 2002; Scarani et al., 2009). The present security analyses provide a device-independent application of quantum entanglement.

In general, consider an application $\mathcal{A}$ where all the legal parties make use of measurement outcomes to complete certain tasks.

Definition 4.1.1 $\mathcal{A}$ is completed in a device-independent manner if all the legal parties use only statistical distributions of the inputs and outputs but do not care about the involved entangled or measurement devices to complete the task.

In a device-independent model, all the parties will complete the task independent of the underlying physical realizations of sources and measurements (Mayers, 1996; Acín et al., 2007; Arnon-Friedman et al., 2018; Vazirani and Vidick, 2019; Schwonnek et al., 2021). It is a strong secure model for various quantum tasks including quantum key distribution, quantum secret sharing, or quantum computations.

4.1.2 Collective Attacks

Consider a modification of the Ekert 1992 protocol (Ekert, 1991). Alice and Bob share a quantum entanglement channel. Alice has three measurements A_0, A_1 and A_2, and Bob has two measurements B_1 and B_2. All measurements have binary outcomes labelled by $a_i, b_j \in \{+1, -1\}$. The raw key is assumed to be extracted from the pair $\{A_0, B_1\}$. Denote the quantum bit error rate (QBER) as $Q = \Pr(a_0 \neq b_1)$. Here, two parties use the measurements A_1, A_2, B_1, and B_2 to evaluate the CHSH correlator (4.3). The goal here is to build a directive relation between $\mathcal{S}$ and Q.

Without loss of generality, suppose that all the marginals are random for each measurement. In the device-independent scenario, Eve is assumed not only to control the entanglement source but also the measurement devices. The only data available to the two parties are the observed statistics of measurement setups and outcomes, which are used to bound Eve's leaking knowledge.

In general, suppose that Alice, Bob, and Eve share a state $|\Psi\rangle_{ABE}$ in the Hilbert space $\mathbb{H}_A^{\otimes n} \otimes \mathbb{H}_B^{\otimes n} \otimes \mathbb{H}_E$, where n is the number of bits in the raw key. The dimension d of the Hilbert spaces $\mathbb{H}_A = \mathbb{H}_B = \mathbb{C}^d$ is unknown to both parties and fixed by Eve. The measurement M_k for the k-th outcome of Alice is defined on the k-th subspace of Alice and chosen by Eve based on the measurement setting A_{j_k} of Alice, as well as all previous settings and outcomes, i.e., $M_k = M(A_{j_k}, \vec{A}_{k-1}, \vec{a}_{k-1})$, where $\vec{A}_{k-1} := (A_{j_1}, \ldots, A_{j_{k-1}})$ and $\vec{a}_{k-1} := (a_{j_1}, \ldots, a_{j_{k-1}})$. Similar definitions hold for Bob.

Here, we focus on collective attacks where Eve applies the same attack to each system of Alice and Bob. Specifically, suppose the total state has the product form $|\Psi\rangle_{ABE} = |\psi\rangle_{ABE}^{\otimes n}$, and the measurements are a function of the current setting only, i.e., $M_k = M(A_{j_k})$. In what follows, we write the measurement $M(A_j)$ as A_j for short.

For collective attacks, the secret key rate r by one-way postprocessing from Bob to Alice is lower bounded by the Devetak-Winter rate (Devetak and Winter, 2005) as

$$r \geq r_{DW} = I(A_0; B_1) - \chi(B_1; E), \tag{4.5}$$

where $I(A_0; B_1) = 1 - h(Q)$ (h is the binary entropy) is the mutual information between Alice and Bob, and $\chi(B_1; E) = S(\rho_E) - \frac{1}{2}\sum_{b_1=\pm 1} S(\rho_{E|b_1})$ denotes the Holevo quantity between Eve and Bob. Note that $\chi(A_0; E) \geq \chi(B_1; E)$. There is an advantage for Alice and Bob to do the classical postprocessing.

To find the optimal attack, we evaluate the maximum of $\chi(B_1; E)$ compatible with the observed statistics in terms of the violation of the CHSH inequality.

Theorem 4.1.1 *After Alice and Bob have symmetrized their marginals, we get*

$$\chi(B_1; E) \leq h\left(\frac{1 + \sqrt{S^2/4 - 1}}{2}\right). \tag{4.6}$$

The idea of the proof is to consider a mixed state $\rho_{AB} = \sum_i p_i \rho_{AB}^i$, together with an ancillary owned by Eve for distinguishing the measurements performed by two parties. Each state ρ_{AB}^i can be taken to be a Bell-diagonal while the measurements in the x-z plane of the Pauli sphere.

Given the bound (4.6), the key rate (4.5) is computed for Q and $\mathcal{S}$. Take $\mathcal{S} = 2\sqrt{2}(1 - 2Q)$ as an example, which is arisen from the maximally entangled EPR state $|\Phi^+\rangle = \frac{1}{\sqrt{2}}(|00\rangle + |11\rangle)$ going through a depolarizing channel with local measurement operators $A_0 = B_1 = Z$, $B_2 = X$, $A_1 = (Z + X)/\sqrt{2}$ and $A_2 = (Z - X)/\sqrt{2}$. The key rate may be computed as a function of Q. In an ideal case, it is equivalent to Ekert's protocol (Ekert, 1991), where Eve's information is determined by the phase error (Shor and Preskill, 2000; Kraus et al., 2005). In this case, it follows that $e_p = 1 - Q - \mathcal{S}/2\sqrt{2}$ and

$$\chi(B_1; E) \leq h(Q + \frac{\mathcal{S}}{2\sqrt{2}})). \tag{4.7}$$

If $e_p = Q$, i.e., $\mathcal{S} = 2\sqrt{2}(1 - 2Q)$, this yields to the critical QBER of 11% (Shor and Preskill, 2000), different from 7.1% in the device-independent scenario.

Device-independent security provides a way to ensure the security of QKD protocols without relying on detailed assumptions about the devices and measurements used in the protocol. Instead, device-independent security allows us to account for the presence of noisy quantum devices or uncontrolled side channels that may be present during the protocol.

In some cases, both parties detectors may fail to register a click, making it difficult to apply the device-independent proof to existing devices. One possible strategy is to replace the absence of a click with a chosen outcome, but this introduces additional assumptions that may not be satisfied in practice.

In Bell tests, additional assumptions such as the fair sampling hypothesis can be used to address this problem. The fair sampling hypothesis assumes that the observed statistics of measurement outcomes are representative of the underlying probability distributions, even when some measurement outcomes are missing. By making this assumption, we can use the observed statistics to infer the presence of entanglement, even in the presence of missing measurement outcomes.

4.1.3 No-Signaling Security

While standard QKD protocols are provably secure against eavesdropping attacks under the assumption of quantum theory, this security relies on the validity of a quantum theory. If quantum theory is not valid, the security of QKD is called into question. This raises concerns about the vulnerability of QKD to attackers who may use post-quantum techniques to extract information from quantum communications.

Post-quantum attacks may not cause the quantum disturbances that are assumed in existing security proofs, making it difficult to ensure the security of QKD against such attackers. To address this, researchers are developing new physical principles that can ensure the security of QKD against post-quantum eavesdroppers. For example, some protocols rely on the hardness of certain mathematical problems, such as the shortest vector problem, to ensure security against attackers with quantum computers.

One of the most common assumptions in quantum information theory is the no-signaling principle, which asserts the impossibility of super-luminal signaling. So far, all well-known experimental physics, including both classical and quantum mechanics, satisfy this principle.

Consider a standard Bell test experiment with two parties. With local measurements, neither party can obtain any information about the other party's measurements without transmitting physical information. In terms of the probability distribution, both classical and quantum theories satisfy the no-signaling principle if one cannot infer the other party's input from the observed outcomes. The general no-signaling conditions for the bipartite scenario are given by

$$\sum_{a} p(a, b|x, y) = \sum_{a'} p(a, b|x', y) = p(b|y), \forall x, x',$$
$$\sum_{b} p(a, b|x, y) = \sum_{b} p(a, b|x, y') = p(a|x), \forall y, y', \tag{4.8}$$

where a and b are outcomes of two parties' measurements, respectively. x, y belong to a set of possible measurements performed by Alice and Bob, respectively. If this equality is not satisfied, then the no-signaling principle is violated. On a technical level, the no-signaling condition implies that tracing over subsets of outputs in a correlation will result in tracing over the corresponding inputs (Brunner et al., 2014). These are relevant in our scenario because it is necessary to verify the observed correlations from a specific state and measurement choices without classical communication.

In the case of QKD, it is necessary to verify the observed correlations from a specific state and measurement choices without classical communication. By verifying that the Bell inequality is violated, it is possible to detect any eavesdropping attempts and ensure the security of QKD.

Assume that Alice and Bob have a noiseless quantum channel and an authenticated classical channel. Consider the following protocol against attacks by post-quantum eavesdroppers. Define the bases

$$X_r = \left\{\cos\frac{r\pi}{2n}|0\rangle + \sin\frac{r\pi}{2n}|1\rangle, -\sin\frac{r\pi}{2n}|0\rangle + \cos\frac{r\pi}{2n}|1\rangle\right\} \tag{4.9}$$

for integer r. For each basis, define outcomes 0 and 1 corresponding respectively to the projections onto the first and second basis states. Let n and m (defined below) be the security parameters. For simplicity, we suppose $m \ll n$. The protocol is going as follows.

S1. Both parties share $N = mn^2$ pairs of maximally entangled state $|\psi_-\rangle = \frac{1}{\sqrt{2}}(|01\rangle - |10\rangle)$;

S2. Both parties independently and randomly choose r_A^i and r_B^i from $\{0, 1, \ldots, n-1\}$, $i = 1, \cdots, N$. They measure the i-th particle in the bases $A_i := X_{r_A^i}$ and $B_i := X_{r_B}$;

S3. After all the measurements are completed, they announce the bases over an authenticated classical channel;

S4. They abort the protocol and restart unless

$$2mn \leq \sum_{i} \sum_{c=-1,0,1} |\{j : A_j = X_i, B_j = X_{i+c}\}|. \tag{4.10}$$

S5. Their outcomes should be kept secret for one random pair with the bases X_i and X_{i+e} for some i and $c = -1, 0$ or 1;

S6. They abort the protocol if a and b satisfy $a = b \oplus 1$ in all the cases, where they choose neighboring or identical bases;
S7. If the protocol is not aborted, they get one secret bit.

Attack model

We describe the power available to post-quantum eavesdroppers.

Assumption 1. Assume that Eve can control the entanglement source which generates the transmitted systems.

Suppose that Eve prepares $2N+1$ systems in a post-quantum state λ, sending N of these systems to Alice, N to Bob, and keeping the first system for herself. From the state λ, we can define measurement probabilities as $P_\lambda(abc|ABE)$, where $A = \{A_1, \cdots, A_N\}$ and $B = \{B_1, \cdots, B_N\}$ are sets of measurement choices for Alice and Bob, $E = \{E_1\}$ is a set of measurement choices for Eve, and a, b, and c are the corresponding measurement outcomes.

The state λ may be no-quantum and nonlocal in the sense that it does not necessarily satisfy the principles of quantum mechanics, but it still satisfies the no-signaling principle, even if the parties cooperate. Thus, for any partition $A = A_1 \cup A_2$, $B = B_1 \cup B_2$ and $E = E_1 \cup E_2$, and measurement choices $\bar{A}_2, \bar{B}_2, \overline{E}_2$, we get that

$$\begin{aligned} &\sum\nolimits_{a_2,b_2,e_2} P_\lambda(a_1a_2, b_1b_2, e_1e_2|A_1A_2, B_1B_2, E_1E_2) \\ &= \sum\nolimits_{a_2,b_2,e_2} P_\lambda(a_1\bar{a}_2, b_1b_2, e_1\bar{e}_2|A_1A_2, B_1B_2, E_1E_2) \end{aligned} \tag{4.11}$$

Here, the attacker Eve may wait until all the operations and communications are finished before performing local measurements.

Assumption 2. The range of measurements available to Eve and their outcome probabilities are time-independent.

In a post-quantum theory, all local measurements on a shared state cannot be used to send signals between the parties in any configuration to prevent the propagation of information to Eve instantaneously. Violating this assumption would allow Eve to gain information about the measurement bases and outcomes, which would compromise the security of any key distribution scheme.

In general, it is impossible to achieve secure key distribution when Eve can violate this assumption, as she would be able to intercept the key without being detected. Therefore, it is crucial to ensure that the post-quantum theory used for key distribution satisfies this assumption to guarantee the security of the scheme.

One can justify the security by extending a standard cryptographic assumption to post-quantum scenarios. The standard analyses of quantum key distribution require that both parties are secure against Eve's scrutiny. Similarly, in the post-quantum case, there is no information about both parties' experiments, the measurements or outcomes propagates to Eve. This means both legal parties can establish secure experiments locally. We can prove that the present protocol is secure against general attacks. Define A_j and B_j to be both parties' measurement bases for the j-th pair. These are generally represented by random variables with probability $1/n$. Define a_j and b_j to be the measurement outcomes. We have

$$t_j = \frac{1}{3n} \sum_{c=-1,0,1} \sum_{i=0}^{n-1} P_\lambda(a_j \neq b_j \mid A_j = X_i, B_j = X_{i+c}) \leq 1 - \frac{2}{3n} \tag{4.12}$$

for a local state λ. This gives a generalized Bell inequality, which is similar to the chained Bell inequality (Braunstein and Caves, 1990).

If there is no eavesdropping, both parties share maximally entangled singlets. This follows a quantum bound as

$$t_j = 1 - O(\frac{1}{n^2}), \forall j, \tag{4.13}$$

which violates the inequality (4.12) for a large enough N. This provides the proof of security. Specially, the violation implies the limited leaking of information to Eve.

Now, we present a lower bound on t_s for the secret pair s, given that both parties have passed the test. Meanwhile, Eve has not used a strategy that may stop the protocol. The lower bound may imply an upper bound on Eve's information. Assume that there is at least one pair for which both parties' measurements are neighboring or identical. Let a random variable s be the index of the pair for recovering a secret bit. The probability P_λ(test pass) that both parties' tests are passed is generally dependent on the prepared post-quantum state λ. In fact, for any λ such that $P_\lambda(\text{test pass}) > \epsilon$, we have that

$$P_\lambda(a_s \neq b_s | \text{test pass}) > 1 - \frac{1}{2mn\epsilon}. \tag{4.14}$$

Combining with the no-signaling condition (4.8) and the chain rule for conditional probabilities it follows that

$$t_s > 1 - \frac{1}{2mn\epsilon}. \tag{4.15}$$

We assume the test is passed. In this case, three parties share the system (4.15). By using the contradiction, the information that Eve gets by local measurements is small. Formally, Eve gets an outcome ϵ_0 with a probability $\delta > 0$ such that

$$P_\lambda(a_s = b, b_s = \bar{b} | A_s = X_k, B_s = X_{k+d}, \epsilon_0) < \frac{1}{2}(1 + \delta') \tag{4.16}$$

for some k and $d = -1, 0$ or 1, where $\delta' > 0$ and $b \in \{0, 1\}$.

So far, we have restricted $m \ll n$. It may be extended for arbitrarily large m. In this case, both parties' security tests should be statistically consistent with quantum predictions. This shows the security against post-quantum attackers.

Beyond these attacks, how to extend the key ratio is important for applications in quantum networks. Different methods of practical quantum key distributions have proposed for long-distance secure fibre communication (Ma et al., 2018; Lucamarini et al., 2018; Wang, 2019; Clivati, 2022; Zhong et al., 2022; Zhou et al., 2023).

4.2 Device Independent Model on Quantum Networks

In this section, we extend the device-independent QKD to general secure applications on quantum networks. For a single entangled state, the violation of a certain Bell inequality allows for device-independent information processing against the quantum even post-quantum eavesdroppers. We propose a device-independent model for secure communication, key distribution, and other applications on quantum networks.

4.2.1 No-Signaling Conditions

No-signaling principles impose special constraints on correlations. It implies that parties sharing the correlation cannot use it alone to communicate. No party can then infer the measurement input choices of other parties based on local information (Beckman et al., 2001).

Denote the i-th input to the measurement device by x_i, and the outcome of measurement by a_i, $i \in \{1, \ldots, n\}$ and n is the number of parties involved. Compliance with the no-signaling conditions can be enforced by imposing a space-like separation between all the measured parties. This may be represented in multipartite scenarios as

$$\sum_{a_i} P(\vec{a}|\vec{x}) = \sum_{a_i} P(\vec{a}|\vec{x}') = P(\vec{a}_i|\vec{x}_i), \quad \forall x_i, \bar{x}_i, \tag{4.17}$$

where $\vec{x}'$ denotes $\vec{x}$ by replacing x_i with x_i', $\vec{x}_i = x_1 \cdots x_{i-1}x_{i+1} \cdots x_n$ and $\vec{a}_i = a_1 \cdots a_{i-1}\ a_{i+1} \cdots a_n$. The set of all correlations that satisfy (4.17) is called no-signaling correlations. The no-signaling conditions may be equivalently expressed in terms of conditional mutual information as follows:

$$I(A_i; \vec{A}_i|\vec{X}_i)_\rho = 0, \forall i = 1, \cdots, n, \tag{4.18}$$

where $\vec{X}_i = X_1 \cdots X_{i-1}X_{i+1} \cdots X_n$, $\vec{A}_i = A_1 \cdots A_{i-1}A_{i+1} \cdots A_n$,

$$\rho_{\vec{A},\vec{X}} = \sum_{\vec{a},\vec{x}} q(\vec{x})P(\vec{a}|\vec{x})|\vec{a}\rangle\langle\vec{a}|_{\vec{A}} \otimes |\vec{x}\rangle\langle\vec{x}|_{\vec{X}}. \tag{4.19}$$

Here, the distribution $P(\vec{a}|\vec{x})$ is a no-signaling correlation, and the conditional mutual information of random variables X, Y, and Z is defined as

$$I(X; Y|Z) := H(X, Z) + H(Y, Z) - H(Z) - H(X, Y, Z) \tag{4.20}$$

with Shannon entropy H. It suffices to take the input distribution q to be uniform. Eq. (4.17) implies an equivalent one as

$$I(\vec{A}_i; A_i|X_i)_\rho = 0, \forall i = 1, \cdots, n. \quad (4.21)$$

In quantum scenarios, the correlation for a given state $\rho_{\tilde{A}_1\cdots\tilde{A}_n}$ is given by the following form:

$$P(\vec{a}|\vec{x}) = \mathrm{Tr}[(\Pi_{a_1}^{x_1} \otimes \cdots \otimes \Pi_{a_n}^{x_n})\rho_{\tilde{A}_1\cdots\tilde{A}_n}], \quad (4.22)$$

where $\Pi_{a_i}^{x_i}$ is a POVM for each x_i. Quantum correlations are a subset of no-signaling correlations. This can easily be seen in the analysis below

$$\begin{aligned}\sum_{a_i} P(\vec{a}|\vec{x}) &= \sum_{a_i} \mathrm{Tr}((\Pi_{a_1}^{x_1} \otimes \cdots \otimes \Pi_{a_n}^{x_n})\rho_{A_1\cdots A_n}) \\ &= \mathrm{Tr}[\otimes_{j\neq i}\Pi_{b_j}^{y_j}\varrho_{\vec{A}_i}] \\ &= P(\vec{a}_i|\vec{x}_i),\end{aligned} \quad (4.23)$$

where $\vec{A}_i = A_1\cdots A_{i-1}\cdots A_{i+1}\cdots A_n$ and $\varrho_{\vec{A}_i}$ denotes the reduced state of $\vec{A}_i$.

Similar to the constraints imposed by no-signaling conditions on QKD, it is sensible to impose no-signaling conditions on the systems that an eavesdropper may possess (Kaur et al., 2017, 2020). No-signaling extensions are extensions of a correlation that obey the above no-signaling constraints. It can be expressed as follows:

$$\sum_{a_i} P(\vec{a}|\vec{x})\rho_E^{\vec{a}\vec{x}} = \sum_{a_i} P(\vec{a}|\vec{x}')\rho_E^{\vec{a}\vec{x}'}, \forall x_i, x_i'. \quad (4.24)$$

Consider a general network. For a device-independent model, eavesdroppers may hold local systems correlated with some sources to recover private information such as measurement outcomes. Here, the eavesdropper possesses a system E that is correlated with the systems possessed by the n parties A_i, and this correlation extends the total state $\rho_{\vec{A}}$ to

$$\rho_{\vec{A}} = \mathrm{Tr}_E\, \rho_{\vec{A}E}. \quad (4.25)$$

A quantum extension of a correlation is defined as follows:

$$\begin{aligned}\rho_{\vec{A}E\vec{X}} &= \sum_{\vec{a},\vec{x}} q(\vec{x})|\vec{a},\vec{x}\rangle\langle\vec{a},\vec{x}|_{\vec{A},\vec{X}} \otimes \mathrm{Tr}_{\vec{A}}[(\Pi_{a_1}^{(x_1)} \otimes \cdots \otimes \Pi_{a_n}^{(x_n)} \otimes \mathbb{1}_E)\rho_{\vec{A}E}] \\ &= \sum_{\vec{a},\vec{x}} q(\vec{x})|\vec{a},\vec{x}\rangle\langle\vec{a},\vec{x}|_{\vec{A},\vec{X}} \otimes P(\vec{a}|\vec{x})\rho_E^{\vec{a}\vec{x}}.\end{aligned} \quad (4.26)$$

4.2.2 *Semi-Device Independent Model*

Consider a general network $\mathcal{N}$ (e.g., Fig. 4.3) consisting of n agents: $\mathsf{A}_1, ..., \mathsf{A}_n$. $\mathcal{N}$ is k-independent if there are k agents $\mathsf{A}_1, \cdots, \mathsf{A}_k$ who have not shared any sources. Let A_i and B share sources $\Lambda_i = \lambda_{j_1} \cdots \lambda_{j_{\ell_i}}$, where Λ_i's satisfy $\cup_{i=1}^n \Lambda_i = \lambda_1 \cdots \lambda_m$. For simplicity, we assume that $\Lambda_i = \lambda_i$. Let $\overline{\lambda} = \lambda_{k+1} \cdots \lambda_m$, and B includes all parties $\mathsf{A}_{k+1}, \cdots, \mathsf{A}_n$. This network could be represented by a directed acyclic graph (DAG) in the device-independent model, as shown in Fig. 4.3b.

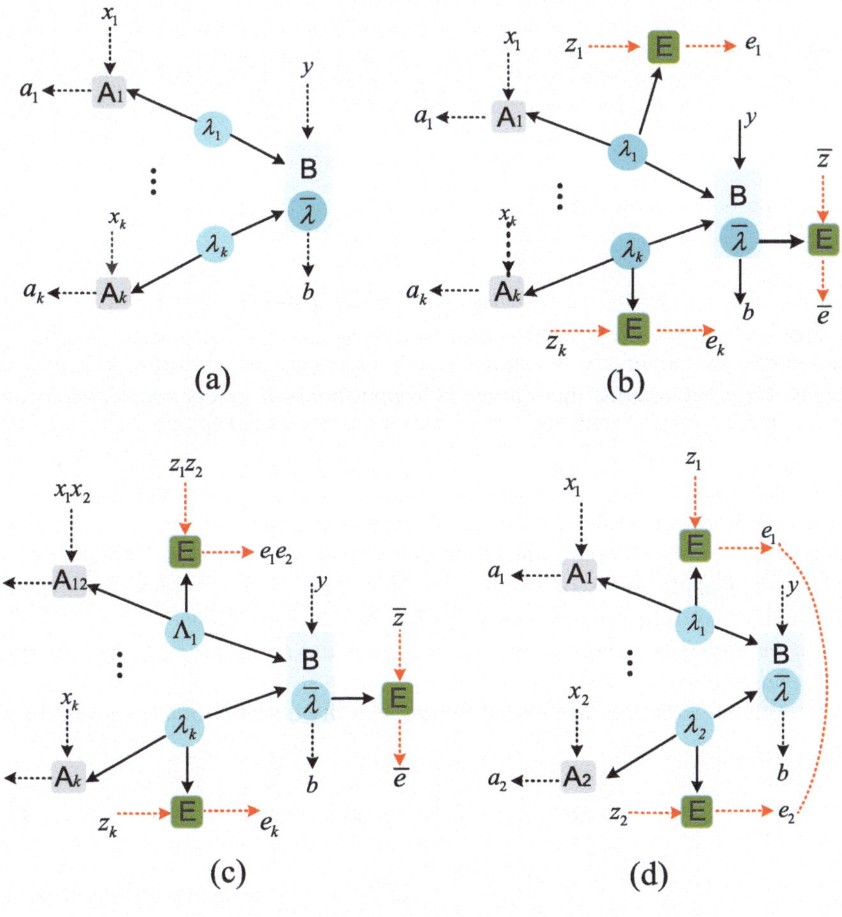

Fig. 4.3 (Color online) **a** A k-independent classical network. B includes all parties $\mathsf{A}_{k+1}, \cdots, \mathsf{A}_n$. **b** A DAG in terms of the device-independent model. An eavesdropper holds some systems each of them is correlated with one source λ_i. **c** An eavesdropper holds some systems some of which are correlated with multiple sources in Λ_i. **d** An eavesdropper holds a system that correlates with λ_1 and λ_2 (represented by red lines)

Assume that an eavesdropper holds some systems, each of which is correlated with one source λ_i. Let z_i and e_i denote the respective input and outcome of the measurement on each eavesdropper's system λ_i, and $\overline{e} = e_{k+1} \cdots e_n$. In general, each agent A_i performs local measurements with dichotomic input, denoted $x_i \in \{0, 1\}$, and obtains dichotomic output, denoted $a_i \in \{-1, 1\}$. Similar results hold for multiple inputs and outputs (Luo, 2018, 2020).

If all sources λ_i are random variables, the classically achievable correlations satisfy the nonlinear inequality (Luo, 2018):

$$\mathcal{R}_k := |I_{n,k}|^{\frac{1}{k}} + |J_{n,k}|^{\frac{1}{k}} \leq 1 \tag{4.27}$$

where $I_{n,k}$ and $J_{n,k}$ are defined as

$$I_{n,k} = \frac{1}{2^k} \sum_{x_i, i \in \{1, \cdots, k\}} \langle A_{x_1} \cdots A_{x_{k+1}} \cdots A_{x_n} \rangle, \tag{4.28}$$

$$J_{n,k} = \frac{1}{2^k} \sum_{x_i, i \in \mathcal{I}} (-1)^{\sum_{j \in \{1, \cdots, k\}} x_j} \langle A_{x_1} \cdots A_{x_k} A'_{x_{k+1}} \cdots A'_{x_n} \rangle, \tag{4.29}$$

and A_{x_j} and A'_{x_j} are different measurement of the party A_j, $j = k+1, \cdots, n$.

Similar to standard device-independent information processing on single entangled states, the adversary is limited to recovering privacy information from legal agents. Here, we assume that the eavesdropper holds m independent systems, each of which is correlated with one of m sources, as shown in Fig. 4.3b.

The eavesdropper's systems can be post-quantum (no-signaling). The output e_i of each eavesdropper's system depends on its input z_i and correlated sources. To complete a general network task, it is reasonable to permit independent agents $\mathsf{A}_1, \cdots, \mathsf{A}_k$ to communicate with each other classically. Unlike in device-independent applications of single entanglement, in the semi-device-independent model, the eavesdropper will satisfy the following assumptions.

Assumption-The eavesdropper cannot correlate the sources shared by all the independent parties.

Denote the variation distance of two probability distributions $\{p(x)\}$ and $\{q(x)\}$ as:

$$D(p, q) = \frac{1}{2} \sum_x |p(x) - q(x)|. \tag{4.30}$$

One way to ensure security against no-signaling eavesdroppers is by limiting their ability to predict the outcomes of agents' devices. This can be achieved by bounding the total variational distance, denoted by $D(P; Q)$, between the distributions $P(\vec{E}|\vec{A}; \vec{X}; \vec{Z})$ and $P(\vec{E}|\vec{Z})$. Another method is by estimating the guessing probability $P(\vec{E} = \vec{A})$. In simple terms, the first approach limits the amount of information that an eavesdropper can infer about agents' outcomes from local measurements, while the second approach limits the probability that the eavesdropper can correctly guess agents' outcomes.

Under this assumption, a violation of the inequality (4.27) provides an upper bound of an eavesdropper's information about the outcomes of independent agents.

Theorem 4.2.1 *The total information recovered by an eavesdropper satisfies the following inequality:*

$$D(P(\vec{e}|a_1\cdots a_k;\vec{x},\vec{z}),\prod_{i=1}^{m}p(e_i|z_i))\leq k(2-\mathcal{R}_k) \tag{4.31}$$

if $\lambda_1,\cdots,\lambda_k$ *are independent.*

Proof. Consider a quantum realization of k-independent network $\mathcal{N}$, where $\mathcal{N}$ consists of generalized EPR states (Einstein et al., 1935) or GHZ states (Greenberger et al., 1989). Each observer A_i performs local measurements. The expectation of quantum correlations among space-like separated observers satisfies the following bound (Luo, 2018):

$$1<|I_{n,k}^{q}|^{\frac{1}{k}}+|J_{n,k}^{q}|^{\frac{1}{k}}\leq\sqrt{2}, \tag{4.32}$$

which violates the inequality (4.27), where $I_{n,k}^{q}$ and $J_{n,k}^{q}$ are the corresponding quantities by quantum correlations.

Consider the conditional distribution $P(a,b,e|x,y,z)$, where a and b are binary random variables and x and y are s-valued ($s\geq 2$), satisfying the no-signaling conditions:

$$\begin{aligned}P(a,b|x,y,z)&=P(a,b|x,y),\\P(a,e|x,y,z)&=P(a,e|x,z),\\P(b,e|x,y,z)&=P(b,e|y,z).\end{aligned} \tag{4.33}$$

It easily implies new no-signaling conditions

$$\begin{aligned}p(a|b,x,y,z)&=p(a|b,x,y),\\p(b|c,x,y,z)&=p(b|c,y,z),\\p(c|a,x,y,z)&=p(c|b,x,z).\end{aligned} \tag{4.34}$$

We only prove $p(a|b,x,y,z)=p(a|b,x,y)$ as an example. It can be obtained from the following equalities

$$\begin{aligned}p(b|x,y,z)p(a|b,x,y,z)&=P(a,b|x,y,z)\\&=P(a,b|x,y)\\&=p(b|x,y)p(a|b,x,y),\end{aligned} \tag{4.35}$$

and $p(b|x,y,z)=p(b|x,y)$, i.e., the no-signaling condition.

From Fig. 4.3b, we get the following conditional independence relations:

$$P(\vec{e}|a_1\cdots a_k;\vec{x},\vec{z})=\prod_{i=1}^{k}p(e_i|\vec{x}_{\lambda_i},\vec{a}_{\lambda_i},z_i)\prod_{j=k+1}^{m}p(e_j|\vec{x}_{\lambda_j},\vec{a}_{\lambda_j},z_j), \tag{4.36}$$

where we have assumed $\lambda_1,\cdots,\lambda_k$ are shared by $\mathsf{A}_1,\cdots,\mathsf{A}_k$. All the other sources $\lambda_{k+1},\cdots,\lambda_m$ are shared by the agents included in B. $\vec{x}_{\lambda_i}$ and $\vec{a}_{\lambda_i}$ denote the respective inputs and outputs of all the agents who have shared the source λ_i, $i=1,\cdots,m$.

Note that all the agents in B are not allowed for classical communications. From the no-signaling conditions (4.33) and (4.34), we rewrite Eq. (4.36) into

$$P(\vec{e}|a_1\cdots a_k;\vec{x},\vec{z})=\prod_{i=1}^{k}p(e_i|\vec{x}_{\lambda_i},\vec{a}_{\lambda_i},z_i)\prod_{j=k+1}^{m}p(e_j|z_j). \tag{4.37}$$

Consider the left side of the inequality (4.31). From Eq. (4.37) it can be decomposed into

$$\begin{aligned}D_{\vec{e}}:&=D\left(P(\vec{e}|a_1\cdots a_k;\vec{x},\vec{z}),\prod_{i=1}^{m}p(e_i|z_i)\right)\\&=\sum_{e_{k+1},\cdots,e_m}D\left(\prod_{i=1}^{k}p(e_i|\vec{x}_{\lambda_i},\vec{a}_{\lambda_i},z_i),\prod_{i=1}^{k}p(e_i|z_i)\right)\prod_{j=k+1}^{m}p(e_j|z_j)\\&=D\left(\prod_{i=1}^{k}p(e_i|\vec{x}_{\lambda_i},\vec{a}_{\lambda_i},z_i),\prod_{i=1}^{k}p(e_i|z_i)\right)\end{aligned} \tag{4.38}$$

from the normalization conditions: $\sum_{e_j}p(e_j|z_j)=1$ for $j=k+1,\cdots,m$. From the triangle inequality $|x-y|\le|x-z|+|z-y|$, it follows

$$\begin{aligned}D_{\vec{e}}\le&\frac{1}{2}\sum_{e_1,\cdots,e_k}p(e_1|\vec{x}_{\lambda_1},\vec{a}_{\lambda_1},z_1)\left|\prod_{i=2}^{k}p(e_i|\vec{x}_{\lambda_i},\vec{a}_{\lambda_i},z_i)-\prod_{i=2}^{k}p(e_i|z_i)\right|\\&+\frac{1}{2}\sum_{e_1,\cdots,e_k}|p(e_1|\vec{x}_{\lambda_1},\vec{a}_{\lambda_1},z_1)-p(e_1|z_1)|\prod_{i=2}^{k}p(e_i|z_i)\\\le&\frac{1}{2}\sum_{e_1,\cdots,e_k}\left|\prod_{j=2}^{k}p(e_j|\vec{x}_{\lambda_j},\vec{a}_{\lambda_j},z_j)-\prod_{j=2}^{k}p(e_j|z_j)\right|\\&+D(p(e_1|\vec{x}_{\lambda_1},\vec{a}_{\lambda_1},z_1),p(e_1|z_1))\end{aligned} \tag{4.39}$$

from the normalization conditions: $\sum_{e_1} p(e_1|\vec{x}_{\lambda_1}, \vec{a}_{\lambda_1}, z_1) = 1$ and $\sum_{e_j} p(e_j|z_j) = 1$ for $j = 2, 3, \cdots, k$. Finally, by iteratively using the inequality (4.39) for it follows that

$$
\begin{aligned}
D_{\vec{e}} &\leq \sum_{i=1}^{k} D(p(e_i|\vec{x}_{\lambda_i}, \vec{a}_{\lambda_i}, z_i), p(e_i|z_i)) \\
&\leq \sum_{i=1}^{k} I_2(P(a_i, b_i|x_i, y_i)),
\end{aligned} \tag{4.40}
$$

where I_2 is from the chained Bell inequality (Chaves et al., 2015) with two measurement settings. The inequality (4.40) is followed from the inequality: $D(p(e|a, x, z), p(e|z)) \leq I_k(P(a, b|x, y))$ (Barrett et al., 2006, 2012; Rabelo et al., 2011) and the following general form

$$
D_{\vec{e}} \leq I_k(P(a_i, a_j|x_i, x_j)) \tag{4.41}
$$

with $I_k(P(a, b|x, y)) = P(a = b|x = 1; y = k) + \sum_{|x-y|=1} P(a \neq b|x; y)$, where all the agents $\mathsf{A}_1, \cdots, \mathsf{A}_k$ and the potential eavesdropper are correlated by one source. The inequality (4.41) can be proved by following the same procedure (Barrett et al., 2012) and the fact that $P(\vec{a}|\vec{x}, z)$ is a conditional probability distribution for given inputs $\vec{x}$ and z.

Now, consider a quantum network $\mathcal{N}$ in which all the agents have binary inputs and outputs (Branciard et al., 2012). Specially, all the independent observers A_i perform separable measurements $A_i^{x_i} = A_{i,0}^{x_i} \otimes A_{i,1}^{x_i}$ while all the other agents B_j included in B perform separable measurements $B_j^{y_j} = B_{j,0}^{y_j} \otimes B_{j,1}^{y_j}$ on local systems. We get

$$
\langle A_1^{x_1} A_2^{x_2} \cdots A_k^{x_k} B^{y} \rangle = \prod_{i=1}^{k} \langle A_i^{x_i} B_i^{y_i} \rangle. \tag{4.42}
$$

From the definitions of $I_{n,k}$ and $J_{n,k}$ it follows that

$$
\begin{aligned}
I_{n,k} &= \frac{1}{2^k} \prod_{i=1}^{k} (\langle A_j^0 B_j^0 \rangle + \langle A_j^1 B_j^0 \rangle), \\
J_{n,k} &= \frac{1}{2^k} \prod_{i=1}^{k} (\langle A_j^0 B_j^1 \rangle - \langle A_j^1 B_j^1 \rangle).
\end{aligned} \tag{4.43}
$$

From Eq. (4.43) and the arithmetic-geometric inequality $(\prod_{i=1}^{n} x_i)^{1/n} \leq \frac{1}{n} \sum_{i=1}^{n} x_i$, we get

$$
\mathcal{R}_k = |I_{n,k}|^{\frac{1}{k}} + |J_{n,k}|^{\frac{1}{k}}
$$

$$\leq \frac{1}{2}\sum_{i=1}^{k}(|\langle A_j^0 B_j^0\rangle + \langle A_j^1 B_j^0\rangle| + |\langle A_j^0 B_j^1\rangle - \langle A_j^1 B_j^1\rangle|)$$
$$:= \frac{1}{2k}\sum_{i=1}^{k} C_2^{A_i B_i}, \tag{4.44}$$

where $C_2^{A_i B_i} := |\langle A_j^0 B_j^0\rangle + \langle A_j^1 B_j^0\rangle| + |\langle A_j^0 B_j^1\rangle - \langle A_j^1 B_j^1\rangle|$ is a special quantity used in the CHSH inequality (Clauser et al., 1969).

By using $\langle AB\rangle = 2p(A = B) - 1$ it follows that

$$I_2(e_i) = 2 - \frac{1}{2}C_2^{A_i B_i}. \tag{4.45}$$

Combining Eqs. (4.44) and (4.45), the right side of the inequality (4.40) is evaluated as

$$\sum_{i=1}^{k} I_2(e_i) = \sum_{i=1}^{k}(2 - \frac{1}{2}C_2^{A_i B_i})$$
$$\leq k(2 - \mathcal{R}_k), \tag{4.46}$$

which has completed the proof. □

A stronger assumption for eavesdropper is to correlate the multiple shared sources in Λ_i into a new variable $\hat{\lambda}_i$, as shown in Fig. 4.3c. If the eavesdropper can correlate sources λ_i, the inequality (4.31) can be extended using a similar proof. One example is shown in Fig. 4.3c, where the correlated sources λ_1 and λ_2 (λ_3 and λ_4) are combined into new variables $\hat{\lambda}_1$ ($\hat{\lambda}_2$). We define $\vec{e} = (e_1e_2, e_3e_4, e_5, \cdots, e_m)$. A similar result holds by replacing the left-hand side of the inequality (4.31) with $D(P(\vec{e}|a_i, i \in \mathcal{I}; \vec{x}, \vec{z}), p(e_1e_2|z_1z_2)\ p(e_3e_4|z_3z_4)\prod_{i=3}^{m} p(e_i|z_i))$.

Similar to the standard QKD with single entanglement, the security of networks using noisy states cannot be strengthened by the nonlinear inequality (4.27). Agents may employ non-separable measurements or different Bell inequalities to prevent information leakage. However, these strategies are ineffective against the strongest eavesdropper, who can correlate all shared sources, as shown in Fig. 4.3d. In this case, the analysis reduces to that of single entanglement (Barrett et al., 2005; Acín et al., 2007; Vazirani and Vidick, 2019).

Example 4.2.1 Consider a long-distance chain network with n independent hidden sources $\lambda_1, \cdots, \lambda_n$. Each space-like separated agent A_i shares some sources λ_j's. The multipartite quantum correlations violate the inequality (4.27) for all bipartite entangled pure states, where $k = \lceil n/2 \rceil$ denotes the number of independent agents, and $\lceil x \rceil$ denotes the smallest integer no less than x. An eavesdropper holds n independent systems correlated with $\lambda_1, \cdots, \lambda_n$ and measures each system with a device z_i, obtaining an outcome e_i. In the experiment, A_i first obtains a_i depending on measurement settings x_i and shared sources, $i = 2, \cdots, n$. Then A_j outputs a_j depending on its input x_j and shared sources, $j = 1, n+1$. z_i and e_i denote the respective input and outcome of each system λ_i by the eavesdropper.

Assuming that all independent agents $\mathsf{A}_1, \mathsf{A}_3, \cdots, \mathsf{A}_{n+1}$ for an even n (or $\mathsf{A}_1, \mathsf{A}_3, \cdots, \mathsf{A}_{n-2}, \mathsf{A}_{n+1}$ for an odd n) cannot communicate with each other, we obtain the upper bound $k(2 - \mathcal{R}_k)$ for the classical correlations. Another approach is to use multiple settings of Bell inequalities (Lee and Hoban, 2018).

Example 4.2.2 Consider a general star network with n independent hidden sources $\lambda_1, \cdots, \lambda_n$ and space-like separated agents A_i and B who share one source λ_i. The inequality (4.27) with $k = n$ is violated by multipartite quantum correlations created by generalized EPR states. For device-independent information processing, an eavesdropper holds n independent systems correlated with λ_i. In the experiment, B outputs b based on its measurement settings y and shared sources, followed by A_i outputting a_i based on x_i and shared sources, $i = 1, \cdots, n$. The eavesdropper's measurement on λ_i has input and outcome denoted by z_i and e_i respectively, and it is assumed that the sources $\lambda_1, \cdots, \lambda_n$ are uncorrelated with the eavesdropper. Theorem 4.2.1 provides an upper bound $k(2 - \mathcal{R}_k)$ on the eavesdropper's information, where $\mathcal{R}_k$ depends on all the independent agents $\mathsf{A}_{i_1}, \cdots, \mathsf{A}_{i_k}$ chosen to construct the nonlinear inequality (4.27). Another approach is to use multiple settings of the Bell inequality (Lee and Hoban, 2018).

4.2.3 *Device Independent Model*

For a powerful attacker, all the involved devices may be correlated in a device-independent model. This implies that the semi-device independent models (Lee and Hoban, 2018; Luo, 2020) should be improved without independent assumption of sources. The independence of sources in these secure scenarios cannot be guaranteed by experiment improvements. We propose a full device-independent model on quantum networks in this subsection.

Consider the connected network $\mathcal{N}$ shown in Fig. 4.4a. In a device-independent model, an eavesdropper can hold local systems correlated with some sources to recover private information of legal parties, such as measurement outcomes in QKD. To ensure security, we use the genuinely multipartite nonlocality of quantum networks (Svetlichny, 1987). Let I and $\overline{I}$ be a bipartition of $\{1, \cdots, n\}$. Denote $M_{x_k(I)}$ as the dichotomic observable of A_k for input index $x_k(I)$, where x_k depends on the set I. Denote $M_I = \prod_{k\in I} M_{x_k(I)}$ and $M_{\overline{I}} = \prod_{k\in \overline{I}} M_{x_k(\overline{I})}$. We can prove the following Bell inequality for any biseparable state (Luo, 2022):

$$\sum_{i=1}^{\lfloor \frac{n}{2} \rfloor} \mathcal{L}_i \leq 2^n\sqrt{2} - 4\sqrt{2} + 2, \tag{4.47}$$

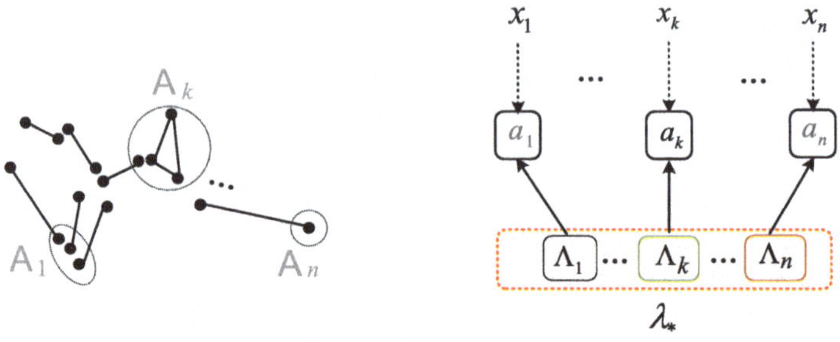

Fig. 4.4 (Color online) **a** A quantum network. **b** A k-independent network in terms of the generalized local hidden variable model. Here, all the independent sources λ_i's may be correlated into one source λ^*

where $\mathcal{L}_i = \sum_{|I|=i} CH_I$ with $CH_I = (M_I + \hat{M}_I)M_{\overline{I}} + (M_I - \hat{M}_I)\hat{M}_{\overline{I}}$, M_S and $\hat{M}_S$ are observables depending on $S = I$ or $\overline{I}$. The largest bound for quantum networks is $2^n\sqrt{2} - 2\sqrt{2}$. Since the inequality (4.47) is linear, it is sufficient to prove the result using product states. This inequality also holds for the following bilocal correlations $P(\vec{a}|\vec{x}) = \sum_I p_I P_I(a_i, i \in I|x_j, j \in I)P_{\overline{I}}(a_i, i \in \overline{I}|x_j, j \in \overline{I})$, where P_I and $P_{\overline{I}}$ may be quantum nonlocal correlations. Using the inequality (4.47), we can obtain the following results.

Theorem 4.2.2 *The predictive power of an eavesdropper to learn the outcomes of legitimate parties satisfies*

$$D_e \leq 2n - \frac{\varpi_q - 2\sqrt{2}(\alpha - \lfloor \alpha \rfloor)}{2\lfloor \alpha \rfloor}, \tag{4.48}$$

where $D_e = D(\prod_{i=1}^n P(e_i|a_i; \vec{x}, z_i), \prod_{i=1}^n P(e_i|z_i))$, $\alpha = \frac{1}{n}(2^n - 1)$, $\varpi_q = \sum_{i=1}^{\lfloor \frac{n}{2} \rfloor} \mathcal{L}_i$, *and* $\mathcal{L}_i$ *is defined in the inequality (4.47) with the estimated quantum correlations.*

Proof We prove the inequality (4.48) for connected quantum networks. Consider a general network $\mathcal{N}$ (Fig. 4.4) with n parties $\mathsf{A}_1, \cdots, \mathsf{A}_n$, who share m sources $\lambda_1, \cdots, \lambda_m$. An eavesdropper may recover the outcome a_i information by local measurements, given that all inputs of $\vec{x}$ are accessible. In secure applications, legal parties do not send out measurement outcomes before detecting the eavesdropper. Therefore, a smart eavesdropper can recover one outcome a_i by prior knowledge of $a_i, \vec{x}, z_i$, i.e., $P(e_i|a_i, \vec{x}, z_i)$. From the no-signaling condition, we get that

$$P(e_i|\vec{x}, z_i) = P(e_i|z_i),\ P(e_i|a_i, \vec{x}, z_i) = P(e_i|a_i, x_i, z_i). \tag{4.49}$$

Denote $D_e := D(\prod_{i=1}^n P(e_i|a_i; \vec{x}, z_i), \prod_{j=1}^n p(e_j|z_j))$ for convenience. From Eq. (4.49), the left side of the inequality (4.48) can be rewritten into

$$D_e \leq \frac{1}{2}\sum_{\vec{e}} p(e_1|a_1;\vec{x},z_1)\left|\prod_{i=2}^{n} p(e_i|a_i;\vec{x},z_i) - \prod_{i=2}^{n} p(e_i|z_i)\right|$$
$$+\frac{1}{2}\sum_{\vec{e}} |p(e_1|a_1;\vec{x},z_1) - p(e_1|z_1)| \prod_{i=2}^{n} p(e_i|z_i). \quad (4.50)$$

Here, the inequality (4.50) is followed from the triangle inequality $|x-y| \leq |x-z| + |z-y|$. Moreover, we have

$$\begin{aligned} D_e \leq & D(p(e_1|a_1;\vec{x},z_1), p(e_1|z_1)) \\ & + D(\prod_{i=2}^{n} p(e_i|a_i;\vec{x},z_i), \prod_{i=2}^{n} p(e_i|z_i)) \quad (4.51) \\ \leq & \sum_{i=1}^{n} D(p(e_i|a_i;\vec{x},z_i), p(e_i|z_i)) \quad (4.52) \\ = & \sum_{i=1}^{n} D(p(e_i|a_i;x_i,x_j,z_i), p(e_i|z_i)) \quad (4.53) \\ \leq & \sum_{i=1}^{n} I_2(P(a_i,a_j|x_i,x_j). \quad (4.54) \end{aligned}$$

The inequality (4.51) is from the equalities $\sum_{e_1} p(e_1|z_1)$=1, and $\sum_{e_j} p(e_j|a_j;\vec{x},z_j)$=1, $j=2,\cdots,n$. The inequality (4.52) is obtained by a similar process of the inequality (4.50) with $n-1$ iterations. Eq. (4.53) is followed from Eq. (4.49), where A_i and A_j shares at least one hidden variable. The inequality (4.54) is obtained by using the inequality of $D(p(e_i|a_i;x_i,x_j,z_i), p(e_i|z_i)) \leq I_2(P(a_i,a_j|x_i,x_j))$ with the chained Bell inequality (Chaves et al., 2015; Barrett et al., 2012; Rabelo et al., 2011).

Consider a quantum network $\mathcal{N}_q$ on which all parties have multiple inputs and binary outcomes. The quantum violation ϖ_{es} of practical quantum correlations is shown as follows

$$\begin{aligned} \varpi_{es} = & \sum_{i=1}^{\lfloor\frac{n}{2}\rfloor}\sum_{I}[\langle(M_I+\hat{M}_I)M_{\overline{I}}\rangle + \langle(M_I-\hat{M}_I)\hat{M}_{\overline{I}}\rangle] \\ \leq & \sum_{i=1}^{\lfloor\frac{n}{2}\rfloor}\sum_{I}[|\langle(M_I+\hat{M}_I)M_{\overline{I}}\rangle| + |\langle(M_I-\hat{M}_I)\hat{M}_{\overline{I}}\rangle|] \\ \leq & \sum_{i=1}^{\lfloor\frac{n}{2}\rfloor}\sum_{j\in I}[|\langle(M_{x_j(I)}+\hat{M}_{x_j(I)})M_{x_{k_j}(\overline{I})}\rangle| \\ & + |\langle(M_{x_j(I)}-\hat{M}_{x_j(I)})\hat{M}_{x_{k_j}(\overline{I})}|] \quad (4.55) \end{aligned}$$

$$=\sum_{i=1}^{\lfloor\frac{n}{2}\rfloor}\sum_{I} CH_{M_jM_{k_j}}. \tag{4.56}$$

Here, the inequality (4.55) has used the inequality of $\langle A\rangle \leq |\langle A\rangle|$. The inequality (4.55) can be proved by separable observables ($M_{x_i} \leq 1$) except for $M_{x_{k_j}(\overline{I})}$ and $\hat{M}_{x_{k_j}(\overline{I})}$. $M_{x_{k_j}}$ and $\hat{M}_{x_{k_j}}$ denote two observables of $\mathsf{A}_{x_{k_j}}$ with $k_j \notin I$ associated with $j \in I$. $CH_{M_jM_{k_j}}$ in Eq. (4.56) is defined by

$$CH_{M_jM_{k_j}} = \langle (M_{x_j(I)} + \hat{M}_{x_j(I)})M_{x_{k_j}(\overline{I})}\rangle + \langle (M_{x_j(I)} - \hat{M}_{x_j(I)})\hat{M}_{x_{k_j}(\overline{I})}\rangle, \tag{4.57}$$

where we have $\langle (M_{x_j(I)} + \hat{M}_{x_j(I)})M_{x_{k_j}(\overline{I})}\rangle \geq 0$ and $\langle (M_{x_j(I)} - \hat{M}_{x_j(I)})\hat{M}_{x_{k_j}(\overline{I})}\rangle \geq 0$ in applications.

We obtain that

$$\varpi_{es} \leq \lfloor\frac{N}{n}\rfloor\sum_{j=1}^{n} CH_{M_jM_{k_j}} + 2\sqrt{2}(\frac{N}{n} - \lfloor\frac{N}{n}\rfloor), \tag{4.58}$$

where $N = 2^n - 1$. From $\langle XY\rangle = 2p(X = Y) - 1$, it follows that

$$I_2(P(a_i, a_j|x_i, x_j)) = 2 - \frac{1}{2}CH_{X_iX_j}. \tag{4.59}$$

Using the inequalities (4.54) and (4.58), we obtain that

$$\begin{aligned} D_e &\leq 2n - \frac{1}{2}\sum_{j=1}^{n} CH_{M_jM_{k_j}} \\ &\leq 2n - \frac{\varpi_{es} - 2\sqrt{2}(n_0 - \lfloor n_0\rfloor)}{2\lfloor n_0\rfloor} \end{aligned} \tag{4.60}$$

with $n_0 = \frac{N}{n}$. This has completed the proof. □

There are two assumptions for eavesdroppers: first, that they have access to all inputs, and second, that they cannot communicate with other parties. If either assumption is violated, then bipartite quantum correlations can be simulated using classical variables. This holds for general quantum networks as well. For example, if a network is k-independent, meaning that there are k parties who do not share entanglement, then the outcome of each party depends on shared classical variables and local measurements. The network can be represented by a directed acyclic graph (DAG) with variables $\lambda_1, \cdots, \lambda_m$. The joint conditional distributions of the outcomes conditional on the inputs are given

$$P_c(\vec{a}|\vec{x}) = \int_{\times_{i=1}^{m}\Omega_i} \prod_{j=1}^{n} P_c(a_j|x_j, \Lambda_j) \prod_{i=1}^{m} d\mu_i(\lambda_i). \tag{4.61}$$

Define a classical correlation as

$$E_c(\vec{x}) = \sum_{\Xi_1} P_c(\vec{a}_k|\vec{x}_k) - \sum_{\Xi_2} P_c(\vec{a}_k|\vec{x}_k), \tag{4.62}$$

where $P_c(\vec{a}_k|\vec{x}_k)$ denote the marginal probability of $\vec{a}_k$ conditional on $\vec{x}_k$, and Ξ_1 denotes all combinations with even number of output -1 while Ξ_2 denotes all combinations with odd number of output -1.

Consider special applications with a joint system in the state ρ, such as multipartite QKD. All the independent parties of $\mathsf{A}_1, \cdots, \mathsf{A}_k$ will finally perform qubit measurements after the others' local measurements and classical communication. The joint conditional probability $P(\vec{a}|\vec{x}_k, x_{k+1}\cdots x_n)$ is defined

$$P_q(\vec{a}|\vec{x}_k, x_{k+1}\cdots x_n) = \mathrm{Tr}(\vec{x}_1 \otimes \cdots \vec{x}_n\rho), \tag{4.63}$$

where $\vec{x}_i$ is a vector under the Pauli basis (X, Y, Z). Similar to Eq. (4.62), define a quantum correlation as

$$E_q(\vec{x}) = \sum_{\Xi_1} P_q(\vec{a}_k|\vec{x}_k) - \sum_{\Xi_2} P_q(\vec{a}_k|\vec{x}_k). \tag{4.64}$$

In the context of a device-independent model shown in Fig. 4.5b, suppose that the eavesdroppers E_i share a common variable Λ_i, replacing all A_i in the local hidden variable model of $\mathcal{N}_q$. The objective is to simulate the quantum correlation (4.64) using classical correlation (4.62). If all measurement inputs are accessible to the eavesdroppers, as shown in Fig. 4.5b, then the quantum correlation (4.64) can be simulated using shared randomness and finite classical communication in a classical simulation. This provides a partial solution to a recent conjecture (Toner and Bacon, 2003; Branciard et al., 2012).

Theorem 4.2.3 *Suppose that eavesdroppers E_i's can access all the measurement inputs $\vec{x}$. They can generate a classical correlation $E_c(\vec{x})$ such that*

$$E_c(\vec{x}) = E_q(\vec{x}) \tag{4.65}$$

by shared randomness and finite classical communication, where $E_c(\vec{x})$ is defined by $E_c(\vec{x}) = \sum_{\Xi_1} P_c(\vec{e}_k|\vec{x}_k) - \sum_{\Xi_2} P_c(\vec{e}_k|\vec{x}_k)$, $P_c(\vec{e}_k|\vec{x}_k)$ denotes the joint distribution of $\mathsf{E}_1, \cdots, \mathsf{E}_k$ conditional on $\vec{x}_k$ and $\vec{e}_k = e_1\cdots e_k$.

The basic idea is to simulate the quantum correlations by combining the tripartite entanglement swapping networks and bipartite entanglement (Luo, 2022; Branciard et al., 2012).

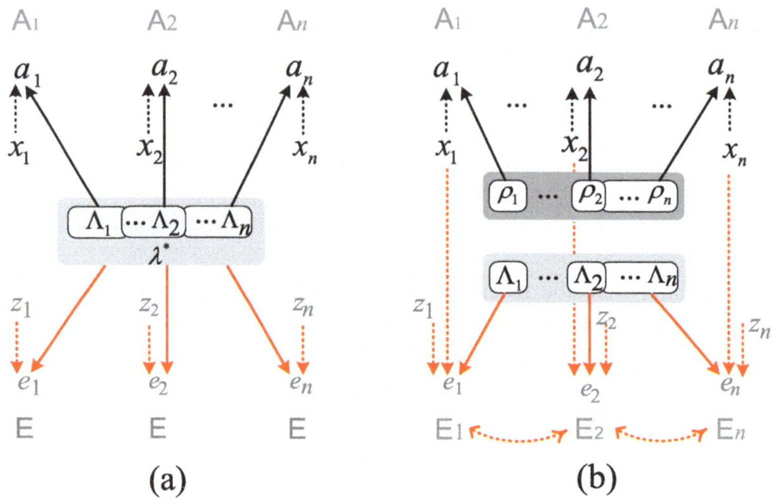

Fig. 4.5 **a** A device-independent model on an n-partite network $\mathcal{N}$ consisting of m sources $\lambda_1, \cdots, \lambda_m$. An eavesdropper E may correlate all the shared sources λ_i's into new source λ^*. **b** A quantum network. A_i shares a system ρ_i consisting of some EPR states and GHZ states with A_j, $i = 1, \cdots, k$, and $j = k+1, \cdots, n$

4.3 Quantum Conference Key Distribution

This section discusses the extension of bipartite QKD to multipartite quantum conference key distribution (QCKD) on networks. In the device-independent model, to achieve conference key agreement, all parties are provided with a share of a multipartite state. Each party has access to a black box that allows for classical interactions and generates an output for each classical input. The objective is to generate a random bit series for all legal parties that is secure against potential eavesdroppers by measuring relevant statistics.

4.3.1 QCKD Based on Parity-CHSH Game

In device-independent quantum conference key distribution, all legal parties can use the measurement outcomes to accomplish specific tasks such as detecting attacks or generating keys. This is independent of the physical implementation of the measurements and shared states. In this subsection, we accomplish the task using the CHSH game (Clauser et al., 1969).

Denote x_i as the i-th measurement input, and its outcome as a_i, $i \in \{1, \ldots, n\}$ and n is the number of parties involved. Define an n-partite parity-CHSH game (Ribeiro and Wehner, 2018) as follows.

Definition 4.3.1 An n-partite parity-CHSH game is a cooperating game where all the parties $\mathsf{A}_1, \cdots, \mathsf{A}_n$ cooperate to win a game respect to the referee. The winning condition is given by the following Boolean function

$$a_1 \oplus a_2 = x_1 \wedge (x_2 \oplus_{i=3}^{n} a_i), \tag{4.66}$$

where $a_1, \ldots, a_n, x_1$, and x_2 are random variables.

For all classical players, the maximal winning probability is $P_{\text{pCHSH}} = \frac{3}{4}$ for any classical strategy. This will be violated by quantum players by sharing some quantum entangled states. It may be written into an equivalent Bell inequality. One example for tripartite case (Holz et al., 2020) is given by

$$\mathcal{B} = \langle O_1^1 O_2^+ O_3 \rangle - \langle O_1^0 O_2^- \rangle \leq 1, \tag{4.67}$$

where $O_i^{\pm} = (O_i^0 \pm O_i^1)/2$, the subscript i refers to the i-th party, and O_i^0 and O_i^1 are observables corresponding to the inputs 0 and 1, respectively.

Based on the n-partite parity-CHSH game, we can construct a QCKD protocol as follows. It is completed by m round experiments (Ribeiro and Wehner, 2018). The main steps are given by repeating the following m times.

S1. The party A_1 randomly picks a Bernoulli random variable T_j taking values in $\{0, 1\}$ such that $\mu := \Pr(T_j = 1) \in [0, 1]$, and sends to other parties A_i's.
S2. If $T_j = 0$, A_i chooses $(X_{1,j}, X_{(2,\ldots,n),j}) = (0, 2, 0, \ldots, 0)$. Otherwise, A_i chooses $X_{i,j} \in \{0, 1\}$.
S3. A_i inputs these values into their local measurement devices and record the outputs as $A_{i,j}$.

Now, all parties communicate their inputs over a classical authenticated channel.

Error Correction-All parties apply an error correction protocol based on the universal hashing (Ribeiro and Wehner, 2018).

Parameter Estimation-If $T_j = 1$ for $j \in \{1, \ldots, m\}$ and all parties win the n-partite parity-CHSH game, they set $C_j = 1$. If they lose the game, they set $C_j = 0$. If $T_j = 0$, they set $C_j = \perp$. Once m rounds are completed, all legal parties can estimate the winning probability for the n-partite parity-CHSH game. If the winning probability is only slightly higher than $\frac{3}{4}$, the protocol is aborted.

Privacy Amplification-All the legal parties apply a privacy amplification protocol to generate the final random bits $L_{\mathsf{A}_1} L_{\mathsf{A}_2} \cdots L_{\mathsf{A}_n}$.

The experiment generates an n-partite correlation $P(\vec{a}_1, \ldots, \vec{a}_m | \vec{x}_1, \ldots, \vec{x}_m)$, where $\vec{a}_i$ denotes the series $\vec{a}_i = a_{i,1} \cdots a_{n,1}$ and $\vec{x}_i = x_{i,1} \cdots x_{n,1}$. The security proof is based on analyzing the correlation $P(\vec{a}_1, \ldots, \vec{a}_m | \vec{x}_1, \ldots, \vec{x}_m)$. In particular, eavesdropping can be detected by estimating the maximum winning probability of n-partite parity-CHSH games similar to Theorem 4.2.2. This provides a straightforward approach for implementing QCKD. In practical applications, all parties may utilize multipartite entanglement such as GHZ states (Greenberger et al., 1989), quantum networks consisting of EPR states (Einstein et al., 1935), or GHZ states with small sizes (Greenberger et al., 1989).

4.3.2 QCKD Based on Network Intrinsic Nonlocality

In this section, we introduce a general form of device-independent QCKD based on network intrinsic nonlocality. In this scenario, the eavesdropper has access to a no-signaling extension of the underlying correlation. This can be reduced to a quantum eavesdropper (Holz et al., 2020; Philip et al., 2021; Horodecki et al., 2022).

Let $n \in \mathbb{Z}^+$, $R \geq 0$, and $\varepsilon \in [0, 1]$. We consider a tripartite example to show the main idea. Let $p(a, b, c|x, y, z)$ be the tripartite correlation of the device shared among Alice, Bob, and Charlie. We define an (n, R, ε) device independent QCKD as follows:

S1. All the legal parties generate the input sequences x^n, y^n, and z^n according to the probability distribution $q(x^n, y^n, z^n)$. For round $j \in \{1, \ldots, n\}$, a_j, b_j, c_j are their outputs correspond to the inputs x_j, y_j, z_j, respectively. The joint distribution for the inputs and outputs is embedded in a classical state as

$$\sigma_{A^nB^nC^n}^{X^nY^nZ^n} = \sum_{\substack{a^n,b^n,c^n \\ x^n,y^n,z^n}} q(x^n, y^n, z^n)P^{(n)}(a^n, b^n, c^n|x^n, y^n, z^n) \times |a^nb^nc^nx^ny^nz^n\rangle\langle a^nb^nc^nx^ny^nz^n|, \tag{4.68}$$

where $P^{(n)}(a^n, b^n, c^n|x^n, y^n, z^n)$ is the n-fold independent and identically distributed extension of $P(a, b, c|x, y, z)$. This state is held by all the legal parties. If an attacker Eve is involved we consider an arbitrary no-signaling extension $\sigma_{A^nB^nC^n}^{X^nY^nZ^nE}$ of $\sigma_{A^nB^nC^n}^{X^nY^nZ^n}$.

S2. Alice has access to a local channel $\mathcal{L}_{A^n\to M_AC_A}$, where C_A is a classical register for public communication, and M_A is a classical memory that is not used for public communication. Eve has a register $\bar{C}_A$ that is a copy of C_A. Similarly, Bob has a channel $\mathcal{L}_{B^n\to M_BC_B}^B$ with a public register C_B and private memory M_B. If the parameters are found to be outside of a predetermined range, the protocol is aborted.

S3. After the parameter estimation and key generation steps, Alice applies the decoding transformation $\mathcal{D}_{M_AC_AC_BC_C\to L_A}^A$ to obtain a final set of random bits L_A. Bob applies the decoding mapping $\mathcal{D}_{M_BC_AC_BC_C\to L_B}^B$ to obtain L_B, while Charlie applies $\mathcal{D}_{M_CC_AC_BC_C\to L_C}$ to obtain L_C. The protocol yields to a state ϱ satisfying

$$\frac{1}{2}\|\rho - \varrho\|_1 \leq \varepsilon, \tag{4.69}$$

where ρ is defined by

$$\rho = 2^{-nR}\sum_{l=1}^{2^{nR}} |l\rangle\langle l|_{L_A} \otimes |l\rangle\langle l|_{L_B} \otimes |l\rangle\langle l|_{L_C} \otimes \varrho_{EX^nY^nZ^n\bar{C}_A\bar{C}_B\bar{C}_C}. \tag{4.70}$$

In what follows, we consider the key rate R.

Definition 4.3.2 A rate R is achievable for a device characterized by a correlation P if there exists an $(n, R-\delta, \varepsilon)$ device independent QCKD for all $\varepsilon \in (0, 1]$, $\delta > 0$, and sufficiently large n.

The device-independent QCKD capacity is denoted by $R^{NS}_{\max}(P)$ and represents the maximum achievable rate (Philip et al., 2021; Horodecki et al., 2022). This definition can also be adapted for a quantum eavesdropper who is restricted by quantum mechanics. The main consideration is the quantum extension $\sigma^{X^nY^nZ^nE}_{A^nB^nC^n}$ of $\sigma^{X^nY^nZ^n}_{A^nB^nC^n}$. We denote the corresponding capacity by $R^{Q}_{\max}(P)$.

The tripartite scheme may be easily extended for multipartite QCKD. This follows a general correlation $P(\vec{a}|\vec{x})$.

To analyze the security we first prove the additivity of tripartite intrinsic nonlocality for quantum networks. This provides indeed an upper bound on the device's independent QCKD capacity. For convenience, we define

$$N(A; B; C) := \sup_{q(x,y,z)} \inf_{\rho^{XYZE}_{ABC}} I(A; B; C|E, X, Y, Z)_\rho. \tag{4.71}$$

Definition 4.3.3 The tripartite intrinsic nonlocality (TINL) of a correlation $P(a, b, c|x, y, z)$ is defined as

$$N(A; B; C)_P := \frac{1}{2} \sup_{q(x,y,z)} \inf_{\rho^{XYZE}_{ABC}} I(A; B; C|E, X, Y, Z)_\rho, \tag{4.72}$$

where $q(x, y, z)$ is a probability distribution for the inputs and $\rho_{ABCXYZE}$ is a no-signalling extension given by

$$\begin{aligned}\rho^{XYZE}_{ABC} = &\textstyle\sum_{a,b,c,x,y,z} q(x, y, z) P(a, b, c|x, y, z) \\ &\times [abcxyz]_{ABCXYZ} \otimes \rho^{abcxyz}_E.\end{aligned} \tag{4.73}$$

The nonlocality will be used to bound the QCKD capacity as follows (Philip et al., 2021; Horodecki et al., 2022).

Theorem 4.3.1 *The device-independent QCKD capacity with no-signaling eavesdropper is bounded by the TINL as*

$$R^{NS}_{\max}(P) \le N(A; B; C)_P. \tag{4.74}$$

Tripartite intrinsic nonlocality and quantum tripartite intrinsic nonlocality vanish for distributions $P(a, b, c|x, y, z)$ having a local hidden variable model, that is,

$$P(a, b, c|x, y, z) = \int_\Omega p(\lambda) p(a|x, \lambda) p(b|y, \lambda) p(c|z, \lambda), \tag{4.75}$$

then $N(A; B; C)_P = N_Q(A; B; C)_P = 0$.

Theorem 4.3.2 *The device-independent QCKD capacity for quantum eavesdropper is bounded by the quantum tripartite intrinsic nonlocality* $N_Q(A; B; C)_P$ *as*

$$R^Q_{\max}(P) \leq N_Q(A; B; C)_P. \tag{4.76}$$

The proof is based on the additivity of tripartite intrinsic nonlocality. In fact, for every tripartite state $\rho^{A_2B_2C_2E}_{A_1B_1C_1}$ we have

$$\begin{aligned} I(A_1A_2; B_1B_2; C_1C_2|E)_\rho = {} & I(A_1; B_1; C_1|EA_2B_2C_2)_\rho + I(A_2; B_2; C_2 \mid E)_\rho \\ & + I(C_1; A_2B_2|EC_2)_\rho + I(A_1; B_2C_2|EA_2)_\rho \\ & + I(B_1; A_2C_2|EB_2)_\rho. \end{aligned} \tag{4.77}$$

Let $P_1(a_1, b_1, c_1|x_1, y_1, z_1)$ and $P_2(a_2, b_2, c_2|x_2, y_2, z_2)$ be no-signaling correlations corresponding to the marginals of the distribution $P(a_1, a_2, b_1, b_2, c_1, c_2|x_1, x_2, y_1, y_2, z_1, z_2)$. Then the intrinsic nonlocality is superadditive, in the sense that

$$N(A_1A_2; B_1B_2; C_1C_2)_P \geq N(A_1; B_1; C_1)_{P_1} + N(A_2; B_2; C_2)_{P_2}, \tag{4.78}$$

where the equality (additive) holds if $P = P_1 \cdot P_2$. Similarly, the superadditive of quantum intrinsic nonlocality is given by

$$N_Q(A_1A_2; B_1B_2; C_1C_2)_P \geq N_Q(A_1; B_1; C_1)_{P_1} + N_Q(A_2; B_2; C_2)_{P_2}, \tag{4.79}$$

where the additive holds if $P = P_1 \cdot P_2$.

Similar proof holds by the additivity of tripartite intrinsic nonlocality (Philip et al., 2021). It may be extended for multipartite intrinsic nonlocality of non-signaling distribution $P(\vec{a}|\vec{x})$ as

$$N(A_1; \cdots; A_M)_P := \frac{1}{M-1} \sup_{q(\vec{x})} \inf_{\rho^{\vec{X}E}_{\vec{A}}} I(A_1; \cdots; A_M|E\vec{X})_\rho, \tag{4.80}$$

where $q(\vec{x})$ is a probability distribution for the inputs of all legal parties and $\rho^{\vec{X}E}_{\vec{A}}$ is a non-signaling extension, of the state shared by all legal parties, given by $\rho_{\vec{A}\vec{X}E} = \sum_{\vec{a},\vec{x}} q(\vec{x})P(\vec{a}|\vec{x})\ [\vec{a}\vec{x}]_{\vec{A}\vec{X}} \otimes \rho^{\vec{a}\vec{x}}_E$. For quantum distribution, we can define

$$N_Q(A_1; \cdots; A_M)_P := \frac{1}{M-1} \sup_{q(\vec{x})} \inf_{\rho^{\vec{X}E}_{\vec{A}}} I(A_1; \cdots; A_M|E\vec{X})_\rho. \tag{4.81}$$

The nonlocality is used to bound the QCKD capacity.

Theorem 4.3.3 *The device-independent QCKD capacity for any no-signaling eavesdropper is bounded by* $N(A_1; \cdots; A_M)_P$ *as*

$$R_{\max}^{NS}(P) \leq N(A_1; \cdots; A_M)_P. \tag{4.82}$$

The main proof is that the capacity of DI-QKD to securely distribute a secret key against a no-signaling eavesdropper is bounded by the multipartite quantum intrinsic nonlocality $N_Q(A_1; \cdots; A_M)_P$, as expressed by the inequality $R_{\max}^{Q}(P) \leq N_Q(A_1; \cdots; A_M)_P$, where $R_{\max}^{Q}(P)$ is the maximum achievable rate of the secret key that can be generated per entangled state by a protocol that uses projective measurements and classical post-processing, and P is the probability of detecting an eavesdropper. The proof follows from the fact that any protocol that generates a secret key with a rate higher than $R_{\max}^{Q}(P)$ must either violate the no-signaling assumption or the multipartite Bell inequality, which would allow Eve to potentially gain information about the secret key without being detected.

Theorem 4.3.4 *The device-independent QCKD capacity for any no-signaling eavesdropper is bounded by* $N_Q(A_1; \cdots; A_M)_P$ *as*

$$R_{\max}^{Q}(P) \leq N_Q(A_1; \cdots; A_M)_P. \tag{4.83}$$

The violation of a Bell inequality can be characterized by the multipartite quantum intrinsic nonlocality, denoted as $N_Q(A_1; \cdots; A_M)_P$. This quantity measures the maximum possible violation of a Bell inequality that involves measurements on the M subsystems, under a certain protocol P. We can prove that $R_{\max}^{Q}(P) \leq N_Q(A_1; \cdots; A_M)_P$ by leveraging the fact that the Bell violation can be quantified by the intrinsic nonlocality. By doing so, we establish a fundamental connection between the security of QCKD and the nonlocal properties of multipartite entangled states, thus providing insights into the underlying physics of quantum communication protocols.

Example 4.3.1 Consider a noisy GHZ state (Greenberger et al., 1989) as

$$\rho_{ABC} = (1-p)|GHZ\rangle\langle GHZ|_{ABC} + \frac{v}{8}\mathbb{1}, \tag{4.84}$$

where $|GHZ\rangle = \frac{1}{\sqrt{2}}(|000\rangle + |111\rangle)$. Alice takes two Pauli measurements Z and X. Bob performs three measurements $(Z - X)/\sqrt{2}$, $(Z + X)/\sqrt{2}$, and σ_z. Charlie performs two measurements Z and X. This leads to a quantum correlation $P(a, b, c|x, y, z)$.

Using the Bell inequality corresponding to the parity-CHSH game (Ribeiro and Wehner, 2018), the parity-CHSH violation $\mathcal{S}$ is given by

$$\mathcal{S} = \sqrt{2}(1-p). \tag{4.85}$$

This means ρ_{ABC} produces a local correlation when the parity-CHSH violation is less than or equal to one or, equivalently when $v \geq 1 - 1/\sqrt{2}$. For local correlations, there is no quantum tripartite intrinsic nonlocality. It means that

$$N_Q(A;\, B;\, C)_{P_{S^v}} \le (1-\alpha(\varepsilon))N_Q(A;\, B;\, C)_{P_{S^\varepsilon}}. \tag{4.86}$$

Consider a trivial extension of $P_{S^v}(a, b, c|x, y, z)$. We obtain

$$N_Q(A;\, B;\, C)_{P_{S^v}} \le \min_{0\le\varepsilon\le v} \sup_{q(x,y,z)} (1-\alpha(\varepsilon))I(A;\, B;\, C)_{P_{S^\varepsilon}}. \tag{4.87}$$

The lower bound is calculated by the winning probability of the parity-CHSH game, given by $v_{win} = \frac{1}{2} + \frac{(1-v)}{2\sqrt{2}}$.

A straightforward approach utilizing the inherent nonlocality can be utilized to assess the security of QCKD in various quantum networks. To extend this method to other networks, it is important to identify a suitable multipartite Bell inequality, such as the parity-CHSH game. This necessitates an exploration of genuinely multipartite nonlocality in the context of multipartite entangled states, which will be elaborated on in the upcoming section.

4.3.3 QCKD on Tripartite Cyclic Networks

In this subsection, we present a device-independent QCKD with three parties using quantum networks. The main idea is based on the Bell inequality (4.47).

Consider a tripartite network consisting of A_1, A_2 and A_3 as shown in Fig. 4.6a. The inequality (4.47) implies a special inequality for tripartite QCKD as

$$\begin{aligned} &A_1B_1C_1 + A_2B_1C_1 + A_1B_2C_2 - A_2B_2C_2 + A_3B_3C_1 \\ &+A_3B_4C_1 + A_4B_3C_2 - A_4B_4C_2 + A_3B_1C_3 + A_3B_1C_4 \\ &+A_4B_2C_3 - A_4B_2C_4 \le 4\sqrt{2} + 2, \end{aligned} \tag{4.88}$$

which holds for all biseparable states. A_i, B_j, and C_k are dichotomic observable of A_1, A_2, and A_3, respectively, $i, k = 1, \cdots, 4$; $j = 1, \cdots, 6$.

Let ρ_{ABC} be the ideal total system of three parties. From the inequality (4.88), each party has four inputs $x, y, z \in \{1, \cdots, 4\}$ and two outputs $a, b, c \in \{0, 1\}$ for each input. The QCKD involves N runs, with each run consisting of a triangle network represented by $\rho^{\otimes N}$. In each round, A_1 (or A_2, A_3) measures A_{x_i} (or A_{y_i}, A_{z_i}) on local particles. The resulting outputs are $a_i, b_i, c_i \in \{0, 1\}$. The parties may select a subset B of size γN for attack testing, which will be completed by public discussions. The goal is to determine the fraction of pairs in B that satisfy inequality (4.88). If the fraction is less than a predefined threshold, the protocol is aborted. Otherwise, the parties publicly reveal their input.

Denote C as all the check rounds with $(x_i, y_i, z_i) \in$ { (1,1,1),(2,1,1), (1,2,2), (2,2,2), (3,3,1), (3,4,1), (4,3,2), (4,4,2), (3,1,3), (3,1,4), (4,2,3),(4,2,4) }. They perform information reconciliation on the rounds in C to obtain the raw key k_a, k_b, k_c.

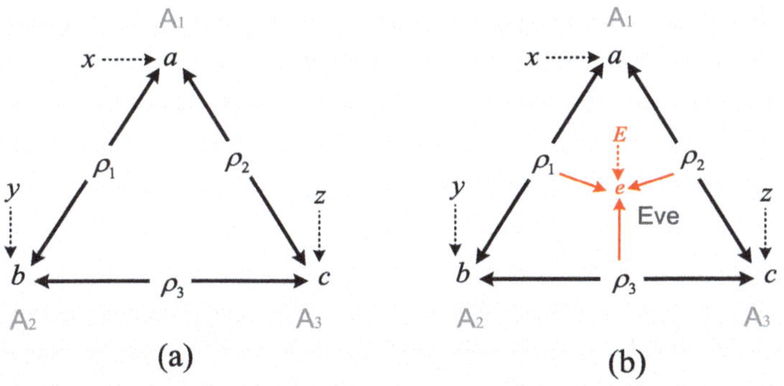

Fig. 4.6 Device independent tripartite QCKD. **a** Tripartite QCKD is based on a triangle network. a, b, and c (x, y, and z) denote the output (input) of A_1, A_2 and A_3, respectively. **b** Device-independent tripartite QCKD. The output e of eavesdropper E depends on its input and the shared systems

These strings are further used to obtain the final keys by error reconciliation and privacy amplification.

In a device-independent QCKD, the information available to a quantum eavesdropper is denoted by E correlated with ρ_{ABC} as shown in Fig. 4.6b. The total state is then extended as ρ_{ABCE} which should satisfy $\mathrm{Tr}_E\rho_{ABCE} = \rho_{ABC}$. Suppose that Eve aims to learn the information of a. For one copy resource, the predictability for Eve is quantified by the guessing probability of a as $P_{gs}(a) = \max_a P(a|x)$. This follows the asymptotic secret-key rate (Masanes et al., 2011; Koenig et al., 2009; Rabelo et al., 2011) as

$$r := -\frac{1}{s}\log_2 P_{gs}(\vec{a}|E) \tag{4.89}$$

In general, we get the following result for the guessing probability.

Theorem 4.3.5 *The guessing probability is bounded by*

$$P_{gs}(a) \leq \frac{1}{2} + \frac{1}{12}\sqrt{72 - \varpi_{es}^2}, \tag{4.90}$$

where ϖ_{es} denotes the quantum violation of the inequality (4.88). Moreover, the asymptotic secret-key rate satisfies

$$r \geq -\log_2(\frac{1}{2} + \frac{1}{12}\sqrt{72 - \varpi_{es}^2}) - H(a|b,c), \tag{4.91}$$

where $P_{gs}(\vec{a}|E)$ denotes the optimal guessing probability. The first term in Eq. (4.91) represents the knowledge of Eve for the privacy amplification, and $H(a|b,c)$ quantifies the information required by error correction.

Proof From the inequality (4.88), the expected quantum violation is given by

$$\begin{aligned}\varpi_{es} = & \sum_{\substack{a_i,b_i,c_i\\ x_i,y_i,z_i\in S_{in}}} [(-1)^{a_1+b_1+c_1+x_1y_1z_1} P(a_1b_1c_1|x_1y_1z_1) \\ & +(-1)^{a_2+c_2+b_2+x_2y_2z_2} P(a_2b_2c_2|x_2y_2z_2) \\ & +(-1)^{a_3+b_3+c_3+x_3y_3z_3} P(a_3b_3c_3|x_3y_3z_3)],\end{aligned} \tag{4.92}$$

where $P(a_ib_ic_i|x_iy_iz_i) = \mathrm{tr}[(A_{x_i}^{a_i}B_{y_i}^{b_i}C_{z_i}^{c_i})\rho]$, $A_{x_i}^{a_i}$, $B_{y_i}^{b_i}$ and $C_{z_i}^{c_i}$ denote the POVM operators of A_1, A_2 and A_3, respectively, and S_{in} denotes the input set satisfying the conditions $x_1 = y_1$, $x_2 = z_2$ and $y_3 = z_3$. From the symmetry of x_i, y_i, and z_i, it follows that

$$\varpi_{es} = 3 \sum_{\substack{a_1,b_1,c_1\\ x_1,y_1,z_1\in S_{in}}} (-1)^{a_1+b_1+c_1+x_1y_1z_1} P(a_1b_1c_1|x_1y_1z_1) \tag{4.93}$$

In what follows, we only consider the case of $a_1 = 0$ and $x_1 = 0$ to prove the inequality (4.90) which is inspired by bipartite QKD scheme (Masanes et al., 2011; Rabelo et al., 2011). Note that the right side of the inequality (4.90) is a concave function of ϖ_{es}. Denote

$$W = \sum_{\substack{a_1,b_1,c_1\\ x_1,y_1,z_1\in S_{in}}} (-1)^{a_1+b_1+c_1+x_1y_1z_1} A_{x_1}^{a_1} B_{y_1}^{b_1} C_{z_1}^{c_1} \tag{4.94}$$

as a tripartite quantum operator. By the semidefinite programming relaxation of the bipartite scheme (Masanes et al., 2011; Rabelo et al., 2011), the inequality (4.90) is equivalent to the following inequality

$$A_0^0 \le \frac{1}{2} + \frac{6}{\sqrt{72-\varpi^2}} - \frac{\varpi}{12\sqrt{72-\varpi^2}} W. \tag{4.95}$$

This can be then rewritten into

$$\frac{1}{2} + \frac{1}{2} A_0 \le \frac{1}{2} + \frac{6}{\sqrt{72-\varpi^2}} - \frac{\varpi}{12\sqrt{72-\varpi^2}} W, \tag{4.96}$$

where $A_0 = A_0^0 - A_0^1$ is Hermitian operator. Hence, it is sufficient to prove the inequality (4.96) for the guessing probability (4.90).

Define four operators L_i's as follows

$$L_1 = -\frac{\sqrt{\gamma}}{2} A_1 + \frac{\varpi\sqrt{\gamma}}{24}(B_0C_0 - B_1C_1) + \frac{\varpi}{48\sqrt{\gamma}}(A_1B_0C_0 + A_1B_1C_1), \tag{4.97}$$

$$L_2 = \frac{\sqrt{\gamma}}{8} - \frac{\varpi}{24\sqrt{\gamma}}A_0 + \frac{1}{8\sqrt{\gamma}}(B_0C_0 + B_1C_1) - \frac{\sqrt{\gamma}}{2}(A_0B_0C_0 + A_0B_1C_1)$$
$$+\frac{(36-\varpi^2)\sqrt{\gamma}}{144}(A_1B_0C_0 - A_1B_1C_1), \tag{4.98}$$
$$L_3 = \frac{\sqrt{\gamma}}{4}(A_0B_0C_0 - A_0B_1C_1) - \frac{1}{8\sqrt{\gamma}}(B_0C_0 - B_1C_1)$$
$$-\frac{(36-\varpi^2)\sqrt{\gamma}}{144}(A_1B_0C_0 + A_1B_1C_1), \tag{4.99}$$

and

$$L_4 = \frac{1}{4\sqrt{\gamma}} - \frac{(36-\varpi^2)\sqrt{\gamma}}{72}A_0 - \frac{\sqrt{\gamma}\varpi}{24}(B_0C_0 + B_1C_1)$$
$$-\frac{\varpi}{48\sqrt{\gamma}}(A_1B_0C_0 - A_1B_1C_1), \tag{4.100}$$

where $\gamma = \frac{3}{\sqrt{72-\varpi^2}}$. We get that

$$\sum_{i=1}^{4} L_i^\dagger L_i = \frac{6}{\sqrt{72-\varpi^2}} - \frac{1}{2}A_0 + \frac{\varpi}{12\sqrt{72-\varpi^2}}W \tag{4.101}$$
$$\geq 0 \tag{4.102}$$

for $\varpi < 6\sqrt{2}$. Here, Eq. (4.101) is from the equalities $A_x^2 = B_y^2 = C_z^2 = \mathbb{1}$, and $[A_x, B_y] = [A_x, C_z] = [B_y, C_z] = 0$. The inequality (4.102) has used the fact that the right side of Eq. (4.101) is positive semidefinite. The inequality (4.96) is then obtained from the concavity of its right side for ϖ with $\varpi \in [4\sqrt{2}+2, 6\sqrt{2}]$.

In what follows, we prove the asymptotic secret-key rate. The length of the real secret k is lower bounded by $H_{min}(A|E) - s_p$ up to the order $\sqrt{s}$, where $H_{min}(A|E)$ denotes the min-entropy of A conditional on Eve's information, $s_p \approx sH(A|B)$ is used for error correction, and the conditional entropy is given by $H(A|B) = -\sum_{a,b} p(a,b)\log_2 p(a|b)$. This means we can get the asymptotic secret-key rate R by evaluating $H_{min}(A|E)$. After the measurement of A_1, the joint state of A_1 and Eve is represented by

$$\rho_{AE} = \sum_{\vec{a}} P(\vec{a}|\vec{x}_r)|\vec{a}\rangle\langle\vec{a}| \otimes \rho_E^{\vec{a}}, \tag{4.103}$$

where $\rho_E^{\vec{a}}$ denotes the reduced state of Eve conditional on the output $\vec{a}$.

Let $H_{min}(\vec{a}|E)$ be the min-entropy (Masanes et al., 2011) of the raw key $\vec{a} = a_1 \cdots a_s$ conditional on Eve's information related to ρ_{AE}. Denote $P(\vec{a}|\vec{x}_r, ez)$ as the probability of $\vec{a}$ conditional on Eve's information. The expected optimal guessing probability for Eve is shown as

$$P_{gs}(\vec{a}|E) = \max_{z} \sum_{e} P(e|z) \max_{\vec{a}} P(\vec{a}|\vec{x}_r, ez). \tag{4.104}$$

Denote $\rho_{ABC|ez}$ as the initial state when Eve measures z and obtains e, where $\rho_{ABC|ez}$ satisfies $\rho_{ABC} = \sum_e p(e|z)\rho_{ABC|ez}$. Denote $A_{\vec{x}_r}^{\vec{a}} = \prod_{i=1}^{s} A_{x_r}^{a_i}$ as POVMs for all outputs $\vec{a}$ conditional on A_1's inputs $\vec{x}_r$. We get that

$$P(\vec{a}|\vec{x}_r, ez) = \mathrm{Tr}[\rho_{ABC|ez}^{\otimes s} A_{\vec{x}_r}^{\vec{a}}], \tag{4.105}$$

where $\vec{x}_r$ denote all the parties' inputs with x_r. By the semidefinite programming relaxation (Masanes et al., 2011; Rabelo et al., 2011), it follows that

$$A_{\vec{x}_r}^{\vec{a}} \leq A(\varpi) = \prod_{i=1}^{s} [\alpha(\varpi)\mathbb{1} + \beta(\varpi)O_i], \tag{4.106}$$

where $\alpha(\varpi) = \frac{1}{2} + \frac{1}{12}\sqrt{72 - \varpi^2} + \frac{\varpi^2}{12\sqrt{72-\varpi^2}}$, $\beta(\varpi) = -\frac{\varpi}{12\sqrt{72-\varpi^2}}$ and O_i is defined by $O_i = \sum_{abc,xyz} g_{xyz}^{abc} A_{x_i}^{a_i} B_{y_i}^{b_i} C_{z_i}^{c_i}$. From the inequality (4.106), we get that

$$\begin{aligned} P_{gs}(\vec{a}|E) &= \max_{z} \sum_{e} p(e|z) \max_{\vec{a}} \mathrm{Tr}[\rho_{ABC|ez}^{\otimes s} A_{\vec{x}_r}^{\vec{a}}] \\ &\leq \max_{z} \sum_{e} p(e|z) \min_{\varpi} \mathrm{Tr}[\rho_{ABC|ez}^{\otimes s} A(\varpi)] \\ &\leq \min_{\varpi}(\alpha(\varpi) + \beta(\varpi)\varpi_{es} + s_{es}^{-1/4})^s && (4.107) \\ &\leq (\frac{1}{2} + \frac{1}{12}\sqrt{72 - \varpi_{es}^2} + s_{es}^{-1/4})^s. && (4.108) \end{aligned}$$

The inequality (4.107) is followed from the result (Masanes et al., 2011; Rabelo et al., 2011) by combining the systems B and C, where s_{es} denotes the number of distilled keys. The inequality (4.108) has used the inequality

$$\min_{\varpi}\{\alpha(\varpi) + \beta(\varpi)\varpi_{es}\} \leq \frac{1}{2} + \frac{1}{12}\sqrt{72 - \varpi_{es}^2}, \tag{4.109}$$

and the concavity. This implies from the inequality (4.108) that

$$\begin{aligned} r &= -\frac{\log_2 P_{gs}(\vec{a}|E)}{s} \\ &\geq -\log_2(\frac{1}{2} + \frac{1}{12}\sqrt{72 - \varpi_{es}^2}) - H(a|b, c). && (4.110) \end{aligned}$$

We have completed the proof. □

The guessing probability and asymptotic secret-key rate for QCKD on triangle networks have been demonstrated by Theorem 4.3.5, which can be readily extended to multipartite QCKD on various quantum networks. The focus here pertains to quantum attackers, as suggested by the inequality (4.47) applicable to biseparable states (Svetlichny, 1987). A potential enhancement involves the generalization of this approach for no-signaling eavesdroppers by examining the Bell inequality in the context of biseparable correlations under the no-signaling principle.

4.3.4 QCKD on Star Networks

In this subsection, we propose a QCKD on star networks, where each party is connected with one party in the center. The main result is based on the result in (Devetak and Winter, 2005).

For a device-independent QCKD, we consider a protocol with n parties: B who possesses one device with two inputs $\{0, 1\}$, and $\mathsf{A}_1, \ldots, \mathsf{A}_{n-1}$ who each possess a device with three inputs $\{0, 1, 2\}$, every input with two outputs.

Let P_0^i, P_1^i be two Pauli matrices, $\forall i \in [n]$. We firstly prove a Bell inequality inspired by the CHSH inequality (Clauser et al., 1969) or the Mermin-Ardehali-Belinksii-Klyshko (MABK) inequality (Mermin, 1990; Ardehali, 1992; Belinskii and Klyshko, 1993). This is defined by recursion as

$$M_2 := \frac{1}{2} F_{\text{CHSH}}(P_0^1, P_1^1, P_0^2, P_1^2), \tag{4.111}$$

$$M_n := \frac{F_{\text{CHSH}}(M_{n-1}, \overline{M_{n-1}}, P_0^n, P_1^N)}{2}, \tag{4.112}$$

where $F_{\text{CHSH}}(P_0^1, P_1^1, P_0^2, P_1^2)$ denotes the CHSH operator (Clauser et al., 1969), and $\overline{M}_l$ is the operator obtained from M_l by replacing P_k^i by P^i, $\forall i \in [l]$, $\forall k \in \{0, 1\}$.

The main steps of the protocol are given as follows:

S1. For every round $i \in [n]$

(a) B produces and distributes an n-partite star quantum network in the total state $\rho_{BA_0 \ldots A_{n-1};i}$, shared among herself and A_i's;
(b) B randomly chooses T_i satisfying $\Pr(T_i = 1) = \mu$, and publicly communicates with others;
(c) If $T_i = 0$, B and A_i's choose $X_i = 0$, $Y_{j,i} = 2$ with any j, and if $T_i = 1$ they all choose X_i, $Y_{j,i} \in \{0, 1\}$ uniformly at random;
(d) B and A_i's record the outputs as B_i', $A_{j,i}'$, $j = 1, \cdots, n-1$.

S2. They communicate publicly the list of bases X_1^N, Y_j^N what they used, $j = 1, \cdots, n-1$.

After obtaining a raw key, B and A_i's will complete the error correction as follows. B sends the classical information O_A to A_i's. During the test rounds $T_i = 1$, A_k sends the classical information $O_{(k)}$, where $k = 1, \cdots, n-1$. If at least one B_k aborts, the protocol is aborted. Otherwise, the raw keys $\tilde{K}_B = B'_i$ and $\tilde{K}_{A_j} = A'_j$, where $j = 1, \cdots, n-1$, are obtained.

Now, they will complete the parameter estimation. For $T_i = 1$, B get $C_i = 1$ by B'_i and his guess on $A'_{j,i}$ if they can win the n-partite MABK game. Otherwise, he sets $C_i = 0$ when they lost. For $T_i = 0$, he gets $C_i = \perp$. He aborts the protocol if $\sum_i C_i < \delta \sum_i T_i$, where $\delta \in (p_{\min}, p_{\max})$.

This final step is privacy amplification. B and the A_i use a privacy amplification scheme to generate final keys K_B, K_{A_j}.

In the protocol, B and A_i randomly choose some rounds to test the violation of the MABK inequality (Mermin, 1990; Ardehali, 1992; Belinskii and Klyshko, 1993). They abort the protocol if the frequency of rounds for winning the MABK game does not reach a threshold δ. As for QKD (Renner, 2008; Portmann and Renner, 2014) the security of QCKD (Epping et al., 2017) will be split into two terms: correctness and secrecy.

Definition 4.3.4 A device-independent QCKD is ϵ_{corr}-correct if all keys, K_B, K_{A_1}, $\ldots$, $K_{A_{n-1}}$ are all identical with probability at least $1 - \epsilon_{corr}$.

Definition 4.3.5 A device-independent QCKD is ϵ_{sec}-secret if the key K_B is ϵ_{sec}-close to a key that Eve is ignorant about. This is formalized as

$$p_{\hat{\Omega}} \left\| \rho_{K_B E|\hat{\Omega}} - \frac{1}{2^l} \mathbb{1}_B \otimes \rho_{E|\hat{\Omega}} \right\|_{\mathrm{tr}} \leqslant \epsilon_{\mathrm{sec}}, \tag{4.113}$$

where $\|\cdot\|_{tr}$ denotes the trace norm, l is the key length, $\hat{\Omega}$ is the event of the protocol not aborting, and $p_{\hat{\Omega}}$ is the probability for $\hat{\Omega}$.

If a protocol is both ϵ_{corr}-correct and ϵ_{sec}-secret then it is ϵ^s-secure for any $\epsilon^s \geqslant \epsilon_{corr} + \epsilon_{sec}$.

In general, an ϵ^s-secure QCKD protocol means that for any possible physical implementation of the protocol, either it aborts with a probability larger than $1 - \epsilon^s$ or ϵ^s-secure in Definitions 4.3.4 and 4.3.5.

In what follows, we estimate the size of a secret key. By combining Definitions 4.3.4 and 4.3.5 and the leftover hashing lemma (Renner, 2008), an ϵ_{sec}-secret key of size is given by

$$l = H_{\min}^{\epsilon}(B'|E) - 2\log_2 \frac{1}{\epsilon_{\mathrm{PA}}} \tag{4.114}$$

for $\epsilon_{sec} > 2\epsilon + \epsilon_{PA}$. Here, $H_{\min}^{\epsilon}(B|E)_\rho$ denotes the conditional smooth min-entropy defined as

$$H_{\min}^{\epsilon}(B|E)_\rho := \sup_{\sigma \in \mathcal{B}(\rho)} H_{\min}(B|E)_\sigma. \tag{4.115}$$

The supremum is taken over all POVMs ϵ-close to ρ in the purified distance. For a classical register B and a quantum state σ, $H_{\min}(B|E)_\sigma$ is the maximum probability with which Eve guesses B. In general, we have

$$H_{\min}(B|E)_\sigma := \sup_{\tau_E} \sup \left\{\lambda : \sigma_{BE} \leqslant 2^{-\lambda} \mathbb{1}_B \otimes \tau_E\right\}, \tag{4.116}$$

where the second supremum is taken over all the states τ_E.

Definitions 4.3.4 and 4.3.5 provide a criterion for the composable security of the device-dependent QKD (Portmann and Renner, 2014). However, it is unknown whether such a criterion is enough for composable security in the device-dependent case. We choose these definitions as the security criteria for device-independent QCKD by the main idea in (Arnon-Friedman et al., 2019).

The length of a secure key is obtained as follows. Denote ϵ'_{EC} as an error parameter of the error correction, ϵ_{PA} as the privacy amplification error probability, and ϵ_{EA} as a security parameter.

Theorem 4.3.6 *The present protocol generates an ϵ^s-secure key with $\epsilon^s \leqslant \epsilon_{PA} + 2(n-1)\epsilon'_{EC} + 2\epsilon + \epsilon_{EA}$ of length:*

$$\begin{aligned} l = & \max_{p_{\min} \leqslant \delta_{opt} \leqslant p_{\max}} \left\{(f(\delta, \delta_{opt}) - \mu)n - \tilde{v}\sqrt{n}\right\} \\ & +3\log_2(1 - \sqrt{1 - \frac{\epsilon^2}{16}}) + 2\log \epsilon_{PA} \\ & - \operatorname{leak}_{EC}(O_B) - \sum_{k=1}^{N-1} \operatorname{leak}_{EC}(O_{(k)}), \end{aligned} \tag{4.117}$$

where ϵ is a smoothing parameter, δ is a secure threshold, $f(\cdot, \delta_{opt})$ is the tangent of the function $\hat{f}(\cdot)$ at the point δ_{opt}, and $\tilde{v}$ is given by

$$\begin{aligned} \tilde{v} = & 2(\log 13 + (\frac{\hat{f}'(p_{opt})}{\mu} + 1))\sqrt{1 - 2\log(\epsilon \cdot \epsilon_{EA})} \\ & +2\log 7\sqrt{-\log(\epsilon_{EA}^2(1 - \sqrt{1 - \epsilon^2/16}))}. \end{aligned} \tag{4.118}$$

The security proof is completed in two main steps. The first is to split the overall entropy of B's string into a sum of the entropy produced on each round of the protocol by the entropy accumulation theorem (Dupuis et al., 2020). The other is to bound the entropy produced in one round by the violation of the n-partite MABK inequality (Mermin, 1990; Ardehali, 1992; Belinskii and Klyshko, 1993).

Theorem 4.3.7 *(MABK-CHSH correspondence). An n-partite MABK inequality with a violation $\mathcal{S}_n > 2^{(n-2)/2}$ can be regarded as the CHSH inequality for a bipartite Bell test with a violation of $\mathcal{S}_2 = \mathcal{S}_n/2^{(n-2)/2}$, where $n-1$ parties are combined into one.*

A device-independent QCKD may be beneficial for certain regimes of noise. Consider the special case of each qubit experiencing independent bit errors at QBER Q. By some numerical evaluations we can show that there are advantageous to use device-independent QCKD instead of $n-1$ times of independent bipartite QKDs (Arnon-Friedman et al., 2019) for the low noise regime. Moreover, for independent bipartite QKDs, each pair shares an independent random bit series. In this case, all the parties will make use of a one-time pad to share one common bit series. This is different from the device-independent QCKD. Of course, the device-independent QCKD is more difficult in experiments. In general, which method should be applied depends on the cost and noisiness of quantum networks over pairwise EPR pairs.

In what follows, we get a lower bound on the length of a secret key. It generally requires estimating the leakages due to error correction in an honest implementation. Let Q be QBER between B and one of the A_i's. By the depolarizing noise, we get

$$\text{leak}_{EC}(O_A) \leqslant (1-\mu)h(Q)n + \mu n + O(\sqrt{n}), \tag{4.119}$$

$$\text{leak}_{EC}(O_{(k)}) \leqslant \mu n + O(\sqrt{n}). \tag{4.120}$$

This follows the asymptotic key rate (Arnon-Friedman et al., 2019) as

$$r_{\text{QCKD}}^{\infty} = 1 - h(\frac{1}{2} + \frac{1}{2}\sqrt{2(1-2Q)^n - 1}) - h(Q) \tag{4.121}$$

when $\mu \to 0$ with $\mu\sqrt{n} \to \infty$.

Instead, suppose that the party B has performed $n-1$ device-independent QKDs to establish a common key with all the A_i's. It follows an asymptotic key rate as (Arnon-Friedman et al., 2019):

$$r_{\text{bQKD}}^{\infty} = \frac{1}{n-1} - \frac{1}{n-1}h(\frac{1}{2} + \frac{1}{2}\sqrt{2(1-2Q)^2 - 1}) - \frac{1}{n-1}h(Q). \tag{4.122}$$

There is a factor of $\frac{1}{n-1}$ because the party B needs n rounds for each of $\mathsf{A}_1, \cdots, \mathsf{A}_n$. While both quantum networks will be generated similarly in a local manner. The present asymptotic key rates (4.121) and (4.122) provide a relatively fair comparison of QCKD and independent QKDs.

4.4 QCKD on Multiplex Quantum Channels

In this section, we extend the QCKD to multiplex quantum channels (MQC), where MQCs connect some parties with one of three roles, i.e., only sender, only receiver, or both sender and receiver. This provides a potential realization of Quantum Internet (Dowling and Milburn, 2003; Wehner et al., 2018).

The standard bipartite QKD may be regarded as a point-to-point (P2P) protocol. The second case is to send messages from multiple parties B_i's to multiple parties C_i's by using an interference channel. It may be generally described as $c_i = f_i(b_1, b_2)$, where b_i denotes the input of B_i and c_i denotes the received message of C_i, and f_i denotes the interference channel. The third one is a multiple access communication where all the messages from B_i's will be sent to one party C by a multiple access channel g, which may be described as $c = g(b_1, b_2)$. The last one is inspired by to broadcast in which each receiver C_i will receive some message (which may be different) from one party B by using the interference channel f_i.

Instead of the special cases, we consider a general case for different inputs and different outputs with one-way or two-way channels. Denote all the senders as $\mathsf{B}_1, \cdots, \mathsf{B}_k$, and their inputs as $B_1, \cdots, B_k$; receivers as $\mathsf{C}_1, \cdots, \mathsf{C}_m$, and their inputs as $C_1, \cdots, C_m$; and parties that are both senders and receivers as $\mathsf{A}_1, \cdots, \mathsf{A}_n$, with respective inputs $A'_1, \cdots, A'_n$ and outputs $A_1, \ldots, A_n$.

Definition 4.4.1 A MQC is defined as $\mathcal{N}_{\vec{A}'\vec{B}\to\vec{A}\vec{C}}$, where $\vec{A} = A_1 \cdots A_n$, $\vec{B} = B_1 \cdots B_k$ and $\vec{C} = C_1 \cdots C_m$. $\vec{A}$: denotes the n-partite partition of $A_1, \cdots, A_n$ and $\vec{A} : \vec{B}$ stands for the partition of $A_1, \ldots, A_n, B_1, \cdots, B_k$. Similar definitions hold for others.

In what follows, we extend the distributed classical states or random bit series to general states. We consider multipartite private states that are the most general states, which can imply the QCKDs (i.e., without distillation) by local measurements (Augusiak and Horodecki, 2009).

Definition 4.4.2 An n-partite state is the private state if it may be represented by

$$\varrho_{\vec{S},\vec{K}} := U_{\vec{S},\vec{K}}(\rho_{\vec{K}} \otimes \omega_{\vec{S}})U^\dagger_{\vec{S},\vec{K}}, \tag{4.123}$$

where $\vec{K} = K_1 \ldots K_n$ denotes the key part, $\vec{S} = S_1 \ldots S_n$ denotes the secure systems unknown to the eavesdropper, ρ is an n-partite quantum network, ω is some density operator, and U is a specifically constructed bipartite unitary operation known as twisting.

This definition gives a faithful private state for QCKD. In applications, there are some errors from noises or attackers' operations. This may follow an approximate state which may be close to the faithful private state. Hence, it might be convenient in an approximate manner. This can be defined as follows.

Definition 4.4.3 A state γ is ϵ-approximate of private state ϱ if it satisfies

$$F(\gamma, \varrho) \geq 1 - \epsilon \tag{4.124}$$

for some fidelity function $F(\cdot)$.

Definition 4.4.4 A ϱ-privacy test corresponding to $\varrho_{\vec{S}\vec{K}}$ is defined as the dichotomic measurement $\{\Pi_{\vec{S}\vec{K}}, \mathbb{1} - \Pi_{\vec{S}\vec{K}}\}$, where $\Pi_{\vec{S}\vec{K}} := U_{\vec{S}\vec{K}}(\varrho_{\vec{K}} \otimes \mathbb{1}_{\vec{S}})U^\dagger_{\vec{S}\vec{K}}$.

Using the properties of fidelity and the form of the test measurement, we arrive at the following proposition.

Proposition 4.4.1 *Given a state $\gamma_{\vec{S}\vec{K}}$ is ε-approximate to the state $\varrho_{\vec{S}\vec{K}}$, i.e., $F(\gamma_{\vec{S}\vec{K}}, \varrho_{\vec{S}\vec{K}}) \geq 1 - \varepsilon$, then the following inequality holds*

$$\mathrm{Tr}[\Pi_{\vec{S}\vec{K}}\gamma_{\vec{S}\vec{K}}] \geq 1 - \varepsilon. \tag{4.125}$$

A simple fact is any biseparable state $\sigma_{\vec{S}\vec{K}} \in \mathrm{BS}(\vec{S}\vec{K}:)$ can never pass any γ-privacy test with probability greater than $1/K$, i.e., $\mathrm{tr}[\Pi_{\vec{S}\vec{K}}\sigma_{\vec{S}\vec{K}}] \leq \frac{1}{K}$. Here, a biseparable state is a probabilistic mixing of bipartite product states. This follows a result that the private states are necessarily genuinely multipartite entangled (GME) in the biseparable model (Svetlichny, 1987).

Theorem 4.4.1 *The private state (4.123) is a GME.*

The main idea to show the result is a multipartite privacy test. Especially, consider a dichotomic measurement $\{\Pi, \mathbb{1} - \Pi\}$ such that any ϵ-approximate multipartite private state (MPS) ρ passes the test with success probability given by $1 - \epsilon$. For any biseparable state Φ_{bs} we cannot pass the privacy test with probability larger than $1/K$, where $\log K$ is the number of QCKD bits achieved by measuring the key part of ϱ. The idea is to construct an MQC by interleaving with LOCC. For example, in a measurement device independent QKD setup, i.e., we do not consider the attacking on sources, A_1 and A_2 send their states to the total measurement device using respective channels $\mathcal{N}^{1,2}$. This kind of MQC may be written into

$$\mathcal{N}_{A_1'A_2' \to A_1A_2} := \mathcal{B}_{X \to A_1A_2} \circ \mathcal{M}_{A_1''A_2'' \to X} \circ \mathcal{N}^1_{A_1' \to A_1''} \otimes \mathcal{N}^2_{A_2' \to A_2''}, \tag{4.126}$$

where $\mathcal{M}_{A_1''A_2'' \to X}$ is the quantum channel for the total measurement, and $\mathcal{B}_{X \to A_1A_2}$ is a classical broadcast channel for sending the measurement outcomes to A_1 and A_2. Other examples of multipartite QKDs over quantum networks depend on key repeaters (Christandl and Müller-Hermes, 2017; Bauml et al., 2015).

Denote E as entanglement and FS as the set of fully separable states. Denote $\mathcal{P}(\cdot) = \sum_{k \in \mathcal{K}} |k\rangle\langle k|(\cdot)|k\rangle\langle k|$ as a projective measurement channel. Suppose $|K_i| = K$ for $i \in [m]$. In this case, we define a conference key state $\varrho^c_{\vec{K}E}$ as

$$\otimes_{i=1}^m \mathcal{P}_{K_i}(\varrho^c_{\vec{K}E}) := \frac{1}{K}\sum_{k \in \mathcal{K}} |k\rangle\langle k|_{K_1} \otimes \cdots \otimes |k\rangle\langle k|_{K_m} \otimes \sigma_E, \tag{4.127}$$

where σ_E is a state of the system E, which may be accessible to Eve.

The generalized divergence of an arbitrary state $\rho_{\vec{A}}$ is defined as (Contreras-Tejada et al., 2019):

$$\mathcal{D}_r(\vec{A}:)_\rho := \inf_{\sigma \in \mathrm{F}(\vec{A}:)} D(\rho_{\vec{A}} \| \sigma_{\vec{A}}), \tag{4.128}$$

when F $\in$ {FS, BS} for $r = \text{E}$ or GME, respectively, and $D(\cdot\|\cdot)$ denotes the divergence of two states. $\mathcal{D}_r$ is an efficient entanglement measure. The entangling power of $\mathcal{N}_{\vec{A}'\vec{B}\to\vec{A}\vec{C}}$ is defined as the maximum possible gain for any input states as

$$\mathcal{D}_r^p(\mathcal{N}) := \sup_{\rho}\{\mathcal{D}_r(\vec{L}\vec{A} : \vec{R} : \vec{P}\vec{C})_{\mathcal{N}(\rho)} - \mathcal{D}_r(\vec{L}\vec{A}' : \vec{R}\vec{B} : \vec{P})_{\rho}\}, \quad (4.129)$$

where the optimization is over all input states $\rho_{\vec{L}\vec{A}'\vec{R}\vec{B}\vec{P}}$.

By generalizing results for P2P channels (Pirandola et al., 2017; Wilde et al., 2017; Christandl and Müller-Hermes, 2017), bidirectional channels (Bauml et al., 2018) or single-letter end-to-end capacities over a general network (Pirandola, 2019), we obtain a divergence-based measure for MQCs.

The discussion above provides upper bounds on the key capacities. Define

$$E_r(\mathcal{N}) := \sup_{\rho\in \text{FS}(\vec{L}\vec{A}':\vec{R}\vec{B})} \mathcal{D}_r(\vec{L}\vec{A} : \vec{R} : \vec{C})_{\mathcal{N}(\rho)}, \quad (4.130)$$

for $r = \text{E}$ or GME, respectively. In the one-shot case of a MQC with classical preprocessing and postprocessing (CPP), we have the following weak converse result.

Proposition 4.4.2 *For any fixed* $\varepsilon \in (0, 1)$, *the achievable region of CPP-assisted secret key agreement over a MQC* $\mathcal{N}$ *satisfies*

$$R_{CPP}^{(1,\varepsilon)}(\mathcal{N}) \le \mathcal{D}_{h,GME}^{\varepsilon}(\mathcal{N}), \quad (4.131)$$

where $\mathcal{D}_{h,GME}^{\varepsilon}(\mathcal{N})$ *is the* ε*-ht (hypothesis-testing) relative entropy of GME of* $\mathcal{N}$ *based on the* ε*-ht divergence (Buscemi and Datta, 2010).*

For many channels uses interleaved by LOCC, we get the following strong converse bound:

$$R_{\text{LOCC}}(\mathcal{N}) \le \mathcal{D}_{\max,E}(\mathcal{N}), \quad (4.132)$$

where $\mathcal{D}_{\max,E}(\mathcal{N})$ is the max-relative entropy of entanglement of $\mathcal{N}$ based on the max-relative entropy (Datta, 2009). □

Another strong result may be followed by the regularized relative entropy in finite-dimensional Hilbert spaces as

$$R_{\text{LOCC}}(\mathcal{N}) \le \mathcal{D}_{R,E}^{\infty}(\mathcal{N}). \quad (4.133)$$

If $\mathcal{N}$ is teleportation simulatable (Pirandola et al., 2017; Bennett et al., 1996), i.e., it can be simulated by a resource state and a LOCC operation, in this case, the bound of $R_{\text{LOCC}}(\mathcal{N})$ reduces to the relative entropy of entanglement of the resource state.

The present techniques allow us to compute the upper bounds of the rates achievable in measurement device-independent QKDs.

Example 4.4.1 Consider a dual-rail scheme based on single photons (Das et al., 2018) for two users. The channels between the users and the relay station can be represented by erasure channels. The QKD capacity is given by

$$R_{\mathrm{LOCC}}(\mathcal{N}^{\mathcal{E}}_{\vec{A}\to\vec{Z}}) = q\eta_1\eta_2, \tag{4.134}$$

where η_i's are the parameters of the erasure channels connecting users to the relay station and q is the probability of success of the Bell measurement. This allows us to trade off the rate distance.

Example 4.4.2 Consider a bidirectional network, where all nodes are connected with their neighbors by a product of P2P channels. Define its spanning tree among any pair of the parties as $\mathcal{T}$. By exploring the lowest rate (Devetak and Winter, 2005) in $\mathcal{T}$, it follows a tighter bound for the achievable rate.

Another application is to estimate the upper bound of the rates of distilled keys from multipartite quantum states. The upper bound of the one-shot distillable conference key is given by

$$R_{\mathrm{D}}^{(1,\varepsilon)}(\gamma) \leq \mathcal{D}^{\varepsilon}_{h,\mathrm{GME}}(\gamma) \tag{4.135}$$

in terms of the hypothesis-testing relative entropy with respect to biseparable states. It may be further extended by the regularized relative entropy for input state γ as

$$R_D(\gamma) \leq \mathcal{D}^{\infty}_{\mathrm{GME}}(\gamma). \tag{4.136}$$

In what follows, we consider the achieve key rate for CPP-assisted protocols. Let $\hat{R}^{\mathcal{N}}_{\mathrm{CPP}}(n, \varepsilon)$ denote the maximum rate R such that the (n, K, ε) QCKD is achievable for any $\mathcal{N}$ using CPP. Denote the ε-ht relative entropy of GME over MQC $\mathcal{N}$ as

$$\mathcal{D}^{\varepsilon}_{h,\mathrm{GME}}(\mathcal{N}) := \sup_{\Psi\in\mathrm{FS}(\vec{L}\vec{A}':\vec{R}\vec{B})} \inf_{\sigma\in\mathrm{BS}(\vec{L}\vec{A}:\vec{R}:\vec{C})} D^{\varepsilon}_{h}(\mathcal{N}(\Psi)\|\sigma). \tag{4.137}$$

Here, it is sufficient to optimize over pure states $\Psi \in \mathrm{FS}(\vec{L}\vec{A}' : \vec{R}\vec{B})$. The following bound holds for the one-shot secret key rate of a MQC.

For the lower bounds on the secure key rates of MQCs achieved by CPP, the main is to use a distributing party (trusted party) who performs the directed QCKD protocol (Devetak and Winter, 2005) with all the remained parties. The achievable rate is the worst case of the achievable rate (Devetak and Winter, 2005).

Theorem 4.4.2 *For a given* $\varepsilon \in (0, 1)$, *the achievable region of CPP-assisted QCKD with one copy of* $\mathcal{N}_{\vec{A}'\vec{B}\to\vec{A}\vec{C}}$ *satisfies*

$$R^{\mathcal{N}}_{CPP}(1, \varepsilon) \leq \mathcal{D}^{\varepsilon}_{h,\mathrm{GME}}(\mathcal{N}). \tag{4.138}$$

We now get converse and strong converse bounds on a LOCC-assisted QCKD over MQC $\mathcal{N}_{\vec{A}'\vec{B}\to\vec{A}\vec{C}}$. For a LOCC-assisted QCKD protocol, the max-relative entropy of entanglement of $\mathcal{N}$ is defined as

$$\mathcal{D}_{\max,E}(\mathcal{N}) := \sup_{\Psi\in\mathrm{FS}(\vec{L}\vec{A}':\vec{R}\vec{B}),\sigma\in\mathrm{FS}(\vec{L}\vec{A}:\vec{R}:\vec{C})} D_{\max}(\mathcal{N}(\Psi)\|\sigma). \tag{4.139}$$

A converse bound can be explored by employing the proof (Das et al., 2020) to the multiplex scenario as

$$\frac{1}{n}\log_2 K \leq \mathcal{D}_{\max,E}(\mathcal{N}) + \frac{1}{n}\log\frac{1}{1-\varepsilon} \tag{4.140}$$

for a fixed $n, K \in \mathbb{N}, \varepsilon \in (0,1)$.

Theorem 4.4.3 *The asymptotic LOCC-assisted key capacity is given by*

$$R_{\mathrm{LOCC}}(\mathcal{N}) \leq \mathcal{D}_E^{\infty}(\mathcal{N}). \tag{4.141}$$

Moreover, the achievable key region from one state $\gamma_{\vec{A}}$ satisfies for any fixed $\varepsilon \in (0,1)$

$$R_{\mathrm{D}}^{(1,\varepsilon)}(\gamma) \leq \mathcal{D}_{h,\mathrm{GME}}^{\varepsilon}(\vec{A})_{\gamma}, \tag{4.142}$$

where $\mathcal{D}_{h,\mathrm{GME}}^{\varepsilon}(\vec{A})_{\gamma}$ is the ε hypothesis-testing relative entropy of GME $\gamma_{\vec{A}}$.

These findings offer a general approach to investigating the conference key ratio of QCKD on quantum networks, where the attainable key is dependent on the specific resource states of the quantum networks. A common approach is to use a singlet state or multipartite GHZ state (Greenberger et al., 1989); however, this method necessitates a lengthy decoherence and storage period for experiments involving a large number of parties. Alternatively, multisource quantum networks comprising EPR states (Einstein et al., 1935) can be utilized, which provides a more adaptable configuration and setup. Moreover, these results may be extended to examine private states in the context of quantum communication (Pirandola et al., 2017; Bennett et al., 1996).

4.5 Secure Communication on Quantum Networks

Deterministic quantum protocols for distributing messages involve the use of quantum entanglement to achieve message distribution. These protocols rely on sharing entangled pairs of qubits between the sender and receiver and then manipulating those qubits to encode the message. By using a deterministic process, the receiver can decode the message with high fidelity, without the need for error correction. This protocol has the potential to enable secure and efficient communication in a range of applications, including cryptography, quantum computing, and telecommunications.

In this section, we present secure communication on quantum networks based on QKD. Different from the standard way by combining QKD and one-time pad secure communication, we may construct a deterministic quantum protocol for distributing message (Beige et al., 2002; Bostroem, 2002; Deng and Long, 2004; Zhou et al., 2020; Qi et al., 2021).

Suppose one party Alice communicates the message $\vec{a} = a_1 \cdots a_N$, where $a_i \in \{0, 1, 2, 3\}$. Alice has a quantum source to generate two-qubit in one of the Bell states $|\phi^{\pm}\rangle$ and $|\psi^{\pm}\rangle$. These are further written into $|\Psi_i\rangle$, $i \in \{0, \cdots, 3\}$. A simple encoding is by $a_i \to |\Psi_{a_i}\rangle$.

Consider four unitary operations $\{\sigma_0 = \mathbb{1}, \sigma_1, \sigma_2, \sigma_3\}$. Note that $(\mathbb{1} \otimes \sigma_i)|\Psi_j\rangle$ is equivalent to $|\Psi_j\rangle$ up to a global phase for some k. This implies an encoding of two bits $b \in \{0, 1, 2, 3\}$ by performing σ_b on one qubit as

$$|\Psi_f\rangle \sim (\mathbb{1} \otimes \sigma_b)|\Psi_a\rangle, \tag{4.143}$$

which is similar to a dense coding protocol (Bennett and Wiesner, 1992).

As far as an eavesdropper Eve is concerned, legal parties will use eavesdropping detection techniques similar to those used in QKD. However, unlike QKD, the encoded states in this case already contain the message. This means that a powerful eavesdropper like Eve could potentially launch a man-in-the-middle attack by intercepting the transmitted particles from Alice. She would then prepare another Bell state $|\Psi_e\rangle$ and distribute it to Bob. In this scenario, Eve would know exactly the operation σ_b performed by Bob.

The man-in-the-middle attack may be detected by a classical discussion on public authentical channels. We present an explicit protocol as follows.

S1. Alice prepares two qubits in the Bell state $|\Psi_{a_n}\rangle$. She keeps one qubit and sends the other to Bob;

S2. After receiving the qubit, Bob switches to control mode with a probability p_c or message mode with $1 - p_c$;

Control mode: Bob and Alice complete the CHSH test (Clauser et al., 1969) for eavesdropping detection;

Message mode: Bob and Alice complete super dense coding (Bennett and Wiesner, 1992) without eavesdropping. Otherwise, they abort the protocol.

Let the probability for a control transfer be p_c. The security can be followed from the standard QKD if Eve attacks all the time. For a random attack, one may asymptotically complete the security analysis. This provides a device-independent model of quantum secure communication. The current bipartite secure communication protocol can be extended to quantum networks consisting of multiple subnetworks. One possible is to take use of multipartite GHZ state (Greenberger et al., 1989). For general networks it can be broken down into subnetworks $\mathcal{N}_1, \cdots, \mathcal{N}_k$, each of which is a densely connected network. However, since these subnetworks may be located far apart, more efficient communication methods beyond QKD and one-time pad are required under reasonable assumptions of secure channels. Instead of focusing on secure communication within each subnetwork, the security of the quantum channels between subnetworks should be ensured, as this long-distance communication may be vulnerable to powerful eavesdroppers. For this reason, device-independent QKD or QCKD may be considered as solutions. Quantum storage can be achieved by adjusting the time delay of the electrical signal (Qi et al., 2021), and the communication process is carried out only after eavesdropping detection. The specific transfer scheme will depend on the network configuration and the task at hand.

4.6 Conclusion

In this chapter, we have extended the bipartite QKD protocol (Bennett and Brassard, 1984; Ekert, 1991) to quantum networks. The traditional security of the bipartite protocol is based on the monogamy relationship between the quantum violation of a specific Bell inequality (Ekert, 1991; Clauser et al., 1969) and the leakage of information from the final keys. This provides a strong key distribution that is device-independent and resistant to post-quantum eavesdroppers. For multipartite applications on quantum networks, similar monogamy relationships are generally required. This is based on the genuinely multipartite nonlocality of the resource states involved. One possible model is the biseparable model (Svetlichny, 1987) or the MABK inequality (Mermin, 1990; Ardehali, 1992; Belinskii and Klyshko, 1993). For further information on the security of these protocols, see recent reviews (Gisin et al., 2002; Scarani et al., 2009; Portmann and Renner, 2022; Scarani et al., 2009).

The no-signaling security in the Sect. 4.1.3 is based on (Brunner et al., 2014). The device-independent model in Sect. 4.2 is from (Luo, 2018, 2020). Quantum conference key distribution on networks in the Sect. 4.3 is based on the result (Ribeiro and Wehner, 2018; Arnon-Friedman et al., 2019; Dupuis et al., 2020; Philip et al., 2021; Horodecki et al., 2022). The main results on multiplex quantum channels are based on the result (Contreras-Tejada et al., 2019; Das et al., 2018). One may read further materials (realistic QKD (Xu et al., 2020) or Micius quantum experiments (Lu et al., 2022)).

References

A. Acín, N. Brunner, N. Gisin, S. Massar, S. Pironio, V. Scarani, Device-independent security of quantum cryptography against collective attacks. Phys. Rev. Lett. **98**, 230501 (2007)

M. Ardehali, Bell inequalities with a magnitude of violation that grows exponentially with the number of particles. Phys. Rev. A **46**, 5375 (1992)

R. Arnon-Friedman, F. Dupuis, O. Fawzi, R. Renner, T. Vidick, Practical device-independent quantum cryptography via entropy accumulation. Nat. Commun. **9**, 459 (2018)

R. Arnon-Friedman, R. Renner, T. Vidick, Simple and tight device-independent security proofs. SIAM J. Comput. **48**, 181 (2019)

R. Augusiak, P. Horodecki, Multipartite secret key distillation and bound entanglement. Phys. Rev. A **80**, 042307 (2009)

J. Barrett, L. Hardy, A. Kent, No-Signaling and Quantum Key Distribution. Phys. Rev. Lett. **95**, 010503 (2005)

J. Barrett, A. Kent, S. Pironio, Maximally nonlocal and monogamous quantum correlations. Phys. Rev. Lett. **97**, 170409 (2006)

J. Barrett, R. Colbeck, A. Kent, Unconditionally secure device-independent quantum key distribution with only two devices. Phys. Rev. A **86**, 062326 (2012)

S. Bauml, M. Christandl, K. Horodecki, A. Winter, Limitations on quantum key repeaters. Nat. Commun. **6**, 1–5 (2015)

S. Bauml, S. Das, M.M. Wilde, Fundamental limits on the capacities of bipartite quantum interactions. Phys. Rev. Lett. **121**, 250504 (2018)

D. Beckman, D. Gottesman, M.A. Nielsen, J. Preskill, Causal and localizable quantum operations. Phys. Rev. A **64**, 052309 (2001)

A. Beige, B.-G. Englert, C. Kurtsiefer, H. Weinfurter, Secure communication with a publicly known key. Acta Phys. Pol. A **101**, 357 (2002)

A.V. Belinskii, D.N. Klyshko, Interference of light and Bell's theorem. Sov. Phys. Usp. **36**, 653 (1993)

C.H. Bennett, G. Brassard, Quantum cryptography: Public key distribution and coin tossing, in *International Conference on Computer System and Signal Processing* (IEEE, New York, 1984), pp. 8

C.H. Bennett, S.J. Wiesner, Communication via one-and two-particle operators on Einstein-Podolsky-Rosen states. Phys. Rev. Lett. **69**, 2881 (1992)

C.H. Bennett, H.J. Bernstein, S. Popescu, B. Schumacher, Concentrating partial entanglement by local operations. Phys. Rev. A **53**, 2046 (1996)

K. Bostroem, Secure direct communication using entanglement. Phys. Rev. Lett. **89**, 187902 (2002)

C. Branciard, N. Brunner, H. Buhrman, R. Cleve, N. Gisin, S. Portmann, D. Rosset, M. Szegedy, Classical simulation of entanglement swapping with bounded communication. Phys. Rev. Lett. **109**, 100401 (2012)

S. Braunstein, C. Caves, Statistical distance and the geometry of quantum states. Ann. Phys. **202**, 22 (1990)

N. Brunner, D. Cavalcanti, S. Pironio, V. Scarani, S. Wehner, Bell nonlocality. Rev. Mod. Phys. **86**, 419 (2014)

F. Buscemi, N. Datta, The quantum capacity of channels with arbitrarily correlated noise. IEEE Trans. Inf. Theory **56**, 1447 (2010)

R. Chaves, R. Kueng, J.B. Brask, D. Gross, Unifying framework for relaxations of the causal assumptions in Bell's theorem. Phys. Rev. Lett. **114**, 140403 (2015)

M. Christandl, A. Müller-Hermes, Relative entropy bounds on quantum, private and repeater capacities. Commun. Math. Phys. **353**, 821 (2017)

J.F. Clauser, M.A. Horne, A. Shimony, R.A. Holt, Proposed experiment to test local hidden variable theories. Phys. Rev. Lett. **23**, 880 (1969)

C. Clivati et al., Coherent phase transfer for real-world twin-field quantum key distribution. Nat. Commun. **13**, 157 (2022)

P. Contreras-Tejada, C. Palazuelos, J.I. de Vicente, Resource theory of entanglement with a unique multipartite maximally entangled state. Phys. Rev. Lett. **122**, 120503 (2019)

S. Das, S. Khatri, J.P. Dowling, robust quantum network architectures and topologies for entanglement distribution. Phys. Rev. A **97**, 012335 (2018)

S. Das, S. Bäuml, M.M. Wilde, Entanglement and secret-key-agreement capacities of bipartite quantum interactions and read-only memory devices. Phys. Rev. A **101**, 012344 (2020)

N. Datta, Max-relative entropy of entanglement, alias log robustness. Inter. J. Quantum Inf. **7**, 475–491 (2009)

F.-G. Deng, G.L. Long, Secure direct communication with a quantum one-time-pad. Phys. Rev. A **69**, 052319 (2004)

I. Devetak, A. Winter, Distillation of secret key and entanglement from quantum states. Proc. Royal Soc. A **461**, 207 (2005)

J.P. Dowling, G.J. Milburn, Quantum technology: the second quantum revolution. Phil. Trans. Royal Soc. London Ser. A. **361**, 1655 (2003)

F. Dupuis, O. Fawzi, R. Renner, Entropy accumulation. Commun. Math. Phys. **379**, 867–913 (2020)

A. Einstein, B. Podolsky, N. Rosen, Can quantum mechanical description of physical reality be considered complete? Phys. Rev. **47**, 777 (1935)

A.K. Ekert, Quantum cryptography based on Bell's Theorem. Phys. Rev. Lett. **67**, 661 (1991)

M. Epping, H. Kampermann, C. Macchiavello, D. Bruß, Multi-partite entanglement can speed up quantum key distribution in networks. New J. Phys. **19**, 093012 (2017)

N. Gisin, G. Ribrody, W. Tittel, H. Zbinden, Quantum cryptography. Rev. Mod. Phys. **74**, 145–195 (2002)

N. Gisin, G. Ribordy, W. Tittel, H. Zbinden, Quantum cryptography. Rev. Mod. Phys. **74**, 145 (2002)

D.M. Greenberger, M.A. Horne, A. Zeilinger, in *Bell's Theorem, Quantum Theory and Conceptions of the Universe*. ed. by M. Kafatos (Kluwer, Dordrecht, 1989), pp.69–72

T. Holz, H. Kampermann, D. Bruß, Genuine multipartite Bell inequality for device-independent conference key agreement. Phys. Rev. Research **2**, 023251 (2020)

K. Horodecki, M. Winczewski, S. Das, Fundamental limitations on the device-independent quantum conference key agreement. Phys. Rev. A **105**, 022604 (2022)

E. Kaur, X. Wang, M.M. Wilde, Conditional mutual information and quantum steering. Phys. Rev. A **96**, 022332 (2017)

E. Kaur, M.M. Wilde, A. Winter, Fundamental limits on key rates in device-independent quantum key distribution. New J. Phys. **22**, 023039 (2020)

R. Koenig, R. Renner, C. Schaffner, The operational meaning of min and max-entropy. IEEE Trans. Inf. Theory **24**, 339 (2009)

B. Kraus, N. Gisin, R. Renner, Lower and upper bounds on the secret-key rate for quantum key distribution protocols using one-way classical communication. Phys. Rev. Lett. **95**, 080501 (2005)

C.M. Lee, M.J. Hoban, Towards device-independent information processing on general quantum networks. Phys. Rev. Lett. **120**, 020504 (2018)

H.K. Lo, H.F. Chau, Unconditional security of quantum key distribution over arbitrarily long distances. Science **283**, 2050 (1999)

C.-Y. Lu, Y. Cao, C.-Z. Peng, J.-W. Pan, Micius quantum experiments in space. Rev. Mod. Phys. **94**, 035001 (2022)

M. Lucamarini, Z.L. Yuan, J.F. Dynes, A.J. Shields, Overcoming the rate-distance limit of quantum key distribution without quantum repeaters. Nature **557**, 400–403 (2018)

M.-X. Luo, Computationally efficient nonlinear bell inequalities for quantum networks. Phys. Rev. Lett. **120**, 140402 (2018)

M.-X. Luo, Nonsignaling causal hierarchy of general multisource networks. Phys. Rev. A **101**, 062317 (2020)

M.X. Luo, Fully device-independent model on quantum networks. Phys. Rev. Research **4**, 013203 (2022)

X. Ma, P. Zeng, H. Zhou, Phase-matching quantum key distribution. Phys. Rev. X **8**, 031043 (2018)

L. Masanes, S. Pironio, A. Acin, Secure device-independent quantum key distribution with causally independent measurement devices. Nat. Commun. **2**, 238 (2011)

D. Mayers, Quantum key distribution and string oblivious transfer in noisy channels, in *Advances in Cryptology: Proceedings of Crypto' 96* (Springer-Verlag, New York, 1996), pp. 343–357

N.D. Mermin, Extreme quantum entanglement in a superposition of macroscopically distinct states. Phys. Rev. Lett. **65**, 1838 (1990)

A. Philip, E. Kaur, P. Bierhorst, M.M. Wilde, Intrinsic non-locality and device-independent conference key agreement (2021). arXiv:2111.02596v2

S. Pirandola, End-to-end capacities of a quantum communication network. Commun. Phys. **2**, 51 (2019)

S. Pirandola, R. Laurenza, C. Ottaviani, L. Banchi, Fundamental limits of repeaterless quantum communications. Nat. Commun. **8**, 15043 (2017)

C. Portmann, R. Renner, Cryptographic security of quantum key distribution (2014). arXiv:1409.3525

C. Portmann, R. Renner, Security in quantum cryptography. Rev. Mod. Phys. **94**, 025008 (2022)

Z. Qi, Y. Li, Y. Huang, J. Feng, Y. Zheng, X. Chen, A 15-user quantum secure direct communication network. Light: Sc. Appl. **10**, 183 (2021)

R. Rabelo, M. Ho, D. Cavalcanti, N. Brunner, V. Scarani, Device-independent certification of entangled measurements. Phys. Rev. Lett. **107**, 050502 (2011)

R. Renner, Security of quantum key distribution. Inter. J. Quantum Inf. **6**, 1–127 (2008)

J. Ribeiro, G. Wehner, Fully device-independent conference key agreement. Phys. Rev. A **97**, 022307 (2018)

V. Scarani, H. Bechmann-Pasquinucci, N.J. Cerf, M. Dusek, N. Lutkenhaus, M. Peev, The security of practical quantum key distribution. Rev. Mod. Phys. **81**, 1301 (2009)

R. Schwonnek, K. Tong Goh, I.W. Primaatmaja, E.Y.-Z. Tan, R. Wolf, V. Scarani, C.C.-W. Lim, Device-independent quantum key distribution with random key basis. Nat. Commun. **12**, 2880 (2021)

P.W. Shor, J. Preskill, Simple proof of security of the BB84 Quantum Key Distribution protocol. Phys. Rev. Lett. **85**, 441 (2000)

G. Svetlichny, Distinguishing three-body from two-body nonseparability by a Bell-type inequality. Phys. Rev. D **35**, 3066 (1987)

B.F. Toner, D. Bacon, the communication cost of simulating Bell correlations. Phys. Rev. Lett. **91**, 187904 (2003)

U. Vazirani, T. Vidick, Fully device-independent quantum key distribution. Commun. ACM. **62**, 133 (2019)

S. Wang et al., Beating the fundamental rate-distance limit in a proof-of-principle quantum key distribution system. Phys. Rev. X **9**, 021046 (2019)

S. Wehner, D. Elkouss, R. Hanson, Quantum internet: A vision for the road ahead. Science **362**, 9288 (2018)

M.M. Wilde, M. Tomamichel, M. Berta, Converse bounds for private communication over quantum channels. IEEE Trans. Inf. Theory **63**, 1792 (2017)

W.K. Wootters, W.H. Zurek, A single quantum cannot be cloned. Nature **299**, 802 (1982)

F. Xu, X. Ma, Q. Zhang, H.-K. Lo, J.-W. Pan, Secure quantum key distribution with realistic devices. Rev. Mod. Phys. **92**, 025002 (2020)

X. Zhong, W. Wang, R. Mandil, H.-K. Lo, L. Qian, Simple multiuser twin-field quantum key distribution network. Phys. Rev. Appl. **17**, 014025 (2022)

L. Zhou, Y.-B. Sheng, G.-L. Long, Device-independent quantum secure direct communication against collective attacks. Sci. Bul. **65**, 12–20 (2020)

L. Zhou, J. Lin, Y. Jing, Z. Yuan, Twin-field quantum key distribution without optical frequency dissemination. Nature Commun. **14**, 928 (2023)

Chapter 5
Distributed Quantum Computations

In distributed quantum computing, the communication between the processors plays a critical role, as it enables the distribution of quantum states and operations among the processors. The key challenge in this paradigm is to implement efficient protocols for distributing quantum information while minimizing the errors that may arise due to noise and other sources of decoherence.

Distributed quantum computing has the potential to enable the development of large-scale quantum networks, which could be used for a variety of applications ranging from secure communication to scientific simulations. Some promising approaches to implementing distributed quantum computing include the use of hybrid classical-quantum networks, quantum error correction codes, and topological qubits.

However, there are still many challenges that need to be addressed before distributed quantum computing becomes a practical reality, including the development of robust quantum hardware, efficient communication protocols, and scalable software tools for programming and optimizing distributed quantum algorithms.

5.1 Quantum Circuits

Distributed quantum computing is a paradigm that holds tremendous promise for enabling large-scale quantum networks and performing highly complex computational tasks. Unlike classical computers, which rely on binary logic gates and Boolean algebra, quantum systems utilize qubit gates and Pauli algebra to undertake intractable computational tasks Deutsch (1989); Nielsen and Chuang (2000). This is due to the unique features of quantum systems such as entanglement or quantum correlation, which do not exist in the classical realm. However, local qubit operations are often insufficient for general global quantum system processing. Therefore, correlated qubits operation has been designed to construct universal quantum logic gates, which can be performed on a small, fixed number of qubits.

M.-X. Luo, *Quantum Networks*,
https://doi.org/10.1007/978-981-97-6226-2_5

Especially, a global unitary transformation of multiple qubits can be synthesized using only two qubits at a time, a result that has no analog in classical reversible logic where three-bit gates are needed to simulate all reversible Boolean functions. Despite these advantages, there are still significant challenges that need to be addressed before distributed quantum computing becomes practical. The development of robust quantum hardware is one such challenge, as well as the implementation of efficient communication protocols to distribute quantum information accurately and minimize errors due to decoherence. In addition, there is a need for scalable software tools for programming and optimizing distributed quantum algorithms. Nonetheless, with continued research and development, distributed quantum computing holds tremendous potential for revolutionizing many fields, from secure communication to scientific simulations.

The concept of a quantum circuit involves an acyclic arrangement of quantum gates connected by wires. The quantum gates serve as representations of quantum operations, while the wires represent the qubits that these gates act upon. An illustration of a quantum circuit can be seen in Fig. 5.1, which features four input qubits and three output qubits.

A quantum circuit can have any number of input and output qubits, represented by integers n and m with $n, m \geq 0$, respectively. By combining the actions of individual gates in a specific manner, such a circuit produces a quantum operation that maps n input qubits to m output qubits. The size of a quantum circuit is defined as the sum of its total number of gates and wires.

A unitary quantum circuit is a type of quantum circuit composed entirely of gates that correspond to unitary quantum operations. To qualify as a unitary quantum circuit, every gate within it must have the same number of input and output qubits. In the field of quantum computing, it is often standard practice to work solely with unitary quantum circuits. These circuits are related to the general model, and the nature of this relationship will be explained shortly.

1. *Toffoli gate*. A Toffoli gate is a three-qubit unitary gate U_{Tf} identified with the unitary matrix T_f defined by the following action on elementary unit vectors:

$$T_f : |a\rangle|b\rangle|c\rangle \mapsto |a\rangle|b\rangle|c \oplus ab\rangle. \tag{5.1}$$

2. *Hadamard gate*. A Hadamard gate is a single-qubit unitary gate U_H identified with the unitary matrix H defined as:

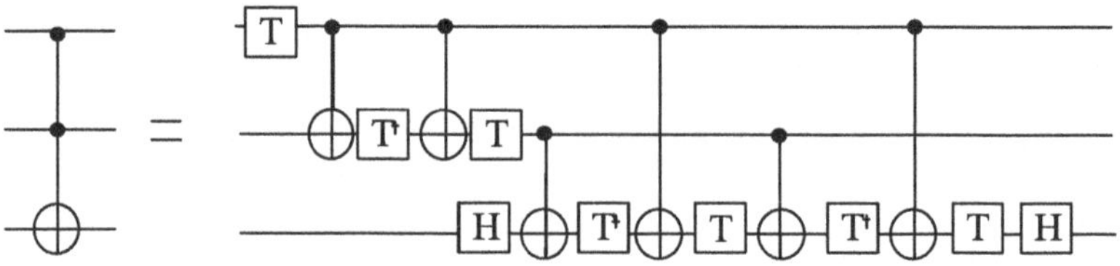

Fig. 5.1 A quantum circuit for Toffoli gate

$$H : |a\rangle \mapsto \frac{1}{\sqrt{2}}|0\rangle + \frac{(-1)^a}{\sqrt{2}}|1\rangle. \tag{5.2}$$

3. *Phase-shift gate.* A Phase-shift gate is a single-qubit unitary gate U_P identified with the unitary matrix P defined as:

$$P : |a\rangle \mapsto (\sqrt{-1})^a|a\rangle. \tag{5.3}$$

4. *T-gate.* The T-gate is a type of single-qubit rotation gate, identified with the unitary matrix T, which is defined as

$$T : |a\rangle \mapsto \exp(\frac{i^a \pi}{4})|a\rangle. \tag{5.4}$$

5. *Ancillary gates.* Ancillary gates are non-unitary gates that produce a single qubit in the state $|0\rangle$ as output, without taking any input.
6. *Erasure gates.* Erasure gates are a type of non-unitary gate that takes a single qubit as input and produce no output. Their operation is represented by the partial trace on the qubit they act upon.

Theorem 5.1.1 *(Universality Theorem). Let U be an arbitrary quantum operation from n qubits to m qubits. Then for every $\epsilon > 0$ there exists a quantum circuit W with n input qubits and m output qubits such that U and W are ϵ-indistinguishable. Moreover, for fixed n and m, the circuit W may be taken to satisfy* $\text{size}(W) = \text{poly}(\log(1/\epsilon))$.

The proof relies on the Solovay-Kitaev theorem which states that for any finite set of quantum gates, i.e., there exists a quantum circuit that approximates any unitary operator to within a desired precision using only those gates Brennen et al. (2005).

Let S be a universal set of quantum gates, meaning that any quantum operation can be expressed as a product of gates from S. We consider the set S_n of all gates acting on n qubits and the set S_m of all gates acting on m qubits. Then, for any $\epsilon > 0$, we can use the Solovay-Kitaev theorem to find two sets of gates $A \subset S_n$ and $B \subset S_m$ such that any gate in S_n can be approximated to within $\epsilon/2$ by a product of gates from A, and any gate in S_m can be approximated to within $\epsilon/2$ by a product of gates from B.

We can now construct the desired circuit W as follows. First, decompose U into a product of gates from S. Then, approximate each gate in the decomposition to within $\epsilon/2$ using the corresponding set of gates from A or B. The resulting circuit W consists of gates from A and B and approximates U to within ϵ.

To estimate the size of W, note that each gate in the decomposition of U can be approximated to within $\epsilon/2$ using at most $\text{poly}(\log(1/\epsilon))$ gates from A or B by the Solovay-Kitaev theorem. Therefore, the size of W is bounded by $\text{poly}(d \log(1/\epsilon))$, where d is the size of the original decomposition of U into gates from S. Since d depends only on n and m, the circuit W can be taken to satisfy $\text{size}(W) = \text{poly}(\log(1/\epsilon))$ for fixed n and m.

5.2 Qudit Circuit

The universal qubit circuit can be extended to include quantum qudit circuit, which utilizes qudits as the unit of memory Brennen et al. (2005); Luo and Wang (2014). By utilizing a d-dimensional state space, qudit systems offer greater flexibility in the storage and processing of quantum information. This increased dimensionality provides various advantages, including improving channel capacity, enhancing the implementation of quantum gates, increasing security for quantum communication protocols, and exploring novel quantum features.

Despite the potential benefits of qudit-based approaches, implementing qudit logic requires more complex primitive qudit gates than their binary counterparts. These gates involve a controlled transformation of all the levels of each qudit, making them necessarily more complex than binary gates. Several general schemes have been proposed for explicitly implementing qudit logic, but they are not yet as well-developed as binary logic.

Define U_d as a family of d-dimensional transforms that maps a known single-qudit state to $|d-1\rangle$ on d-dimensional Hilbert space $\mathbb{H}$ as

$$U_d : \sum_{j=0}^{d-1} \alpha_j |j\rangle \mapsto |d-1\rangle. \tag{5.5}$$

Similar to the qubit case, U_d does not unique in terms of d complex parameters $\alpha_0, \cdots, \alpha_{d-1}$.

U_d can realize by deterministic unitary transformations at most $d-1$ steps. In detail, U_d may be decomposed into

$$U_d = (X_{d-1}(a_{d-1}, b_{d-1})X_{d-2}(a_{d-2}, b_{d-2}) \cdots X_1(a_1, b_1))^{-1} \tag{5.6}$$

with

$$X_j(x, y) = \begin{pmatrix} \mathbb{1}_{j-1} & & & \\ & x & y & \\ & -y & x & \\ & & & \mathbb{1}_{d-j-1} \end{pmatrix} \tag{5.7}$$

and $a_j = \alpha_{j-1}/b_j, b_j = b_{j-1}/\sqrt{\sum_{i=j}^{d-1} \alpha_i^2}, b_0 = 1$. The new primitive transformations $X_j(x, y)$ are easily implemented in physics with only one parameter such as linear ion trap model and linear optics with multiport.

Define a d-dimensional phase gate Z_d as

$$Z_d(\theta) = \sum_{j=0}^{d-1} e^{i(1-\mathrm{sgn}(d-1-j))\theta} |j\rangle\langle j|, \tag{5.8}$$

where sgn denotes the sign function. $\{Z_d, X_d\}$ with $X_d = \{X_j(x, y)\}$ are sufficient to simulate all single-qudit unitary transforms. Each primitive gate can be implemented by controlling 2 complex parameters. Define the controlled-qudit gate

$$C_2[R_d] = \begin{pmatrix} \mathbb{1}_{d^2-d} & \\ & R_d \end{pmatrix} \tag{5.9}$$

acting in d^2-dimensional basis of two qudits, where R_d represents either X_d or Z_d. The identity matrix $\mathbb{1}_{d^2-d}$ acts on the states $\{|0, 0\rangle, \cdots, |d-2, d-1\rangle\}$, and R_d acts on the remaining d states $\{|d-1, 0\rangle, \cdots, |d-1, d-1\rangle\}$.

Denote $C_k[R_d]$ as

$$C_k[R_d] = \begin{pmatrix} \mathbb{1}_{d^k-d} & \\ & R_d \end{pmatrix} \tag{5.10}$$

which is acted on the d^k-dimensional basis of k-qudit state. It transforms the last qudit with the operation R_d conditional on the first $k-1$ qudits being in $|d-1, \cdots, d-1\rangle$.

Consider an N-dimensional unitary transform U acting on an n-qudit state, where $U \in \mathbb{SU}(d^n)$. Each U can be decomposed into combinations of $C_k[U_d]$ and $C_k[P_d]$ (the controlled permutation P_d). Here, $C_m[R_d]$ with $m > 2$ can be built from the two-qudit gates $C_2[R_d]$, similar to the qubit scenarios Luo and Wang (2014). For $d > 2$, the circuit uses $r = \lceil (m-2)/(d-2) \rceil$ auxiliary qudits ($\lceil x \rceil$ represents the smallest integer greater than x), where k is the remainder of $(m-2)/(d-2)$. This implies a general result for quantum qudit computation Luo and Wang (2014).

Theorem 5.2.1 *The set $\{X_d, Z_d, C_2[R_d]\}$ is universal for quantum qudit computation.*

5.3 Distributed Quantum RAM Model

Assuming non-signaling and tomographic locality (single-wire measurements), consider a theory of n gbits, where individual gbits are characterized by a $(d \geq 2)$-dimensional Bloch ball state space and are subject to the single-gbit transformation group $\mathrm{SO}(d)$. Additionally, suppose that the global transformations form a closed connected matrix group $\mathcal{G}$. Therefore, quantum computation is the only reversible circuit model for which bits can be represented as balls.

The typical theoretical representation of quantum algorithms differs significantly from how quantum computers are expected to be built in practice. Quantum algorithms are commonly described using the quantum circuit model, which permits any number of simultaneous two-qubit gates for an N-qubit circuit at a given time-step. However, practical implementations of quantum computers are anticipated to be predominantly local in two or three dimensions and feature only a few long-range connections, unlike classical computers that operate on von Neumann architectures with a small number of cores sharing a vast random access memory (RAM).

In some cases, the circuit model may also prove inadequate for our needs. An analog to the classical distributed RAM model or the concept of RAM would be useful for algorithm design. Therefore, the quantum distributed RAM model may permit simultaneous queries to shared quantum RAM in addition to any step of the circuit model. This enables to design of new quantum algorithms for parallel database searching, as well as solving problems like element distinctness and collision finding. Another improvement is to show that quantum computers can access memory in a parallel and unrestricted way. These allow distributed quantum computers to implement any quantum algorithm presented in terms of a quantum circuit. In this model, each processor acts on a few well-functioning qubits connected to a small number of other memory sites, possibly through long-range interactions. Quantum memory does not need to be in a single place but can be distributed among many processors, where a memory site could even be a single qubit.

5.3.1 Quantum Data Transferring

To transfer a quantum state stored in a qubit across a network $\mathcal{G}$, one method is to use a series of SWAP gates. Another way is by teleportation assisted by entanglement Cleve and Buhrman (1997). In this section, we present an algorithm to permute qubits on a given network $\mathcal{G}$. The goal is to transfer qubits stored in nodes of $\mathcal{G}$. Here, a qubit transferring is unitary map V_N on the register $|x_1, x_2, \ldots, x_N\rangle$, where the node j has stored $|x_i\rangle$. V_N can permutate $(1, \ldots, N) \mapsto (j_1, \ldots, j_N)$.

A sorting network is composed of binary comparators on T wires where each wire represents one of T elements. The comparators do not change the inputs but with new orders depending on comparison requirements. A simple fact is one can sort deterministically any finite elements even on any connected network $\mathcal{G}$ with a lower-depth network. Take a cyclic network $\mathcal{G}$ as an example, one way is by using a bitonic sorting network with 6 comparators in-depth 3 if $\mathcal{G}$ has 4 vertexes.

Theorem 5.3.1 *Suppose that $\mathcal{G}$ is a connected network that has N nodes, there exists an algorithm with a depth of $O(D_{\mathcal{G}})$, where $D_{\mathcal{G}}$ is the depth of a reversible sorting network over $\mathcal{G}$ with N inputs.*

As a classical reversible sorting network can be used as the primary subroutine for both the data-transferring and the parallel memory access. This fact implies a quantum extension of the classical reversible sorting network for proof.

Example 5.3.1 Consider the 1D Nearest Neighbour Network . The sorting network is the bubble/insertion sort. The depth of the sorting network is $O(N)$.

Example 5.3.2 Consider the 2D Nearest Neighbour Network where each node is connected to the closest adjacent nodes(s). The depth of the sorting network is $O(\sqrt{N})$.

Example 5.3.3 Consider the hypercube network with the valency $O(\log N)$. The depth of the sorting network is $O(\log^2 N)$ by Bitonic sorting Knuth (1998), or $O(\log N)^{6100}$ by AKS sorting Ajtai et al. (1983).

An ancilla bit will be added to each comparator for ensuring reversible sorting. The flipping operation of the ancilla will be used to identify the changes of a comparator. A swapping conditional on the ancilla will be applied. One way to construct a reversible comparator is based on binary tree Thapliyal et al. (2010) for lexicographic sorting. Its sorting operation is defined by:

$$S_N : |0\rangle \otimes_{k=1}^{N} |x_k\rangle \mapsto |\sigma\rangle \otimes_{k=1}^{N} \left|x_{\sigma(k)}\right\rangle . \tag{5.11}$$

Here, $\sigma \in \mathrm{Sym}(N)$ denotes a permutation, and the first register consists of the $s(N)$ sorting-ancilla bits. Its width is $O(b)$ and depth is $O(\log b)$. It generally requires $\Omega(N \log N)$ comparators for N elements.

Now, we discuss the quantum data-moving circuit with *formatting*, *sorting*, and *permutation*. After the permutation, the formatting and sorting are reversible. This process can be represented by

$$V_N = F_2 \circ S_{2N}^{-1} \circ P \circ S_{2N} \circ F_1 . \tag{5.12}$$

Here, all logical gates satisfy the locality condition of the network. Each step generally requires depth and additional ancilla. SWAP and X gates are used to realize F without ancilla. Its depth is only 1. It can be represented by

$$F_1 : |\vec{x}\rangle \otimes_{i=1}^{N} |0, 0, 0\rangle_{2i} \;\; |0, 0, 0\rangle_{2i+1} \mapsto \left|\vec{0}\right\rangle \otimes_{i=1}^{N} |i, x_i, a\rangle_{2i} \;\; |j_i, 0, q\rangle_{2i+1} \tag{5.13}$$

Each *packet*, i.e., ancilla registers, contains an *address* with $\lceil \log_2 N \rceil$ bits, a *flag* with 1 bit, and a *data* with d bits. q denotes question and a is for answer.

We aim to sort all packets assisted by address and flag. This will be completed by using the following subroutine S_{2N} as:

$$S_{2N} \;:\; |0\rangle \otimes_{i=1}^{N} |i, x_i, a\rangle_{2i} \;\; |j_i, 0, q\rangle_{2i+1} \mapsto |\tau\rangle \otimes_{i=1}^{N} |i, 0, q\rangle_{2i} \;\; |i, x_i, a\rangle_{2i+1} \tag{5.14}$$

where τ is the permutation with $\sigma(2i + 1) = 2i$ and $\sigma(2j_i) = 2i + 1$. The lexicographical order will be used to sort packet series. The total depth depends on the sorting network and comparator. The inverse of S_{2N} is easily followed by reverting this mapping.

The final subroutine is to transfer the sorted data to the original registers and clearing ancillas. Similar to the transformation F we can define it as

$$F_2 : \left|\vec{0}\right\rangle \otimes_{i=1}^{N} |i, 0, a\rangle_{2i} \;\; \left|j_i, x_{j_i}, q\right\rangle_{2i+1} \mapsto \left|\vec{x}_j\right\rangle \otimes_{i=1}^{N} |0, 0, 0\rangle_{2i} \;\; |0, 0, 0\rangle_{2i+1} . \tag{5.15}$$

To record which-way information, we need $O(D_{\mathcal{G}})$ qubits at each node. However, there may not be sufficient local space in low-valency graphs. One example is 1D nearest neighbor network Hirata et al. (2011). This can be resolved by computing all the moves by the classical sorting before the quantum algorithm. The depth is also $O(D_{\mathcal{G}})$.

5.3.2 Reversible Sorting Networks

The circuit model allows for long-range interactions between qubits, but this is impossible because of fault tolerance. In this section, we show that reversible sorting networks can embed any circuit into physical hardware in an asymptotical optimal manner.

The scheme assumes N quantum processors, with L qubits memory per processor, which is connected in a network $\mathcal{G}$. For a circuit with width T, If every two-qubit gate will be performed in the same or neighboring processor, each processor can then implement a single gate in each logic step. The goal is to minimize depth for implementing the circuit, subject to constraints on the network $\mathcal{G}$. The total cost in the worst case is defined by $\max_C D_p(C')/D_p(C)$ over all circuits C with the same width. Normally, the scheduling task is similar to exploring sorting networks associated with a given network. For sorting $2N$ bit series with length $\lceil \log_2 N \rceil + 2$ over $\mathcal{G}$, let $D_{\mathcal{G}}$ be the depth of the optimal circuit.

Theorem 5.3.2 *For a given network $\mathcal{G}$ with N nodes, there is an algorithm that can implement arbitrary circuits with a depth overhead $O(D_{\mathcal{G}})$ in a distributed quantum model with $O(\log N)$ qubits per processor.*

Theorem 5.3.2 provides an algorithm for minimizing the depth overhead factor to implement worst-case circuits by embedding them into a physical network. Note sorting networks can sort N elements with $O(\log N)$ depth Ajtai et al. (1983); Paterson (1990). This implies a simplified version of the AKS sorting network Paterson (1990). The following theorem shows up to logarithmic factors that this algorithm is optimal.

Theorem 5.3.3 *The lower bound for implementing any circuit with width $O(N \log N)$ on a distributed quantum computer is given by $\Omega(\frac{D_{\mathcal{G}}}{\log N \log \log N})$.*

A lower bound is found for an addition circuit on a k-dimensional nearest neighbor graph Choi and Van Meter (2011). This can be translated into implementing an additional circuit on a network. In other words, the minimum cost for a k-D nearest neighbor network is proportional to $N^{1/k}$.

5.3.3 Accessing Quantum Memory

Transferring quantum data according to a compile-time permutation has been considered for efficient distributed quantum computing. The algorithm can be modified to permutate quantum data with SWAP gates. This section will generalize the idea further to address the same indices.

Here, we examine the concept of accessing quantum memory without oracles.

Definition 5.3.1 A unitary that can access single data is a mapping $F_{1,N}$ given by

$$U_{1,N} : |j\rangle |y\rangle |\vec{x}\rangle \mapsto |j\rangle \left|y \oplus x_j\right\rangle |\vec{x}\rangle , \tag{5.16}$$

where $\oplus$ is bitwise addition.

This definition can be extended to multiple pieces of data. It is given by

$$U_{N,N} : \left|\vec{j}\right\rangle |\vec{y}\rangle |\vec{x}\rangle \mapsto \left|\vec{j}\right\rangle |\vec{y} \oplus \vec{x}\rangle |\vec{x}\rangle . \tag{5.17}$$

Theorem 5.3.4 *For any circuit which can complete $U_{1,N}$ on N registers with d qubits per register, it has the width $\Omega(Nd)$ and depth $\Omega(\log N)$.*

As any different pairs of control qubits cannot overlap, it follows there are at least $\binom{N}{2}$ computational paths. The total number of qubits for formatting computational paths is $\Omega(\log N)$. Additionally, the width is lower bound by the number of data registers.

Theorem 5.3.5 *There exists a set of circuits realizing $U_{N,N}$ with the width $O(N(\log N + d))$ and depth $O(\log N \log(d \log N))$.*

The proof relies on the fact that each gate can be performed using a polynomial number of elementary gates. Specifically, each controlled-U gate can be realized using $O(d)$ two-qubit gates and the Hadamard gate can be implemented using $O(1)$ single-qubit gates. The circuit width is then given by $O(N(\log N + d))$. Moreover, each level contains $N/2$ non-interacting pairs of control qubits. This implies the total depth $O(\log N \log(d \log N))$. The extensions of these results are generally based on the quantum circuit model for decomposing the logic gates, parallel RAM for accessing the data , and network configuration for distributing implementation Van Meter et al. (2009); Van Meter and Oskin (2006); Beals et al. (2013). In applications, one may trade off the cost with the node number and the connection degree of the network.

5.3.4 Distributed Control Systems

Distributed control systems (DCSs) present a viable solution for connecting smaller quantum devices in a network, allowing for flexibility in hardware configurations and the ability to scale by adding more devices DiAdamo et al. (2021); Hespanha et al. (2007); Yang (2006). These systems, commonly used in various applications, can be either centrally or decentralized controlled Mahmoud and Xia (2019), resembling the potential networking of distributed quantum computers.

To efficiently use a network of distributed quantum computers, developing robust communication protocols for efficient and reliable communication and control between the quantum devices is essential. In this section, we consider quantum systems with classical control and a separated quantum central processing node (QCPN) consisting of a three-layered system with inputs and outputs to a communication network through a classical computer. The interfaces with the network and controls the programmable gates based on the control instructions from the network. We assume there is a layered architecture with the network. Qubit measurements and other classical messages are transmitted back to the network via a reversed path.

We discuss two different network configurations and elaborate on how these systems could be implemented in practice. We list the communication requirements needed for performing distributed quantum computations and explore available protocols to achieve these requirements under two scenarios: one with a centralized controller and classical information and quantum entanglement-based communication to devices.

The first scenario we consider for distributed quantum computing is illustrated in Fig. 5.2. In this scenario, there is a single controller with many control units that behaves according to the instructions that are fed from this controller. The QCPN systems are connected to the controller via a classical network, and they are further connected allowing them to generate entanglement among themselves.

The proposed network scheme for quantum computers involves a centralized control unit that can access classical computers with static IP addresses and controls

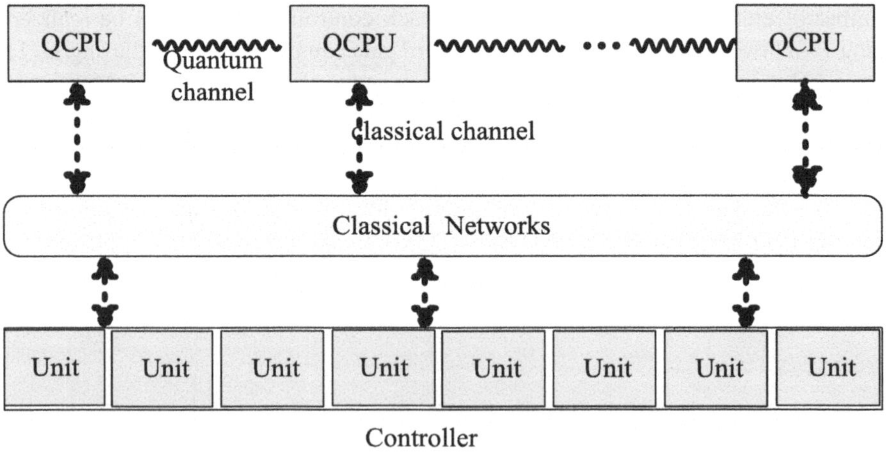

Fig. 5.2 Networked quantum control system with a centralized controller. Assume the network between the QCPNs is connected according to quantum channels assisted by classical channels

the synchronization between quantum computer processing units (QCPUs). This scheme is suitable for small and medium-sized networks, but a distributed control system is more suitable for larger networks where multiple nodes are involved. In a decentralized control system, resources are requested from different parties, and communication between QCPUs is facilitated by a low-latency protocol, as shown in Fig. 5.3.

To execute instructions synchronously on networked quantum hardware, we propose a temporal operation schedule with precise timestamps for execution. These schedules can then be sent to each QCPU in the network with a time to begin execution. As quantum gates have an upper bound for how long they take to execute, we can use this information when generating the schedule. We assume that all gates have a known execution time, and latency times for classical communication and entanglement generation are also known.

Definition 5.3.2 (Distributed Quantum Algorithm Scheduling) Produce a schedule for executing gates in a distributed circuit with multiple gate layers. Each layer has a set of gates that operate on specific qubits, and each gate has a specified execution time for each processing unit in the system. The schedule must follow certain constraints.

In the given problem scenario, with synchronized clocks among all nodes, classical communication, and entanglement occur simultaneously for both sending and receiving parties. Each qubit can undergo only one operation per time instance during the gate time, with no overlapping gates. Additionally, controlled operations require

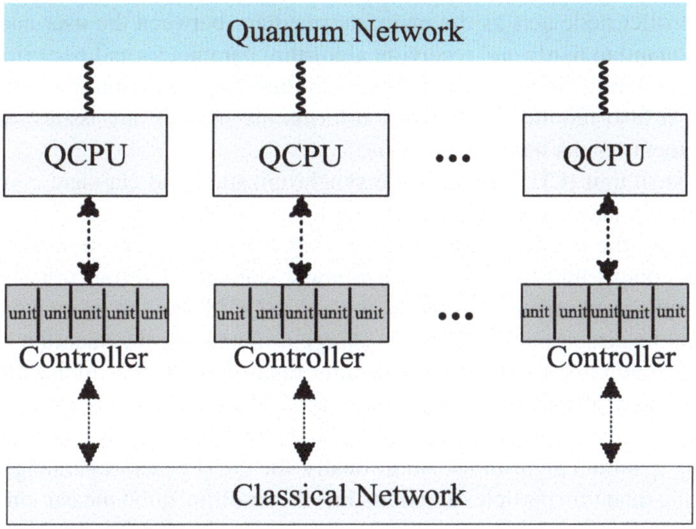

Fig. 5.3 Internetworked distributed QCPNs with controllers. There are independent controllers with different controlling units that may control a set of QPCNs. The quantum networks will be used for entanglement routing. The controller is between the QCPNs and the classical network

both qubits to be available for performing the control gate, meaning ongoing gate operations on one of the qubits are not allowed.

To address this problem, a temporal planning and constraint programming approach is adopted Booth et al. (2018). However, this approach differs in that it assumes worst-case gate times which include the possibility of swapping qubits to their nearest neighbor if needed, and re-mapping the index of the qubit so it does not have to be swapped back. No constraint programming is used at the end of each layer of gates, assuming each qubit is free to be operated on without swapping.

To run a distributed quantum algorithm on a distributed quantum computer there is classical communication for sending computation instructions to QCPUs and control Units between QCPUs. Clock synchronization may be used. In general, it may require certification steps to ensure the cooperative execution of distributed algorithms by multiple QCPUs.

When using a decentralized controller, additional protocols are necessary to ensure that all parties can execute distributed quantum algorithms and perform nonlocal operations synchronously. In this situation, the user has no control over the quantum hardware, so a protocol is needed to guarantee that the user's instructions are executed as specified.

Cflow is a system that uses nodes in a graph to control information flow within a hardware network. The nodes contain instructions on what to do when an input message is received. Here, we describe the highest level nodes used and their logic for conducting a distributed α-VQE with a single centralized controller. Hardware-specific nodes can be added within these high-level nodes in the future. Figure 5.4 shows a Cflow graph on top of the hardware, with three nodes:

- The controller node acts as the main intermediary between the user and the distributed quantum hardware, receiving algorithm parameters and performing tasks such as validating circuits, distributing Hamiltonians, and sending instructions to QCPUs. It also monitors for errors, informs the user of any issues, and sends measurement results back for analysis.
- The classical unit (CU) handles time synchronization and classical communication between quantum circuit processing units (QCPUs), receiving timing information from the controller and coordinating communication. It works with the quantum control unit to collect measurement statistics and transmit them to the corresponding quantum circuit node, with each QCPU having its own paired CU.
- The quantum circuit unit (QCU) is in charge of managing the quantum hardware, executing gate operations, and performing measurements on qubits. It interacts with the classical unit (CU) to receive classical control information for nonlocal operations and communicates with the controller to receive instructions for executing quantum algorithms. Additionally, the QCU generates entanglement by controlling quantum particles and transmits the resulting qubit measurements back to the controller.

In a decentralized controller setting, the CGraph is divided among the QCPUs, and each QCU has its own CGraph, with intercommunication managed within each

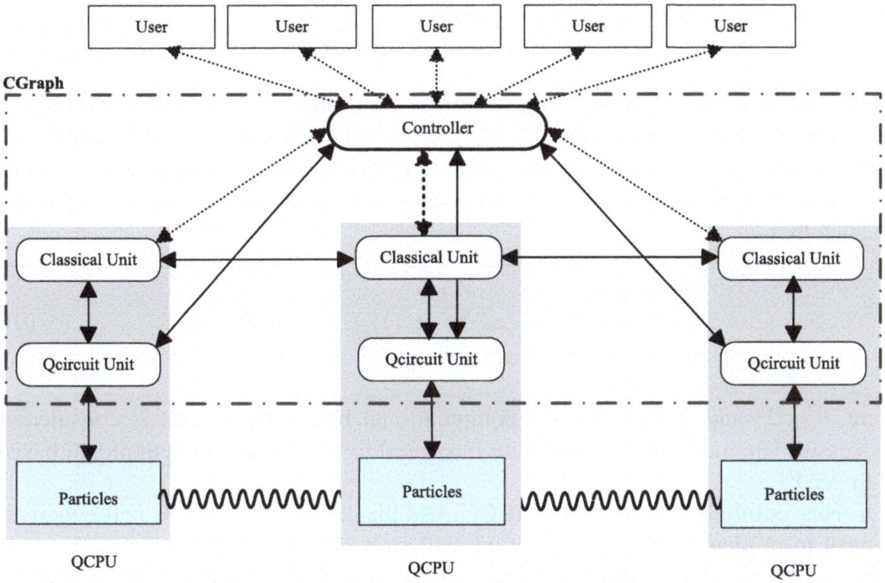

Fig. 5.4 CGraph laid out on top of distributed QCPUs. The CGraph controls the distributed computations performed by various QCPUs. Different users may input special parameters to one controller for setting up computation tasks

CGraph. Agreed-upon contracts ensure secure communication without malicious parties. Each remote QCU has its controller and time reference node, allowing for different clock hardware and synchronization protocols. The distributed QRAM model extends classical RAM to quantum computing, enabling efficient simulation and analysis of quantum algorithms by utilizing classical inputs and outputs with quantum memory for parallel processing and quantum computation. While powerful in theory, its implementation relies on advancements in quantum hardware and software.

5.3.5 Linear Combining Quantum Operations

The linear combining of quantum operations allows for the representation of complex quantum computations as combinations of simpler operations, enabling more efficient resource utilization. This technique is particularly advantageous for implementing quantum computing on remote servers while preserving privacy and security.

To implement a unitary U_T, which operates on a d-dimensional target subspace, we can represent it as a sum of gates V_j with complex coefficients α_j. Especially, consider U_T to be expressed in the form,

$$U_T = \sum\nolimits_{j=0}^{n-1} \alpha_j V_j, \tag{5.18}$$

where the V_j are gates acting on a d-dimensional target (T) subspace, and the α_j are complex coefficients satisfying $\sum_{j=0}^{n-1} |\alpha_j|^2 = 1$. When controlled-$V_j$ gates are available, we can implement the unitary U_T in a probabilistic manner. In the circuit, the initial state of the k-qubit control register (C) encodes the complex coefficients α_j such that

$$|\phi\rangle_C = \sum_{j=0}^{n-1} \alpha_j \, |j\rangle_C \,, \tag{5.19}$$

Here, $n = 2^k$ and j represent the computational basis. The circuit is considered successful when all control qubits are measured to be 0 in the computational basis at the end.

Linear combination operations (LCC) use black box unitaries W and coherent control to implement a target quantum evolution on a $(n \times d)$-dimensional target subspace T, which decomposes it into n d-dimensional subspaces. Subspace-swap operations, controlled by qubits in C, are used to exchange basis elements between the 0th and jth subspaces and a sum operation is performed using $V_j^{(j)}$ on each subspace. The initial state for T is a superposition state given by

$$|\Psi_{ext}\rangle_T = \sum_{j=0}^{d-1} \beta_j |j\rangle_T. \tag{5.20}$$

When control qubits are measured to be 0, the target evolves according to $\sum \alpha_j V_j^{(0)} \, |\Psi_{ext}\rangle_T$, with a success probability $1/n$.

Cartan's KAK decomposition Theorem allows for the explicit representation of any two-qubit unitary operation in $\mathbb{SU}(4)$ as a linear combination of four tensor products of single-qubit gates. Similarly, an n-qubit unitary operation in $\mathbb{SU}(2^n)$ can be decomposed into a tensor product of n single-qubit gates using this method, although this may not always be efficient due to the potential for exponentially many terms Khaneja and Glaser (2001). As a result, the success probability of LCC may be exponentially small for general unitary operations.

However, for certain families of unitary operations, such as controlled-unitary gates, the linear decomposition method can be highly efficient. For instance, the n-qubit controlled-unitary gate CU can be decomposed as $\frac{\mathbb{1}+Z}{2} \otimes \mathbb{1} + \frac{\mathbb{1}-Z}{2} \otimes U$, where U is an $n-1$-qubit operation Zhou et al. (2011). With only one control qubit required, a high success probability can be achieved when implementing this operation.

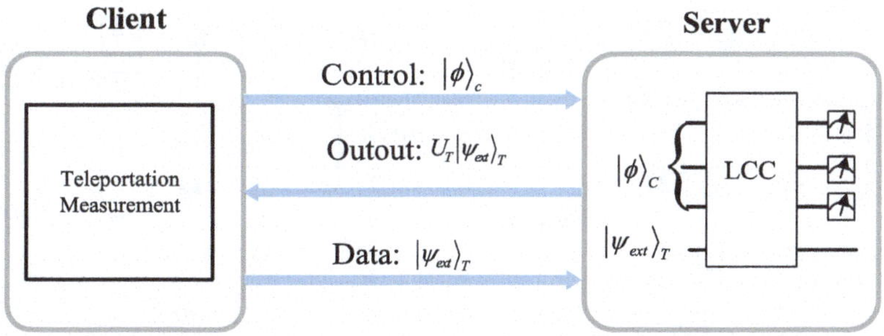

Fig. 5.5 Protocol for remote quantum processing

5.3.6 Remote Quantum State Control

The LCC, as described above, enables quantum information processing through a client-server model, as shown in Fig. 5.5. The computational resources V_j are provided by the server, and the client configures the α_j's to encode the algorithm. These parameters are encoded into the control state $|\phi\rangle_C$ and sent remotely from the client to the server using a (multi-)qubit teleportation protocol with generalized Bell measurements. The control state has k qubits and k EPR channels must be shared between the client and server to enable the teleportation of this state Qiang et al. (2017).

Similarly, to teleport the computational input $|\Psi_{ext}\rangle_T$ from the client to the server and the computational output back to the client, $\lceil \log_2 d \rceil$ EPR channels are needed for each step (d is defined previously) Bennett et al. (1993); Chen et al. (2006). To initiate the computation, the client requests the server to run the LCC repeatedly on the EPR channels while resetting them as needed. When the LCC succeeds, the server notifies the client and performs teleportation measurements on the LCC output and corresponding EPR channels. Finally, the client performs teleportation measurements on $|\phi\rangle_C$ and $|\Psi_{ext}\rangle_T$ (and their corresponding EPR channels). Upon successful completion of all LCC and teleportation steps, $U_T\,|\Psi_{ext}\rangle_T$ is returned to the client.

To ensure the security of the client's computation, the protocol keeps the client's control state $|\phi\rangle_C$ hidden from the server. In the case where the client sends a one-qubit control state to implement a linear combination of two quantum operations UU and VV, the corresponding circuit is shown in Fig. 5.6a. Here, we assume that UU and VV are not black-box operations and ignore the teleportation of the input state for the computation. The circuit starts with the initial state $\frac{1}{\sqrt{2}}\,|0\rangle_1\,(|0\rangle_2\,|0\rangle_3 + |1\rangle_2\,|1\rangle_3)\,|\varphi\rangle_4$.

If the server follows the protocol, they run the LCC until it succeeds, at which point qubit 3 (the local control qubit) is measured to be 0" on the computational basis. The state of the remaining qubits becomes $\frac{1}{\sqrt{2}}(|0\rangle_1\,|0\rangle_2\,U\,|\varphi\rangle_4 + |0\rangle_1\,|1\rangle_2\,V\,|\varphi\rangle_4)$. When the client performs the quantum teleportation and measures qubits 1 and 2 to be 0"

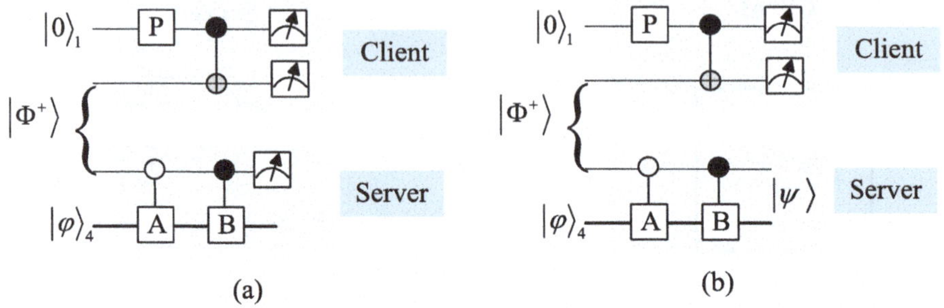

Fig. 5.6 Secure quantum processing with one-qubit control. **a** An EPR state is shared between the client and the server. **b** The server does not measure the local control qubit

in the computational basis, the state of the remaining qubit immediately becomes $(\alpha U + \beta V)\,|\phi\rangle_4$. Throughout the process, the server cannot detect the control state, encoded in qubit 1 by the client's local operation PP, as they need to measure the local control qubit (qubit 3) before the client configures the control.

In the case where the server does not measure the local qubit before the teleportation, the circuit evolves as shown in Fig. 5.6b. When the client measures qubits 1 and 2 to be "0", the state of the remaining qubits becomes $\alpha\,|0\rangle_3\,U\,|\varphi\rangle_4 + \beta\,|1\rangle_3\,V\,|\varphi\rangle_4$ (denoted as $|\Psi\rangle$). The question is whether the server can extract the information of the control state $|\phi\rangle_C = \alpha\,|0\rangle + \beta\,|1\rangle$ without being detectable by the client.

To achieve this, the server needs to extract $|\phi\rangle_C$ and output the correct computation result $(\alpha U + \beta V)\,|\varphi\rangle_4$ to the client. In other words, the server must find an operation U_s that satisfies $(\alpha\,|0\rangle + \beta\,|1\rangle)(\alpha U + \beta V)\,|\varphi\rangle = U_s(\alpha\,|0\rangle\,U\,|\varphi\rangle + \beta\,|1\rangle\,V\,|\varphi\rangle)$. However, this operation U_s cannot exist for unknown parameters α and β because it would violate the no-cloning theorem by allowing copying of an unknown quantum state Wootters and Zurek (1982); Dieks (1982).

Nevertheless, an attacker could still generate a copy of the control state with imperfect fidelity, using a universal quantum cloning machine (UQCM) Bužek and Hillery (1996); Gisin and Massar (1997), even with a single copy of the control state. Such cloning attacks are difficult to prevent as they could be disguised as channel loss, and thus could lead to information leakage about the client's computation.

In applications such as Shor's factorization algorithm and Grover's search algorithm, only a few runs of the protocol may be needed for the client to obtain the result. However, in this scenario, the server or an attacker could potentially gain partial information about the control state using a UQCM. For applications requiring numerous runs, multiple copies of the control state would need to be sent by the client, increasing the risk of the server gaining complete information about the control state through methods like quantum state tomography.

5.3.7 *Noisy Control State*

To improve security, the client can use decoy states. For a computation with control state $\rho = |\phi\rangle\langle\phi|$, the client generates a decoy-state ρ_m according to

$$\rho_m = \frac{1+v}{n}\mathbb{1} - v\rho, \tag{5.21}$$

where n is the number of dimensions of ρ, and $0 < v \leq 1/(n-1)$. The client sends the control state ρ with probability $v/(1+v)$ and the decoy state ρ_m with probability $1/(1+v)$ on each run of the protocol. By discarding the output states corresponding to the decoy states, the client keeps only the correct ones for further applications.

From the server's perspective, the received state becomes

$$\frac{v}{1+v}\rho + \frac{1}{1+v}\rho_m = \frac{1}{n}\mathbb{1}. \tag{5.22}$$

Since the state $\frac{1}{n}\mathbb{1}$ has maximal entropy (=$\log n$), the server does not know the received states at all. Therefore, using decoy states improves security by making it impossible for the server to determine which runs correspond to the actual computation.

In certain applications such as Shor's factorization and Grover's search, the client can directly verify the results of quantum computations, while for others, such as large simulations, direct verification is not possible. However, the client can still monitor the computation process by decomposing components into verifiable single-qubit gates and randomly sending basis states to the server, allowing for diagnosis of correct execution while maintaining control state privacy.

The client can delegate a computation to a server while keeping it private by encoding the control state in a single qubit and sending it to the server. The server applies the operation without learning any information about the control state. To prevent attacks, the client can perform quantum verification by randomly sending basis states and checking the output, combined with a strategy to prevent the control state from being measured. These techniques enable secure quantum computation in resource-limited scenarios.

5.4 Distributed Shor Factoring Algorithm

We introduce a distributed version of Shor's quantum factoring algorithm that operates on a distributed quantum model. This allows multiple small-scale quantum computers to collaborate and simulate the performance of a large-scale quantum computer.

5.4.1 Quantum Fourier Transform

The quantum Fourier transform (QFT) is a crucial component in quantum computing, used in algorithms like Shor's and Grover's. It maps a quantum state to another state, encoding frequency components, and its efficient implementation relies on techniques such as modular arithmetic and circuit optimization.

In general, the QFT is defined on standard basis states as follows,

$$|j\rangle \rightarrow \frac{1}{\sqrt{2^n}} \sum_{k=0}^{2^n-1} e^{2\pi ijk/2^n} |k\rangle , \tag{5.23}$$

where n is the number of qubits. Instead of decomposition circuits for local QFT Nielsen and Chuang (2000); Cleve et al. (1998), here implements a distributed version of the Fourier transformation by replacing a controlled R_k with non-local controlled R_k, when necessary. This requires a distributed swap gate that enables the teleportation of qubits between two computers as needed. It allows multiple small-scale quantum computers to collaborate and simulate the performance of a large-scale quantum computer. Our method presents an alternative approach for implementing the QFT on a distributed quantum network model, where small quantum computers work together to perform complex computations.

We can implement a distributed Fourier transformation for $2n$ qubits with two quantum computers where the gate R_k is defined as: $R_k = \text{diag}(1, e^{2\pi i/2^k})$, where $k \in \{2, 3, \cdots\}$ determines the rotation angle of the gate. The distributed Fourier transformation uses non-local controlled R_k gates instead of locally controlled R_k gates. Unlike local QFT, in the distributed version, the qubits are divided into two groups and transformed separately. The first group contains the most significant n qubits, while the second group contains the other qubits, with each group located on a different computer or node in the network. This may extend to multiple quantum computers.

To perform the computation, each group applies their respective QFTs independently, followed by a controlled swap gate that entangles each group. Finally, all groups apply their inverse QFTs to complete the transformation.

5.4.2 Phase Estimation Algorithm

The phase estimation algorithm (PEA) is a quantum algorithm that accurately estimates the phase of an eigenvector of a unitary operator. It is used as a subroutine in various quantum algorithms and relies on controlled rotations and inverse Fourier transform. PEA can provide exponential speedup for certain problems but requires error correction due to sensitivity to errors in ancilla preparation and measurements.

Definition 5.4.1 Let U be an n-qubit unitary transformation. PEA is to estimate the phase θ_t of corresponding givenvalue $e^{2\pi i\theta_t}$.

To solve this problem, it may construct two quantum registers Cleve and Buhrman (1997). A unitary transformation $c_m(U)$ will be designed on both registers as follows:

$$c_m(U) : |k\rangle |\psi\rangle \rightarrow |k\rangle U^k |\psi\rangle , \tag{5.24}$$

where $|k\rangle$ and $|\psi\rangle$ denotes respectively the state of the first and second register.

The phase estimation algorithm can be described as follows:

Input: U and $|\psi_t\rangle$
Output: An estimate of θ_t.

(1) Let $|r_1\rangle |r_2\rangle = |0\rangle |\psi_t\rangle$.
(2) $|r_1\rangle |r_2\rangle = (H^{\otimes m} \otimes \mathbb{1}) |r_1\rangle |r_2\rangle$.
(3) $|r_1\rangle |r_2\rangle = c_m(U) |r_1\rangle |r_2\rangle$.
(4) $|r_1\rangle |r_2\rangle = (QFT^{-1} \otimes \mathbb{1}) |r_1\rangle |r_2\rangle$.
(5) j = the result of measuring $|r_1\rangle$.
(6) Output $j/2^m$.

Step (1) initializes the registers to $|0\rangle |\psi_t\rangle$ using input $|\psi_t\rangle$. In step (2), the first register undergoes a Hadamard transformation, resulting in the state $\sum_{k=0}^{2^m-1} |k\rangle |\psi_t\rangle /\sqrt{2^m}$. Step (3) applies the controlled unitary operator $c_m(U)$ to the registers, which involves applying U conditioned on the first register being in state $|k\rangle$ for all k. This transforms the registers into $\sum_{k=0}^{2^m-1} e^{2\pi i k\theta_t} |k\rangle |\psi_t\rangle /\sqrt{2^m}$. Step (4) applies the inverse QFT to the registers, resulting in the state $|j\rangle |\psi_t\rangle$. In step (5), measuring the first register provides the value of j, where $\theta_t = j/2^m$.

In many cases, the exact value of an eigenvalue θ_t may be unknown beforehand, and it is desirable to obtain an estimate of its value using quantum computations. The inverse QFT in step (4) of the phase estimation algorithm provides an m-bit approximation of θ_t, denoted by $j/2^m$, with high probability.

It is important to note that although θ_t may not have the form of $j/2^m$, the inverse QFT in step (4) provides the best m-bit approximation of θ_t. Specifically, the probability of obtaining an m-bit approximation that is off by more than $1/2$ is bounded above by $4/\pi^2$ Cleve et al. (1998). This implies that the precision of the estimated eigenvalue increases as m increases. The algorithm has various applications in quantum computation, including quantum simulation, factoring large numbers, and implementing quantum algorithms such as the quantum Fourier transform.

5.4.3 Order Finding Algorithm

The order finding problem is a fundamental problem in computational number theory that involves finding the period of a function $f(x) = a^x \bmod N$ for integers a and N. The period of this function determines the order of a modulo N, which is the smallest integer r such that $a^r \equiv 1 \bmod N$. The order-finding problem has important applications in cryptography, including the factorization of large composite numbers using Shor's algorithm.

Definition 5.4.2 Given a positive integer N and an integer a relatively prime to N, find the smallest positive integer r such that $a^r = 1 \mod N$.

Consider the transformation M_a, which maps an n-qubit register $|x\rangle$ to $|ax \mod N\rangle$, where a and N are integers. By defining $\omega = e^{2\pi i/r}$ for some integer period r, we can construct eigenvectors $|\psi_t\rangle$ for each t using

$$|\psi_t\rangle = \frac{1}{\sqrt{r}} \sum_{s=0}^{r-1} \omega^{-st} \left|a^s\right\rangle. \tag{5.25}$$

These eigenvectors have corresponding eigenvalues ω^t. It follows that M_a applied to $|\psi_t\rangle$ results in $\omega^t |\psi_t\rangle$. By approximating θ_t as t/r for each t, we can use the phase estimation algorithm to find the period r of a if we have an eigenvector $|\psi_1\rangle$ and knowledge of how to construct $c_m(M_a)$.

The Shor factoring algorithm is a quantum algorithm that efficiently determines the order of an integer modulo N using a quantum computer Shor (1994). It utilizes the phase estimation subroutine to estimate eigenvalues of the unitary operator performing modular exponentiation. These estimated eigenvalues, which are rational numbers with denominators as powers of 2, are then used as potential values of the order. The classical part of the algorithm employs techniques like continued fractions to determine the actual order. Shor's algorithm and the order-finding problem showcase the power of quantum computation, leading to advancements in quantum information and computation, error-correcting codes, quantum communication protocols, and quantum complexity theory.

To implement the unitary transformation $c_m(M_a)$ using the method of repeated squaring, we can express the contents of the first register in binary as $k_{m-1}k_{m-2}\ldots k_1k_0$, and then compute $M_a^k = \prod_{i=0}^{m-1}(M_a^{2^i})^{k_i}$ by implementing each term $(M_a^{2^i})^{k_i}$ as a controlled $M_a^{2^i}$ with control qubit $|k_i\rangle$. Since a is a constant integer and $M_{a^{2^i}} = M_a^{2^i}$ for all $0 \leq i < m$, we can precompute the values of a^{2^i} classically and use the same technique used to implement M_a to implement $M_{a^{2^i}}$.

For a given polynomial-time function f we can construct a unitary transformation F that maps $|x\rangle|0\rangle$ to $|x\rangle|f(x)\rangle$. However, the complete definition of F also includes ancillary qubits that contain information necessary for F to be reversed, computed by a function g.

To construct M_a, we first implement binary adders: binary full adders denoted as $\mathrm{BFA}_a : |c, b, 0\rangle \mapsto \left|a \oplus b \oplus c, b, c'\right\rangle$ and binary half adders denoted as BHA_a :

$|c, b\rangle \mapsto |a \oplus b \oplus c, b\rangle$. Here, $|c\rangle$ and $|c'\rangle$ are input and output carries, respectively, and a is a classical bit. For every classical integer a with n bits, FA_a is a unitary transformation on n-qubit full adder. It transforms the state $|b, 0, c\rangle$ to $|b, s, c'\rangle$, where $|s\rangle$ is an n-qubit register and $s = a + b + c \mod 2^n$, c and c' are input and output carries. To create HA_a, we replace the last $\text{BFA}_{a_{n-1}}$ with a $\text{BHA}_{a_{n-1}}$, which only requires $n - 1$ input ancillary qubits with an initial state of $|0\rangle$. When an input carry qubit $|c\rangle$ is included, HA_a becomes a $2n$-qubit unitary transformation.

Implementing the n-qubit adder AN_a modulo N will take use of FA_a and HA_a. If $a + b < N$, then $a + b \mod N = a + b \mod 2^n$. Otherwise, $a + b \mod N = a + b + 2^n - N \mod 2^n$. The transformation AN_a is computed by adding $|b\rangle$ with the classical number $a + 2^n - N$ modulo 2^n. If the carry is not set, subtract $2^n - N$. The resulting transformation AN_a is given by

$$\text{AN}_a : |b, 0, 0, 0\rangle \mapsto |b, s, c, a + b \mod N\rangle, \tag{5.26}$$

where $s = b + a + 2^n - N \mod 2^n$ and c is the carry. The adder A_a is defined as

$$A_a = \text{XAN}^r_{-a} \cdot \text{SWAP} \cdot \text{XAN}_a, \tag{5.27}$$

which maps $|b\rangle$ to $|a + b \mod N\rangle$. The transformation A_a is an n-qubit transformation with $3n + 1$ ancillary qubits.

To construct M_a, which maps $|x\rangle$ to $|ax\rangle$, we define an $2n$-qubit unitary transformation

$$\text{MF}_a : |x, 0\rangle \rightarrow |x, ax \mod N\rangle. \tag{5.28}$$

Express ax as $\sum i = 0^{n-1} ax_i 2^i$, where $x_{n-1} \dots x_1 x_0$ is the binary representation of x. For each $0 \le i < n$, the term $ax_i 2^i$ can be implemented using control-A_{a2^i}, where $|x_i\rangle$ is the control qubit and A_{a2^i} transforms $|b\rangle$ to $|b + a2^i \mod N\rangle$. As $a2^i$ is constant for each i, we can compute each $a2^i$ using a classical computer. This allows to implementation of MF_a using the method of repeated squaring.

5.4.4 Distributed Shor Factoring Algorithm

Assuming that we have a quantum computer network consisting of $(n + c)$-qubit devices, where $n = \log N$, with c extra qubits available for each device to be used either as channel qubits or ancillary qubits. We will show that the value of c is independent of the input number N. By choosing $m = 2n$, we need $7n + 1$ qubits for this particular implementation.

To perform a distributed quantum computation of the XANDER adder, we begin by splitting the control register of $c_m(M_a)$, denoted by $|k\rangle$, into two groups of n-qubits each. These two groups are placed on separate computers, with each qubit $|k_i\rangle$ in the groups remotely controlling the transformation $M_a^{2^i}$. Another computer is assigned to hold the control register of MF_a, represented by $|x\rangle$, where each qubit $|x_j\rangle$ remotely controls A_a.

The XANDER adder includes the transformation XAN_a as a part of AN_a. This generally requires four computers. For XAN_a an n-qubit input register is labeled by $|b\rangle$ and an n-qubit output register is labeled by $|a+b\rangle$. As XAN_a uses $2n+1$ ancillary qubits, the computation is evenly divided among the four computers. Each computer computes 1/4 of each register and has two extra carry qubits used for FA and FA'. The implementation of a distributed control FA involves distributing two control qubits to the target qubit-holding computer. The double control is then locally executed.

For each FA' and the single HA, there are three controlling qubits. One qubit comes from the first register $|k\rangle$, another from the register $|x\rangle$, and the last from the output carries a bit of FA_{a+a^n-N}. To implement a non-local three-control gate, all three control qubits are distributed to the target computer, where the control gate is locally executed with three control qubits.

Since each computer has $n/4$ of the intermediate output register and the final output register, implementing distributed COPY that is a bitwise copy is simple by applying local CNOT gates. However, COPY still requires remote control by two qubits from register $|k\rangle$ and $|x\rangle$.

Likewise, each machine owns $n/4$ qubits of both the input register and output register. The distributed SWAP can be locally implemented on each machine while being remotely controlled by two qubits from the $|k\rangle$ and $|x\rangle$ registers.

5.4.5 Communication Complexity

The communication complexity of distributing a quantum algorithm refers to the amount of information exchanged between different quantum devices during its execution. Quantum communication involves transmitting quantum states, which is more challenging due to noise and decoherence.

Minimizing communication is crucial for efficient execution, achieved through careful algorithm design and communication protocols. In the case of distributing the Shor algorithm, the communication complexity includes the number of entangled pairs and classical bits exchanged between devices. The optimal cost for a non-local operation is one EPR pair and two classical bits, allowing the estimation of communication overhead by counting non-local control gates and teleportation circuits.

The communication overhead for a distributed implementation of circuit F can be more precisely estimated as $O(Ln^2)$, where L is the number of non-local control gates implemented in the distributed implementation of circuit F. This takes into account the fact that the cost of each non-local control gate can be shared among its elementary gates. For distributing the Shor algorithm, there are 8 non-local control gates per AN_a. Four non-local control circuits are sufficient to implement COPY or SWAP. This implies $L = O(mn + m^2)$ for Shor algorithm.

Similarly, denote K as the number of teleportation circuits implemented in the distributed implementation of circuit F. It is easy to check $K = O(mn)$. This implies the communication complexity of distributing Shor algorithm is $L + K = O(mn + m^2)$.

5.5 Distributed VQE

The short coherence time of qubits without error correction poses a challenge for implementing deep circuits in near-term quantum computing. However, variational quantum algorithms, such as the variational quantum eigensolver (VQE), combine classical and quantum components to effectively execute shallow circuits and find the minimum eigenvalue of a chemical Hamiltonian Peruzzo et al. (2014).

The accelerated variational quantum eigensolver (AVQE) is a fundamentally new approach to solving for the ground state of a Hamiltonian. The distributing AVQE algorithm opens the door for many researchers to access this powerful tool by using near-term devices.

5.5.1 *Scheduling Hamiltonian*

To express the Hamiltonian H of a molecular electronic system, one can employ a series of Pauli matrices that comprises a polynomial number with respect to the system size. This can be represented mathematically as:

$$H = \sum_i a_i P_i. \tag{5.29}$$

Here, $P_i \in \{\mathbb{1}, X, Y, Z\}^{\otimes n}$ denotes a Pauli string that constitutes a tensor product of qubit n Pauli operators, which may be the identity, and $a_i \in \mathbb{R}$ denotes the coefficient.

Example 5.5.1 Under the Jordan-Wigner transformation, the Hamiltonian of molecule H_2 can be decomposed into the sum of Pauli strings as:

$$\begin{aligned} H = & -\frac{J}{2}(Z_0 Z_1 - X_0 X_1 - Y_0 Y_1) + \frac{1}{2}(Z_0 + Z_1) \\ & + \frac{1}{4R}(X_0 - X_1 + Y_0 - Y_1 - 2Z_0 - 2Z_1), \end{aligned} \tag{5.30}$$

where X, Y, Z are the Pauli matrices that represent spin operators for two hydrogen atoms, J is the coupling constant, and R is the distance between the two hydrogen atoms.

Example 5.5.2 Consider a LiH molecule with a bond length of 1.547 angstroms. Under the Born-Oppenheimer approximation, the Hamiltonian can be expressed in the second-quantized form as follows:

$$\begin{aligned} H = & -3.3283\mathbb{1} + 0.9437Z_0 - 0.9437Z_1 + 0.1813Z_0Z_1 \\ & +0.4149X_0X_1 + Y_0Y_1 - 0.0747X_0X_1 - Y_0Y_1 + 0.1686Z_2Z_0 \\ & -0.1686Z_2Z_1 + 0.0456Z_2Z_0Z_1 + 0.0456(Z_2X_0X_1 + Z_2Y_0Y_1 \\ & -Z_2X_0Y_1 - Z_2Y_0X_1), \end{aligned} \tag{5.31}$$

where $\mathbb{1}$ denotes the identity operator, X_i, Y_i, Z_i denotes the Pauli operators acting on the i-th qubit, and the electronic structure of the molecule is calculated for a fixed nuclear configuration. The specific values of the coefficients will depend on the basis set and level of theory used for the electronic structure calculation.

To efficiently use distributed quantum computing, a parallelized and distributed approach for expectation value estimation is preferred. The motivation behind this strategy is based on the linear nature of obtaining $\langle\psi|H|\psi\rangle$, which permits breaking the summation into separate parts. For each component, an n qubit Ansatz must be developed to estimate each $\langle\psi|P_i|\psi\rangle$ independently, and then evaluate $\langle\psi|H|\psi\rangle$ by rejoining the outcomes. The distributed quantum processing unit (QPU) architecture necessitates allocating the qubits appropriately so that Ansatz states can be prepared for all P_i in the summation. Subsequently, the coefficients can be combined to arrive at a final value for $\langle\psi|H|\psi\rangle$.

We need to allocate qubits for an α-VQE computation, where we have N qubits, and we will perform α-QPE on one additional qubit. We may split the Ansatz between M QPUs and need to reserve qubits for entanglement between QPUs. The worst case is when a three-qubit control gate is split between QPUs during α-QPE. Use the method of cat-entangling and disentangling, where we reserve two qubits from each QPU for entanglement. Determine the minimum number of qubits needed for this computation.

Definition 5.5.1 (Ansatz Distribution Problem): Given a Hamiltonian H and a collection of m QPUs, output an estimated expectation $\langle\psi|H|\psi\rangle$ for a given Ansatz $|\psi\rangle$.

To create an ansatz, if P_i is divided among two QPUs, two qubits from each QPU must be assigned for nonlocal operations to prepare the Ansatz $|\psi\rangle$ across multiple QPUs. Additionally, one qubit should be kept aside for α-QPE. The problem's resolution provides a distributed schedule that allows running over the distributed system to approximate $\langle\psi|H|\psi\rangle$.

The ansatz distribution problem involves finding the optimal distribution of ansatz parameters for a set of p Pauli strings on m QPUs, such that the expectation values of all Pauli strings are minimized. A simple greedy algorithm for solving this problem could be as follows:

- Initialize each parameter distribution as uniform over $[0, 2\pi]$.
- For each QPU in sequence, compute the expectation value of each Pauli string using the current parameter distribution.
- Identify the Pauli string with the highest deviation from its target expectation value, and select the corresponding parameter distribution to adjust.
- Perturb the selected parameter distribution by a small amount in the direction that reduces the deviation of the target expectation value (use a gradient descent step).

This algorithm is greedy in the sense that it only considers the locally-optimal adjustment at each step, without considering the global optimality of the entire distribution. However, it can provide a reasonable solution in practice and does not require any complex optimization methods. This algorithm executes with run time on the order of $O(pm^2)$. One example is by using a Pauli string as $[XZ, ZY, ZZ]$ with two QPUs. The optimal parameter distributions are given by: [4.5639, 4.0438], [0.5616, 2.3561], [4.6833, 1.5708] for QPU 1, and [3.7163, 2.6779], [1.0035, 0.4142], [1.6992, 3.9270] for QPU 2 by using PennyLane, a quantum machine learning library.

As another approach to solving the ansatz distribution problem, we use constraint programming. The tradeoff with constraint programming is that setting up a collection of constraints is generally straightforward forward but solving constraint problems on a finite domain is generally NP-complete, trading simplicity for time. Define the variables: Create a variable for each ansatz that represents its distribution. A general outline for constraint programming algorithm is given by:

- Assign a domain of possible values for each variable, representing the range of potential distributions for each ansatz.
- Specify the constraints that limit the valid combinations of distributions.
- Use a heuristics-based algorithm such as backtracking to search through the possible solutions.
- Verify that the solution satisfies all constraints.

Adding more constraints is straightforward relative to implementing the same constraints greedily. This provides an easy way for solving the ansatz distribution problem.

5.5.2 Distributing α-VQE

In this section, we adapt the proposed α-VQE from Wang et al. (2019) to a distributed system. Instead of statistical sampling, the α-VQE employs α-QPE as a subroutine, which is more efficient when $\alpha > 0$ but requires the ability to run deeper circuits on quantum hardware. The section focuses on combining the results Eisert et al. (2000) by utilizing nonlocal control gates over separated QPNs.

The QPE algorithm is a crucial component of various popular quantum algorithms, including the Shor algorithm. It is utilized to determine the phase of a quantum state $|\psi\rangle$ that results from applying a specific unitary operation U to it

Kitaev (1997). Here, $|\psi\rangle$ is an eigenstate of U, and QPE aims to estimate the phase ϕ in $U|\psi\rangle = e^{2i\pi\phi}|\psi\rangle$ with high probability.

Distributed a**-QPE**

Input: A schedule S providing the qubit mapping of an ansatz of size n for p Pauli strings on m QPUs. The vector constants for each Pauli string $A = [a_1, \cdots, a_n]$. The circuits for c-QPE $U = [U_1, \cdots, U_n]$. The Pauli operators $P = [P_1, \cdots, P_n]$ associated with each U.

Output: The value of the estimated expectation value $\alpha - QPE(S, A, U, P)$.

- for $r \in S$, initalize $|S(r)|$ length array;
- for $p \in S(r)$, estimate $|\langle\psi(\lambda)|P_p|\psi(\lambda)\rangle|$ away from 0 and 1; Estimate $\langle\psi(\lambda)|P_p|\psi(\lambda)\rangle$ with statistical sampling using constant distribution circuits using the ansatz distribution p.

The distributed a-VQE algorithm is a modified version of QPE called α-QPE, which uses a free parameter α chosen based on circuit depth. α-QPE estimates the eigenphase in $U|\phi\rangle = e^{\pm i\phi}|\phi\rangle$, where U is a rotation operator that rotates an ansatz $|\psi\rangle$ by an angle ϕ in the plane spanned by $\{|\psi\rangle, P|\psi\rangle\}$, and P is a Pauli string. The algorithm executes a circuit to obtain a value E, performs rejection sampling to produce a posterior distribution for ψ, and repeats until sufficient accuracy is reached. Finally, it recovers $|\langle\psi|P|\psi\rangle|$ using the relation $|\langle\psi|P|\psi\rangle| = \cos(\phi/2)$ and mechanisms to recover the sign of $\langle\psi|P|\psi\rangle$ are provided in Wang et al. (2019).

One example is using the reflection operator $\Pi := \mathbb{1} - 2|0\rangle\langle 0|$ and $U := R\Pi R^\dagger P R\Pi R^\dagger P^\dagger$ to estimate the eigenphase in $U|\phi\rangle = e^{\pm i\phi}|\phi\rangle$. With an estimated E rejection sampling is performed to produce a normal posterior distribution for ψ, which is sampled to repeat the process until sufficient accuracy is reached. Finally, $|\langle\psi|P|\psi\rangle|$ is recovered using the relation $|\langle\psi|P|\psi\rangle| = \cos(\phi/2)$.

Example 5.5.3 With regards to the chemical H_2 and its electronic molecular Hamiltonian with two electrons and two active orbitals, a Bravyi-Kitaev transformation necessitates four qubits to estimate the Hamiltonian. The resulting form of the Hamiltonian is then as follows

$$H = \sum_{i=1}^{15} a_i P_i. \tag{5.32}$$

We will use four ansatzes across three QPUs in at least four rounds to perform α-VQE, with arbitrary qubit rotations represented by $R(\lambda_1, \lambda_2, \lambda_3)$ gates. A distributed version of the reflection operation is needed to support the distributed ansatz, and a circuit for the control part of c-U, c-Π, is required for running α-QPE and α-VQE.

Example 5.5.4 Consider the Hamiltonian of CO_2 with 16 electrons and 16 molecular orbitals. By using the Bravyi-Kitaev transformation we need 20 qubits (10 spin-orbitals). The Hamiltonian is then expressed as a linear combination of tensor products of Pauli operators on these qubits. The explicit expression for the CO_2 Hamiltonian in the second quantization is:

$$H = \sum_{p,q,r,s} h_{p,q,r,s} a_p^\dagger a_q^\dagger a_r a_s + \frac{1}{2} \sum_{p,q,r} h_{pqqp} a_p^\dagger a_q^\dagger a_r a_r \\ + \frac{1}{2} \sum_{p,q,r,s} h_{pqrp} a_p^\dagger a_q^\dagger a_r a_s + \text{const}, \tag{5.33}$$

where $a^\dagger$ and a are fermionic creation and annihilation operators, respectively, and h_{pqrs} are the integrals defining the Hamiltonian. Using the Jordan-Wigner or Bravyi-Kitaev transformations, it can be mapped to a sum over tensor products of Pauli operators acting on qubits as

$$H = \sum_{i=1}^{4^{20}} a_i P_i. \tag{5.34}$$

For this large Hamiltonian, one may use a larger number of QPUs to perform distributed α-VQE.

5.6 Distributed Grover Searching Algorithm

The Grover quantum searching algorithm is a powerful quantum computing algorithm that can search an unsorted database of N items in $O(\sqrt{N})$ time, offering a polynomial speedup compared to classical algorithms Grover (1996). By repeatedly applying the Grover operator to a superposition of all possible input states, the algorithm amplifies the amplitude of the target item, making it highly probable to find the desired item in large datasets.

The versatility of the Grover algorithm Grover (1996) and its extension to amplitude amplification Brassard et al. (2002) is a significant advantage. These algorithms can be used to solve various problems, such as the element distinctness Buhrman et al. (2005); Ambainis (2007), the collision finding Brassard et al. (1998), and NP-hard tree search problems Cerf et al. (2000). This is because amplitude amplification can be applied to improve the efficiency of these algorithms. In this section, we will revisit the Grover quantum searching algorithm and the element distinctness problem and address a challenge raised by Grover and Rudolph Grover and Rudolph (2004). In this section, we present distributed ideas using the distributed quantum model.

5.6.1 Searching an Unstructured Database

Grover searching algorithm Grover (1996) is designed to identify solutions s for a given function $f : \{1, \ldots, N\} \to 0, 1$ satisfying $f(s) = 1$. An oracle in the form of a unitary operator U_f, which acts on a two-register state $|x\rangle\,|y\rangle$ by transforming it to $|x\rangle\,|y \oplus f(x)\rangle$, can identify solutions. Grover's algorithm requires $\Theta(\sqrt{N/M})$ calls to U_f and outputs one of the M possible solutions uniformly at random with a success probability of $1 - O(M/N)$.

When searching unstructured databases, the Grover algorithm constructs the oracle using $U_{1,N}$ and a function that checks whether a database entry is a solution. In a more general case, where the inputs come from a database $X = \{x_j : j \in [N]\}$ of bit strings of length d, and we seek solutions $s \in \{1, \ldots, N\}$ such that $\alpha(x_s) = 1$, Grover algorithm evaluates the function α using the unitary operator $U_\alpha : \left|x_j\right\rangle |y\rangle \mapsto \left|x_j\right\rangle \left|y \oplus \alpha(x_j)\right\rangle$ after loading x_j into the computational memory.

There are two scenarios for the database-one, where one is stored as a quantum state $|\vec{x}\rangle$, and another is a classical database whose indices can be accessed through superposition. Regardless of the type of database, once the relevant entry has been retrieved, we apply a function evaluation step to determine if it is a solution to α. This leads to the construction of the oracle $\mathcal{O}_\alpha$ that takes the form $\mathcal{O}_\alpha : |j\rangle \mapsto (-1)^{\alpha(x_j)}\,|j\rangle$, where x_j is the database entry indexed by j.

When accessing a quantum state database, the initial state of the quantum computer is set to $|j\rangle\,|0\rangle\,|y\rangle\,|\vec{x}\rangle$, where $|j\rangle$ represents the index of the database item to retrieve, $|0\rangle$ is used to load the memory, and $|y\rangle$ is utilized as the target register. The first two registers are used to perform the memory look-up operation, while the last two registers contain the quantum state $|\vec{x}\rangle$ that represents the database. To access the relevant database entry, we apply the single memory look-up unitary transformation $U_{(1,N)}$ to the state $|j\rangle\,|0\rangle\,|y\rangle\,|\vec{x}\rangle$.

In some situations, it needs to search pairs or r-tuples of database elements. An oracle $\mathcal{O}_\alpha$ can be designed for Grover's algorithm using $U_{(1,N)}$ and the unitary $U_\alpha : \left|\vec{x_j}\right\rangle |b\rangle \mapsto \left|\vec{x_j}\right\rangle \left|b \oplus \alpha(\vec{x_j})\right\rangle$, where $\vec{j} = (j^1, \ldots, j^r) \in [N]^r$ and $\vec{x_j} = (x_{j^1}, \ldots, x_{j^r}) \in X^r$. The initial state is $\left|\vec{j}\right\rangle \left|\vec{0}\right\rangle |b\rangle\,|x_1, \ldots, x_N\rangle$, and the oracle construction proceeds similarly to the $r = 1$ case. The quantum circuit implementing Grover algorithm with $\mathcal{O}_\alpha$ as an oracle makes $\Theta(\sqrt{N^r/M})$ calls to $\mathcal{O}_\alpha$, where M is the number of solutions. The circuit width and depth are then $\Omega(Nd)$ and $\Omega(\sqrt{N^r/M}(r \log N + D_\alpha))$, respectively.

After developing an efficient algorithm for parallel memory look-ups, we investigate the use of the unitary $U_{(N,N)}$ in conjunction with up to N functions as oracles for Grover's algorithm. Suppose we have up to N functions that take r database elements as inputs, $\alpha_i : X^r \to \{0, 1\}$ for $i \in [N]$, and we aim to find solutions $\vec{s_i} = (s_i^1, \cdots, s_i^r) \in [N]^r$ such that $\alpha_i(x_{s_i^1}, \ldots, x_{s_i^r}) = 1$ for $i \in [N]$. We assume that we have circuits for the unitaries U_{α_i}, where $\left|\vec{x_j}\right\rangle |b\rangle \mapsto \left|\vec{x_j}\right\rangle \left|b \oplus \alpha_i(\vec{x_j})\right\rangle$, and we assume that the depth and width of these circuits are polynomial functions of $\log N$,

d, and r. We consider each node to have size d, which is $\Omega(\log N)$, and we treat the function as a black box.

Given these assumptions, the following theorem asserts that an algorithm can discover a solution for each function α_i using the same quantum circuit size (up to a polynomial factor in $\log N$ and r) required to find only one solution using $U_{(1,N)}$ as an oracle. Although the algorithm can be expressed entirely in terms of the circuit model, it is convenient to discuss it in terms of quantum processors and memory locations. Since we know that the distributed quantum RAM model can be emulated by the circuit model with a $O(\log N)$ overhead, the presented algorithm becomes a simple generalization of the Grover algorithm. The number of processors, N, is equal to the number of items in the database. The cost of quantum memory will be equivalent to that of quantum processors Grover and Rudolph (2004).

Theorem 5.6.1 *An efficient quantum algorithm exists that, for each $i \in [N]$, can either output the solution index $\vec{s}_i \in [N]^r$ where $\alpha_i(\vec{x}_{s_i}) = 1$, or indicate that there are no solutions. The algorithm has a high probability of success, and it can be realized using a quantum circuit with width $\tilde{O}(N)$ and depth $\tilde{O}(\sqrt{N^r/M})$, where M is the number of solutions.*

The main idea of the proof is to determine the value M, which represents the minimum number of solutions for each function α_i. For the case $r = 1$, the algorithm organizes the Hilbert space across N processors to perform N parallel Grover searches. A circuit is defined to implement the unitaries U_α in parallel. Grover's algorithm calls $O(\sqrt{N/M})$ iterations of the unitary $U_{(N,N)}U_\alpha U_{(N,N)}$. This approach finds a solution for each α_i with a circuit of width $O(N)$ and depth $O(\sqrt{N/M}(\log N \log d + D_\alpha))$. For $r > 1$, the indices and memory data are replaced by r-tuples.

Grover's algorithm makes $O(\sqrt{N^r/M})$ calls to $\mathcal{O}_\alpha$, resulting in a circuit with width $\tilde{O}(N)$ and depth $\tilde{O}(\sqrt{N^r/M})$.

To handle the unstructured database, the efficient parallel memory look-up algorithm is utilized to access its elements. This enables the interleaving of Grover algorithm steps with database lookups, allowing for N parallel Grover searches regardless of the structure. This approach eliminates the need to explicitly represent the data in a structured manner and results in a circuit with a width of $O(N)$, making it feasible for large-scale applications.

5.6.2 Parallel Element Distinctness Problem

In this section, we present a quantum algorithm for the element distinctness problem, offering an exponential speedup over classical algorithms. This algorithm solves the problem in $O(\sqrt{N})$ time, making it a promising candidate for demonstrating the potential of quantum computing in practical applications.

Definition 5.6.1 The problem involves determining whether a given function f, which maps binary strings of length n to binary strings of length n, has distinct input strings i and j that produce the same output string $f(i) = f(j)$.

The following discussion focuses on the parallel memory look-up algorithm inspired by the algorithm Buhrman et al. (2005), where $N = 2^n$ determines the problem size. In classical computing, the problem can be solved using $O(N)$ function calls if $ST = O(N^2)$, where S and T represent available memory and time, respectively. However, the Grover algorithm does not improve the number of function queries because $\Theta(N^2)$ pairs require $\Theta(N)$ queries. Instead, a quantum algorithm can solve this problem Ambainis (2007) with $O(N^{2/3})$ queries and requires memory proportional to $\tilde{O}(N^{2/3})$, i.e., $S = T = \tilde{O}(N^{2/3})$. This algorithm is optimal since $\Omega(N^{2/3})$ queries are also required.

If we regard f as something computed by a small circuit (say poly(log N)), we can achieve the same performance as the algorithm in Ambainis (2007) through a more naive approach Grover and Rudolph (2004). This approach involves dividing the N^2 search space into N^2/S blocks, performing a Grover search in each one that takes $\tilde{O}(\sqrt{N^2/S})$ time, resulting in a time-space trade-off of $ST^2 = \tilde{O}(N^2)$. This trade-off includes both the special case $S = T = \tilde{O}(N^{2/3})$ Ambainis (2007) and the naive Grover search $S = O(\text{poly}(\log N))$, $T = \tilde{O}(N)$.

Theorem 5.6.2 *Given a* poly$(\log N)$ *size circuit for computing the function f, the Element Distinctness problem can be solved in space $\tilde{O}(S)$ and time $T = \tilde{O}(N/S)$ for any $S \le N$.*

The main idea of the proof is to randomly select S inputs $x_1, \ldots, x_S$ to query and store them, along with their corresponding outputs, in the distributed memory. Then, using the AKS sorting network, we sort the stored outputs in time $O(\log N)$ and check if any pair of them are equal. If so, we output the indices of the colliding elements. Otherwise, we use the sorted list to construct a function g_L, which checks for collisions between the remaining $N - S$ inputs and the stored inputs. We can split the remaining $N - S$ inputs into S collections of size $O(N/S)$ and use the Multi-Grover algorithm to perform S parallel searches. This procedure requires $\Theta(\sqrt{N/S})$ calls to g_L, and the total depth is $\tilde{O}(\sqrt{N/S})$.

We will succeed (with constant probability) in finding a collision if one of the colliding elements was in $x_1, \ldots, x_S$, which happens with probability $\Theta(S/N)$. To boost the probability to $\Theta(1)$, we can use amplitude amplification and repeat the routine $\Theta(\sqrt{N/S})$ times. Therefore, the total runtime is $T = \tilde{O}(N/S)$, and we have achieved a time-space trade-off of $ST = \tilde{O}(N)$.

Example 5.6.1 Suppose there is a set of $N = 1000000$ elements and we want to find out if there are any duplicates. Using the algorithm Ambainis (2007), we can solve the Element Distinctness problem for $N = 1000000$ in time $O(3000)$. The present distributed algorithm requires $O(10000)$ runtime with 100 quantum parallel RAMs.

5.6.3 Distributed Collision Finding Problem

The collision finding problem, which aims to find a pair of inputs that produce the same output for a given function, is a fundamental problem in computer science with applications in cryptography, data mining, and database management. Classical algorithms for this problem require $O(N)$ evaluations of the function, but quantum algorithms have been developed that can solve it in sub-linear time, offering a potential speedup.

Definition 5.6.2 $f : [N] \to [N]$ is an efficiently-computable function with the promise of being either 1-1 or 2-1.

A simple query algorithm can be described as follows.

- Initialize two registers, A and B, with n qubits each, where n is the number of bits in the message.
- Apply a Hadamard gate to each qubit in register A to put it into a superposition state.
- Apply a quantum oracle that maps $|x, y\rangle$ to $|x, y + f(x)\rangle$ where $f(x)$ is a function that maps input x to output y.
- Apply a second Hadamard gate to register A.
- Measure the states of registers A and B and compare them. If they are the same, output the message x.

This procedure may be repeated until a collision is found. A collision is found when two different inputs x and x' produce the same output y, i.e., $f(x) = f(x')$. In this case, output x and x'.

In terms of distributed quantum computation scenarios, it may be solved with $ST = \tilde{O}(\sqrt{N})$ using two possible approaches. The first approach involves randomly selecting $O(\sqrt{N})$ elements and applying Element Distinctness to solve the problem. The second approach involves using the algorithm proposed in Brassard et al. (1998) with slight modifications, whereby S processors with shared memory are utilized with the look-up algorithm. This implies that the problem can be solved by performing parallel Grover searches using S processors, with each search conducted on a space of size N/S^2, in contrast to the standard Grover search which involves conducting a single search on a space of size N/S. By dividing the space into smaller portions, each search is performed on a reduced dataset, thereby significantly reducing the computation time required to solve the problem. This means it is possible to solve distributed quantum computation problems with considerably lower computational resources, thereby promoting scalability and efficiency.

5.7 Conclusion

This chapter has introduced DQC which is a field of research in quantum computing that focuses on the use of multiple, separated quantum devices to perform computations. The aim is to leverage the power of quantum computing to solve problems that are beyond the capabilities of classical computers Meter et al. (2008); Ying and Feng (2009), while also addressing the practical challenges of building larger quantum computers Andres-Martinez (2018); Andres-Martinez and Heunen (2019); van Houte et al. (2020); Ghodsollahee et al. (2021); Sarvaghad-Moghaddam and Zomorodi (2021); Zhao et al. (2021); Chatterjee et al. (2022); Cheng et al. (2023).

DQC poses significant challenges, including the need to synchronize the different qubits used in the computation, control the interactions between the devices, and manage the noise and errors that arise in quantum systems. This chapter has discussed various topics related to DQC including the quantum circuit model Deutsch (1989); Brennen et al. (2005); Luo and Wang (2014), distributed RAM model Cirac et al. (1997); Van Meter et al. (2009), quantum information processing by remote quantum control Qiang et al. (2017) and distributing Shor algorithm Beals et al. (2013), distributing accelerated variational quantum eigensolver DiAdamo et al. (2021), distributed Grover's search algorithm Grover and Rudolph (2004) and some applications Brassard et al. (1998); Buhrman et al. (2005); Ambainis (2007). This has shown the power of distributed quantum devices for solving large-scale tasks.

References

M. Ajtai, J. Komlos, E. Szemeredi, An $O(n \log n)$ sorting network, in *Proc. 15th annual ACM symposium on Theory of computing* (1983), pp. 1–9

A. Ambainis, Quantum walk algorithm for element distinctness. SIAM J. Comput. **37**, 210–239 (2007)

P. Andres-Martinez, Automated distribution of quantum circuits. Theoret. Comput. Sci. **410**, 2489–2510 (2018)

P. Andres-Martinez, C. Heunen, Automated distribution of quantum circuits via hypergraph partitioning. Phys. Rev. A **100**(3), 032308 (2019)

R. Beals, S. Brierley, O. Gray, A.W. Harrow, S. Kutin, N. Linden, D. Shepherd, M. Stather, Efficient distributed quantum computing. Proc. Royal Society A **469**, 20120686 (2013)

C.H. Bennett, G. Brassard, C. Crépeau, R. Jozsa, A. Peres, W.K. Wootters, Teleporting an unknown quantum state via dual classical and Einstein-Podolsky-Rosen channels. Phys. Rev. Lett. **70**, 1895 (1993)

K.E.C. Booth, M. Do, J.C. Beck, E.G. Rieffel, D. Venturelli, J. Frank, Comparing and integrating constraint programming and temporal planning for quantum circuit compilation, in *Proc. 28th Int. Conf. Automated Planning Scheduling* (2018), pp. 366–374

G. Brassard, P. Hoyer, and A. Tapp, Cryptology Column–Quantum algorithm for the Collision problem, *Third Latin American Symp. on Theoretical Informatics (LATIN'98)*, pp. 163-169, 1998

G. Brassard, P. Hoyer, M. Mosca, A. Tapp, Quantum amplitude amplification and estimation. Contemp. Math. **305**, 53–74 (2002)

G. Brennen, D. OLeary, S. Bullock, Criteria for exact qudit universality. Phys. Rev. A **71**, 052318 (2005)

H. Buhrman, C. Durr, M. Heiligman, P. Hoyer, F. Magniez, M. Santha, R. de Wolf, Quantum algorithms for element distinctness. SIAM J. Comput. **34**, 1324–1330 (2005)

V. Bužek, M. Hillery, Quantum copying: Beyond the no-cloning theorem. Phys. Rev. A **54**, 1844 (1996)

N. Cerf, L. Grover, C. Williams, Nested quantum search and structured problems. Phys. Rev. A **61**, 032303 (2000)

T. Chatterjee, A. Das, S.I. Mohtashim, A. Saha, A. Chakrabarti, Qurzon: A Prototype for a Divide and Conquer-Based Quantum Compiler for Distributed Quantum Systems. SN Computer Science **3**, 323 (2022)

P.-X. Chen, S.-Y. Zhu, G.-C. Guo, General form of genuine multipartite entanglement quantum channels for teleportation. Phys. Rev. A **74**, 032324 (2006)

X. Cheng, X. Chen, K. Cao, P. Zhu, S. Feng, Z. Guan, Optimization of the transmission cost of distributed quantum circuits based on merged transfer. Quantum Inf. Process. **22**, 187 (2023)

B.-S. Choi, R. Van Meter, On the effect of quantum interaction distance on quantum addition circuits. ACM J. Emerg. Tech. Com. Sys. **7**, 11 (2011)

J. Cirac, P. Zoller, H. Kimble, H. Mabuchi, Quantum state transfer and entanglement distribution among distant nodes in a quantum network. Phys. Rev. Lett. **78**, 3221 (1997)

R. Cleve, H. Buhrman, Substituting quantum entanglement for communication. Phys. Rev. A **56**, 1201 (1997)

R. Cleve, A. Ekert, C. Macchiavello, M. Mosca, Quantum algorithms revisited. Proc. Royal Soc. A **454**, 339–354 (1998)

D. Deutsch, Quantum computational networks. Proc. R. Soc. A **425**, 73–90 (1989)

S. DiAdamo, M. Ghibaudi, J. Cruise, Distributed quantum computing and network control for accelerated vqe (2021) arXiv:2101.02504

D. Dieks, Communication by EPR devices. Phys. Lett. A **92**, 271 (1982)

J. Eisert, K. Jacobs, P. Papadopoulos, M.B. Plenio, Optimal local implementation of nonlocal quantum gates. Phys. Rev. A **62**, 052317 (2000)

I. Ghodsollahee, Z. Davarzani, M. Zomorodi, P. Piawiak, M. Houshmand, M. Houshmand, Connectivity matrix model of quantum circuits and its application to distributed quantum circuit optimization. Quantum Inf. Process. **20**, 235 (2021)

N. Gisin, S. Massar, Optimal quantum cloning machines. Phys. Rev. Lett. **79**, 2153 (1997)

L. Grover, A fast quantum mechanical algorithm for database search, *Proc. 28th Annual ACM STOC*, pp.212-219, 1996

L. Grover, T. Rudolph, How significant are the known collision and element distinctness quantum algorithms? Quantum Inf. & Comput. **4**, 201–206 (2004)

J.P. Hespanha, P. Naghshtabrizi, Y. Xu, A survey of recent results in networked control systems. Proc. IEEE. **95**, 138–162 (2007)

Y. Hirata, M. Nakanishi, S. Yamashita, Y. Nakashima, An efficient conversion of quantum circuits to a linear nearest neighbor architecture. Quant. Inf. Comput. **11**, 142 (2011)

N. Khaneja, S.J. Glaser, Cartan decomposition of $SU(2^n)$ and control of spin systems. Chem. Phys. **267**, 11–23 (2001)

A.Y. Kitaev, Quantum computations: algorithms and error correction. Russian Math. Surv. **52**, 1191 (1997)

D. Knuth, *The Art of Computer Programming*, vol. 3. (Addison-Wesley, 1998)

M.-X. Luo, X. Wang, Universal quantum computation based on qudits. Sci. China Phys. Mech. Astr. **57**, 1712–1717 (2014)

M.S. Mahmoud, Y. Xia, *Networked Control Systems: Cloud Controland Secure Control* (Butterworth-Heinemann, Oxford, U.K., 2019)

V. Meter, W. Munro, K. Nemoto, K.M. Itoh, Arithmetic on a distributed-memory quantum multicomputer, ACM. J. Emerg. Technol. Comput. Syst. JETC **3**, 2 (2008)

M. Nielsen, I. Chuang, *Quantum Computation and Quantum Information* (Cambridge University Press, 2000)

M. Paterson, Improved sorting networks with $O(\log N)$ depth. Algorithmica **5**, 75 (1990)

A. Peruzzo, J. McClean, P. Shadbolt, M.-H. Yung, X.-Q. Zhou, P.J. Love, A. Aspuru-Guzik, J.L. O'Brien, A variational eigenvalue solver on a photonic quantum processor. Nat. Commun. **5**, 4213 (2014)

X. Qiang, X. Zhou, K. Aungskunsiri, H. Cable, J.L. O'Brien, Quantum processing by remote quantum control. Quantum Sci. Tech. **2**, 045002 (2017)

M. Sarvaghad-Moghaddam, M. Zomorodi, A general protocol for distributed quantum gates. Quantum Inf. Process. **20**, 265 (2021)

P.W. Shor, Algorithms for quantum computation: discrete logarithms and factoring, in *Proceedings 35th Annual Symposium on Foundations of Computer Science*, IEEE Comput. Soc. Press (1994), pp. 124–134

H. Thapliyal, N. Ranganathan, R. Ferreira, Design of a comparator tree based on reversible logic, in *Proc. 10th IEEE Conference on Nanotechnology (IEEE-NANO)* (2010), pp. 1113–1116

R. van Houte, J. Mulderij, T. Attema, I. Chiscop, F. Phillipson, Mathematical formulation of quantum circuit design problems in networks of quantum computers. Quantum Inf. Process. **19**, 141 (2020)

R. Van Meter, M. Oskin, Architectural implications of quantum computing technologies. ACM J. Emerg. Tech. Com. Sys. **2**, 31 (2006)

R. Van Meter, W. Munro, K. Nemoto, K. Itoh, Arithmetic on a distributed-memory quantum multicomputer. ACM J. Emerg. Tech. Com. Sys. **3**, 2 (2009)

D. Wang, O. Higgott, S. Brierley, Accelerated variational quantum eigensolver. Phys. Rev. Lett. **122**, 140504 (2019)

W.K. Wootters, W.H. Zurek, A single quantum cannot be cloned. Nature **299**, 802 (1982)

T.C. Yang, Networked control system: a brief survey. IEEE Proc.-Control Theory Appl. **153**, 403–412 (2006)

M. Ying, Y. Feng, An algebraic language for distributed quantum computing. IEEE Trans. Comput. **58**, 728–743 (2009)

X. Zhao, B. Zhao, Z. Wang, Z. Song, and X. Wang, Practical distributed quantum information processing with LOCCNet, *npj Quantum Inf.***7**, 159 (2021)

X.Q. Zhou, T.C. Ralph, P. Kalasuwan, M. Zhang, A. Peruzzo, B.P. Lanyon, J.L. O'Brien, Adding control to arbitrary unknown quantum operations. Nat. Commun. **2**, 413 (2011)

Chapter 6
Distributed Quantum Games

In this chapter, we present a distributed game on quantum networks. A nonlocal game is generally described as multiple, space-separated players interacting with a referee. Communication between players is forbidden. This kind of game allows all players to determine a joint strategy from complete knowledge of input distribution and predicate conditions for winning the game. The corresponding implementation procedure is generally featured by a generalized variable model for classical sources or a quantum model with entangled sources. One fundamental problem is to determine whether quantum mechanics is superior to classical theory in terms of specific nonlocal tasks. These evaluations are equivalent to solving special optimization problems from Tsirelson's reductions.

6.1 CHSH Game

Clauser-Horne-Shimony-Holt (CHSH) game aims to distributed evaluate Boolean equation (Clauser et al. 1969):

$$x \oplus y = a \,\text{AND}\, b, \tag{6.1}$$

where a_i and x_i are the respective binary input and output of two players. It is formally defined as follows.

Definition 6.1.1 (*CHSH game*) Alice and Bob play the following game with a referee:

- A referee chooses $x, y \in \{0, 1\}$ according to probability distribution $P(x, y)$.
- The referee sends Alice x and Bob y.
- Alice responds with $a \in \{0, 1\}$ and Bob responds with $b \in \{0, 1\}$.

Both players win the game if and only if the output x, y satisfies the equality (6.1).

M.-X. Luo, *Quantum Networks*,
https://doi.org/10.1007/978-981-97-6226-2_6

In a nonlocal game, there is a specification of an input or question set $\mathcal{X}$ and a possible set of responses or answers, $\mathcal{Y}$, to questions in $\mathcal{X}$. The payoff is defined according to special compatible rules, that is, payoff functions $\mathcal{F}$. With these settings, one can define the expected payoff as follows:

$$\wp_\omega = \mathbb{E}_{(x,y)\in\mathcal{X}\times\mathcal{Y}}[\mathcal{F}(x, y) = \text{true}]. \tag{6.2}$$

Here, the term "true" refers to the compatibility of inputs and outputs, otherwise, it is considered a failure. All the players cannot communicate with each other during the experiment but can make use of local shared resources of randomness or entanglement. This kind of distributed game may be further classified according to the rules of all players. One type is a cooperative game in which most of the players jointly compete with one referee. The other type is a game where different groups may compete with each other.

As for the CHSH game, the payoff function is defined by

$$F(x, y, a, b) = \begin{cases} 1, & \text{if } x \oplus y = a \text{ AND } b; \\ 0, & \text{otherwise} . \end{cases} \tag{6.3}$$

This implies an average winning probability given by

$$\begin{aligned} \wp_{\text{CHSH}} &= \mathbb{E}_{(x,y)\in\mathcal{X}\times\mathcal{Y}} F(x, y, a, b) \\ &= \sum_{x,y,a,b} F(x, y, a, b) P(a, b|x, y) P(x, y), \end{aligned} \tag{6.4}$$

where $P(a, b|x, y)$ is a conditional probability distribution that corresponds to the output strategy of both players.

To evaluate the winning probability for different strategies, the key is to find a special distribution $P(a, b|x, y)$. For classical strategies, assume that both players share a random variable λ. The optimal conditional probability allows the following decomposition:

$$P(a, b|x, y) = \int d\mu(\lambda) P(a, b|\lambda, x, y) p(\lambda|x, y). \tag{6.5}$$

This is achieved by using the Bayes' rule, where $p(\lambda|x, y)$ is a conditional probability distribution. From the no-signaling principle, due to no communication, we have the following decomposition:

$$P(a, b|\lambda, x, y) = p(a|\lambda, x) p(b|\lambda, y). \tag{6.6}$$

Moreover, $p(\lambda|x, y)$ is generally independent of inputs, which means the shared correlations are before the game. To sum up, the distribution $p(a, b|x, y)$ allows a hidden variable model:

$$P(a, b|x, y) = \int d\mu(\lambda)p(a|\lambda, x)p(b|\lambda, y)p(\lambda). \tag{6.7}$$

The winning probability (6.4) may be optimized with respect to all classical strategies. From linearity, it is sufficient to consider the deterministic output. For the CHSH game, the maximal winning probability for classical players is $\frac{3}{4}$ by outputting the same for any inputs.

In quantum scenarios, both players Alice and Bob may share a quantum state ρ_{AB} on the Hilbert space $\mathbb{H}_A \otimes \mathbb{H}_B$. They perform local POVM measurements depending on the inputs x and y. Denote Alice's POVM measurement as $\{M_x^a\}$, and Bob's POVM measurement as $\{N_y^b\}$. From the Born rule, the conditional probability distribution $P(a, b|x, y)$ for outputs a and b is given by

$$P(a, b|x, y) = \text{Tr}(\rho M_x^a \otimes N_y^b). \tag{6.8}$$

For a special case of inputs with uniform probability, i.e., the balanced CHSH game, the difference between the winning and losing probabilities is equal to

$$\wp_{\text{win}} - \wp_{\text{loss}} = \frac{1}{4}\text{Tr}(\rho\Gamma_{AB}), \tag{6.9}$$

where Γ_{AB} is the CHSH operator (Clauser et al. 1969) defined as

$$\Gamma_{AB} = A_0 \otimes B_0 + A_0 \otimes B_1 + A_1 \otimes B_0 - A_1 \otimes B_1. \tag{6.10}$$

It can be checked that $|\Gamma_{AB}|_\infty \leq 2\sqrt{2}$ with the matrix norm $|X|_\infty$ of an operator X. This implies an optimal winning probability of $\cos^2 \frac{\pi}{8}$. Specially, if both players share a maximally entangled Einstein-Podolsky-Rosen (EPR) state $\frac{1}{\sqrt{2}}(|00\rangle + |11\rangle)$, Alice performs a measurement of $Z^{x\oplus 1}X^x$ and gets the outcome a, and Bob performs a measurement of $\frac{1}{\sqrt{2}}(Z + (-1)^y X)$ and gets the outcome b. This quantum strategy follows the optimal winning probability beyond classical strategies. This provides a notable way for witnessing the nonlocality of the EPR state.

Theorem 6.1.1 *The optimal winning probability of the CHSH game is* 0.75 *with classical players while* $\cos^2 \frac{\pi}{8}$ *with quantum players.*

There are many interesting nonlocal games other than the CHSH game. XOR games are the most well-studied (Ambainis et al. 2012). The global task is the XOR of the outcomes of all players. Other games include the Kochen-Specker game (Kochen and Specker 1967), non-zero sum games (Eisert et al. 1999), graph isomorphism game (Atserias et al. 2019), Bayesian game (Brunner and Linden 2013), and non-cooperating game (Pappa et al. 2015).

The key to achieving a quantum advantage is the nonlocal correlations generated by local measurements on the shared entangled states. Similar correlations are not achievable by remote players using local strategies with any shared classical resources. These tasks show the important role of quantum entanglement in information processing (Brunner et al. 2014).

6.2 Graphic Games

An m-player nonlocal cooperating game consists of n players, denoted $\mathsf{A}_1, \ldots, \mathsf{A}_m$, and a referee (Luo 2019), as shown in Fig. 6.1. All the players agree on a joint strategy beforehand but cannot communicate with each other during the game.

Definition 6.2.1 (*Distributed cooperating games*) In a nonlocal game with m players, the referee first chooses n questions $x_1, \ldots, x_m$ from a finite set $\mathcal{X} := X_1 \times \cdots \times X_m$ according to a known distribution $P(\vec{x})$. The referee then sends each player A_i, $i \in [m]$, their respective question x_i. The players must reply with answers $y_1, \ldots, y_m$ chosen from a finite set $\mathcal{Y} := Y_1 \times \cdots \times Y_m$ to the referee. Finally, the referee determines whether all the players win or lose the game according to a payoff function $F : \mathcal{X} \times \mathcal{Y} \to \{0, 1\}$. The outcome is a win if $F(\vec{x}, \vec{y}) = 1$, and it is a loss if $F(\vec{x}, \vec{y}) = 0$, where $\vec{x} = x_1 \cdots x_m$, and $\vec{y} = y_1 \cdots y_m$.

The optimal average winning probability of all players, known as the nonlocal value of the game, is defined as follows:

$$\varpi_c = \sup_{\Omega} \sum_{\vec{x},\vec{y},\lambda} P(\vec{x})\mu(\lambda)F(\vec{x}, \vec{y})P(\vec{y}|\vec{x}, \lambda), \tag{6.11}$$

where $\mu(\lambda)$ is the probability measure of the hidden classical resource λ and satisfies $\sum_\lambda \mu(\lambda) = 1$. $P(\vec{y}|\vec{x}, \lambda)$ denotes the joint conditional probability depending on the shared randomness λ. The supremum is taken over all possible classical spaces Ω of hidden variables λ.

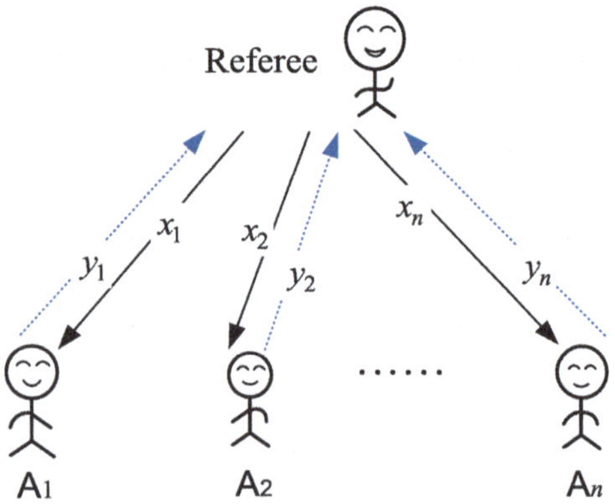

Fig. 6.1 Cooperating nonlocal game. A referee chooses questions x_k from a finite set according to a prior distribution $P(\mathbf{x})$ and sends them to the players. Each player A_k outputs an answer y_k based on the received question x_k and the shared resource, without communicating with the other players. The referee then evaluates a payoff function with all outputs and inputs

Finding the nonlocal value ϖ_c of multipartite games is generally a difficult task. However, due to the linearity of ϖ_c with respect to all probability distributions, it is sufficient to use deterministic strategies for all players, i.e., $p(\lambda) = 1$ for some λ. This simplifies the search for the optimal strategy and reduces the problem of finding the maximum winning probability achievable using a specific deterministic strategy.

In quantum scenarios, players are allowed to share various entangled states and perform quantum measurements, which are described by positive-operator valued measurements (POVMs) $\{M_{y_1}^{x_1}\}, \ldots, \{M_{y_m}^{x_m}\}$ that depend on the received questions $x_1, \ldots, x_m$. These POVMs satisfy normalization conditions $\sum_{y_i} M_{y_i}^{x_i} = \mathbb{1}$ for $i = 1, \ldots, m$, where $\mathbb{1}$ is the identity operator. Let ρ be the shared entangled state. The optimal winning probability for all quantum players to play a cooperating nonlocal game is defined as follows:

$$\varpi_q = \sup \sum_{\vec{x}, \vec{y}} P(\vec{x}) F(\vec{x}, \vec{y}) \mathrm{Tr}[\otimes_{j=1}^m M_{y_j}^{x_j} \rho], \tag{6.12}$$

where the supremum is taken over all possible quantum states and local POVMs. Note that each linear Bell-type inequality is equivalent to a cooperating nonlocal game (Silman et al. 2008). The nonlocal value ϖ_q is related to the Tsirelson bound of linear Bell inequality. A central problem in nonlocality theory is to find computable good bounds for the nonlocal value ϖ_q. Unfortunately, it is QMA-hard to approximate the nonlocal value of ϖ_q.

6.2.1 Graphic Games on Quantum Networks

The CHSH game is a novel approach to investigating the quantum nature of bipartite entanglement. It would be further intriguing to characterize multiple entangled states shared by observers separated in space in a network scenario. While linear testing of single entanglements enables the creation of equivalent nonlocal games, designing nonlocal games for multi-source quantum networks requiring nonlinear or semiquantum testing remains a challenge. To tackle this issue, we suggest a general technique for constructing nonlocal games from graphs containing only vertices.

The multi-source network $\mathcal{N}_q$ comprises independent EPR states shared among different observers. Let $\mathcal{G}_q$ be graphically represented in this network, with each node representing an entanglement. For any graph, players can be assigned subgraphs according to their inputs. Thus, two observers sharing an entangled state corresponds to two players sharing a vertex, where each player has vertices of the same color. Consistency conditions can be applied to the common vertices shared by different players to characterize these nonlocal games. Specifically, the graphic game is defined as follows.

Definition 6.2.2 An m-partite graphic game is a six-tuple: $(\mathcal{G}, \mathcal{A}, \mathcal{V}, \mathcal{X}, \mathcal{Y}, F)$. $\mathcal{G}$ is a general graph with vertex set V (the edges are not required in this application). $\mathcal{A}$

denotes the set of all A_i's and referee. $\mathcal{V}$ denotes the set of all the vertex sets $V^{x_i} \subseteq V$ owned by players, where V^{x_i} denotes the set of vertices assigned to the player A_i according to the input question x_i. Assume that V^{x_i}'s satisfy the following conditions: $V^{x_i} \cap V^{x_j} = \emptyset$ for $x_i = x_j = 1$ with $1 \leq i < j \leq m$, where n is a parameter satisfying $1 \leq n < m$. $\mathcal{X}$ denotes the set of binary problems $x_1, \ldots, x_m \in \{0, 1\}$. $\mathcal{Y}$ denotes the set of all the assignments on vertices (as outputs) by players. $F : \mathcal{X} \times \mathcal{Y} \rightarrow \{0, 1\}$ is a payoff function depending on all the inputs and outputs of players. A graphic game is implemented as follows:

Input-The referee randomly chooses binary question x_i according to the distribution $\{p, 1 - p\}$, and sends to the player A_i, $i = 1, \ldots, m$.
Output-The player A_i assigns an integer $y_i \in \{\pm 1\}$ on each vertex in the set V^{x_i}, and sends all the assignments to the referee secretly.
Winning conditions-All the players win the graphic game, i.e., $F(\vec{x}, \vec{y}) = 1$, if and only if their assignments satisfy the following consistency conditions:

(a) $S^{x_i} = 1$ for $i = n + 1, \ldots, m$;
(b) $S^{x_i;x_j} = -1$ when $x_i = x_j = 1$, or 1 otherwise, for $1 \leq i \leq n < j \leq m$;
(c) $S^{x_i;x_j} = S^{x_j;x_i} = 1$ for $n < i < j \leq m$;

where S^{x_i} denotes the product of all the assignments of the player A_i, and $S^{x_i;x_j}$ denotes the product of all the assignments by the player A_i on the vertices shared with the player A_j.

The consistency conditions are crucial for most nonlocal games. Quantum correlations derived from entangled resources can satisfy these requirements and achieve higher winning probabilities than all classical achievable correlations. In the proposed graphical game, n serves as a variable parameter ranging from 1 to m while identifying players whose assignments satisfy consistency conditions. To win the nonlocal game, all players must produce outputs satisfying consistency conditions. This graphical game resembles a nonlinearly constrained nonlocal computation, like the CHSH game depicted in conditions (a)–(c).

For a graph $\mathcal{G}$ with N vertices, there are $2^N - 1$ nontrivial subgraphs, which implies $O(4^{mN})$ different graphic games. It is difficult to evaluate or approximate the nonlocal values for all these games. Therefore, it would be interesting if some restricted games can be distinguished, allowing for a more efficient study of nonlocality in multi-source quantum networks.

Example 6.2.1 As an example, let us consider the graph $\mathcal{G}$ shown in Fig. 6.2a. In Fig. 6.2b, there are two players, each holding one vertex independent of inputs, and their outputs are $1 - 2x_1$ and $1 - 2x_2$. Since there is no consistency requirement for two players with $m = 1$, the optimal winning probability for classical players is $\varpi_c = 1$. Therefore, there is no quantum advantage, and we have $\varpi_q = \varpi_c = 1$.

For the game in Fig. 6.2c, the player A_1 holds a different vertex v_{x_1+1} according to the input x_1, while A_2 owns the entire graph independent of the input. In this case, the two players have to satisfy the consistency requirements in conditions (b) and (c),

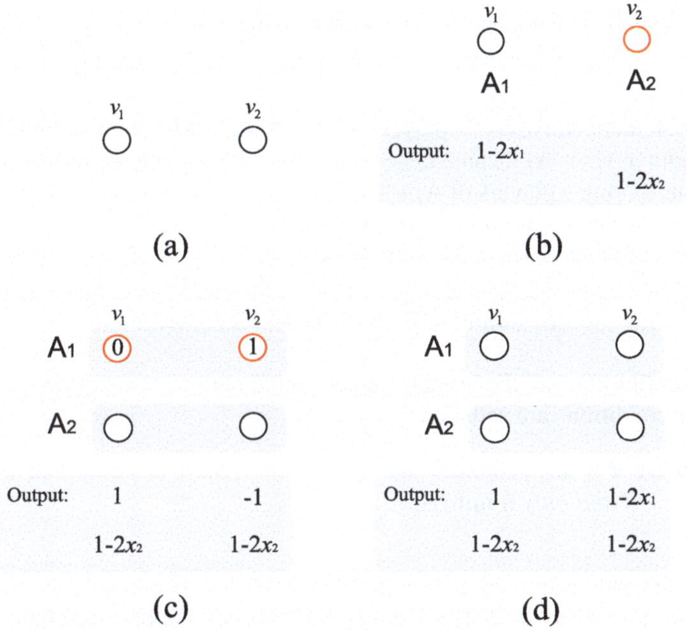

Fig. 6.2 A graphic nonlocal game. **a** A graph with two vertices. **b** A graphic game with two players. The player A_i holds the vertex v_i independent of the input, $i = 1, 2$. **c** A graphic game with two players. The player A_1 holds the vertex v_{x_1} depending on its input x_1 while the player A_2 owns the vertices v_1, v_2 independent of the input. **d** A graphic game with two players. The player A_i holds the vertexes v_1, v_2 independent of the input, $i = 1, 2$. The players can share classical or quantum resources

i.e., $y_{1;A_1} = y_{1;A_2} = 1$ and $y_{2;A_1} = y_{2;A_2} = -1$, where $y_{i;A_j}$ denotes the assignment on vertex v_i by player A_j. Suppose A_1 assigns $1 - 2x_1$ to the shared vertex according to the input x_1, and A_2 assigns the same value of $1 - 2x_2$ to two vertices. With these assignments, the graphic game is equivalent to the CHSH game, which implies a strict quantum advantage for two players who share an EPR state. A similar result holds for the graphic game shown in Fig. 6.2d.

To explain the main result, some definitions will be introduced beforehand.

Definition 6.2.3 Consider a graphic game with n players $\mathsf{A}_1, ..., \mathsf{A}_m$. For each i with $1 \leq i \leq n$, $\mathcal{A}_i$ denotes the set of players A_j's $(n + 1 \leq j \leq m)$ who share vertices with A_i for each input $x_i x_j$, i.e.,

$$\mathcal{A}_i = \{\mathsf{A}_j | V^{x_i} \cap V^{x_j} \neq \emptyset, \forall x_i, x_j = 0, 1\}. \tag{6.13}$$

Denote $\mathcal{A}_i^s$ as the set of players as follows:

$$\mathcal{A}_i^s = \{(\mathsf{A}_i, \mathsf{A}_{i_1}, \ldots, \mathsf{A}_{i_{s-1}})|\mathsf{A}_{i_j} \in \mathcal{A}_i,$$
$$V^{x_i} \cap (\cap_{j=1}^{s-1} V^{x_{i_j}}) \neq \emptyset, \forall x_i, \ldots, x_{i_{s-1}} = 0, 1\}, \quad (6.14)$$

where each element consists of s players $\mathsf{A}_i, \mathsf{A}_{i_1}, \ldots, \mathsf{A}_{i_{s-1}}$ who share common vertices for each input $x_i x_{i_1} \cdots x_{i_{s-1}}$, and $2 \leq s \leq n-m$. $\mathcal{A}_i^s$ characterizes the consistency requirements among s players of A_j's and A_i. In what follows, we omit the case that all the players have no common vertex because of $\varpi_c = \varpi_q$.

Let I_i be an integer associated with $\mathcal{A}_i^s$ as:

$$I_i = \min\{s|\mathcal{A}_i^s \neq \emptyset\}, \forall i = 1, \ldots, m. \quad (6.15)$$

With these definitions, all the graphic games with quantum advantages when proper consistency conditions are satisfied.

Theorem 6.2.1 *For any graphic game with* $\varpi_c \neq 1$, *there is a quantum advantage, i.e.,* $\varpi_q > \varpi_c$, *if and only if* $\min\{I_1, \ldots, I_m\} = 2$.

Proof The first step is to reduce the winning probability, inspired by the unbalanced CHSH game (Lawson et al. 2010). The key is to separate the optimal classical bound from Tsirelson's bound of the expected winning probability. Tsirelson proved that the optimal nonlocal values or quantum bounds are achievable for generalized linear games using maximally entangled EPR states. Thus, from the equivalence of linear games and Bell inequalities, it is sufficient to prove $\varpi_c < \varpi_q$ with the optimal bounds ϖ_c and ϖ_q when $I_1 = 2$. It is only necessary to prove that there are quantum strategies for all players such that their joint winning probability is larger than the classical upper bound of ϖ_c. It is very challenging to find the achievable optimal quantum bound ϖ_q.

For input question series $\vec{x}$, denote the product of assignments on the common vertices $V^{x_i} \cap V^{x_j}$ with respect to two players A_i and A_j as:

$$\zeta_{x_i,x_j}^{\vec{y}_{ij}} = \prod_{\forall v \in V^{x_i} \cap V^{x_j}} y_i(v), \quad (6.16)$$

$$\zeta_{x_i,x_j}^{\vec{y}_{ij}} = \prod_{\forall v \in V^{x_i} \cap V^{x_j}} y_j(v), \quad (6.17)$$

where the outputs of the player A_i with respect to the player A_j are respectively defined by $\vec{y}_{ij} = y_i(v_{j_1}) y_i(v_{j_2}) \cdots y_i(v_{j_s})$ for $v_{j_1}, \ldots, v_{j_s} \in V^{x_i}$, which depend on the input x_i. The consistency condition (c) is equivalent to the following equations:

$$\zeta_{x_i,x_j}^{\vec{y}_{ij}} = \zeta_{x_j,x_i}^{\vec{y}_{ji}}, \forall m < i < j \leq n. \quad (6.18)$$

Hence, the valuation function $F(\vec{x}, \vec{y})$ can be rewritten into

$$F(\vec{x}, \vec{y}) = \prod_{i,j} \delta(\zeta_{x_i,x_j}^{\vec{y}_{ij}} = \zeta_{x_i,x_j}^{\vec{y}_{ji}}), \tag{6.19}$$

where $\vec{y}_i = \vec{y}_{i1}\vec{y}_{i2}\cdots\vec{y}_{im}$ denote the outputs of the player A_i, and $\delta(\cdot)$ denotes the characteristic function, i.e., $\delta(x) = 1$ if x is true and 0 for other cases.

In what follows, assume that all the quantum resources are EPR states. For given inputs $\vec{x}$, denote the measurement of the player A_i as $M_{x_i}^{\vec{y}_i}$, $i = 1, \ldots, m$. The joint probability is given by $P(\vec{y}|\vec{x}) = \mathrm{Tr}(\otimes_{i=1}^m M_{x_i}^{\vec{y}_i}\rho)$. So, any strategy satisfying the consistency condition (6.18) can imply the average winning probability as follows:

$$P_{\vec{x}} = \sum_{\vec{y}} F(\vec{x}, \vec{y})\langle(\otimes_{i=1}^n M_{x_i}^{\vec{y}_i})\rangle \tag{6.20}$$

$$= \sum_{\vec{y}} \prod_{i<j} \langle \delta(\zeta_{x_i,x_j}^{\vec{y}_i} = \zeta_{x_i,x_j}^{\mathbf{y}_j})(M_{x_i}^{\vec{y}_{ij}} \otimes M_{x_j}^{\vec{y}_{ji}})\rangle \tag{6.21}$$

$$= \prod_{i<j} \sum_{\vec{y}_{ij},\vec{y}_{ji}} \langle \delta(\zeta_{x_i,x_j}^{\vec{y}_{ij}} = \zeta_{x_i,x_j}^{\vec{y}_{ji}})(M_{x_i}^{\vec{y}_{ij}} \otimes M_{x_j}^{\vec{y}_{ji}})\rangle. \tag{6.22}$$

Equation (6.20) follows from the definition of the conditional probability $P(\vec{y}|\vec{x})$. In Eq. (6.21), it has taken use of the equalities: $M_{x_j}^{\vec{y}_{ji}} M_{x_j}^{\vec{y}_{ji}} = M_{x_j}^{\vec{y}_{ji}}$ for the projection operators $M_{x_j}^{\vec{y}_{ji}}$, $i, j = 1, \ldots, m$. Equation (6.22) is from the swapping of two operations by using the orthogonal conditions: $M_{x_i}^{\vec{y}_{ij}} M_{x_i}^{\vec{y}'_{ij}} = \mathbf{0}$ with $\vec{y}_{ij} \neq \vec{y}'_{ij}$ for any x_i, i, j, where $M_{x_i}^{\vec{y}_{ij}}$ and $M_{x_i}^{\vec{y}'_{ij}}$ are projection operators of the subsystem owned by the player A_i.

Note that $\delta(\zeta_{x_i,x_j}^{\vec{y}_{ij}} = \zeta_{x_i,x_j}^{\vec{y}_{ji}}) = \frac{1}{2}(1 + \zeta_{x_i,x_j}^{\vec{y}_{ij}}\zeta_{x_i,x_j}^{\vec{y}_{ji}})$ for any $i, j = 1, 2, \ldots, m$. It follows that

$$P_{\vec{x}} = \prod_{i<j} \frac{1}{2}\langle(\mathbb{1} + \hat{M}_{x_i,x_j} \otimes \hat{M}_{x_j,x_i})\rangle, \tag{6.23}$$

where $\hat{M}_{x_i,x_j} = \sum_{\vec{y}_{ij}} \zeta_{x_i,x_j}^{\vec{y}_{ij}} M_{x_i}^{\vec{y}_{ij}}$ denotes the successful measurement of the player A_i on the system shared with the player A_j. The total win probability is given by:

$$\sum_{\vec{x}} P(\vec{x})P_{\mathbf{x}} = \sum_{\vec{x}} p_* P(\vec{x}) \prod_{l=1}^{k} \prod_{j=s_l}^{n_l} \frac{1}{2}\langle \mathbb{1} + \hat{M}_{x_{i_l},x_j} \hat{M}_{x_j,x_{i_l}}\rangle. \tag{6.24}$$

Equation (6.24) holds for classical hidden resources by simple evaluations under the no-signaling principle, where the goal is to evaluate the maximal winning probability over all the resources.

For the quantum model, it is sufficient to use EPR states as shared resources, i.e., each pair of players A_i and A_j can share some EPR states. From the assumption of the consistency condition, it is only necessary to consider the outputs of each pair of players who have common subgraphs. For other pairs, the consistency condition can be easily satisfied by setting their outputs to be 1 for all vertices. Therefore, it is reasonable to denote $p*$ in Eq. (6.24) as the success probability of all pairs without common subgraphs. This means that all the players can win the game with a constant success probability regardless of their input distributions and shared resources. Here, we assume that for each $j = 1, 2, \ldots, n$, all pairs of players A_{i_j} and Alice$_t$ have common vertices for $t = s_j, s_j + 1, \ldots, m_j$, where s_j and m_j satisfy $1 < s_1 < m_1 < s_2 < m_2 < \cdots < s_n < m_n < m$.

For the sufficient condition, due to the requirements of the consistency conditions (a) and (b) for the first player A_1, there exist four common vertex sets $V^{x_1x_j}$ for each pair of players A_1 and A_j satisfying that

$$\hat{M}_{x_1,x_j} = (-1)^{x_1}\hat{M}_{x_1,1-x_j}, \tag{6.25}$$

$$\hat{M}_{x_j,x_1} = \hat{M}_{x_j,1-x_1}, \tag{6.26}$$

for binary inputs x_1, x_j, where Eq. (6.26) follows from the non-signaling condition.

Now, assume that there is no more than two players sharing vertices, i.e., $I_1 = 2$. From Eqs. (6.24)–(6.26), the total winning probability can be rewritten into:

$$\begin{aligned}\sum_{\vec{x}} P(\vec{x})P_{\vec{x}} = p_*(p\prod_{i=s_1}^{n_1}\frac{1}{2}\langle \mathbb{1} + p\hat{M}_{x_1=0,x_i}\hat{M}_{x_i=0,x_1} + q\hat{M}_{x_1=0,x_i}\hat{M}_{x_i=1,x_1}\rangle \\ +q\prod_{i=s_1}^{n_1}\frac{1}{2}\langle \mathbb{1} + p\hat{M}_{x_1=1,x_i}\hat{M}_{x_i=0,x_1} - q\hat{M}_{x_1=1,x_i}\hat{M}_{x_i=1,x_1}\rangle) \\ \times\prod_{i=2}^{k}(\prod_{j=s_i+1}^{n_i}\frac{1}{2}\langle \mathbb{1} + p\hat{M}_{x_i=0,x_j}\hat{M}_{x_j=0,x_i} + q\hat{M}_{x_i=0,x_j}\hat{M}_{x_j=1,x_i} \\ +p\hat{M}_{x_i=1,x_j}\hat{M}_{x_j=0,x_i} - q\hat{M}_{x_i=1,x_j}\hat{M}_{x_j=1,x_i}\rangle].\end{aligned} \tag{6.27}$$

In Eq. (6.27), for the simplicity we assume $i_j = j$ for $j = 1, 2, \ldots, k$, $s_1 = k + 1$ and $s_j = m_{j-1} + 1$ for $j = 2, 3, \ldots, k$. Moreover, Eq. (6.27) follows from the independence assumption of input distributions, where $\{p, q = 1 - p\}$ denotes the input distribution of each player. The following proof is based on the unbalanced CHSH game, that is, for an unbalanced CHSH game with the input probability distribution $\{p_{00}, p_{01}, p_{10}, p_{11}\}$, the quantum advantage exists if the following conditions satisfy

- $(\frac{1}{p_{10}} - \frac{1}{p_{11}})^2 - (\frac{1}{p_{00}} + \frac{1}{p_{01}})^2 < 0$ if $\min\{p_{10}, p_{11}\} \le \min\{p_{00}, p_{01}\}$;
- $(\frac{1}{p_{00}} - \frac{1}{p_{01}})^2 - (\frac{1}{p_{10}} + \frac{1}{p_{11}})^2 < 0$ if $\min\{p_{10}, p_{11}\} > \min\{p_{00}, p_{01}\}$.

This has proved the sufficient condition.

The next is to prove the necessary condition. Specially, there is no quantum advantage when there are more than two players who share vertices. The proof is based on reductions (6.16)–(6.24). Consider the special case where l players A_1, ..., A_l have common vertices. For convenience, let us assume that $l = 3$. For the classical resources we have

$$\begin{aligned}
\varpi_c = & \max_{\lambda,\{M_{x_i}^{\hat{y}_i}\},\forall i} \sum_{\vec{x}} P(\vec{x}) P_{\vec{x}} \\
= & \max_{\lambda,\{M_{x_i}^{\hat{y}_i}\},\forall i} (\sum_{\vec{x}} p_* P(\vec{x}) (\frac{1}{2}\langle \mathbb{1} + \hat{M}_{x_1,2}\hat{M}_{x_2,1}\rangle) \\
& \times (\frac{1}{2}\langle \mathbb{1} + \hat{M}_{x_1,3}\hat{M}_{x_3,1}\rangle)(\frac{1}{2}\langle \mathbb{1} + \hat{M}_{x_2,3}\hat{M}_{x_3,2}\rangle)),
\end{aligned} \tag{6.28}$$

where p_* is the success probability of all the pairs without the consistency requirements.

Note that all players $\mathsf{A}_1, \ldots, \mathsf{A}_l$ share the same vertex set. Denote $A_0 := \hat{M}_{x_1=0,2} = \hat{M}_{x_1=0,3}$, $A_1 := \hat{M}_{x_1=1,2} = \hat{M}_{x_1=1,3}$, $B_0 := \hat{M}_{x_2=0,1} = \hat{M}_{x_2=0,3}$, $B_1 := \hat{M}_{x_2=1,1} = \hat{M}_{x_2=1,3}$, $C_0 := \hat{M}_{x_3=0,1} = \hat{M}_{x_3=0,2}$, $C_1 := \hat{M}_{x_3=1,1} = \hat{M}_{x_3=1,2}$. It follows that

$$\varpi_c = \begin{cases} p_*(1-2p+3p^2-p^3), & \text{for } 0 < p \leq \frac{1}{2}; \\ p_* p(1+p-p^2), & \text{for } \frac{1}{2} \leq p < 1. \end{cases} \tag{6.29}$$

In what follows, it needs to prove that the quantum bound ϖ_q derived from quantum measurements on any quantum resources satisfying $\varpi_q = \varpi_c$. The basic idea is from the consistence of classical bound and quantum bound of generalized CHSH operator $CHSH_{AB} + \frac{1}{2}\mathcal{L}_{AB} = \frac{1}{2}(3p^2 A_0 B_0 + 3pq A_0 B_1 + p^2 q A_1 B_0 - 3q^3 A_1 B_1)$ for $p \geq q$, or $CHSH_{AB} + \frac{1}{2}\mathcal{L}_{AB} = \frac{1}{2}(3p^2 A_0 B_0 + 3pq A_0 B_1 + 3q^3 A_1 B_1 - p^2 q A_1 B_0)$ for $p \leq q$. A similar result holds for other cases. □

Analyzing general Cube games with general input distributions is difficult. However, for the special case of independent distribution, where the referee randomly chooses binary question $x_i \in \{0, 1\}$ according to a given distribution $\{p, 1-p\}$, Theorem 6.2.1 can be applied. Here, each player assigns 1 or -1 to their own vertices on an $n-1$-dimensional hyperplane defined by $\{(q_1, \ldots, q_m)|q_i = x_i\}$ and sends it to the referee. All players win the game if and only if their assignments satisfy the specific consistent requirements.

From Theorem 6.2.1, it follows that the cubic game has no quantum advantage over all classical strategies when $m \geq 3$, where $I_1 = m$. This result differs from a recent result with relaxed consistency conditions (Eisert et al. 1999).

Another simple extension with $n \geq 1$ is defined as follows: the player A_i owns the local subgraph defined by $\{(q_1, \ldots, q_m)|q_1 = \cdots = q_n = x_i\}$ for $i = 1, \ldots, m$, and $\{(q_1, \ldots, q_m)|q_j = x_j\}$ for $j = n+1, \ldots, m$. Note that any two players $\mathsf{A}_1, \ldots, \mathsf{A}_n$ have no common subgraph for all inputs. It is straightforward to check that this generalization has a quantum advantage when $n = m - 1$.

6.2.2 A Boolean Satisfiability Problem

A Boolean satisfiability problem (SAT) aims to check some given equations (Schaefer 1978). The decision problem is of central importance in theoretical computer science, complexity theory, and algorithmic theory. Interestingly, each graphic game is equivalent to a distributive SAT problem. For example, consider the game shown in Fig. 6.3a. Let y_{i,x_j} be the assignment on vertex v_i by player A_j according to input x_j, where $i = 1, \ldots, 8$ and $j = 1, 2, 3$. The winning conditions are equivalent to Boolean equations with 48 variables.

Theorem 6.2.1 shows that there is no quantum advantage from shared entangled resources in the game. Generally, for an n-player graphic game involving N vertices, there are $O(mN)$ conditions that should be checked. Theorem 6.2.1 provides an intuitive completion for distributive SAT problems with different resources using the graphic game model. In applications, an equivalent graphic game should be found for special decision problems. One possible way is to reduce the required clauses first, even if it is hard. Then, a graphic game can be defined clause-by-clause.

6.2.3 Witnessing Multi-source Quantum Networks

The potential applications of a single-source Bell network can be greatly extended by multi-source quantum networks. However, due to the non-convexity of quantum correlations, it is challenging to verify distributive entangled states on a large scale. One effective approach is to use nonlinear Bell-type inequalities. Interestingly, the current model provides another way to verify general quantum networks. Specifically, if each

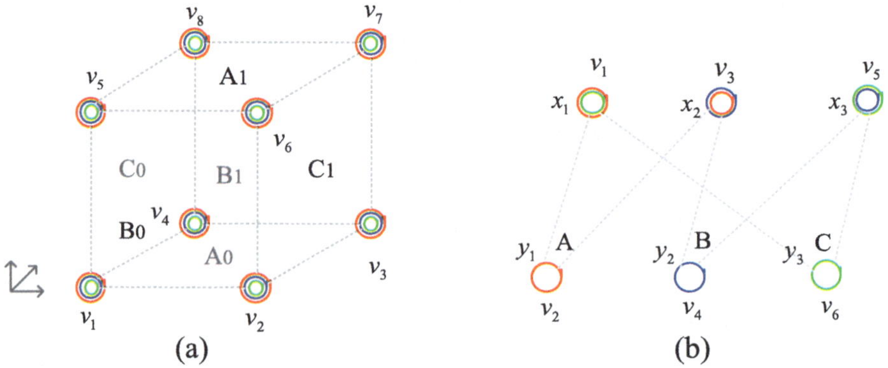

Fig. 6.3 Some graphic games. **a** Cubic game. Each player owns all the vertices on a 2-dimensional plane (denoted by A_i, B_j, C_k) with different colors depending on the input. **b** The equivalent graphic game of the GYNI game. Each player owns three vertices of one subgraph (denoted by A, B, C) with different colors. Here, the edges with dot lines are used for visualization

vertex in a graph represents an EPR state, then any multi-source quantum network consisting of EPR states can be represented by a graph with multiple vertices. The assumption that each pair of players shares an EPR state can be shown schematically by sharing a vertex in terms of the graphic game. To achieve a quantum advantage when defining graphic games, significant testing is required. Theorem 6.2.1 implies the following result:

Theorem 6.2.2 *For any quantum network $\mathcal{N}_q$ consisting of EPR states shared by m observers, there are nonlocality witnesses in terms of the graphic game if $\mathcal{N}_q$ is k-independent with $k \geq 2$.*

To verify quantum networks, all shared vertices V^{x_i} must be independent of inputs. The equivalent reduction shows that any k-independent network is equivalent to a generalized star-type network. The proof of Theorem 6.2.2 follows easily from the corresponding graphic games, where $\min_i I_i = 2$ because no more than two players share vertices.

As an illustration, consider a star network where Bob shares one vertex v_i with each of the players A_i, independent of their inputs for $i = 1, \ldots, m$. Assuming $n = 1$ (or $n = m$), a quantum advantage exists due to $I_1 = 2$. Similarly, in a chain network, each vertex v_i is shared by two adjacent players A_i and A_{i+1}. Different values of m can be defined by selecting players who do not share vertices (which are independent in quantum networks). Quantum advantages can be achieved in these graphic games. In general, these graphic games provide a means for the first linear testing of multi-source quantum networks composed of EPR states.

An intuitive way to demonstrate the relationship between I_i and quantum advantage is through $\mathcal{A}_i^s$, which represents the number of players who satisfy the consistency requirements regardless of their inputs. When a network is divided into n subnetworks according to players Alice$_i$ ($i = 1, \ldots, n$) who do not share vertices, I_i characterizes the topology of the network. Specifically, $I_i = 2$ signifies that the network comprises star or chain subnetworks. For $I_i = k > 2$, the network consists of various cyclic subnetworks. In such cases, Theorem 6.2.1 suggests that there is no useful linear testing scheme for cyclic networks, similar to the nonlinear method proposed in Luo (2018).

6.3 Guess Your Neighbour's Input Game

The Guess Your Neighbour's Input (GYNI) game has been recently put forward to differentiate multipartite quantum correlations from classical correlations (Almeida et al. 2010). This game can be portrayed equivalently as a graphic game, as demonstrated in Fig. 6.3b. The initial consistency requirements are substituted by particular output strategies for all players, where the input is designated as the vertex v_{2i-1} for players $\mathsf{A}_i, i = 1, 2, 3$. This form of graphic game does not provide any quantum advantage for any quantum resources. It can be considered a relaxed graphic game that fulfills partial consistency conditions. In this paper, we establish a more general result.

Definition 6.3.1 (*Generalized GYNI*) Consider a given m-party quantum network $\mathcal{N}$ with m nodes.

- Input: The referee randomly chooses binary question $x_i \in \{0, 1\}$ according to a given distribution $P(x_1, \ldots, x_m)$, and sends x_i to the player Alice$_i$.
- Output: Each player outputs $y_i = 0$ or 1 to the referee.
- Win: All the players win the game if $y_i = f_i(\vec{x})$ for all i's. Here, f_i denotes some function of the input bit series $\vec{x}$.

As an example, $f_i(\vec{x})$ can be regarded as the restriction of the product of all the assigned values in the graphic games. In this case, we can prove the following theorem.

Theorem 6.3.1 *There is no quantum advantage for a multipartite nonlocal game if* $\mathcal{F} : \vec{x} \mapsto (f_1(\vec{x}), f_2(\vec{x}), \ldots, f_m(\vec{x}))$ *is injective.*

Proof We show that the optimal classical and quantum winning strategies are identical for any prior distribution p of the inputs. Note that there is a simple classical strategy for achieving a winning probability

$$\varpi_c \geq \max_{\vec{x}}\{P(\vec{x}) + P(\overline{\vec{x}})\}, \tag{6.30}$$

where $\overline{\vec{x}}$ denotes the negation of the input string $\vec{x}$, $\overline{\vec{x}} = (\overline{x}_1, \overline{x}_2, \ldots, \overline{x}_m)$ with $\overline{x}_i = x_i \oplus 1$, and $\oplus$ denotes addition modulo 2. This strategy is based on the following simple observation from the assumption of $\mathcal{F}$.

- Let $\vec{x}^*$ be an arbitrary string. If $\vec{x} \neq \vec{x}^*, \overline{\vec{x}^*}$, then it follows that

$$\begin{aligned} (f_1(\vec{x}^*), \ldots, f_m(\vec{x}^*)) &\neq (f_1(\vec{x}), \ldots, f_m(\vec{x})), \\ (f_1(\overline{\vec{x}^*}), \ldots, f_m(\overline{\vec{x}^*})) &\neq (f_1(\vec{x}), \ldots, f_m(\vec{x})). \end{aligned} \tag{6.31}$$

Indeed, if this was not the case, we would have that for any i, either $x_i \neq y_i$ or $x_{i+1} = y_{i+1}$. But this would in turn imply that either $\vec{x} = \vec{x}^*$ or $\vec{x} = \overline{\vec{x}^*}$, which is a contradiction with the assumption.

Let a classical strategy be specified by the string $\vec{x}^*$, where each party sends out the bit $y_i = f_i(\vec{x}^*)$ if it has received the input x_i^*, and outputs $y_i = f_i(\overline{\vec{x}^*})$ if it has received $\overline{x_i^*}$. This obviously follows that $P(y_i = f_i(\vec{x}^*)|\vec{x}^*) = 1$ and $P(y_i = f_i(\overline{\vec{x}^*})|\overline{\vec{x}^*}) = 1$. On the other hand, $P(y_i|\vec{x}) = 0$ for all $\vec{x} \neq \vec{x}^*, \overline{\vec{y}}$. Indeed, from the observation (6.31), there exists an index i such that $x_i = x_i^*$, but for which $x_{i+1}^* \neq x_{i+1}$. Since $\mathcal{F}$ is injective, it follows that

$$(f_1(\vec{x}^*), f_2(\vec{x}^*), \dots, f_m(\vec{x}^*)) \neq (f_1(\vec{x}), f_2(\vec{x}), \dots, f_m(\vec{x})), \tag{6.32}$$
$$(f_1(\overline{\vec{x}^*}), f_2(\overline{\vec{x}^*}), \dots, f_m(\overline{\vec{x}^*})) \neq (f_1(\vec{x}), f_2(\vec{x}), \dots, f_m(\vec{x})). \tag{6.33}$$

The winning probability of this classical strategy equals to $P(\vec{x}^*) + P(\overline{\vec{y}})$, which yields Eq. (6.30) if we take $\vec{x}^*$ to be $P(\vec{x}^*) + P(\overline{\vec{x}^*}) = \max_{\vec{x}}(P(\vec{x}) + P(\overline{\vec{x}}))$.

We will now prove that there cannot exist a superior quantum strategy, and consequently, a superior classical strategy. In the most general quantum protocol, the players share an entangled state ρ, and they perform projective measurements on their subsystems, contingent on their inputs x_i. They output their measurement results y_i. Let $M_{y_i}^{x_i}$ refer to the projection operator related to output y_i for input x_i. The probability that all players produce the correct output is given by:

$$\begin{aligned} P(f_1(\vec{x}), \dots, f_m(\vec{x})|\vec{x}) &= \langle M_{f_1}^{x_1} \otimes \cdots \otimes M_{f_m}^{x_m} \rangle \\ &= \mathrm{Tr}[(M_{f_1}^{x_1} \otimes \cdots \otimes M_{f_m}^{x_m})\rho], \end{aligned} \tag{6.34}$$

The average winning probability is

$$\varpi_q = \max_{\rho, \forall \{M_{y_i}^{x_i}\}} \sum_{\vec{x}} P(\vec{x}) \langle M_{\vec{x}} \rangle, \tag{6.35}$$

where $M_{\vec{x}}$ is given by $M_{\vec{x}} = M_{f_1}^{x_1} \otimes \cdots \otimes M_{f_m}^{x_m}$ for short, and $\langle \cdot \rangle$ is associated with some fixed quantum system ρ. Note that the operators $M_{\vec{x}}$ satisfy the following properties

$$M_{\vec{x}}^2 = M_{\vec{x}}, \tag{6.36}$$
$$M_{\vec{x}} M_{\vec{x}^*} = 0 \text{ if } \vec{x} \neq \vec{x}^*, \overline{\vec{x}^*}. \tag{6.37}$$

The first property follows from the fact that $M_{\vec{x}}$s are projection operators. The second follows from the orthogonality relations: $M_{y_i}^{x_i} M_{\overline{y}_i}^{x_i} = 0$ and observation in Eq. (6.31). Note that protocols involving mixed states or general measurements can all be represented in the above form by expanding the dimensionality of the initial state.

Using Eqs. (6.36) and (6.37), we show that $\varpi_q = \max_{\rho, \forall \{M_{y_i}^{x_i}\}} \sum_{\vec{x}} p(\vec{x}) M_{\vec{x}} \leq \varpi_c$, where the inequality should be understood as an operator inequality, i.e., $A \leq B$ means that $\langle A \rangle \leq \langle B \rangle$ for all ρ. Note that we cannot assume $P(\vec{x}) + P(\overline{\vec{x}}) = \varpi_c$, but $P(\vec{x}) + P(\overline{\vec{x}}) \leq \varpi_c$ for all $\vec{x}$ since the normalization condition of $\sum_{\vec{x}} P(\vec{x}) = 1$. We have

$$\begin{aligned}\Delta &= (\sqrt{\varpi_c} - \sum_{\vec{x}} \alpha_{\vec{x}} M_{\vec{x}})^2 + \frac{1}{2}\sum_{\vec{x}} (\beta_{\vec{x}} M_{\vec{x}} - \beta_{\bar{\vec{x}}} M_{\bar{\vec{x}}})^2 \\ &= \varpi_c - 2\sum_{\vec{x}} (\varpi_c - P(\bar{\vec{x}})) M_{\vec{x}} + \sum_{\vec{x}} \alpha_{\vec{x}}^2 M_{\vec{x}} + \sum_{\vec{x}} \alpha_{\vec{x}} \alpha_{\bar{\vec{x}}} M_{\vec{x}} M_{\bar{\vec{x}}} \\ &+ \frac{1}{2}\sum_{\vec{x}} \frac{p(\vec{x})P(\bar{\vec{x}})}{\varpi_c} (M_{\vec{x}} + M_{\bar{\vec{x}}} - M_{\vec{x}} M_{\bar{\vec{x}}} - M_{\bar{\vec{x}}} M_{\vec{x}}),\end{aligned} \tag{6.38}$$

where $\alpha_{\vec{x}} = \sqrt{\varpi_c} - P(\bar{\vec{x}})/\sqrt{\varpi_c}$ and $\beta_{\vec{x}} = \sqrt{P(\vec{x})P(\bar{\vec{x}})/\varpi_c}$. Equation (6.38) is followed from Eq. (6.36). And then, we have

$$\begin{aligned}\Delta &= \varpi_c - \sum_{\vec{x}} \varpi_c M_{\vec{x}} + \sum_{\vec{x}} (\varpi_c - P(\vec{x}) - P(\bar{\vec{x}})) M_{\vec{x}} M_{\bar{\vec{x}}} \\ &+ \sum_{\vec{x}} \frac{P(\bar{\vec{x}})(P(\vec{x}) + P(\bar{\vec{x}}))}{\varpi_c} M_{\vec{x}}.\end{aligned} \tag{6.39}$$

In Eq. (6.39), we have taken use of the commutate of projection operators, i.e., $M_{\vec{x}} M_{\bar{\vec{x}}} = M_{\bar{\vec{x}}} M_{\vec{x}}$, and identity $\sum_{\vec{x}} \frac{P(\bar{\vec{x}})P(\vec{x})}{\varpi_c} M_{\bar{\vec{x}}} = \sum_{\vec{x}} \frac{P(\vec{x})P(\bar{\vec{x}})}{\varpi_c} M_{\vec{x}}$.

$$\Delta \le \varpi_c - \sum_{\vec{x}} \varpi_c M_{\vec{x}} + \sum_{\vec{x}} P(\bar{\vec{x}}) M_{\vec{x}} + \sum_{\vec{x}} (\varpi_c - P(\vec{x}) - P(\bar{\vec{x}})) M_{\vec{x}} M_{\bar{\vec{x}}}. \tag{6.40}$$

Equation (6.40) follows from the inequality : $P(\vec{x}) + P(\bar{\vec{x}}) \le \varpi_c$ which is derived from the inequality (6.30).

$$\begin{aligned}\Delta &\le \varpi_c - \sum_{\vec{x}} (\varpi_c - P(\bar{\vec{x}})) M_{\vec{x}} + \sum_{\vec{x}} (\varpi_c - P(\vec{x}) - P(\bar{\vec{x}})) M_{\vec{x}} M_{\bar{\vec{x}}} \\ &\le \varpi_c - \sum_{\vec{x}} (\varpi_c - P(\bar{\vec{x}})) M_{\vec{x}} + \sum_{\vec{x}} (\varpi_c - P(\vec{x}) - P(\bar{\vec{x}})) M_{\vec{x}} \end{aligned} \tag{6.41}$$

$$= \varpi_c - \sum_{\vec{x}} P(\vec{x}) M_{\vec{x}}. \tag{6.42}$$

Equation (6.41) follows from the operator inequality of $M_{\vec{x}} M_{\bar{\vec{x}}} \le M_{\vec{x}}$. From Eq. (6.30), it follows that

$$\max_{\rho, \forall \{M_{y_i}^{x_i}\}} \sum_{\vec{x}} P(\vec{x}) M_{\vec{x}} \le \varpi_c. \tag{6.43}$$

Note that any POVM measurements associated with non-commuting operators can be realized using projection measurements in extended state space. The proof presented above is also independent of quantum resources, which explains why the same bound applies to both classical and quantum payoffs. This fact also means that the inequality $\sum_{\vec{x}} P(\vec{x})P(y_1 = f_1(\vec{x}), \ldots, y_m = f_m(\vec{x})|\vec{x}) \leq \varpi_c$ can be interpreted as a Bell inequality, with its local and quantum bounds coinciding.

Example 6.3.1 Take $m = 3$ as an example. All the permutations are given by the permutation group $S_3 = \{(1), (1, 2), (1, 3), (2, 3), (1, 2, 3), (1, 3, 2)\}$, where (1) denotes the identity operator, (i, j) denotes the permutation of $i \to j$ and $j \to i$, and (i, j, k) denotes the permutation of $i \to j$, $j \to k$ and $k \to i$. Now, let's define the expected output of three parties as $g(i, j, k)$ for the corresponding inputs i, j, k where $g \in S_3$. The games with $(f_1, f_2, f_3) = (i, j, k)$ are previous games of guessing the neighbors' inputs. Another example is when f_i are any injective mappings that go beyond the permutation group S_m, meaning they are functions that map from the input space $\{0, 1\}^m$ to the output space $\mathbb{R}^m$.

6.4 Cooperating Games on MAC

In this section, we consider a multiple-player cooperating game on a multiple access channel (MAC). A MAC is a communication system where multiple users share a common communication medium, and each user can transmit information simultaneously. While all the players can agree on strategies before starting the game, they are not allowed to communicate after they have received the questions.

A general MAC is shown in Chapter 3. The central object in this section is a MAC $\mathcal{N}_G$ defined in terms of a distributed game $\mathcal{G} = (\mathcal{X}_1, \ldots, \mathcal{X}_m, \mathcal{Y}_1, \ldots, \mathcal{Y}_m, F)$ as follows.

Definition 6.4.1 The input alphabets of m senders, $\mathsf{A}_1, \ldots, \mathsf{A}_m$, are the question-answer sets $\mathcal{X}_1 \times \mathcal{Y}_1, \ldots, \mathcal{X}_m \times \mathcal{Y}_m$, respectively, and the output alphabet of $\mathcal{N}_G$ is $\mathcal{X}_1 \times \cdots \times \mathcal{X}_m$. If $\mathsf{A}_1, \ldots, \mathsf{A}_m$ win the game, i.e., $(x_1, y_1, \ldots, x_m, y_m) \in W$ with a winning set $W \subseteq \times_{i=1}^m \mathcal{X}_i \times \mathcal{Y}_i$, the channel is noiseless and outputs the question pair $(x_1, \ldots, x_m)$ to the receiver. If they lose the game, i.e., $(x_1, y_1, \ldots, x_m, y_m) \notin W$, the channel outputs a question pair $(\hat{x}_1, \ldots, \hat{x}_m)$ drawn uniformly at random from $\mathcal{X}_1 \times \cdots \times \mathcal{X}_m$.

More formally, the MAC $\mathcal{N}_G : (\mathcal{X}_1 \times \mathcal{Y}_1) \times \cdots \times (\mathcal{X}_m \times \mathcal{Y}_m) \to \mathcal{X}_1 \times \cdots \times \mathcal{X}_m$ is defined as

$$\mathcal{N}_G(\hat{x}_1, \ldots, \hat{x}_m | x_1, y_1; \cdots; x_m, y_m) = \begin{cases} \delta_{x_1\hat{x}_1}\delta_{x_2\hat{x}_2}\cdots\delta_{x_m\hat{x}_m}, & \text{if } (x_1, x_2, \ldots, y_m, y_m) \in W, \\ (|\mathcal{X}_1||\mathcal{X}_2|\cdots|\mathcal{X}_m|)^{-1}, & \text{otherwise.} \end{cases} \tag{6.44}$$

The channel construction is inspired by Refs. (Leditzky et al. 2020; Quek and Shor 2017), where the authors used a similar construction based on the CHSH game (Clauser et al. 1969) for an interference channel comprised of two senders and two receivers. Our findings pertain to the capacity region $\mathcal{C}(\mathcal{N})$ of a MAC $\mathcal{N}$, which encompasses all rate pairs $(R_1, \ldots, R_m)$ for which sender i can transmit information reliably at a rate of R_i to the corresponding receiver. For the special case of $m = 2$, prior research (Ahlswede 1971; Liao 1972) has demonstrated that the convex hull of all pairs (R_1, R_2) that satisfy the following conditions is the capacity region of the MAC:

$$R_1 \leq I(A; Z|B), \tag{6.45}$$
$$R_2 \leq I(B; Z|A), \tag{6.46}$$
$$R_1 + R_2 \leq I(AB; Z), \tag{6.47}$$

for some product distribution $\pi_A \pi_B$ on $\mathcal{A} \times \mathcal{B}$.

Theorem 6.4.1 *If a game $\mathcal{G}$ does not admit a good strategy, $\omega_U(\mathcal{G}) < 1$, then the sum rate $\sum_{i=1}^{m} R_i$ of the MAC $\mathcal{N}_G$ is strictly bounded away from the maximal value $\sum_{i=1}^{m} \log |\mathcal{X}_i|$.*

Although all of A_i's players can agree on a strategy before the game, they are not permitted to communicate during the game. A *deterministic strategy* for Alice is a pair of deterministic functions $f_i : \mathcal{X}_i \to \mathcal{Y}_i$ for $i = 1, \ldots, n$. A *probabilistic strategy* is a probabilistic mixture of deterministic strategies. For a given probability distribution $\pi : P \to [0, 1]$ on the promised question set $P \subset \mathcal{X}_1 \times \cdots \times \mathcal{X}_2$, we define $\omega(\mathcal{G}, \pi)$ as the maximum winning probability achievable using probabilistic strategies. If the given probability distribution on P is the uniform distribution π_U, we use the shorthand $\omega_U(\mathcal{G}) \equiv \omega(\mathcal{G}, \pi_U)$. It is important to note that the maximum winning probability can always be attained at an extreme point, i.e., a deterministic strategy. A strategy that achieves $\omega(\mathcal{G}, \pi) = 1$ is referred to as a *good* strategy.

Proof In what follows, we consider discrete memoryless multiple access channels without feedback, as multiple access channels (MAC). An m-sender MAC is a tuple $(\mathcal{A}_1, \ldots, \mathcal{A}_m, \mathcal{Z}, N(z|a_1, \ldots, a_m))$, where $\mathcal{A}_i$'s are the input alphabets for sender A_i, respectively, $\mathcal{Z}$ is the output alphabet for the single receiver B, and $N(z|a_1, \ldots, a_n)$ is a probability distribution for all pairs $(a_1, \ldots, a_m) \in \mathcal{A}_1 \times \cdots \times \mathcal{A}_m$.

An $(R_1^{(n)}, \ldots, R_m^{(n)}, n)$-code is a tuple $(\mathcal{M}_1, \ldots, \mathcal{M}_m, a_1^n, \ldots, a_m^n, \hat{z}^n)$, where:

- $\mathcal{M}_i$ are message sets with $|\mathcal{M}_i| = 2^{nR_i^{(n)}}$ for $i = 1, \ldots, m$;
- $a_i^n : \mathcal{M}_i \to \mathcal{A}_i^n$ are encoding functions;
- $\hat{z}^n : \mathcal{Z}^n \to \mathcal{M}_1 \times \cdots \times \mathcal{M}_m \cup \{e\}$ is a decoding function, with e an arbitrary error message.

Assume a uniform distribution over the messages $(M_1, \ldots, M_m)$; in particular, the codewords $a_i^n(M_i)$ are independent for any i. The average probability of error is defined as

$$P_e^{(n)} = \Pr[(\hat{M}_1, \ldots, \hat{M}_m) \neq (M_1, \ldots, M_m)], \tag{6.48}$$

where the probability mass function (pmf) of $(\hat{M}_1, \ldots, \hat{M}_m)$ is given by $\hat{z}^n(N^{\times n}(z^n|a_1^n(M_1), \ldots, a_m^n(M_m)))$.

A rate pair $(R_1, \ldots, R_m)$ is achievable if there exists a sequence of codes $\{(R_1^{(n)}, \ldots, R_m^{(n)}, n)\}_{n\in\mathbb{N}}$ such that $\liminf_{n\to\infty} R_i^{(n)} = R_i$ and $\lim_{n\to\infty} Pe^{(n)} = 0$. The capacity region $\mathcal{C}(\mathcal{N})$ of the MAC $(\mathcal{A}_1, \ldots, \mathcal{A}_m, \mathcal{Z}, N(z|a_1, \ldots, a_m))$ is the closure of the set of all achievable rate pairs $(R_1, \ldots, R_m)$. Consider the sum-capacity $S(\mathcal{N})$ defined as $S(N) := \sup\{\sum_{i=1}^m R_i | (R_1, \ldots, R_m) \in \mathcal{C}(\mathcal{N})\}$.

The capacity region of an m-sender MAC $(\mathcal{A}_1, \ldots, \mathcal{A}_m, \mathcal{Z}, N(z|a_1, \ldots, a_m))$ has a single-letter characterization (Ahlswede 1971; Liao 1972).

Let $(A_1, \ldots, A_m)$ be a pair of discrete random variables jointly distributed according to the pmf $p_{A_1} \cdots p_{A_m}$, and let Z be the channel output random variable with conditional pmf $N(z|a_1, \ldots, a_m)$. Define $\mathcal{R}(A_1, \ldots, A_m)$ as the set of rate pairs $(R_1, \ldots, R_m)$ satisfying

$$\begin{aligned} R_i &\leq I(A_i; Z|\mathbf{A}_i), i \in [m], \\ \sum_{i=1}^m R_i &\leq I(A_1, \ldots, A_m; Z), \end{aligned} \tag{6.49}$$

where $\mathbf{A}_i$ denotes all random variables except for A_i. Then the capacity region $\mathcal{C}(\mathcal{N})$ of the MAC $(\mathcal{A}_1, \ldots, \mathcal{A}_m, \mathcal{Z}, N(z|a_1, \ldots, a_m))$ is the convex hull of the union of the regions $\mathcal{R}(A_1, \ldots, A_m)$ over all product distributions $p_{A_1} \cdots p_{A_m}$:

$$\mathcal{C}(\mathcal{N}) = \text{conv}\left(\bigcup \{\mathcal{R}(A_1, \ldots, A_m) \colon (A_1, \ldots, A_m) \sim p_{A_1} \cdots p_{A_m}\}\right). \tag{6.50}$$

This has proved the result. □

6.4.1 MACs and Entanglement Assistance

In this section, we consider coding strategies for classical MACs assisted by entanglement shared between the two senders. Let $(\mathcal{A}_1, \ldots, \mathcal{A}_m, \mathcal{Z}, N(z|a_1, \ldots, a_m))$ be a MAC, and let each pair of two senders A_i and A_j share an entangled state $|\psi\rangle_{S_{A_i} S_{A_j}} \in \mathbb{C}^d \otimes \mathbb{C}^d$, where the d-dimensional quantum systems S_{A_i} with $|S_{A_i}| = d$ are with senders A_i. We then consider the following coding scenario.

Let $A_1, \ldots, A_m$ be random variables taking values in finite alphabets $\mathcal{A}_1, \ldots, \mathcal{B}_1$ for sender $\mathsf{A}_1, \ldots, \mathsf{A}_m$, respectively. Depending on the value $a_1 \in \mathcal{A}_1$ of A_1, the first sender selects a positive operator-valued measure (POVM) $L^{(a_1)} = \{L_{\bar{a}_1}^{(a_1)}\}_{\bar{a}_1 \in \bar{\mathcal{A}}_1}$ with

$$\begin{aligned} L_{\bar{a}_1}^{(a_1)} &\geq 0, \text{ for all } a_i \in \mathcal{A}_i, \\ \sum_{\bar{a}_1 \in \bar{\mathcal{A}}_1} L_{\bar{a}_1}^{(a_1)} &= \mathbb{1}_d, \text{ for all } a_1 \in \mathcal{A}_1. \end{aligned} \tag{6.51}$$

Here, $\mathcal{A}_i$ is some finite alphabet, and $\mathbb{1}_d$ denotes the identity operator on $\mathbb{C}^d$.

Likewise, for $a_i \in \mathcal{A}_i$ the sender A_i selects a POVM $M^{(a_i)} = \{M^{(a_i)}_{\bar{a}_i}\}_{\bar{a}_1 \in \bar{\mathcal{A}}_i}$ for some finite alphabet $\mathcal{A}_i$ satisfying $M^{(a_i)}_{\bar{a}_i} \geq 0$ and $\sum_{a_i \in \mathcal{A}_i} M^{(a_i)}_{\bar{a}_i} = \mathbb{1}_d$ for all $\bar{a}_i \in \mathcal{A}_i$.

Upon drawing $(a_1, \ldots, a_m)$, the senders measure their respective half of the entangled state $|\psi\rangle_{S_{A_i} S_{A_j}}$ using the measurements $L^{(a_i)}$ and $M^{(a_j)}$, producing a correlation

$$P(a'_1, \ldots, a'_m | a_1, \ldots, a_m) = \langle \psi | L^{(a_1)}_{\bar{a}_1} \otimes \cdots \otimes M^{(a_m)}_{\bar{a}_m} | \psi \rangle. \tag{6.52}$$

Finally, the senders can each post-process their measurement outcomes $a'_1, \ldots,$ a'_m together with their inputs $a_1, \ldots, a_m$ to produce inputs $\vec{a}_1, \ldots, \vec{a}_m$ to the MAC N, which we summarize in a function $f_1(\vec{a}_1 | a_1, a'_1) \cdots f_m(\vec{a}_m | a_m, a'_m)$. In total, we have the classical channel

$$E(\vec{a}_1, \ldots, \vec{a}_m | a_1, \ldots, a_m) = \prod_{i=1}^{m} f_i(\vec{a}_i | a_i, a'_i) P(a'_1, \ldots, a'_m | a_1, \ldots, a_m), \tag{6.53}$$

where the correlation P is obtained from measuring the shared entangled state $|\psi\rangle_{S_{A_1} \cdots S_{A_m}}$ through Eq. (6.52).

If the senders to draw a_i independently from a product distribution $p_{A_1}(a_1) \cdots p_{A_m}(a_m)$, then $\mathcal{N} \circ \mathcal{E}$, with $\mathcal{E}$ as defined in Eq. (6.53), can again be interpreted as a MAC with input alphabet $\mathcal{A}_1 \times \cdots \times \mathcal{A}_m$ and output alphabet $\mathcal{Z}$. This prompts us to define the *d-entanglement-assisted achievable rate region* of a classical MAC $\mathcal{N}$ as

$$\mathcal{C}^{(1)}_{ea,d}(\mathcal{N}) := \bigcup_{E} \{\mathcal{C}(\mathcal{N} \circ \mathcal{E})\}, \tag{6.54}$$

where $\mathcal{C}(\cdot)$ is the capacity region of an ordinary MAC defined in Eq. (6.50), and the union is overall classical channels $\mathcal{E}$ as in Eq. (6.53) defined in terms of the following data:

- an entangled state $|\psi\rangle_{S_{A_1} \cdots S_{A_m}} \in \mathbb{C}^d \otimes \cdots \otimes \mathbb{C}^d$;
- arbitrary finite alphabets $\mathcal{A}_1, \ldots, \mathcal{A}_m$ and $\mathcal{A}'_1, \ldots, \mathcal{A}'_m$;
- POVMs $L^{(a_i)} = \left\{ L^{(a_i)}_{\bar{a}_i} \right\}_{\bar{a}_i \in \bar{\mathcal{A}}_i}$ for $a_i \in \mathcal{A}_i$, defined on $\mathbb{C}^d$;
- post-processing $f_i \colon \mathcal{A}_i \times \mathcal{A}_i \ni (a_i, a'_i) \mapsto \mathbf{a}_i \in \mathcal{A}$.

We define the corresponding *achievable sum rate* as

$$\mathcal{S}^{ea,d}(\mathcal{N}) := \sup \left\{ \sum_{i=1}^{m} R_i : (R_1, \ldots, R_m) \in \mathcal{C}^{(0)}_{ea,d}(\mathcal{N}) \right\}. \tag{6.55}$$

The coding Theorem for unassisted MACs implies that the region $\mathcal{C}^{(0)}_{ea}(\mathcal{N})$ is achievable, and hence a natural inner bound on the true entanglement-assisted capacity region of an MAC. We expect that the latter is given by the regularized formula

$$\mathcal{C}_{ea,d}(\mathcal{N}) = \overline{\bigcup_{n \in \mathbb{N}} \frac{1}{n} \mathcal{C}^{(0)}_{ea,d}(\mathcal{N}^{\times n})}, \tag{6.56}$$

where $\overline{X}$ denotes the closure of a set X. We define the *entanglement-assisted achievable rate region*

$$\mathcal{C}^{(1)}_{ea} := \bigcup_{d \in \mathbb{N}} \mathcal{C}^{(1)}_{ea,d}(\mathcal{N}), \tag{6.57}$$

which is achievable by the two senders sharing entanglement on quantum systems of arbitrarily large but finite local dimensions.

The main object of our study is the following construction of a classical multiple access channel, which is presented in terms of a non-local game. This construction is inspired by a similar construction of an interference channel, which involved two senders and two receivers, in terms of the CHSH game as presented in Quek and Shor (2017).

Given a promise-free non-local game $\mathcal{G} = (\mathcal{X}_1, \ldots, \mathcal{X}_m, \mathcal{Y}_1, \ldots, \mathcal{Y}_m, W)$, we define the classical MAC $\mathcal{N}_G : (\mathcal{X}_1 \times \mathcal{Y}_1) \times \cdots \times (\mathcal{X}_m \times \mathcal{Y}_m) \to \mathcal{X}_1 \times \cdots \times \mathcal{X}_m$ as

$$\begin{aligned} &\mathcal{N}_G(\hat{x}_1, \ldots, \hat{x}_m | x_1, y_1; \cdots; x_m, y_m) \\ &:= \begin{cases} \delta_{x_1\hat{x}_1}\delta_{x_2\hat{x}_2} \cdots \delta_{x_m\hat{x}_m}, \text{ if } (x_1, y_1, \ldots, x_m, y_m) \in W; \\ \prod_{i=1}^m |\mathcal{X}_i|^{-1}, \text{ otherwise.} \end{cases} \end{aligned} \tag{6.58}$$

In this particular construction, we link a sender in the MAC scenario with an input alphabet $\mathcal{X}_i \times \mathcal{Y}_i$ for each player of the game, where $i = 1, \ldots, m$. If both senders provide a question-answer tuple $(x_1, y_1, \ldots, x_m, y_m) \in W$ that wins the non-local game $\mathcal{G}$, the channel produces the question pair $(x_1, \ldots, x_m)$ as output; otherwise, the channel produces a question pair selected uniformly at random. For $i = 1, \ldots, m$, we denote the random variables corresponding to the questions for Alice and Bob by $X_i \sim \pi_{X_i}$, the random variables corresponding to the answers by $Y_i \sim p_{Y_i|X_i}\pi_{X_i}$, and the random variable corresponding to the output of the channel $\mathcal{N}_G$ defined in Eq. (6.58) as Z, which takes values in $\mathcal{X}_1 \times \cdots \times \mathcal{X}_m$.

The capacity region of MAC is determined by a product distribution $\prod_{i=1}^m p_{X_iY_i}(x_i, y_i)$ on the set of inputs to N. For the MAC defined in Eq. (6.58), we can interpret this input distribution as follows: given a product probability distribution $\pi(x_1, \ldots, x_m) = \prod_{i=1}^m \pi_{X_i}(x_i)$ on the question set, the players produce answers y_i to the game according to the probabilistic strategy $\prod_{i=1}^m p_{Y_i|X_i}(y_i|x_i)$, on which they agreed upon before beginning the game. This connection allows us to relate the sum-rate capacity of the channel $\mathcal{N}_G$ to the winning probability $\omega(\mathcal{G}, \pi)$ in the following manner:

Proposition 6.4.1 *Consider a promise-free nonlocal game* $\mathcal{G} = (\mathcal{X}_1, \ldots, \mathcal{X}_m, \mathcal{Y}_1, \ldots, \mathcal{Y}_m, W)$, *where* $\pi(x_1, \ldots, x_m) = \prod_{i=1}^m \pi_{X_i}(x_i)$ *is a probability distribution on the questions set* $\mathcal{X}_1 \times \cdots \times \mathcal{X}_m$, *and* $\prod_{i=1}^m p_{Y_i|X_i}(y_i|x_i)$ *is a probabilistic strategy for* $\mathsf{A}_1, \ldots, \mathsf{A}_m$. *Let* $\mathcal{N}_G$ *be the MAC defined in terms of* $\mathcal{G}$ *according to Eq. (6.58).*

Let X_i, Y_i, Z be the random variables corresponding to the questions, answers, and channel output, respectively, as described above. We can then write:

$$I(X_1Y_1\cdots X_mY_m; Z) = H(Z) - p_L(\log|\mathcal{X}_1| + \cdots + \log|\mathcal{X}_m|), \qquad (6.59)$$

where $p_L = \sum_{(x_1,y_1,\ldots,x_m,y_m)\notin W} P(x_1, y_1, \ldots, x_m, y_m)$ denotes the losing probability given the distribution $\pi(x_1, \ldots, x_m)$ on the questions and the strategy $\prod_{i=1}^m p_{X_iY_i}(x_i, y_i)$.

Proof We first expand the mutual information $I(X_1Y_1\cdots X_mY_m; Z)$ as

$$I(X_1Y_1\cdots X_mY_m; Z) = H(Z) - H(Z|X_1Y_1\cdots X_mY_m). \qquad (6.60)$$

Setting $d = |\mathcal{X}_1|\cdots|\mathcal{X}_m|$ and recalling that $W \subset \mathcal{X}_1 \times \mathcal{Y}_1 \times \cdots \times \mathcal{X}_m \times \mathcal{Y}_m$ is the winning set for G, the conditional entropy can be expressed as

$$\begin{aligned} H(Z|X_1Y_1\cdots X_mY_m) &= \sum_{x_1,y_1,\ldots,x_m,y_m} P(x_1, y_1, \ldots, x_m, y_m) \\ &\quad \times H(Z|X_1 = x_1, Y_1 = y_1, \ldots, X_m = x_m, Y_m = y_m) \\ &= \sum_{(x_1,y_1,\ldots,x_m,y_m)\notin W} P(x_1, y_1, \ldots, x_m, y_m) \\ &\quad \times H(Z|X_1 = x_1, Y_1 = y_1, X_2 = x_2, Y_2 = y_2) \\ &= \log d \sum_{(x_1,y_1,\ldots,x_m,y_m)\notin W} P(x_1, y_1, \ldots, x_m, y_m) \\ &= p_L \log d, \end{aligned} \qquad (6.61)$$

where the second equality follows since for $(x_1, y_1, \ldots, x_m, y_m) \in W$ the channel $\mathcal{N}_G$ outputs $(x_1, \ldots, x_m)$ deterministically, and hence $H(Z|X_1 = x_1, Y_1 = y_1, \ldots, X_m = x_m, Y_m = y_m) = 0$. This has proved the result. □

Proposition 6.4.2 *For a promise-free non-local game $\mathcal{G} = (\mathcal{X}_1, \mathcal{Y}_1, \ldots, \mathcal{X}_m, \mathcal{Y}_m, W)$ with $\omega_U(\mathcal{G}) < 1$, let $\mathcal{N}_G$ be the MAC (6.58). For any $0 < \delta < -\log\omega_U(\mathcal{G})$, there exists an $\varepsilon > 0$ such that:*

$$\begin{aligned} I(X_1Y_1\cdots X_mY_m; Z) \leq \max\{&(1-\varepsilon)(\log|\mathcal{X}_1| + \cdots + \log|\mathcal{X}_m|), \\ &\log|\mathcal{X}_1| + \cdots + \log|\mathcal{X}_m| - \delta\}. \end{aligned} \qquad (6.62)$$

For a given $\delta > 0$ the maximal value of ε is given by the ε^* satisfying

$$\frac{\delta + h(\varepsilon^*)}{1-\varepsilon^*} = \delta(\varepsilon^*\|1 - \omega_U(\mathcal{G})), \qquad (6.63)$$

where $h(x) = -x\log x - (1-x)\log(1-x)$ denotes the binary entropy and $\delta(x\|y) := x\log\frac{x}{y} + (1-x)\log\frac{1-x}{1-y}$ denotes the binary relative entropy.

Proof Set $d = |\mathcal{X}_1| \cdots |\mathcal{X}_m|$. For bounding the sum rate capacity $I(X_1Y_1 \cdots X_mY_m; Z)$ away from the maximal value $\log d$, we can assume without loss of generality that the losing probability $p_L = 1 - \omega(\mathcal{G}, \pi)$ for A_i and B is strictly positive, $p_L > 0$. In case $p_L = 0$, the probability distribution π on $\mathcal{X}_1 \times \cdots \times \mathcal{X}_m$ necessarily has supported suppπ strictly contained in $\mathcal{X}_1 \times \cdots \times \mathcal{X}_m$, as, by assumption, the game $\mathcal{G}$ cannot be won with certainty on receiving any one of the full set $\mathcal{X}_1 \times \cdots \times \mathcal{X}_m$. Hence, we have

$$I(X_1Y_1 \cdots X_mY_m; Z) \leq \log|\mathrm{supp}\pi| \leq \log(d-1) < \log d. \tag{6.64}$$

This is because all A_i's had to lose on at least one question pair.

Moreover, we assume, without loss of generality, that $p_L \leq 1 - \omega_U(\mathcal{G})$, since $p_L > 1 - \omega_U(\mathcal{G})$ and $\omega_U(\mathcal{G}) < 1$ imply that

$$I(X_1Y_1 \cdots X_mY_m; Z) < \omega_U(\mathcal{G}) \log d < \log d. \tag{6.65}$$

Therefore, we may assume that $0 < p_L \leq 1 - \omega_U(\mathcal{G})$ for the remainder of the proof.

We prove the assertion of the theorem by contradiction. To this end, assume that

$$H(Z) \geq \log d - \delta \tag{6.66}$$

for some $0 < \delta < -\log \omega_U(\mathcal{G})$. Define a random variable W by

$$W = \begin{cases} 1, & \text{if all } \mathsf{A}_i\text{'s win the game;} \\ 0, & \text{if all } \mathsf{A}_i\text{'s lose the game,} \end{cases} \tag{6.67}$$

by taking values 1 and 0 with probability $1 - p_L$ and p_L, respectively. By the non-negativity of conditional entropy and Eq. (6.66) we have

$$H(W) + H(Z|W) = H(ZW) \geq H(Z) \geq \log d - \delta. \tag{6.68}$$

By expanding the left-hand side of Eq. (6.68) we get

$$h(p_L) + (1 - p_L)H(X_1 \cdots X_m) + p_L \log d \geq \log d - \delta, \tag{6.69}$$

which follows

$$H(X_1 \cdots X_m) \geq \log d - \gamma \tag{6.70}$$

with $\gamma := \frac{\delta + h(p_L)}{1 - p_L}$. From $D(\pi_{X_1} \cdots \pi_{X_m} \| \pi_U) = \log d - H(X_1 \cdots X_m)$, we obtain that

$$\gamma \geq D(\pi_{X_1} \cdots \pi_{X_m} \| \pi_U). \tag{6.71}$$

Let $Q = Q_{Y_1|X_1} \cdots Q_{Y_m|X_m}$ be the optimal probabilistic strategy for $\mathcal{G}$ given the distribution $\pi_{X_1} \cdots \pi_{X_m}$ on the questions, and denote by q_L the losing probability of the same strategy Q with questions drawn uniformly at random. Let $\chi_W : \mathcal{X}_1 \times \mathcal{Y}_1 \times \cdots \times \mathcal{X}_m \times \mathcal{Y}_m \to \{0, 1\}$ be the characteristic function of the winning set $W \subset \mathcal{X}_1 \times \mathcal{Y}_1 \times \cdots \times \mathcal{X}_m \times \mathcal{Y}_m$. Applying the data-processing inequality with respect to $\chi_W \circ Q$ to Eq. (6.71), we obtain

$$\gamma \geq \delta(p_L \| q_L) \geq \delta(p_L \| 1 - \omega_U(\mathcal{G})), \tag{6.72}$$

where $\delta(x \| y)$ denotes the binary relative entropy, and the second inequality follows from the monotonicity of $y \mapsto \delta(x \| y)$ for $y \geq x$ and the fact that $q_L \geq 1 - \omega_U(\mathcal{G})$.

The function $\gamma(x) = \frac{\delta + h(x)}{1-x}$ is monotonically increasing for positive x, and $\lim_{x \to 0} \gamma(x) = \delta < -\log \omega_U(\mathcal{G})$. Moreover, $x \mapsto \delta(x \| 1 - \omega_U(\mathcal{G}))$ is monotonically decreasing for $x \in [0, 1 - \omega_U(\mathcal{G})]$, and $\lim_{x \to 0} \delta(x \| 1 - \omega_U(\mathcal{G})) = -\log \omega_U(\mathcal{G})$. Hence, there exists an $\varepsilon > 0$ such that Eq. (6.72) is violated for all $p_L < \varepsilon$, which means that we either have $p_L \geq \varepsilon$ or the assumption Eq. (6.66) is false. In the first case, we obtain that

$$I(X_1 Y_1 \cdots X_m Y_m; Z) = H(Z) - p_L \log d \leq (1 - \varepsilon) \log d. \tag{6.73}$$

In the second case, we get that

$$I(X_1 Y_1 \cdots X_m Y_m; Z) < (1 - p_L) \log d - \delta < \log d - \delta. \tag{6.74}$$

For a given $\delta > 0$, the maximal value of ε is given by the ε^* satisfying

$$\frac{\delta + h(\varepsilon^*)}{1 - \varepsilon^*} = \delta(\varepsilon^* \| 1 - \omega_U(\mathcal{G})). \tag{6.75}$$

This has completed the proof. □

In applications, the optimal (minimal) upper bound may be obtained by optimizing the right-hand side of Eq. (6.62) over $\delta \in (0, -\log \omega_U(\mathcal{G}))$ and computing ε via Eq. (6.63).

Theorem 6.4.1 facilitates a clear distinction between the capacity region of the unaided MAC $\mathcal{N}_G$ and the achievable rate region of entanglement-assisted strategies for distributed games $\mathcal{G}$ having good quantum tactics and $\omega_U(\mathcal{G}) < 1$. In entanglement-assisted methodologies, the two senders positioned at different locations can gauge shared entangled states to generate the input probability distribution. As a result of the quantum correlations inherent in the entangled state, the probability distribution obtained may no longer be of product form. Depending on the specific game, this entangled strategy could potentially enable the players to secure a certain win, thereby allowing for the transmission of maximal information through the MAC.

6.4.2 *Magic Square Game*

A well-known example of cooperating games is the *magic square game* $\mathcal{G}_{MS}$, in which two players, Alice and Bob, receive questions r and c labeling the rows and columns of a 3×3-matrix, respectively. They answer with bit strings of length 3, and win if (a) the parity of Alice's row is even, (b) the parity of Bob's column is odd, and (c) the bit strings agree in the overlapping cell.

It is impossible to fill the entire square in a manner that satisfies both parity constraints mentioned above. If such a filling did exist, the parity of the whole square would be even as per (a), and odd as per (b), leading to a contradiction. This reasoning can be employed to demonstrate that there is no infallible classical winning strategy for A and B, and in fact, $\omega_U(\mathcal{G}_{MS}) \leq \frac{8}{9}$ (Brassard et al. 2005). In terms of the MAC defined using the magic square game, Theorem 6.4.1 presents an upper bound on the achievable sum rate of 3.13694, which is lower than the maximum value of $2 \log 3 \approx 3.17$.

On the other hand, there is a well-known perfect quantum strategy in which A and B make measurements on two maximally entangled states shared between them (Brassard et al. 2005). Drawing the questions $r, c \in \{0, 1, 2\}$ uniformly at random and using this perfect quantum strategy allows A and B to code at rates $R_1 = R_2 = \log 3$, and hence achieve the perfect rate pair $(\log 3, \log 3)$ not achievable by any classical strategy.

The result is similar in spirit to (Quek and Shor 2017), where the CHSH inequality was employed to establish a distinction between classical and quantum tactics for an interference channel comprising two senders and two receivers. In contradistinction to (Quek and Shor 2017), the divergence between classical methodologies and entanglement-backed methodologies manifested above is relevant to any non-local game G having $\omega_U(\mathcal{G}) < 1$, and where impeccable quantum tactics exist. These games are dubbed "pseudo-telepathy games" (Brassard et al. 2005), and each game belonging to this category leads to a differentiation for the corresponding MAC between the unaided capacity area and the entanglement-assisted area.

6.5 Dynamic Zero-Sum Games

Nash theory, which forms the foundation of non-cooperative games (Fudenberg and Tirole 1991), is centered around identifying the optimal strategies for all participants in a manner that ensures that none of them are inclined towards breaking the equilibrium. Additionally, if the total gains remain zero for all combinations of strategies, it is classified as a zero-sum game (Zhang et al. 2017). Otherwise, it is categorized as a non-zero-sum game. While much of the existing literature has concentrated on cooperative games such as the Bell game (Clauser et al. 1969), we present a novel dynamic zero-sum quantum game that exhibits Nash equilibrium, while the corresponding classical game is characterized by unfairness (Cheng and Luo 2021).

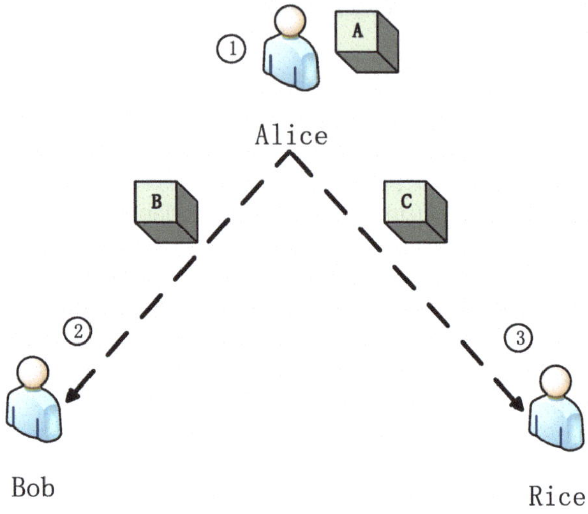

Fig. 6.4 A schematic tripartite dynamic zero-sum classical game G

6.5.1 Game Model

We first present some definitions as follows.

Definition 6.5.1 The dynamic zero-sum game means that the actions of different players have a sequence, the later participant can observe the former's actions, and the sum of payments for all players is zero for any combination of strategies.

Definition 6.5.2 A fair game means that the game does not favor any player. The games in this paper are all zero-sum games, thus the fairness in this paper means everyone's average gain is zero.

Definition 6.5.3 The asymptotically fair game means that under the given initial conditions, the game may not be fair, but with the increase of variable parameters, the degree of deviation from the fair game becomes smaller and smaller and the game is fair when the variable parameter approaches infinity.

Inspired by the bipartite game (Zhang et al. 2017), we introduce a tripartite game G, illustrated in Fig. 6.4. In this game, we assume that the actions of different participants occur in a sequence, where the latter participant can observe the former's action. The present model consists of four stages, as follows:

S1. Alice randomly selects one of three black boxes, namely $\diamond A$, $\diamond B$, and $\diamond C$, and places a ball inside it. Alice then sends box $\diamond C$ to Rice, and box $\diamond B$ to Bob.
S2. Bob chooses to either open box $\diamond B$ or request Alice to open box $\diamond A$, but he does not take any further action.
S3. Rice selects her strategy and acts. She either opens box $\diamond C$ or allows Alice to open box $\diamond A$. If Rice opens box $\diamond C$ and finds no ball inside, the game proceeds to the fourth stage.

S4. It is Bob's turn to take action depending on the strategy he has chosen in S2. He either opens box $\diamond B$ or asks Alice to open box $\diamond A$.

Moreover, for any strategy combination, we assume that the total payment of all participants is zero. In the following, we will explore this kind of game with two settings with different gains.

The classical game tree for G can be easily followed (Cheng and Luo 2021). The game's winning rules are as follows:

W1 Bob and Rice win if Rice chooses to open box $\diamond A$ in S3 and finds the ball.
W2 Alice wins if Rice does not find the ball in box $\diamond A$ in S3.
W3 Rice wins if she opens box $\diamond C$ in S3 and finds the ball.
W4 Bob and Rice win if Bob opens box $\diamond A$ in S4 and finds the ball.
W5 Alice wins if Bob opens box $\diamond A$ in S4 and does not find the ball.
W6 Bob wins if he opens box $\diamond B$ in S4 and finds the ball.
W7 Alice wins if Bob opens box $\diamond B$ in S4 and does not find the ball.

Now, for convenience, we define the probabilities as follows. Denote P_1 as the probability that Alice puts the ball into $\diamond A$. Denote P_2 as the probability that Alice puts the ball into $\diamond B$ or $\diamond C$. Denote $P_{\diamond C}$ as the probability that Rice chooses to open $\diamond C$. Denote $P_{\diamond B}$ as the probability that Bob chooses to open $\diamond B$. Here, we assume that Alice has an equal probability to put the ball into $\diamond B$ or $\diamond C$. It follows that $P_1 + 2P_2 = 1$, with $P_{\diamond B}, P_{\diamond C} \in [0, 1]$.

From winning rules W1–W7 of G, it is easy to get the winning probability P_{Rice_1} for Rice from W3 (i.e., Rice finds the ball after opening $\diamond C$) is given by

$$P_{Rice_1} = P_2 P_{\diamond C}. \tag{6.76}$$

Moreover, the winning probability for Bob from W6 (i.e., Rice chooses to open $\diamond C$ and does not find the ball, but Bob finds the ball after opening $\diamond B$) is given by

$$P_{Bob_1} = P_2 P_{\diamond C} P_{\diamond B}(1 - P_2). \tag{6.77}$$

For Alice, there are three subcases W2, W5, and W7 for winning. It follows that

$$\begin{aligned} P_{Alice} = 1 - P_1 + P_{\diamond B} P_{\diamond C} P_2^2 - P_{\diamond B} P_{\diamond C} P_2 - P_{\diamond C} P_2 \\ + P_{\diamond B} P_{\diamond C} P_1 + P_{\diamond C} P_1 P_2 - P_{\diamond B} P_{\diamond C} P_1 P_2. \end{aligned} \tag{6.78}$$

The winning probability for Bob from W1 and W4 is given by

$$P_{Bob_2} = P_1 - P_{\diamond B} P_{\diamond C} P_1 - P_{\diamond C} P_1 P_2 + P_{\diamond B} P_{\diamond C} P_1 P_2. \tag{6.79}$$

The same result holds for Rice from W1 and W4, i.e., $P_{Rice_2} = P_{Bob_2}$.

In this game G, the players' strategies are analyzed to showcase the main idea. Alice faces a dilemma as she does not know which box Rice or Bob would choose to open before she places the ball. If Alice puts the ball in one of the three boxes with a higher probability, she may end up losing the game. Therefore, there is a tradeoff for Alice in choosing her strategy, which is represented by the probability P_1.

On the other hand, Bob faces uncertainty as he does not know which box Rice would choose to open when he gives the probability $P_{\diamond B}$. Hence, he needs to consider a known parameter P_1 given by Alice and an unknown parameter $P_{\diamond C}$ given by Rice while choosing his strategy.

A similar analysis can be applied to Rice's strategy. Rice needs to choose her parameter based on what she knows before she takes action.

6.5.2 Tripartite Classical Game

It is a well-known fact that players will aim to maximize their interests in a non-cooperative game. In this section, we introduce the first game G_1 with the gain setting provided in Table 6.1. Our goal is to demonstrate the lack of fairness in this game with classical resources.

From Table 6.1, the average gain of Rice is given by

$$\begin{aligned} G_{Rice} &= 2P_{Rice_1} - P_{Bob_1} - P_{Alice} + \eta P_{Rice_2} \\ &= 3P_2 P_{\diamond C} + (\eta + 1)P_1 - (\eta + 1)P_1 P_{\diamond B} P_{\diamond C} \\ &\quad -(\eta + 1)P_1 P_2 P_{\diamond C} + (\eta + 1)P_1 P_2 P_{\diamond B} P_{\diamond C} - 1. \end{aligned} \tag{6.80}$$

From Eq. (6.80), if $\frac{\partial G_{Rice}}{\partial P_{\diamond C}} < 0$, then G_{Rice} is a decreasing function. In this case, Rice has to set $P_{\diamond C} = 0$ to maximize her gains. If $\frac{\partial G_{Rice}}{\partial P_{\diamond C}} > 0$, i.e., G_{Rice} is an increasing function, Rice will choose $P_{\diamond C} = 1$ to maximize her gains. Moreover, when $\frac{\partial G_{Rice}}{\partial P_{\diamond C}} = 0$, i.e., G_{Rice} is a constant, $P_{\diamond C}$ can be any probability.

Similar to Eq. (6.80), we get that the average gain of Bob is given by

Table 6.1 The gain settings of players in the first classical game G_1. The payoff of Rice (or Bob, or Alice) is represented by $g_{c(b,a)}$. The expression $\mathsf{R}_{succ}(\diamond C)$ implies that Rice has found the photon after opening box $\diamond C$. Similarly, $\mathsf{B}_{succ}(\diamond B)$ corresponds to Bob finding the photon by opening box $\diamond B$. The notation $\mathsf{R/B}_{fail}(\cdot)$ denotes the scenario where neither Rice nor Bob finds the photon. The term $\mathsf{R/B}_{succ}(\diamond A)$ pertains to the situation where either Rice or Bob succeeds by the opening box $\diamond A$. In the first classical game, the parameter η is greater than or equal to 1

Cases\gains	g_c	g_b	g_a
$\mathsf{R}_{succ}(\diamond C)$	2	−1	−1
$\mathsf{B}_{succ}(\diamond B)$	−1	2	−1
$\mathsf{R/B}_{fail}(\cdot)$	−1	−1	2
$\mathsf{R/B}_{succ}(\diamond A)$	η	1	$-\eta - 1$

Table 6.2 The values of $P_{\diamond B}$ and $P_{\diamond C}$. $P_{\diamond B(\diamond C)}$ depends on the different cases in the game, where $X_1 = (\eta + 8 - \sqrt{\eta^2 + 4\eta + 52})/(2\eta + 2)$, $\Delta_1 = [0, 3/(2\eta + 5))$, $\Delta_2 = [3/(2\eta + 5), X_1)$, $\Delta_3 = [X_1, 3/(\eta + 1)]$, and $\Delta_4 = (3/(\eta + 1), 1]$

η	$1 \le \eta \le 2$		$\eta > 2$		
P_1	$P_1 \in \Delta_1$	$P_1 \in \Delta_2$	$P_1 \in \Delta_2$	$P_1 \in \Delta_3$	$P_1 \in \Delta_4$
$P_{\diamond B}$	1	X_0	$\frac{3P_2-(\eta+1)P_1P_2}{P_1(\eta+1)(1-P_2)}$	X_0	[0, 1]
$P_{\diamond C}$	1	0	1	0	0

$$
\begin{aligned}
G_{Bob} &= -P_{Rice_1} + 2P_{Bob_1} - P_{Alice} + P_{Bob_2} \\
&= 3P_{\diamond B}P_{\diamond C}P_2 - 3P_{\diamond B}P_{\diamond C}P_2^2 + 2P_1 - 2P_{\diamond B}P_{\diamond C}P_1 \\
&\quad -2P_{\diamond C}P_1P_2 + 2P_{\diamond B}P_{\diamond C}P_1P_2 - 1.
\end{aligned} \tag{6.81}
$$

There are several cases to maximize G_{Bob}. All the results are given in Table 6.2.

It is easy to write the expression for G_{Alice} according to Table 6.1 as follows

$$
\begin{aligned}
G_{Alice} &= -P_{Rice_1} - P_{Bob_1} + 2P_{Alice} - (\eta + 1)P_{Rice_2} \\
&= -3P_{\diamond C}P_2 - 3P_{\diamond B}P_{\diamond C}P_2 + 3P_{\diamond B}P_{\diamond C}P_2^2 \\
&\quad +(\eta + 3)P_{\diamond C}P_1P_2 + (\eta + 3)P_{\diamond B}P_{\diamond C}P_1 \\
&\quad -(\eta + 3)P_1 - (\eta + 3)P_{\diamond B}P_{\diamond C}P_1P_2 + 2.
\end{aligned} \tag{6.82}
$$

For the case of $0 \le P_1 < \frac{3}{2\eta+5}$, it follows from Table 6.2 that $P_{\diamond C} = 1$ and $P_{\diamond B} = 1$. Equation (6.82) is then rewritten into

$$
G_{Alice} = 3P_2^2 - 6P_2 + 2, \tag{6.83}
$$

where $P_2 = \frac{1-P_1}{2}$. It is easy to prove that G_{Alice} achieves the maximum when $P_1 = \frac{3}{2\eta+5}$. Denote $P_1 = \frac{3}{2\eta+5} - \epsilon$, where ϵ is a small constant satisfying $\epsilon > 0$. It follows from Eq. (6.83) that

$$
G_{Alice} = \frac{-\eta^2 + 4\eta + 23}{(2\eta + 5)^2} + O(\epsilon). \tag{6.84}
$$

In general, we can get Table 6.3. Here, $\frac{-\eta^2+4\eta+23}{(2\eta+5)^2} + O(\epsilon) > \frac{\eta+1}{2\eta+5}$ for $1 \le \eta < 2$, and $\frac{-\eta^2+4\eta+23}{(2\eta+5)^2} + O(\epsilon) < \frac{\eta+1}{2\eta+5}$ for $\eta \ge 2$. Thus, Alice will choose $P_1 = \frac{3}{2\eta+5} - \epsilon$ to maximize her gain when $1 \le \eta < 2$, while $P_1 = \frac{3}{2\eta+5}$ when $\eta \ge 2$.

We find that $P_1 = \frac{3}{2\eta+5} - \epsilon$ for $1 \le \eta < 2$. Using induction, we know from Table 6.2 that $P_{\diamond B} = 1$ and $P_{\diamond C} = 1$. From Eqs. (6.80), (6.81), and (6.82), the expressions for G_{Alice}, G_{Bob}, and G_{Rice} in terms of η are presented in Table 6.4.

Theorem 6.5.1 *The tripartite game* G_1 *is unfair for any* $\eta \ge 1$.

A similar result holds for other gain distributions as Table 6.5.

Table 6.3 G^m_{Alice} denotes the maximum of G_{Alice}, and $P_{\max}$ denotes the corresponding point of P_1

P_1	$0 \leq P_1 < \frac{3}{2\eta+5}$	$\frac{3}{2\eta+5} \leq P_1 \leq 1$
$P_{\max}$	$\frac{3}{2\eta+5} - \epsilon$	$\frac{3}{2\eta+5}$
G^m_{Alice}	$\frac{-\eta^2+4\eta+23}{(2\eta+5)^2} + O(\epsilon)$	$\frac{\eta+1}{2\eta+5}$

6.5.3 *Quantizing Zero-Sum Games*

In this section, we have quantized the classical game depicted in Fig. 6.4, and we have discovered that the quantum game also has four stages, as illustrated in Fig. 6.5. The main difference between classical and quantum games is that the former involves placing a ball into three regular black boxes, while the latter entails placing a particle into three quantum boxes. In the classical game, Bob and Rice can strategically allow Alice to open the box $\diamond A$ to prevent her from putting the ball into it, thereby concealing it from Bob and Rice. However, in the quantum game, this issue can be avoided by agreeing on the committed state before the game commences. Alice, Bob, and Rice agree on the photon state that Alice should set, and Alice has three quantum boxes, namely $\diamond A$, $\diamond B$, and $\diamond C$, to store the photon. The photon's state in the boxes is represented by $|a\rangle$, $|b\rangle$, and $|c\rangle$. The quantum game comprises four stages, namely S1-S4.

Table 6.4 The average gains of players in the first classical game G_1

Gain\\η	$1 \leq \eta < 2$	$\eta \geq 2$
G_{Alice}	$\frac{-\eta^2+4\eta+23}{(2\eta+5)^2} + O(\epsilon)$	$\frac{1+\eta}{2\eta+5}$
G_{Bob}	$\frac{-\eta^2-5\eta-13}{(2\eta+5)^2} + O(\epsilon)$	$\frac{1-2\eta}{2\eta+5}$
G_{Rice}	$\frac{\eta-2}{2\eta+5} + O(\epsilon)$	$\frac{\eta-2}{2\eta+5}$

Table 6.5 The gain settings of the second game G_2. Here, $g_{c(b,a)}$ denotes the gain of Rice (or Bob, or Alice). Assume that $\mu \geq 2$ is in this second game

Cases\gains	g_c	g_b	g_a
$\mathsf{R}_{succ}(\diamond C)$	μ	-1	$1-\mu$
$\mathsf{B}_{succ}(\diamond B)$	-1	2	-1
$\mathsf{R/B}_{fail}(\cdot)$	-1	-1	2
$\mathsf{R/B}_{succ}(\diamond A)$	1	1	-2

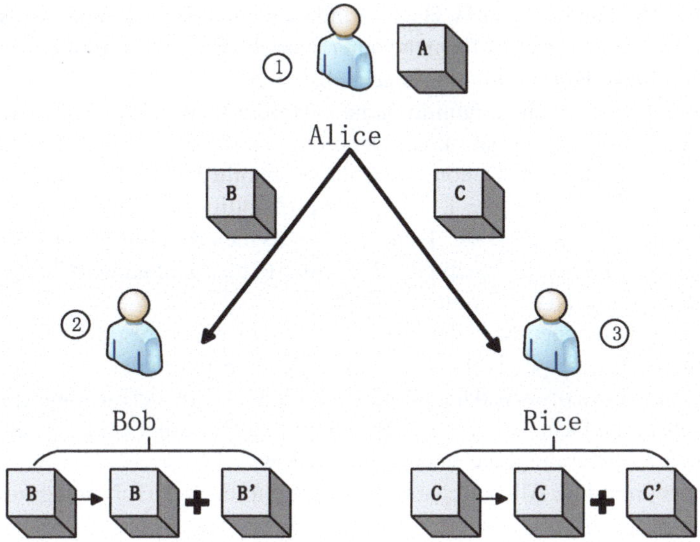

Fig. 6.5 The schematic model of the tripartite dynamic zero-sum quantum game. Here, three parties use a single photon to complete the game while the classic game uses a ball in the game given in Fig. 6.4

S1 Alice randomly places the single photon into one of three quantum boxes: $\diamond A$, $\diamond B$, and $\diamond C$. Alice sends box $\diamond C$ to Rice and sends $\diamond B$ to Bob.
S2 Bob provides his chosen strategy and opens box $\diamond B$.
S3 Rice selects her strategy and takes action, i.e., she opens $\diamond C$. If neither Rice nor Bob finds the photon, the game proceeds to the fourth stage.
S4 Bob (Rice) requests that Alice send him(her) box $\diamond A$ to verify whether the state of the photon prepared by Alice is the same as the committed state.

Additionally, we assume that the total payment of all participants is zero for any strategy combination and that the latter participant can observe the former's action. In the following sections, we will analyze two settings of this game with different gains.

The quantum game tree consists of several stages (Cheng and Luo 2021). In the stages of G, the rules for winning the game are as follows:

W1 Rice wins if she finds the photon after opening $\diamond C$.
W2 Bob wins if he finds the photon after opening $\diamond B$.
W3 Bob and Rice win if neither of them finds the photon but Alice is not honest.
W4 Alice wins if neither of them finds the photon and Alice is honest.

We are examining a dynamic zero-sum quantum game that has the same parameters as classical games but with a slight variation in the interpretation of each symbol. In quantum games, $\mathsf{R}_{succ}(\diamond C)$ denotes the scenario where Rice finds the photon upon opening the box $\diamond C$. Similarly, $\mathsf{B}_{succ}(\diamond B)$ corresponds to Bob finding the photon by opening box $\diamond B$. The notation $\mathsf{R/B}_{fail}(\cdot)$ indicates that neither Rice nor Bob finds the photon, and Alice is truthful. On the other hand, $\mathsf{R/B}_{succ}(\diamond A)$ implies that neither Rice nor Bob finds the photon, but Alice is dishonest.

The winning rules of the quantum game is similar to W1–W4. For convenience, we define the following probabilities. Denote α_1 as the probability that Alice puts the photon into the box $\diamond A$. Denote α_2 as the probability that Alice puts the photon into box $\diamond B$ or box $\diamond C$. Denote $1-\gamma$ as the probability that Rice chooses to open the box $\diamond C$ when she received $\diamond C$. Denote $1-\beta$ as the probability that Bob chooses to open the box $\diamond B$ when he received $\diamond B$. Similar to classical game (Cheng and Luo 2021), we have $\alpha_1 + 2\alpha_2 = 1$ and $\gamma, \beta \in [0, 1]$.

From quantum perspectives, the box of $\diamond A$, $\diamond B$, or $\diamond C$ corresponds to a quantum state $|a\rangle, |b\rangle$ or $|c\rangle$, respectively. If a player opens a particular box (e.g., $\diamond A$), it implies that they have observed the photon in the state $|a\rangle$ by performing a projection measurement in the basis $|a\rangle, |b\rangle, |c\rangle$. Based on these assumptions, we can develop an experimental quantum game as described below.

Alice's preparation. Alice prepares the single photon in the following supposition state

$$|\psi\rangle = \sqrt{\alpha_1}|a\rangle + \sqrt{\alpha_2}|b\rangle + \sqrt{\alpha_2}|c\rangle, \tag{6.85}$$

where $|a\rangle, |b\rangle$ and $|c\rangle$ can be realized by using different paths, $\alpha_1 + 2\alpha_2 = 1$ and $\alpha_1 \in [0, 1]$ is a parameter controlled by Alice.

Bob's operation. Bob splits the box $\diamond B$ into box $\diamond B$ and box $\diamond B'$ according the following transformation

$$|b\rangle \mapsto \sqrt{1-\beta}|b\rangle + \sqrt{\beta}|b'\rangle, \tag{6.86}$$

where $\beta \in [0, 1]$ is a parameter controlled by Bob.

Rice's operation. Rice splits the box $\diamond C$ into two parts $\diamond C$ and $\diamond C'$ according to the following transformation

$$|c\rangle \mapsto \sqrt{1-\gamma}|c\rangle + \sqrt{\gamma}|c'\rangle, \tag{6.87}$$

where $\gamma \in [0, 1]$ is a parameter controlled by Rice.

Alice has the option to increase the probability of the photon appearing in box $\diamond A$ by selecting a large value of α_1, as she would in classical games. This will correspondingly lower the chances of Bob and Rice finding the photon. However, this approach may lead to Alice losing the game with a high probability during the verification stage. The same logical analysis applies to the other players too. Hence, it is imperative to determine reasonable parameters for all players. We will elaborate on the process involved in the subsequent section.

Suppose that Alice prepares the photon in the following state

$$|\psi\rangle = \sqrt{\alpha_1}|a\rangle + \sqrt{\alpha_2}|b\rangle + \sqrt{\alpha_2}|c\rangle \tag{6.88}$$

using path encoding. When Bob (Rice) receives its box $\diamond B(\diamond C)$ and splits it into two parts, the final state of the photon is given by

$$
\begin{aligned}
|\psi_0\rangle = & \sqrt{\alpha_1}|a\rangle + \sqrt{\alpha_2(1-\beta)}|b\rangle + \sqrt{\alpha_2\beta}|b'\rangle \\
& + \sqrt{\alpha_2(1-\gamma)}|c\rangle + \sqrt{\alpha_2\gamma}|c'\rangle.
\end{aligned} \tag{6.89}
$$

From Eq. (6.89), the probability that Rice finds the photon in $\diamond C$ (using single photon detector) is

$$
P_{Rice_1} = |\langle c|\psi_0\rangle|^2 = \alpha_2(1-\gamma). \tag{6.90}
$$

Moreover, the probability that Bob finds the photon in $\diamond B$ is given by

$$
P_{Bob_1} = |\langle b|\psi_0\rangle|^2 = \alpha_2(1-\beta). \tag{6.91}
$$

If neither Rice nor Bob finds a photon, the state in Eq. (6.89) will collapse into

$$
|\psi_1\rangle = \frac{1}{\sqrt{\alpha_1 + \alpha_2\beta + \alpha_2\gamma}}(\sqrt{\alpha_1}|a\rangle + \sqrt{\alpha_2\beta}|b'\rangle + \sqrt{\alpha_2\gamma}|c'\rangle). \tag{6.92}
$$

If Alice did prepare the photon in the committed state $|\phi\rangle = \sqrt{\omega_1}|a\rangle + \sqrt{\omega_2}|b\rangle + \sqrt{\omega_2}|c\rangle$ initially, it is easy to prove that the state at this stage should be

$$
|\psi_C\rangle = \frac{1}{\sqrt{\omega_1 + \omega_2\beta + \omega_2\gamma}}(\sqrt{\omega_1}|a\rangle + \sqrt{\omega_2\beta}|b'\rangle + \sqrt{\omega_2\gamma}|c'\rangle), \tag{6.93}
$$

where $\omega_1 + 2\omega_2 = 1$. By performing a projection measurement on $|\psi_1\rangle$, Rice or Bob gets an ideal state $|\psi_C\rangle$ with the probability

$$
|\langle\psi_1|\psi_C\rangle|^2 = \frac{\left(\sqrt{\alpha_1\omega_1} + x\sqrt{\alpha_2\omega_2}\right)^2}{(\alpha_1 + \alpha_2 x)(\omega_1 + \omega_2 x)}, \tag{6.94}
$$

where $x = \beta + \gamma$.

The probability that neither Rice nor Bob finds the photon when Alice did prepare the photon in the committed state $|\phi\rangle$ is given by

$$
\begin{aligned}
P_{Alice} &= (1 - P_{Rice_1} - P_{Bob_1})|\langle\psi_1|\psi_C\rangle|^2 \\
&= \frac{(\sqrt{\alpha_1\omega_1} + x\sqrt{\alpha_2\omega_2})^2}{\omega_1 + \omega_2 x}.
\end{aligned} \tag{6.95}
$$

Moreover, the probability that neither Rice nor Bob finds the photon but Rice or Bob detects the forge state prepared by Alice is denote by P_{Rice_2} or P_{Bob_2}, which is given by

$$
\begin{aligned}
P_{Rice_2} &= (1 - P_{Rice_1} - P_{Bob_1})(1 - |\langle \psi_1 | \psi_C \rangle|^2) \\
&= -\frac{\left(\sqrt{\alpha_1 \omega_1} + x\sqrt{\alpha_2 \omega_2}\right)^2}{\omega_1 + \omega_2 x} + \alpha_1 + \alpha_2 x
\end{aligned}
\tag{6.96}
$$

with $P_{Rice_2} = P_{Bob_2}$ from the winning rule W3.

6.5.4 Zero-Sum Quantum Games

We introduce the quantum implementation of the first game G with the gain setting shown in Table 6.1. Here, each symbol has a slightly different meaning, i.e., in quantum game, $\mathsf{R}_{succ}(\diamond C)$ means that Rice finds the photon after opening the box $\diamond C$. $\mathsf{B}_{succ}(\diamond B)$ means that Bob finds the photon after opening the box $\diamond B$. $\mathsf{R/B}_{fail}(\cdot)$ means that neither Rice or Bob finds the photon and Alice is honest. $\mathsf{R/B}_{succ}(\diamond A)$ means that neither Rice or Bob finds the photon but Alice is not honest.

Denote G_{Rice} as Rice's average gain. We can easily get G_{Rice} according to Table I as follows

$$
\begin{aligned}
G_{Rice} &= 2P_{Rice_1} + \eta P_{Rice_2} - P_{Bob_1} - P_{Alice} \\
&= -\frac{(\eta + 1)(\sqrt{\alpha_1 \omega_1} + x\sqrt{\alpha_2 \omega_2})^2}{\omega_1 + \omega_2 x} \\
&\quad + (\eta - 2)\alpha_2 \gamma + (\eta + 1)\alpha_2 \beta + \eta \alpha_1 + \alpha_2,
\end{aligned}
\tag{6.97}
$$

where $x = \beta + \gamma$.

Each participant has perfect knowledge before the game reaches this stage, i.e., each participant is exactly aware of what the previous participant may have done. From $\frac{\partial G_{Rice}}{\partial \gamma} = 0$, we get

$$
\begin{aligned}
x_1^* &= -\frac{\sqrt{\omega_1(\eta + 1)}\left|\sqrt{\alpha_1 \omega_2} - \sqrt{\alpha_2 \omega_1}\right|}{\omega_2 \sqrt{3\alpha_2}} - \frac{\omega_1}{\omega_2}, \\
x_2^* &= \frac{\sqrt{\omega_1(\eta + 1)}\left|\sqrt{\alpha_1 \omega_2} - \sqrt{\alpha_2 \omega_1}\right|}{\omega_2 \sqrt{3\alpha_2}} - \frac{\omega_1}{\omega_2}.
\end{aligned}
\tag{6.98}
$$

If Alice chooses $\alpha_1 < \omega_1$, Alice has no benefit if she chooses $\alpha_1 < \omega_1$. It follows that $\alpha_1 \geq \omega_1$ and $\alpha_2 \leq \omega_2$. Hence, from Eq. (6.98) we get

$$
\begin{aligned}
x_1^* &= -\frac{\sqrt{\omega_1(\eta + 1)}(\sqrt{\alpha_1 \omega_2} - \sqrt{\alpha_2 \omega_1})}{\sqrt{3}\sqrt{\alpha_2}\omega_2} - \frac{\omega_1}{\omega_2}, \\
x_2^* &= \frac{\sqrt{\omega_1(\eta + 1)}(\sqrt{\alpha_1 \omega_2} - \sqrt{\alpha_2 \omega_1})}{\sqrt{3}\sqrt{\alpha_2}\omega_2} - \frac{\omega_1}{\omega_2}.
\end{aligned}
\tag{6.99}
$$

From Eq. (6.97)–(6.99), G_{Rice} gets the maximum value at $x = 0$ for $x_2^* \leq 0$, or gets the maximum value at $x = 2$ for $x_2^* \geq 2$, or gets the maximum value at $x = x_2^*$ for $0 < x_2^* < 2$.

Denote G_{Bob} as the average gain of Bob. From Table 6.1, we get G_{Bob} as

$$\begin{aligned} G_{Bob} &= 2P_{Bob_1} + P_{Bob_2} - P_{Rice_1} - P_{Alice} \\ &= -\frac{2(\sqrt{\alpha_1\omega_1} + x\sqrt{\alpha_2\omega_2})^2}{(\omega_1 + \omega_2 x)} + \alpha_1 + \alpha_2 - \alpha_2\beta + 2\alpha_2\gamma, \end{aligned} \tag{6.100}$$

where $x = \beta + \gamma$.

Similar to Eqs. (6.98)–(6.100), we can get

$$\begin{aligned} x_3^* &= -\frac{\sqrt{2\omega_1}(\sqrt{\alpha_1\omega_2} - \sqrt{\alpha_2\omega_1})}{\omega_2\sqrt{3\alpha_2}} - \frac{\omega_1}{\omega_2}, \\ x_4^* &= \frac{\sqrt{2\omega_1}(\sqrt{\alpha_1\omega_2} - \sqrt{\alpha_2\omega_1})}{\omega_2\sqrt{3\alpha_2}} - \frac{\omega_1}{\omega_2}, \end{aligned} \tag{6.101}$$

from $\frac{\partial G_{Bob}}{\partial \beta} = 0$.

From Eqs. (6.100)–(6.101), G_{Bob} gets the maximum value at $x = 0$ for for $x_4^* \leq 0$, or gets the maximum value at $x = 2$ for $x_4^* \geq 2$, or gets the maximum value at $x = x_4^*$ for $0 < x_4^* < 2$. From $\eta \geq 1$, we get $x_2^* \geq x_4^*$.

Similarly, we can get a detailed analysis of Rice. The values of β^* and γ^* depend on x_i^* in the first quantum game such that G_{Bob} achieves the maximum.

Denote G_{Alice} as the average gain of Alice. From Table 6.1, it is easy to evaluate G_{Alice} as

$$\begin{aligned} G_{Alice} &= 2P_{Alice} - (\eta + 1)P_{Rice_2} - P_{Bob_1} - P_{Rice_1} \\ &= (\eta + 3)\frac{(\sqrt{\alpha_1\omega_1} + x\sqrt{\alpha_2\omega_2})^2}{\omega_1 + \omega_2 x} - \eta\alpha_2 x - 1 - \eta\alpha_1, \end{aligned} \tag{6.102}$$

where $x = \beta + \gamma$.

From Eqs. (6.97) and (6.100) we get that

$$\begin{aligned} G_{Alice} &= \max\{G_{Alice_1}, G_{Alice_2}, G_{Alice_3}, G_{Alice_4}, G_{Alice_5}\}, \\ G_{Bob} &= -\frac{2(\sqrt{\alpha_1\omega_1} + x\sqrt{\alpha_2\omega_2})^2}{\omega_1 + \omega_2 x} + 2\alpha_2\gamma - \alpha_2\beta - \alpha_2 + 1, \\ G_{Rice} &= -\frac{(\eta + 1)\left(\sqrt{\alpha_1\omega_1} + x\sqrt{\alpha_2\omega_2}\right)^2}{\omega_1 + \omega_2 x} \\ &\quad +(\eta - 2)\alpha_2\gamma + \eta\alpha_2\beta + \eta\alpha_1 + 2\alpha_2. \end{aligned} \tag{6.103}$$

If $G_{Alice_1} = \max\{G_{Alice_1}, G_{Alice_2}, G_{Alice_3}, G_{Alice_4}, G_{Alice_5}\}$, we obtain that $\alpha_1 = p_1$, and $\beta = \gamma = 0$. The average gains of three players are evaluated using η and ω_1. Here, for each $\eta \geq 1$, the gains of Alice, Bob, and Rice can tend to zero by changing ω_1. From simple simulations (Cheng and Luo 2021), we get that the degree of deviation from the fair game is very small even if the game is not completely fair. The game converges to the fair game when $\eta \to \infty$. To sum up, we get the following theorem.

Theorem 6.5.2 *The quantum game* G *is asymptotically fair.*

It has shown that two present quantum games are asymptotically fair. Interestingly, these games can be easily changed to biased versions, by choosing different η and ω_1. These kinds of schemes may be applicable in gambling theory. Similar to the bipartite scheme (Zhang et al. 2017), a proof-of-principle optical demonstration may be followed for each scheme.

6.6 Prisoner's Dilemma Games

The growing focus on evolutionary games on multilayer networks has generated interest, yet the potential of quantum games within this framework has yet to be investigated, making it a fascinating area for exploration across various fields. Here, we present a model for classical and quantum prisoner's dilemma games on interconnected networks composed of two layers, which simulate two distinct interactive environments and mechanisms.

Prisoner's dilemma game is a simple and paradigmatic metaphor to study cooperation between unrelated individuals. It may be formalized as follows (Smith 1982; Szabó and Fath 2007).

Definition 6.6.1 (*Prisoner's dilemma (PD) game*) In the PD game, two players have the option to select cooperation (C) or defection (D). The payoff obtained by a defector when encountering a cooperator is determined by the temptation of defection $T = b$, while the payoff for mutual cooperation is $R = 1$. In addition, the penalty for mutual defection is defined as $P = 0$, whereas the sucker's payoff $S = 0$ is the reward given to a cooperator when meeting a defector. It is important to note that $1 < b < 2$.

The defection is the unique Nash equilibrium and evolutionarily stable strategy (ESS). The payoff matrix is expressed as

$$F(b) = \begin{pmatrix} 1 & 0 \\ b & 0 \end{pmatrix}. \tag{6.104}$$

The quantum prisoner's dilemma game is an expansion of the classical PD game that incorporates quantum strategies and quantum entanglement (Eisert et al. 1999). One straightforward method is to permit each player's quantum strategy through the "superposition" of cooperation and defection strategies in the shape of a unitary operator as

$$U(\theta, \phi) = \begin{pmatrix} e^{i\phi} \cos \frac{\theta}{2} & \sin \frac{\theta}{2} \\ -\sin \frac{\theta}{2} & e^{-i\phi} \cos \frac{\theta}{2} \end{pmatrix} \tag{6.105}$$

with $0 \leq \theta \leq \pi$ and $0 \leq \phi \leq \pi/2$. Specifically, the identity operator $U(0, 0)$ and bit-flip operator $U(\pi, 0)$ correspond to the classical cooperation and defection strategies, respectively.

As a special version of QPD, one may consider three strategies: cooperation ($C = U(0, 0)$), defection ($D = U(\pi, 0)$), and super cooperation ($Q = U(0, \pi/2)$), respectively (Li and Yong 2014). The payoff matrix is shown as

$$F(\varphi) = \begin{pmatrix} 1 & 0 & \cos^2 \varphi \\ b & 0 & b\cos^2 \varphi \\ \cos^2 \varphi & b\cos^2 \varphi & 1 \end{pmatrix} \tag{6.106}$$

where φ is an entanglement degree. When $\varphi = \pi/2$, the quantum strategy Q represents the only Nash Equilibrium that achieves Pareto efficiency. Additionally, if Q faces a C, there is no quantum advantage. However, as φ increases, Q attains progressively higher payoffs when playing against D, ultimately emerging as a dominant strategy relative to D.

In an alternative model, a co-evolutionary approach to interconnected networks is employed. First, two groups of players construct identical networks of interdependent nodes, $\mathcal{G}_1$ and $\mathcal{G}_2$, where each node on $\mathcal{G}_1$ is linked to a node in $\mathcal{G}_2$. The interdependent structure is described by external links connecting the two networks. During the evolutionary process, one network features classical PD while the other is the QPD. The evolution of both PD versions on the interconnected networks can be simulated via a probability simulation procedure. Initially, each player in the PD game is randomly assigned as either C or D, while the players in the QPD game are randomly assigned as one of three roles: C, D, or Q. Following this, each player from either network obtains its cumulative payoff by engaging in a game with all linked neighbors on the same network (layer).

More specifically, a player on PD just plays games with its neighbors on PD while the same restriction is on QPD. The utility of each player may be computed according to

$$U_x = f(P_x, P_{x'}), U_{x'} = f(P'_x, P_x), \tag{6.107}$$

where x is a node on PD and x' is the coupled node on QPD, and P_x and $P_{x'}$ are payoffs of x and x' received from each network, respectively. f denotes a general coupling functionality. In this model, a straightforward approach is to use a linear combination of two values P_x and $P_{x'}$ with a ratio γ. During each simulation step, players update their strategies asynchronously. One possible method to achieve this is by adopting the strategy s_y from player y with a probability $W(s_y \to s_x) = 1/g(U_x, Y_y, \delta)$, where δ represents the uncertainty related to the strategy adoption process, and g is a positive function with a lower bound. This procedure may be repeated so that each player has an opportunity to change their strategies. Afterward, the interconnected networks proceed to the next round of evolution. This provides a general method for analyzing the co-evolutionary model.

When $\varphi = 0$, the QPD reduces to the classical PD. In this scenario, the evolution of both networks is primarily affected by the coupling functionality. In a linearly coupled network, the evolution is equivalent to that of PD on two-layer networks with a specific coupled parameter (Deng et al. 2016). In general, the level of entanglement φ, coupled parameter, and transition probability jointly impact the overall payoffs. Additionally, even a simple linear coupled network can lead to novel strategies in different networks (Li and Yong 2014). This necessitates detailed simulations to explore such dynamics.

6.7 Conclusion

Since the seminal work in Silman et al. (2008), nonlocal games/distributed games have been extensively studied in the literature and continue to attract significant interest. A detailed introduction to nonlocal games (resp. distributed games) can be found in Attiya and Welch (1998), Brassard et al. (2005), Doherty et al. (2008), Qin and Yao (2021) or others.

In this chapter, we solely consider nonlocal games over quantum networks. Such assumptions are crucial in proving Theorems 6.2.1 and 6.3.1. Otherwise, one may generally require a moving qubit (e.g., Theorem 6.5.2). Such games can be realized in a distributed manner by combining them with quantum teleportation (Bennett et al. 1993). Interested readers can find additional materials in the references, such as Fudenberg and Tirole (1991), Eisert et al. (1999), Ambainis et al. (2012), Brunner and Linden (2013), Pappa et al. (2015). The case with $m = 2$ of Theorem 6.4.1 is proven in Leditzky et al. (2020). The proofs for Theorems 6.2.1 and 6.3.1 can be found in Luo (2019), while the proofs for Theorems 6.5.1 and 6.5.2 are provided in Cheng and Luo (2021).

Another interesting field of study is the investigation of the relationship between the complexity of nonlocal games and the arithmetical hierarchy, which is a classification of languages based on the complexity of arithmetical formulas used to define them. One objective is to explore special complexities for multi-prover games using entangled resources, as seen in further references such as Doherty et al. (2008), Mousavi et al. (2022), Bavarian et al. (2017), Fitzsimons et al. (2019), Ito and Vidick (2012), Regev and Vidick (2015), Goldenberg et al. (1999). Another objective is to identify quantum advantages, as seen in Clauser et al. (1969), Cheng and Luo (2021), Luo (2019). The graph of the involved networks can greatly impact the evolution of such games (see further Refs. Smith (1982); Szabó and Fath (2007)).

References

R. Ahlswede, Multi-way communication channels, in *Second International Symposium on Information Theory: Tsahkadsor* (Armenia, USSR, 1971), pp. 23–52

M.L. Almeida, J.-D. Bancal, N. Brunner, A. Acín, N. Gisin, S. Pironio, Guess your neighbor's input: a multipartite nonlocal game with no quantum advantage. Phys. Rev. Lett. **104**, 230404 (2010)

A. Ambainis, A. Backurs, K. Balodis, D. Kravcenko, R. Ozols, J. Smotrovs, M. Virza, Quantum strategies are better than classical in almost any XOR game, in *International Colloquium on Automata, Languages, and Programming* (Springer, Berlin, Heidelberg, 2012), pp. 25–37

A. Atserias, et al., Quantum and non-signalling graph isomorphisms. J. Combin. Theory, Ser. B **136**, 289–328 (2019)

H. Attiya, J. Welch, *Distributed Computing: Fundamentals, Simulations, and Advanced Topics* (McGraw-Hill Publishing Company, UK, 1998)

M. Bavarian, T. Vidick, H. Yuen, Hardness amplification for entangled games via anchoring, in *Proceedings of the 49th Annual ACM SIGACT Symposium on Theory of Computing* (2017), pp. 303–316

C.H. Bennett, G. Brassard, C. Crépeau, R. Jozsa, A. Peres, W.K. Wootters, Teleporting an unknown quantum state via dual classical and Einstein-Podolsky-Rosen channels. Phys. Rev. Lett. **70**, 1895 (1993)

G. Brassard, A. Broadbent, A. Tapp, Quantum pseudo-telepathy. Found. Phys. **35**, 1877–1907 (2005)

N. Brunner, N. Linden, Connection between Bell nonlocality and Bayesian game theory. Nat. Commun. **4**, 2057 (2013)

N. Brunner, D. Cavalcanti, S. Pironio, V. Scarani, S. Wehner, Bell nonlocality. Rev. Mod. Phys. **86**, 419 (2014)

H.-M. Cheng, M.-X. Luo, Tripartite dynamic zero-sum quantum games. Entropy **23**, 154 (2021)

J.F. Clauser, M.A. Horne, A. Shimony, R.A. Holt, Proposed experiment to test local hidden-variable theories. Phys. Rev. Lett. **23**, 880 (1969)

X. Deng, Q. Zhang, Y. Deng, Z. Wang, A novel framework of classical and quantum prisoner's dilemma games on coupled networks. Sci. Reps. **6**, 23024 (2016)

A. Doherty, Y.-C. Liang, B. Toner, S. Wehner, The quantum moment problem and bounds on entangled multi-prover games, in *Proceedings of the 23rd IEEE Conference on Computational Complexity* (2008), pp. 199–210

J. Eisert, M. Wilkens, M. Lewenstein, Quantum games and quantum strategies. Phys. Rev. Lett. **83**, 3077 (1999)

J. Fitzsimons, Z. F. Ji, T. Vidick, H. Yuen, Quantum proof systems for iterated exponential time, and beyond, in *Proceedings of the 51st Annual ACM SIGACT Symposium on Theory of Computing (STOC 2019)* (2019), pp. 473–480

D. Fudenberg, J. Tirole, *Game Theory* (MIT Press, 1991)

L. Goldenberg, L. Vaidman, S. Wiesner, Quantum gambling. Phys. Rev. Lett. **82**, 3356 (1999)

T. Ito, T. Vidick, A multi-prover interactive proof for NEXP sound against entangled provers, in *2012 IEEE 53rd Annual Symposium on Foundations of Computer Science* (2012), pp. 243–252

S. Kochen, E.P. Specker, The problem of hidden variables in quantum mechanics. Math. Mec. **17**, 59–87 (1967)

T. Lawson, N. Linden, S. Popescu, Biased nonlocal quantum games, arXiv:1011.6245 (2010)

F. Leditzky, M.A. Alhejji, J. Levin, G. Smith, Playing games with multiple access channels. Nat. Commun. **11**, 1497 (2020)

A. Li, X. Yong, Entanglement guarantees emergence of cooperation in quantum prisoner's dilemma games on networks. Sci. Reps. **4**, 6286 (2014)

H. Liao, Multiple access channels, *Phd thesis* (1972)

M.-X. Luo, A new nonlocal game on quantum networks. npj Quantum Inf. **5**, 91 (2019)

M.-X. Luo, Computationally efficient nonlinear bell inequalities for quantum networks. Phys. Rev. Lett. **120**, 140402 (2018)

H. Mousavi, S.S. Nezhadi, H. Yuen, Nonlocal games, compression theorems, and the arithmetical hierarchy, in *STOC 2022: Proceedings of the 54th Annual ACM SIGACT Symposium on Theory of Computing* (2022), pp. 1–11

A. Pappa, N. Kumar, T. Lawson, M. Santha, S. Zhang, E. Diamanti, I. Kerenidis, Nonlocality and conflicting interest games. Phys. Rev. Lett. **114**, 020401 (2015)

M. Qin, P. Yao, Nonlocal games with noisy maximally entangled states are decidable. SIAM J. Comput. **50**, 134592 (2021)

Y. Quek, P.W. Shor, Quantum and superquantum enhancements to two sender, two-receiver channels. Phys. Rev. A **95**, 052329 (2017)

O. Regev, T. Vidick, Quantum XOR games. ACM Trans. Comput. Theory **7**, 1–43 (2015)

T.J. Schaefer, The complexity of satisfiability problems, in *Proceedings of the Tenth Annual ACM Symposium on Theory of Computing* (1978), pp. 216–226

J. Silman, S. Machnes, N. Aharon, On the relation between Bell's inequalities and nonlocal games. Phys. Lett. A **372**, 3796 (2008)

J.M. Smith, *Evolution and the Theory of Games* (Cambridge University Press, 1982)

G. Szabó, G. Fath, Evolutionary games on graphs. Phys. Rep. **446**, 97–216 (2007)

P. Zhang, X.-Q. Zhou, Y.-L. Wang, P.-J. Shadbolt, Y.-S. Zhang, H. Gao, F.-L. Li, J.L. O'Brien, Quantum gambling based on Nash-equilibrium. npj Quantum Inf. **3**, 24 (2017)

Chapter 7
Secure Quantum Cloud Computing

Cloud computation refers to the delivery of computing resources, including software, storage, and processing power, over the Internet or a private network. These resources are hosted in remote data centers operated by third-party providers and can be accessed on-demand via a web browser or API. Cloud computation provides several benefits, including scalability, flexibility, cost savings, and improved reliability. Users can quickly scale up or down their computing resources based on their needs, without the need for significant upfront investment in hardware or infrastructure. They can also access a wide range of software applications, tools, and platforms without having to install or manage them locally.

Quantum cloud computing is an emerging technology that combines the power of quantum computing with the scalability and accessibility of cloud computing. It allows users even classical clients to access quantum resources and run quantum algorithms on remote quantum computers through a cloud-based interface, potentially opening up new possibilities for solving complex problems in various fields such as finance, cryptography, and material science.

In this chapter, we present secure distributive applications in quantum network scenarios where the client can obtain the computation result by just running a remote protocol a few times. This allows especially one classical client to delegate its classical task to a quantum server. A further protocol can even show the blindness and verifications of delegated tasks. This provides a general method for secure applications on hybrid quantum Internet.

7.1 Measurement-Based Quantum Computing

Measurement-based quantum computing is a model of quantum computation that relies on the measurement of entangled qubits to perform operations rather than manipulating individual qubits through quantum gates. The initial state of the system

M.-X. Luo, *Quantum Networks*,
https://doi.org/10.1007/978-981-97-6226-2_7

is prepared by entangling a large number of qubits in a highly entangled state called a cluster state (Deutsch 1989; Nielsen and Chuang 2000; Raussendorf and Briegel 2001; Raussendorf et al. 2003). Measurements are then performed on individual qubits in the cluster, which causes changes in the entanglement between qubits and generates the final output state. This approach to quantum computing has shown promise for certain types of problems, such as simulating quantum systems and solving optimization problems, and may offer advantages over gate-based quantum computing in terms of error correction and fault tolerance.

In measurement-based quantum computing (MBQC), the computation is achieved by iteratively measuring an entangled resource state in specific bases, with adaptive measurements being a possibility (Raussendorf and Briegel 2001; Raussendorf et al. 2003). MBQC relies on a specific type of resource state, which is typically created through entangling a large number of $|+\rangle$-state vectors with controlled-Z gates ($CZ = |0\rangle\langle 0| \otimes \mathbb{1} + |1\rangle\langle 1| \otimes Z$). The resulting entangled state forms the basis for performing quantum computations by measuring individual qubits in the cluster. These measurements generate correlations between the different qubits and ultimately lead to the final output state. There are two requirements for the resource state:

1. The (family of) resource states is computationally universal.
2. There is an efficient classically-computable function f that determines how measurement bases have to be adapted, depending on previous measurement results.

The first requirement means that it can perform any quantum computation with an appropriate measurement pattern. This condition ensures that the resource state can be used as a general-purpose tool for quantum computation. The second dictates that there must exist an efficient classically-computable function, denoted f, that determines how measurement bases should be adapted depending on previous measurement results. This is necessary because the choice of measurement bases affects the correlations generated between the different qubits. By adjusting the measurement bases during the computation, one can create the desired correlations and ultimately obtain the desired output state.

In addition, the resource state needs to be updated to include any quantum input or output involved in the computation. Quantum input must be entangled with the resource state using CZ gates, while the unmeasured part of the resource state remains available for generating the output state. By meeting these requirements, the family of resource states can be used to perform a wide variety of quantum computations using the principles of MBQC.

7.1.1 Cluster States and Graph States

Cluster states and graph states are two types of resource states used in MBQC. These states can be defined by associating a graph with each state, where the vertices of the graph correspond to qubits (Raussendorf et al. 2003; Hein et al. 2006). The corresponding graph state is generated by preparing each qubit in the $|+\rangle = \frac{1}{\sqrt{2}}(|0\rangle + |1\rangle)$ state and applying a controlled Pauli Z operation CZ on every pair of qubits whose vertices are connected by an edge in the graph.

Cluster states are a sub-class of graph states that have an underlying graph structure consisting of an n-dimensional square grid. In contrast, the entanglement structure of graph states can be more flexible, which means they may require fewer qubits to implement a one-way quantum computation compared to cluster states.

The use of graphs to define these states provides a powerful tool for understanding and analyzing their properties. By manipulating the structure of the graph, it may be possible to optimize the resource state's performance for specific computational tasks. The construction of cluster and graph states provides a valuable framework for realizing measurement-based quantum computing and exploring its potential for solving complex problems.

To use graphs as a characterization for a large class of interaction patterns in quantum computing, certain constraints must be placed on the two-particle unitaries involved.

(1) Since the graph does not establish an ordering of the edges, all two-particle unitaries U_{ab} must commute: $[U_{ab}, U_{bc}] = 0$ for all $a, b, c \in V$, where V is the set of vertices in the graph.
(2) For undirected graphs, the unitaries should be symmetric: $U_{ab} = U_{ba}$ for all $a, b \in V$.
(3) If the edges are not specified by weights, all qubits should interact through the same two-particle unitary U: U_{ab} for all $a, b \in V$.

For qubits, condition (1) is already sufficient to restrict particle analysis to the case where the qubits interact according to the same Ising interaction, such as $H^I_{ab} = Z^a Z^b$. By imposing these constraints on the graphs, it may be possible to gain insights into the entanglement structure of the resource state and optimize its performance for specific computational tasks.

Proposition 7.1.1 *Under an appropriate local basis, any set of commuting two-particle unitaries gives an Ising interaction pattern*

$$U^I_{ab}(\varphi_{ab}) := e^{-i\varphi_{ab} Z^a Z^b}. \tag{7.1}$$

Proof Denote the controlled phase gate

$$U_{ab}(\varphi_{ab}) := e^{-i\varphi_{ab} H_{ab}} \tag{7.2}$$

with $H_{ab} := |1\rangle^a\langle 1| \otimes |1\rangle^b\langle 1|$ as the elementary two-qubit interaction. The corresponding interaction strength now is $\varphi_{ab} = \pi$, it follows that

$$U_{ab}(\varphi_{ab}) = e^{-i\frac{\varphi_{ab}}{4}} e^{i\frac{\varphi_{ab}}{4}Z^a} e^{i\frac{\varphi_{ab}}{4}Z^a} U^I_{ab}(\frac{\varphi_{ab}}{4}). \tag{7.3}$$

The phase gate corresponds to the Ising interaction up to some additional $\frac{\pi}{4}$-rotations around the z-axes at each qubit. For simple graphs, i.e., $\varphi_{ab} = \pi$, we find that

$$U_{ab} := U_{ab}(\pi) = P^a_{z,+} \otimes \mathbb{1}^b + P^a_{z,-} \otimes Z^b. \tag{7.4}$$

The choice of $\varphi_{ab} = \pi$ implies that the state vector

$$U_{ab}|+\rangle_a|+\rangle_b = \frac{1}{\sqrt{2}}(|0\rangle_a|+\rangle_b + |1\rangle_a|-\rangle_b) \tag{7.5}$$

is maximally entangled (*Bell state*), meanwhile, $U^2_{ab} = \mathbb{1}_{ab}$ or $U_{ab} = U^\dagger_{ab}$. Consequently, the phase gate U_{ab} creates as well as deletes the edge $\{a, b\}$ in a graph $\mathcal{G}$ depending on whether the edge is already contained in the graph $\mathcal{G}$ or not. □

Definition 7.1.1 Let $\mathcal{G} = (V, E)$ be a graph where V is the set of vertices and E is the set of edges connecting pairs of vertices. The graph state $|G\rangle$ that corresponds to the graph $\mathcal{G}$ is the pure state with state vector

$$|G\rangle = \prod_{\{a,b\}\in E} U_{ab}|+\rangle^V, \tag{7.6}$$

where U_{ab} is the controlled phase gate applied on qubits a and b, and $|+\rangle = \frac{1}{\sqrt{2}}(|0\rangle + |1\rangle)$ is the $+1$ eigenstate of the Pauli X operator.

In other words, the state $|G\rangle$ is obtained by preparing every qubit in the $|+\rangle$ state and applying a controlled-Z gate between each pair of qubits corresponding to an edge in the graph. The preparation of $|G\rangle$ is shown as follows:

- Prepare the state $|+\rangle$ at each vertex.
- Apply U_{ab} to all adjacent vertices a and b in $\mathcal{G}$.

Since U_{ab} is the unitary two-qubit operation applied to vertices a and b, which can add or remove an edge between them in the graph. The initial state of the preparation procedure, with a state vector given by $|+\rangle^V$, can be thought of as the graph state corresponding to the empty graph, where there are no edges present.

Proposition 7.1.2 *Given a graph $\mathcal{G} = (V, E)$, a graph state vector $|G\rangle$ is a unique, common eigenvector in $(\mathbb{C}^2)^V$ that satisfies the following set of independent, commuting observables:*

$$K_a = X^a Z^{N_a} := X^a \prod_{b\in N_a} Z^b, \tag{7.7}$$

where N_a denotes the set of neighbors of vertex a. The eigenvalues of the correlation operators K_a are all equal to $+1$ for every $a \in V$.

The Abelian subgroup $\mathcal{S}$ of the local Pauli-group $\mathcal{P}^V$ generated by the set $\{K_a|a \in V\}$ is called the stabilizer of the graph state. The stabilizer provides a way to characterize and analyze the structure of the graph state, as well as its usefulness for different quantum computing tasks. By analyzing the stabilizer and the associated graph structure, it may be possible to optimize the performance of the graph state for specific computational needs.

Proof For any vertex $a \in V$, the operator X^a anticommutes with all other X^b operators for $b \neq a$. Similarly, the operator Z^b anticommutes with all other Z^c operators for $c \in N_b$, where N_b denotes the set of neighbors of vertex b in the graph.

Consider the product of correlation operators K_a for all $a \in V$. Using the above observations, we can see that this product commutes with every X^b and Z^c operator since they either commute or anticommuting depending on whether they act on the same vertex or a neighboring vertex. This implies that the product of correlation operators K_a commutes with every element of the local Pauli-group $\mathcal{P}^V$, which includes all possible products of X and Z operators on the vertices of the graph.

Since $|G\rangle$ is a graph state vector, it is invariant under all elements of the local Pauli-group $\mathcal{P}^V$, and therefore must be an eigenvector of the product of correlation operators K_a with eigenvalue ± 1. However, since the eigenvalues of K_a are known to be +1 for all $a \in V$, it follows that $|G\rangle$ must be an eigenvector of K_a with eigenvalue +1 for all $a \in V$. □

Proposition 7.1.3 *Given a graph state vector $|G\rangle$, the set of states $|W\rangle = Z^W|G\rangle$ is basis for $(\mathbb{C}^2)^V$. The states $|W\rangle$ are the eigenstates for the correlation operators K_a according to different eigenvalues W_a where $W = (W_1, \cdots, W_N)$, i.e., $K_a|W\rangle_G = (-1)^{W_a}|W\rangle$.*

Proof Note that applying Z^W to $|G\rangle$ results in a phase factor $(-1)^{\sum_{a=1}^{N} W_a N_a}$ for each vertex a, where N_a is the set of neighbors of vertex a. Using the definition of K_a, we see that this phase factor is precisely $(-1)^{W_a}$ for vertex a. Therefore, $|W\rangle$ is an eigenstate of K_a with eigenvalue $(-1)^{W_a}$. □

The projector onto the graph state has a direct representation in terms of the corresponding stabilizer $\mathcal{S}$:

$$|G\rangle\langle G| = \frac{1}{2^N} \sum_{\sigma \in \mathcal{S}} \sigma \tag{7.8}$$

Proposition 7.1.4 *Local complementation of a graph $\mathcal{G} = (V, E)$ at some vertex $a \in V$ produces an LC-equivalent graph state denoted by $|\tau_a(\mathcal{G})\rangle$. This state vector can be obtained by applying a local Clifford unitary $U_a^{\tau}(\mathcal{G})$ to the original graph state $|G\rangle$, as follows:*

$$|\tau_a(G)\rangle = U_a^{\tau}(G)|G\rangle, \tag{7.9}$$

where

$$U_a^{\tau}(G) = e^{-i\frac{\pi}{4}X^a} e^{i\frac{\pi}{4}Z^{N_a}} \propto \sqrt{K_a} \tag{7.10}$$

and N_a denotes the set of neighbors of a in the graph. The operator $\sqrt{K_a}$ is defined as the positive square root of the correlation operator $K_a = X^a Z^{N_a}$, which generates the stabilizer group of the original graph state $|G\rangle$.

Proof The local complementation at vertex a replaces the set of neighbors N_a with its complement $\overline{N}_a$, while leaving the rest of the graph unchanged. This changes the adjacency matrix of the graph but preserves the set of independent commuting observables that generate the stabilizer group of the original graph state $|G\rangle$.

Consider the local Clifford unitary $U_a^{\tau}(\mathcal{G})$ given by Eq. (7.10). This unitary can be written as a product of two operators, namely:

$$U_a^{\tau}(\mathcal{G}) = e^{-i\frac{\pi}{4}X^a} e^{i\frac{\pi}{4}Z^{N_a}} = \prod_{b \in N_a} CZ_b e^{-i\frac{\pi}{4}X^a}, \tag{7.11}$$

where CZ_b denotes the controlled-Z gate between qubits a and b, and the product is taken in any order.

Applying $U_a^{\tau}(\mathcal{G})$ to the original graph state $|G\rangle$ results in a new state vector $|\tau_a(\mathcal{G})\rangle$ that is invariant under all elements of the stabilizer group of $\mathcal{G}$. Moreover, the action of $U_a^{\tau}(\mathcal{G})$ on each qubit corresponds to a series of local Pauli operations that transform the correlation operators K_a and K_b for each neighbor $b \in N_a$ of vertex a.

By analyzing these transformations, it can be shown that the new graph state vector $|\tau_a(\mathcal{G})\rangle$ is equivalent to the original graph state $|G\rangle$ up to local complementation at vertex a. More precisely, the two states are related by a product of local Pauli operations and a permutation of qubits, which corresponds to the change in the adjacency matrix resulting from the local complementation operation. □

Two graph states $|G\rangle$ and $|G'\rangle$ are LC-equivalent if and only if the corresponding graphs are related by a sequence of local complementations, i.e. $G' = \tau_{a_1} \circ \cdots \circ \tau_{a_n}(G)$ for some $a_1, \ldots, a_n \in V$.

Example 7.1.1 Consider an N-qubit Greenberger-Horne-Zeilinger state

$$|GHZ\rangle = \frac{1}{\sqrt{2}}(|0\rangle^{\otimes N} + |1\rangle^{\otimes N}). \tag{7.12}$$

It can be represented as either the star graph or the complete graph. This is apparent by applying Hadamard unitaries $H^{V \setminus a}$ to all but one qubit a in the GHZ state, which transforms it into the star graph state with a as the central qubit. Further application of the LC-unitary U_a^{τ} for the LC-rule τ_a converts the star graph state into the complete graph state. Thus, both the star graph states with different central vertices and the complete graph state are LC-equivalent representations of the GHZ state.

Definition 7.1.2 A family Ψ is considered a universal resource for MBQC if, for every n-qubit state $|\phi\rangle$, there exists a state $|\psi\rangle \in \Psi$ on m qubits with $m \geq n$, such that the transformation $|\psi\rangle \mapsto |\phi\rangle|+\rangle^{\otimes m-n}$ can be achieved deterministically by local operations and classical communication (LOCC), denoted symbolically as $|\psi\rangle \geq_{LOCC} |\phi\rangle$.

Theorem 7.1.1 *A set of states Ψ is a universal resource for MBQC if and only if all 2D cluster states $|C_{k\times k}\rangle$ (for all k) can be prepared from the set Ψ by LOCC.*

The main idea of the proof is as follows. Suppose that Ψ is a universal resource for MBQC. Given a $k \times k$ 2D cluster state $|C_{k\times k}\rangle$, we can prepare it from the GHZ state $|GHZ\rangle = \frac{1}{\sqrt{2}}(|0\rangle^{\otimes k^2} + |1\rangle^{\otimes k^2})$ using local Pauli operations and CNOT gates. Since $|GHZ\rangle$ can be prepared from Ψ by LOCC, it follows that $|C_{k\times k}\rangle$ can also be prepared from Ψ by LOCC.

Moreover, suppose that all 2D cluster states $|C_{k\times k}\rangle$ (for all k) can be prepared from Ψ by LOCC. Given an n-qubit state $|\phi\rangle$, we can prepare a $k \times k$ 2D cluster state $|C_{k\times k}\rangle$ with $k \geq \sqrt{n}$, and use LOCC to transform $|C_{k\times k}\rangle$ to $|\phi\rangle|+\rangle^{\otimes k^2-n}$. Since $|\phi\rangle|+\rangle^{\otimes k^2-n}$ can be prepared from $|C_{k\times k}\rangle$ by LOCC, it follows that $|\phi\rangle$ can be prepared from Ψ by LOCC. Therefore, Ψ is a universal resource for MBQC.

Theorem 7.1.2 *The graph states corresponding to the hexagonal, triangular, and Kagome lattices are universal.*

The main idea is to prove that any n-qubit state can be prepared from these graph states by LOCC. It is easy to show the hexagonal, triangular, and Kagome lattices can be transformed into each other by LOCC, and that they can also be used to prepare arbitrary stabilizer states with a polynomial overhead in the number of qubits. Therefore, it suffices to show that any stabilizer state can be transformed into any arbitrary state by LOCC using these lattice states.

Note that any stabilizer state can be written as a product of single-qubit Pauli operators and stabilizer generators. Since single-qubit operations can be performed locally, it remains to show how to transform the stabilizer generators into arbitrary states.

For the hexagonal, triangular, and Kagome lattices, the stabilizer generators correspond to plaquette operators, which can be measured collectively using adaptively chosen Pauli measurements on the individual qubits. By choosing the measurement basis according to the outcome of previous measurements, one can deterministically prepare any desired stabilizer state from the lattice state using LOCC.

The MBQC model of computation is a framework for quantum computation based on the principle of gate teleportation. In this model, quantum computation is achieved by preparing an entangled resource state and performing adaptive measurements on individual qubits according to a prescribed measurement pattern.

It was shown in various works (Raussendorf and Briegel 2001; Raussendorf et al. 2003), that MBQC can be used to implement universal quantum computing using a certain type of entangled states represented by graphs and one-qubit measurements. Specifically, it has been demonstrated that certain graph states, such as those corresponding to the hexagonal, triangular, and Kagome lattices, are universal resources for MBQC and can be used to implement any quantum computation that can be done with a circuit model. This correspondence between the two models can be described using the tools of measurement calculus (Raussendorf et al. 2003), which provides a formal framework for analyzing the computational power of different types of measurements and resource states in MBQC.

7.1.2 Single-Qubit Measurements and Rotations

Single-qubit measurements in various bases are essential in one-way quantum computation and can be conveniently described using a Bloch sphere picture. Specifically, each projective single-qubit measurement represented by (θ, ϕ) corresponds to an observable $U_z(\phi + \frac{\pi}{2})U_x(\theta)ZU_x(-\theta)U_z(-\phi - \frac{\pi}{2})$.

A rotation around the k axis ($k \in \{x, y, z\}$) by angle ϕ can be written

$$U_k(\phi) = e^{-\frac{i\phi}{2}\sigma_k}. \tag{7.13}$$

We will use the notation $X \equiv X$, etc. in the rest of this chapter. We also adopt standard notation for the eigenstates of Z and X as $\{|0\rangle, |1\rangle\}$ and $\{|+\rangle, |-\rangle\}$.

A measurement with angles (θ, ϕ) corresponds to a measurement of the observable $U_z(\phi + \frac{\pi}{2})U_x(\theta)ZU_x(-\theta)U_z(-\phi - \frac{\pi}{2})$. One way of implementing such a measurement is to apply the single-qubit unitary $U_x(-\theta)U_z(-\phi - \frac{\pi}{2})$ to the qubit before measuring it in the computational basis.

Example 7.1.2 An unknown state $|\psi\rangle = \alpha|0\rangle + \beta|1\rangle$ is prepared for a qubit, while a second qubit is prepared in the state $|+\rangle = \frac{1}{\sqrt{2}}(|0\rangle + |1\rangle)$. The two qubits are subjected to a CZ operation, resulting in a state of the form $\frac{1}{\sqrt{2}}(\alpha|0\rangle|+\rangle + \beta|1\rangle|-\rangle)$. The first qubit is then subjected to a measurement in the basis $\frac{1}{\sqrt{2}}(|0\rangle \pm e^{i\phi}|1\rangle)$, denoted as $(\pi/2, \phi)$. This measurement yields a unit vector in the $x - y$ plane, at an angle ϕ to the x axis. The outcome of the measurement, represented by the bit $m \in 0, 1$, determines the state of the second qubit up to a global phase, given by $X^m HU_z(\phi)|\psi\rangle$. This transfer of information between the two qubits preserves coherence and results in a unitary transformation $X^m HU_z(\phi)$ on the first qubit. The angle of rotation in the transformation is determined by the measurement basis and a Pauli transformation X is applied when the measurement outcome is -1. The randomness of the measurement outcomes implies that any desired unitary can only be implemented up to known Pauli transformations in one-way quantum computation, which is a characteristic feature of this protocol.

Any single-qubit rotation in $\mathbb{SU}(2)$ can be decomposed as a product of three rotations $U_z(\gamma)$, $U_x(\beta)$ and $U_z(\alpha)$. Thus, by repeating the simple two-qubit protocol three times, any arbitrary single-qubit rotation may be obtained (up to an extra Hadamard). Two one-way patterns are combined as one would expect, the output qubit(s) of one pattern becomes the input qubit(s) of the next. The main issue in connecting patterns is to track the effect of the Pauli by-product operators which have accumulated due to the previous measurements.

7.1.3 A Heisenberg Picture

Consider a collection of 2-level systems (qubits), or qubit graph, denoted by the vertices V of a graph $\mathcal{G} = (V, E)$. After initial preparation in a state vector $|\Psi\rangle$, the qubits undergo interaction by means of a graph $\mathcal{G}$ representing a given interaction pattern. Specifically, each edge $\{a, b\} \in E$ causes the qubits of the two adjacent vertices a and b to interact under a (non-local) unitary $U_{ab} = e^{-i\varphi_{ab} H_{ab}}$. Here, H_{ab} represents the interaction Hamiltonian and φ_{ab} denotes the coupling strength or interaction time with appropriate physical units.

To comprehend the one-way patterns responsible for implementing unitary transformations in the one-way model, we employ the stabilizer formalism. In doing so, describe the logical action of a one-way pattern in a Heisenberg picture (Gottesman 1998). Typically, the Schrodinger picture is used to depict the time evolution of quantum systems. This reflects temporal alterations in the system through modifications to the state vector or density matrix. For example, unitary evolution $|\psi\rangle \mapsto U(t)|\psi\rangle$ or $\rho \mapsto U(t)\rho U(t)^\dagger$ are common occurrences. In contrast, observable quantities, such as Pauli observables X, Y, and Z, remain unchanged over time.

On the other hand, in the Heisenberg picture, time-evolution proceeds entirely through physical observables, which evolve according to $O(t) \mapsto U(t)^\dagger O U(t)$. Consequently, states and density matrices do not undergo any temporal modifications.

The *Heisenberg picture*, also referred to as a Heisenberg representation of quantum computation (Gottesman 1998), combines aspects of both Schrodinger picture and Heisenberg picture. To introduce this concept, we will use the example of a single-qubit density matrix $\rho(t)$ that evolves in time. In the Schrodinger picture, we can represent ρ at time $t = 0$ as a linear combination of the n-qubit Pauli-group operators, with real parameters a, b, c, and d defining the state.

At time t, ρ is transformed by a unitary $U(t)$. In the Schrodinger picture, one would describe this transformation in terms of matrix element changes or changes in the parameters a, b, c, and d to $a(t), b(t)$, etc. However, we can also rewrite the equation as $\rho(t) = U(t)\rho U(t)^\dagger = a\mathbb{1} + bU(t)XU(t)^\dagger + cU(t)YU(t)^\dagger + dU(t)ZU(t)^\dagger$ and introduce new time-evolving observables $\bar{X}(t)$, $\bar{Y}(t)$, and $\bar{Z}(t)$ to express it as a linear combination of these observables as $\rho(t) = a\mathbb{1} + b\bar{X}(t) + c\bar{Y}(t) + d\bar{Z}(t)$.

The time-evolution of the observables $\bar{X}(t)$, $\bar{Y}(t)$, and $\bar{Z}(t)$ captures the time-evolution of ρ, while the parameters a, b, c, and d remain constant. Since $\bar{Y}(t) = i\bar{X}(t)\bar{Z}(t)$, specifying $\bar{X}(t)$ and $\bar{Z}(t)$ completely determines the evolution $U(t)$. Therefore, an n-qubit unitary is defined by the evolution of $\bar{X}(t)_a$ and $\bar{Z}(t)_a$ for each qubit a.

In analogy to the (standard) Heisenberg picture, where physical observables evolve, we call this a *Heisenberg picture*. The observables $\bar{X}(t)$, $\bar{Y}(t)$, and $\bar{Z}(t)$ define the logical basis in which ρ is expressed.

Example 7.1.3 Consider a Hadamard $U(t) = H$. This is represented in the Heisenberg picture through $\bar{X}(t) = Z$ and $\bar{Z}(t) = X$.

Example 7.1.4 Consider the representation of the SWAP gate. We obtain that $\bar{X}_1(t) = X_2$ and $\bar{X}_2(t) = X_1$, similarly for the Z variables.

In the one-way quantum computer model, the evolution of quantum states is driven by single-qubit measurements and is therefore discrete. This means that we can suppress the time labeling and focus on the logical evolution of the system. In this context, the Heisenberg picture is a useful tool to describe the encoding of quantum information.

In addition to density matrices, pure state vectors can also be represented in the Heisenberg picture, where the logical basis states play a central role. These are the joint eigenstates of the logical operators $\bar{Z}(t)_a$, with phase relations determined by $\bar{X}(t)_a$.

To encode a state $|\psi\rangle = \alpha|0\rangle + \beta|1\rangle$, we apply a unitary transformation U to map it to a new set of logical basis states, $|0'\rangle$ and $|1'\rangle$. This transformation keeps the coefficients α and β constant while changing the logical basis vectors.

7.1.4 Non-Pauli Measurements

The method outlined above does not provide a way to deal with non-Pauli measurements that are defined by measurement directions other than the x, y, or z-axis. However, these measurements can still be handled within the stabilizer formalism. Consider the state $|\psi\rangle$ stabilized by operator $Z \otimes X$. By rearranging the stabilizer eigenvalue equations, non-Pauli unitary operations that stabilize the sub-space can be generated. For example, consider the state $|\psi\rangle$ stabilized by the operator $Z \otimes X$.

The stabilizer equation can be rewritten as a set of equations that lead to the following unitary transformation: $\exp(i\frac{\phi}{2}(Z \otimes \mathbb{1} - \mathbb{1} \otimes X))$. This unitary operation stabilizes $|\psi\rangle$ and implies that $U_z(\phi) \otimes \mathbb{1}|\psi\rangle = \mathbb{1} \otimes U_x(\phi)|\psi\rangle$.

This technique can be used to understand any one-way pattern solely within the stabilizer formalism, and it was used in Raussendorf et al. (2003) to design and verify many of the gate patterns. Thus, the effect of non-Pauli measurements in a one-way quantum computation can always be understood as the implementation of a generalized rotation $\exp(i\frac{\phi}{2}W)$, where W is any n-qubit Pauli group operator.

It is worth noting that any one-way computation can be transformed into a quantum circuit by using quantum gates to prepare the resource state. This requires one two-qubit gate per bond for cluster and graph resource states, making it efficient. Similarly, a one-way computer can simulate any quantum circuit using a two-dimensional cluster state as the resource state. This is also polynomially efficient, as the required size of the cluster scales with the size of the circuit.

Theorem 7.1.3 *One-way computation and circuit model are equivalent.*

The main idea is as follows. First, let us consider an algorithm that can be computed using a circuit. A circuit consists of gates connected by wires, where each gate takes some inputs and produces some outputs. We can represent each gate as a function that takes its input bits and produces output bits using fixed logic (Deutsch 1989; Brennen et al. 2005; Luo and Wang 2014), such as AND or XOR operations. By composing these functions, we can express the entire circuit as a single function that maps input bits to output bits.

To perform the one-way computation of this algorithm, we can execute the gate functions in sequence, starting from the first gate and proceeding through the layers of gates until we reach the final output. At each step, we feed the output bits of the previous gate as input bits to the next gate. Because we only allow information to flow in one direction (from input to output), we call this method one-way computation.

Conversely, let us consider an algorithm that can be computed using one-way computation. We can represent this algorithm as a sequence of functions, where each function takes some input bits and produces some output bits, similar to the gates in a circuit. However, unlike a circuit, these functions cannot be evaluated in parallel because the output of each function depends on the output of the preceding function.

To compute this algorithm using a circuit, we can create a circuit with a series of layers, where each layer corresponds to one function in the one-way computation sequence. Each layer consists of gates that implement the logic of the corresponding function, with the input bits connected to the inputs of the gates and the output bits connected to the outputs of the gates. The outputs of one layer are fed as inputs to the next layer, just like in the one-way computation sequence.

7.2 Secure Delegated Quantum Computation

Delegated quantum computation refers to the concept of quantum computations to a remote quantum computer while keeping the input, output, and computation confidential. This approach can be useful in scenarios where a client does not have access to a powerful quantum computer or lacks expertise in quantum algorithms (Childs 2001). Instead, they can delegate the computation to a trusted third party with the necessary resources and expertise. The delegated quantum computation protocol ensures that the client's data remains secure by using techniques such as quantum encryption and blind quantum computing (Arrighi and Salvail 2006; Dupuis et al. 2010). It has potential applications in fields such as cryptography, machine learning, and drug discovery.

7.2.1 Measurement-Based Blind Quantum Computing

MBQC works by first selecting an appropriate entangled state. Then, single qubits are measured, and correction operators are applied to the rest of the qubits based on the measurement outcomes. This approach is well-suited for delegating quantum computations where the client only provides quantum inputs and instructs the server on the right measurements to perform. The server then takes on the creation of a large entangled state.

We define two variables: $Z(\theta)$, which equals $|0\rangle\langle 0| + e^{i\theta}|1\rangle\langle 1|$, and $|+_\theta\rangle$, which equals $Z(\theta)\,|+\rangle = \frac{1}{\sqrt{2}}(|0\rangle + e^{i\theta}\,|1\rangle)$. Although angles in $[0, 2\pi)$ are discussed below, if we aim for approximate universality, we can restrict ourselves to using the set of angles $\Theta = \{\frac{k\pi}{4}\}_{k\in\{0,\cdots,7\}}$.

The computation is determined by a graph $\mathcal{G} = (V, E)$, consisting of input and output vertices $I, O \subseteq V$, default measurement angles $\{\phi(i)\}_{i\in O^c}$ for non-output qubits, and a flow function $f : O^c \to I^c$. Each vertex in V corresponds to a qubit in the state $|+\rangle$, while each edge in E corresponds to an entangling operator CZ.

Let $S_X(i)$ and $S_Z(i)$ be respectively the X and Z dependency-sets for qubit i. These sets and the order of measurement are given by the flow. The measurement results from $s(j)$ for qubit $j \in S_X(i) \cup S_Z(i)$ induce Pauli corrections on qubit i. These corrections are equivalent to measuring qubit i with a corrected angle $\phi'(i) = (-1)^{s_X(i)}\phi(i) + \pi s_Z(i)$, where $s_X(i) = \bigoplus_{j\in S_X(i)} s(j)$ and $s_Z(i) = \bigoplus_{j\in S_Z(i)} s(j)$.

After measuring all qubits in O^c according to these angles, the server returns all output qubits $i \in O$ to the client. The client performs the final Pauli correction $Z^{s_Z(i)} X^{s_X(i)}$ on each output qubit.

7.2.2 *Qubit Inputs*

We present a protocol that allows a client to have a server carry out a quantum computation for her such that the client's inputs, outputs, and computation remain perfectly private, and where she does not require any quantum computational power or memory (Reichardt et al. 2013). The client only needs to be able to prepare single qubits randomly chosen from a finite set

$$\mathcal{S} := \{\frac{1}{\sqrt{2}}(|0\rangle + e^{i\theta}\,|1\rangle)|\theta = 0, \frac{\pi}{4}, \ldots, \frac{7\pi}{4}\}, \tag{7.14}$$

and send them to the server, which has the balance of the required quantum computational resources.

The present protocol is interactive: after the initial preparation of quantum states, the client is only required to prepare single qubits from a finite set and communicate with the server classically to instruct measurements for the computation. The protocol supports both classical and quantum inputs and outputs and includes an authentication step to ensure no interference from the server.

Suppose Alice has a unitary operator U implemented with a pattern on a brickwork state $\mathcal{G}_{n\times m}$, with measurements given as multiples of $\pi/4$. The pattern could be designed in MBQC or from a circuit construction. Each qubit $|\psi_{x,y}\rangle \in \mathcal{G}_{n\times m}$ has a measurement angle $\phi_{x,y}$, a set of X-dependencies $D_{x,y}$, and a set of Z-dependencies $D'_{x,y}$. The dependency sets are obtained via the flow construction. During execution, the actual measurement angle $\phi'_{x,y}$ depends on previous measurement outcomes.

Specifically, $\phi'_{x,y}$ is modified by $(-1)^{s^X_{x,y}}\phi_{x,y} + s^Z_{x,y}\pi$, where $s^X_{x,y}$ is the parity of all measurement outcomes for qubits in $X_{x,y}$, and $s^Z_{x,y}$ is the parity of all measurement outcomes for qubits in $Z_{x,y}$.

Assume that Alice's input to the computation is built into U, i.e., Alice wishes to compute $U|0\rangle$, her input is classical and the first layers of U may depend on it.

1. For $x = 1, \cdots, n$; $y = 1, \cdots, m$, Alice prepares $|\psi_{x,y}\rangle \in_R \{|+_{\theta_{x,y}}\rangle = \frac{1}{\sqrt{2}}(|0\rangle + e^{i\theta_{x,y}}|1\rangle) \mid \theta_{x,y} = 0, \pi/4, \cdots, 7\pi/4\}$ and sends the qubits to Bob.
2. Bob creates an entanglement from received qubits, according to their indices, by applying CZ between the qubits to create a brickwork state $\mathcal{G}_{n\times m}$.
3. For $x = 1, \cdots, n$; For $y = 1, \cdots, m$:

   (a) Alice computes $\phi'_{x,y}$ with $s^X_{0,y} = s^Z_{0,y} = 0$. Alice chooses $r_{x,y} \in_R \{0, 1\}$ and computes $\delta_{x,y} = \phi'_{x,y} + \theta_{x,y} + \pi r_{x,y}$. Alice transmits $\delta_{x,y}$ to Bob.
   (b) Bob measures in the basis $\{|+_{\delta_{x,y}}\rangle, |-_{\delta_{x,y}}\rangle\}$. Bob transmits the result $s_{x,y} \in \{0, 1\}$ to Alice.
   (c) If $r_{x,y} = 1$ above, Alice flips $s_{x,y}$; Otherwise, she does nothing.

The universality of the protocol can be followed from the universality of the brickwork state (defined below) for MBQC.

Definition 7.2.1 A brickwork state $\mathcal{G}_{n\times m}$ with $m \equiv 5 \pmod 8$, is an nm-qubit entanglement defined by

(1) Prepare all qubits in state $|+\rangle$ with an index (i, j), where i represents the column ($i \in [n]$) and j represents the row ($j \in [m]$).
(2) Apply CZ gate on qubits (i, j) and $(i, j+1)$ for each row where $1 \leq j \leq m-1$.
(3) Apply a CZ gate on qubits (i, j), $(i+1, j)$, $(i, j+2)$, and $(i+1, j+2)$ for every odd row i and each column $j \equiv 3 \pmod 8$.
(4) Apply CZ on qubits (i, j), $(i+1, j)$, $(i, j+2)$ and $(i+1, j+2)$ for each column $j \equiv 7 \pmod 8$ and each even row i,

Theorem 7.2.1 *The brickwork state $\mathcal{G}_{n\times m}$ is universal for quantum computation. It requires single-qubit measurements under the angles $\{0, \pm\pi/4, \pm\pi/2\}$, and measurements can be done layer-by-layer.*

Proof It is sufficient to prove that any single-qubit unitary gate can be approximated to arbitrary precision using only single-qubit measurements under the angles $\{0, \pm\pi/4, \pm\pi/2\}$ and layer-by-layer measurements on the brickwork state.

Note that any single-qubit unitary gate can be decomposed into a sequence of rotations around the x, y, and z axes by Euler's rotation theorem. Each rotation can then be implemented using measurements under the angles $\{0, \pm\pi/4, \pm\pi/2\}$.

To implement these rotations on the brickwork state, it takes use of the fact that measurements can be done layer-by-layer. One can start by measuring the first layer of qubits in the x basis, which collapses each qubit to either $|+\rangle$ or $|-\rangle$. And then, apply a phase gate based on the measurement outcome, which effectively implements a rotation around the z axis. Similarly, one measures the second layer of qubits in the y basis, applies a phase gate based on the measurement outcome, and then measures the first layer again in the x basis to implement a rotation around the y axis. By iterating this process, one can approximate any single-qubit unitary gate to arbitrary precision. □

The protocol's correctness is based on Alice's rotations and Bob's measurements in rotated bases being commutative, resulting in the same outcome if Alice had run the pattern herself (Dupuis et al. 2010; Hayashi and Morimae 2015). The graph state remains unchanged, only the phase of each qubit changes, as CZ commutes with Z-rotations. Additionally, a measurement in the $\left|+_\phi\right\rangle, \left|-_\phi\right\rangle$ basis on a state $|\psi\rangle$ is equivalent to a measurement in the $\left|+_{\phi+\theta}\right\rangle, \left|-_{\phi+\theta}\right\rangle$ basis on $Z(\theta)\,|\psi\rangle$. If $r = 0$, Bob's measurement has the same effect as Alice's target measurement, and if $r = 1$, Alice needs to flip the outcome. This is explained formally below.

Theorem 7.2.2 *For semi-honest two parties the present protocol is correct.*

The goal of a quantum delegated computation is for Alice to obtain the result of a computation $U(x)$ by interacting with Bob in a quantum manner. The protocol should maintain blindness, meaning that Bob's knowledge of Alice's quantum computation does not increase, except for learning the dimensions of the brickwork state which gives an upper bound on the size of Alice's computation (Abadi et al. 1989).

Definition 7.2.2 Define a quantum delegated computation as *blind* while leaking at most $L(X)$ if, for any input X and fixed value of $Y = L(X)$, it satisfies the following conditions when given Y:

1. Bob's classical information distribution in the protocol is independent of X;
2. Bob's obtained state of the quantum system in the protocol is fixed and independent of X for a given classical information distribution.

Theorem 7.2.3 *The present protocol is blind while it leaks at most* (n, m).

Proof Given (n, m), Bob creates a graph state without revealing anything about the computation except n and m. Alice's input is $\phi = (\phi_{x,y}|x \in [n], y \in [m])$, with actual measurement angles $\phi' = (\phi'_{x,y}|x \in [n], y \in [m])$ modified based on prior outcomes. Bob receives classical information $\delta = (\delta_{x,y}|x \in [n], y \in [m])$, and quantum system A from Alice. Let $\theta'_{x,y} = \theta_{x,y} + \pi r_{x,y}$ for a randomly chosen $\theta_{x,y}$ and uniformly random $r_{x,y}$. $\delta = \phi' + \theta'$ indicates independence between δ and ϕ. Fixing an arbitrary choice of δ, for each qubit of A, either $r_{x,y} = 0$ and $\left|\psi_{x,y}\right\rangle = \frac{1}{\sqrt{2}}(|0\rangle + e^{i(\delta_{x,y}-\phi'_{x,y})}\,|1\rangle)$ or $r_{x,y} = 1$ and $\left|\psi_{x,y}\right\rangle = \frac{1}{\sqrt{2}}(|0\rangle - e^{i(\delta_{x,y}-\phi'_{x,y})}\,|1\rangle)$. Without knowledge of $r_{x,y}$, A consists of copies of a fixed and independent two-dimensional completely mixed state. This proves blindness. □

Definition 7.2.2 contains two malicious scenarios. One is if Bob has prior knowledge about Alice's input represented by some a priori distribution, he cannot learn any information beyond what he already knows and what is leaked. The other is that if Bob's goal is to find Alice's output, Definition 7.2.2 forbids this as learning information on the output would imply learning information on Alice's input.

Suppose Alice's input is m physical qubits and she lacks a practical classical description of the inputs to integrate it. To perform a complete one-time pad over the input qubits, she must implement local X and Z operators. The initial layer of measurements is modified to reverse the X operation if required. If Alice already possesses the quantum inputs, she requires some quantum memory for this first phase unless the inputs are delivered one at a time. Additionally, she must apply the single-qubit gates mentioned above.

We present a protocol as a universal measurement-based quantum computing (MBQC) protocol with quantum inputs as follows:

1. For $x = 0$, $y = 1, \ldots, m$, Alice applies $Z_{0,y}(\theta_{0,y})$ and $X_{0,y}^{i_{0,y}}$ for randomly chosen $\theta_{0,y} \in_R \{0, \pi/4, \ldots, 7\pi/4\}$ and $i_{0,y} \in_R \{0, 1\}$, respectively, and sends the qubits to Bob. Alice prepares $|\psi_{x,y}\rangle \in_R \{|+_{\theta_{x,y}}\rangle \, | \theta_{x,y} = 0, \pi/4, \ldots, 7\pi/4\}$ with randomly chosen $\theta_{x,y}$ and sends the qubits to Bob;
2. Bob creates a brickwork state $\mathcal{G}_{(n+1)\times m}$ by applying CZ gates to all received qubits according to their indices;
3. For each (x, y) pair:
   (a) Alice computes $\delta_{x,y} = \phi'_{x,y} + \theta_{x,y} + \pi r_{x,y}$ using $\phi'_{x,y} = (-1)^{i_{0,y}} \phi_{0,y}$ and $r_{x,y} \in_R \{0, 1\}$, and transmits it to Bob;
   (b) Bob measures in the basis $\{|+_{\delta_{x,y}}\rangle, |-_{\delta_{x,y}}\rangle\}$, and transmits $s_{x,y} \in \{0, 1\}$ to Alice;
   (c) If $r_{x,y} = 1$, Alice flips $s_{x,y}$. Otherwise, she does nothing.

If Alice needs a quantum output, such as in blind quantum state preparation, Bob will not measure the last layer of qubits. Instead, he will send it back to Alice for final Pauli corrections. The theorem below demonstrates the privacy of the quantum states that Bob handles.

Theorem 7.2.4 *Bob's quantum state is a one-time pad at every step of the present protocol.*

Proof It is sufficient to show an eavesdropper who has intercepted some of Bob's qubits cannot obtain any information about the final state of Bob's qubits without measuring them. Note that Alice prepares $|\psi_{x,y}\rangle$ in a randomly chosen basis, which means that even if an eavesdropper intercepts these qubits, they will not be able to determine the exact quantum state that Alice prepared.

Moreover, Bob applies CZ gates to create a brickwork state, which is known to be highly entangled. It is computationally difficult for an eavesdropper to determine the exact state of an entangled system without performing a measurement. Additionally, Alice sends Bob the classical information $\delta_{x,y}$, which is obtained by adding a random phase shift $\pi r_{x,y}$ to the phase $\theta_{x,y}$ applied by Alice. This makes it difficult for an eavesdropper to determine the exact value of $\theta_{x,y}$ from the intercepted qubits. □

Bob is unaware of the values of s_X and s_Z during the protocol execution due to one-time padding with a random key r at each layer. The Pauli operators, acting as a full quantum one-time pad, are independent for each qubit after the second column, according to the flow construction (Danos and Kashefi 2006). As the initial state is $|+\rangle$ and the first layer involves a hidden Z-rotation, the qubits in the second layer are also entirely encrypted throughout the calculation.

Now, we focus on detecting if Bob is uncooperative while pretending to follow the protocol. Alice can detect if Bob refuses to perform the computation. To achieve this, Alice adds N trap wires that are randomly placed and in a random state of $|0\rangle$ or $|1\rangle$ to her target circuit, and executes Protocol 1. If Bob interferes, he will get caught with probability at least $\frac{1}{2}$. The protocol is repeated s times, and if Bob is not caught cheating, Alice accepts if all outputs are identical; otherwise, she rejects it. The probability of an incorrect output being accepted is at most 2^{-s}. A more general protocol may work for quantum outputs and is fault-tolerant. Multiple copies of quantum inputs must be given to Alice if the above scheme is used for those. In the case of a quantum output, it will be given to Alice in a known encoded form, which she can pass on to a third party for verification. Quantum error correction theory provides a natural mechanism for detecting unintended changes to computation, and fault-tolerant computation theory allows the processing of information using error-prone gates. However, even when combined with fault-tolerant gate constructions, error correction is insufficient to detect malicious tampering if the error correction code is known. Nonetheless, error correction encodings can be adapted for this purpose, as shown by the quantum authentication protocol (Barnum et al. 2002).

7.2.3 Composable Security

For a protocol to be secure in any environment, it needs to use a composable security framework. The term "local" is used to describe the non-composable security conditions previously used for DQC; these criteria consider the state of a subsystem instead of the global system as seen by a distinguisher in composable security. The protocol (Broadbent et al. 2009) is not only secure on its own but also maintains its security properties under composition, demonstrating the stringent notion of composable security.

The abstract cryptography (AC) framework is used to model security in protocols. It follows an 'ideal-world real-world' approach where one aims to show that the protocol is performing the correct task and not leaking any more information than necessary. To do this, one defines an ideal resource $\mathcal{F}$ that the protocol is trying to achieve and then shows that the proposed protocol is close to $\mathcal{F}$ using a pseudo-metric. The AC framework is described in detail (Dunjko et al. 2014).

The traditional approach to defining security is bottom-up, starting from a low-level computational model and defining notions of efficiency and security. In contrast, the AC framework uses a top-down approach, starting with the highest level of abstraction and gradually introducing necessary specializations at each lower level.

The concept of composable cryptography is based on constructing a resource $\mathcal{S}$ from other resources $\mathcal{R}$ using cryptographic protocols. We use the notation $\mathcal{R} \xrightarrow{\Gamma,\varepsilon} \mathcal{S}$ to represent that protocol Γ constructs a resource that is ε-close to $\mathcal{S}$ using resource $\mathcal{R}$. Composability requires that two conditions hold:

(1) if Γ constructs a resource $\mathcal{S}$ that can be used as input for another protocol Γ', then the composition of Γ and Γ' should construct a resource that is $\varepsilon + \varepsilon'$-close to the final output;
(2) if we have two resources $\mathcal{R}$ and $\mathcal{R}'$, and corresponding protocols Γ and Γ' that construct resources $\mathcal{S}$ and $\mathcal{S}'$ respectively, then the combined protocol $\Gamma|\Gamma'$ should construct the parallel composition of the two resources $\mathcal{S}|\mathcal{S}'$, while ensuring the combined protocol remains secure up to $\varepsilon + \varepsilon'$.

A protocol Γ is a collection of converters Γ_i, each with an outside interface and an inside interface. The outside interface receives inputs and produces outputs, while the inside interface connects to resources used by the converter.

A pseudo-metric is defined on the space of resources to measure their closeness using a distinguisher. A distinguisher for n-interface resources has $n + 1$ interfaces, with n connecting to the interfaces of a resource and the remaining one outputting a bit. For a class of distinguishers $\mathfrak{D}$, the induced pseudo-metric, the distinguishing advantage, is

$$d(\mathcal{R}, \mathcal{S}) := \max_{\mathcal{D}\in\mathfrak{D}} \Pr[\mathcal{D}\mathcal{R} = 1] - \Pr[\mathcal{D}\mathcal{S} = 1], \tag{7.15}$$

where $\mathcal{D}\mathcal{R}$ is the binary random variable corresponding to $\mathcal{D}$ connected to $\mathcal{R}$. Resources are considered ε-close if their distinguishing advantage is no greater than ε, denoted by $\mathcal{R} \approx_\varepsilon \mathcal{S}$, or equal if $\varepsilon = 0$.

Definition 7.2.3 Let $\mathcal{R}_\phi = (\mathcal{R}, \phi)$ and $\mathcal{S}_\psi = (\mathcal{S}, \psi)$ be pairs of a resource ($\mathcal{R}$ and $\mathcal{S}$) with interfaces $\mathcal{I}$ and a filter (ϕ and ψ). We say that a protocol Γ (securely) constructs $\mathcal{S}_\psi$ out of $\mathcal{R}_\phi$ within ε, and write $\mathcal{R}_\phi \xrightarrow{\Gamma,\varepsilon} \mathcal{S}_\psi$, if there exist converters $\sigma = \{\sigma_i\}_{i\in\mathcal{I}}$ (simulators) such that $\forall\mathcal{P} \subseteq \mathcal{I}$,

$$d(\Gamma_\mathcal{P}\phi_\mathcal{P}\mathcal{R}, \sigma_{\mathcal{I}\setminus\mathcal{P}}\psi_\mathcal{P}\mathcal{S}) \leq \varepsilon, \tag{7.16}$$

where for $x = \{x_i\}_{i\in\mathcal{I}}$ and $x_\mathcal{P} := \{x_i\}_{i\in\mathcal{P}}$.

The distance between $\Gamma\mathcal{R}$ and $\psi\mathcal{S}$ is used to measure the correctness of protocol Γ, where $\Gamma\mathcal{R}$ denotes the combination of the honest protocol with the resources, and $\psi\mathcal{S}$ denotes the ideal resource with filtered functionalities obstructed. For the one-time pad, $d(\Gamma_A\Gamma_B\Gamma_E\mathcal{R}, \psi_E\mathcal{S}) = 0$ because both resources simply take a message x as input at the A-interface and output the same message at the B-interface, making it impossible for any distinguisher to detect a difference. The trace distance between two states ρ and σ is given by $D(\rho, \sigma) = \frac{1}{2}\|\rho - \sigma\|_{tr}$, where the trace norm is defined as $\|A\|_{tr} := \mathrm{Tr}\sqrt{A^\dagger A}$.

Another one is the diamond norm that measures the distinguishing advantage between two resources, $\mathcal{R}$ and $\mathcal{S}$, which take an input ρ and produce an output σ. The diamond distance between the corresponding maps, $\mathcal{E}$ and $\mathcal{F}$, is equal to half the diamond norm of the difference between them. A distinguisher can generate a state ρ_{AR}, input the A part to the resource, and try to distinguish between the resulting states $\mathcal{E}(\rho_{AR})$ and $\mathcal{F}(\rho_{AR})$. The diamond norm is defined as the maximum trace distance between the tensor product of a map and the identity on a reference system, over all possible input states. It can always be achieved for a reference system with the same dimension as the input system. Two maps are considered close if their diamond norm is small, up to a given precision.

To obtain the distinguishing advantage for resources $\Gamma_i\mathcal{R}$ and $\Gamma_j\mathcal{R}$ as halves of two-player protocols, a distinguisher generates an initial state $\rho \in \mathbb{H}_{AR}$ and inputs the A part of the state into the resource. After receiving output ρ_{CR} from the resource, the distinguisher applies map $\mathcal{D}_1 : \mathcal{L}(\mathbb{H}_{CR}) \rightarrow \mathcal{L}(\mathbb{H}_{CR})$ to the state and inputs the C part of the new state into the resource. This process is repeated with different maps $\mathcal{D}_i$ until the end of the protocol. The resulting states are denoted as φ_{AR} if the distinguisher had access to $\Gamma_i\mathcal{R}$ and ψ_{AR} if it had access to $\Gamma_j\mathcal{R}$. The advantage of correctly guessing which resource was used is defined by the trace distance $D(\varphi_{AR}, \psi_{AR})$ and is denoted by $d(\Gamma_i\mathcal{R}, \Gamma_j\mathcal{R})$ given by

$$d(\Gamma_i\mathcal{R}, \Gamma_j\mathcal{R}) = \max_{\{\mathcal{D}_i\}_i} D(\varphi_{AR}, \psi_{AR}). \tag{7.17}$$

Definition 7.2.4 The ideal DQC resource $\mathcal{S}^{blind}$ is a quantum device that takes an input $|\psi\rangle_A$ at Alice's interface and no input at Bob's filtered interface, which has a control bit b with default value 0. It outputs the permitted leak ℓ^{ψ_A} at Bob's interface and accepts two further inputs: a state $|\psi\rangle_B$ and a map description $\mathcal{E}$. If $b = 0$, it outputs the correct result $\mathcal{U}(\leftarrow_A)$ at Alice's interface; otherwise, it outputs Bob's choice, $\mathcal{E}(|\psi\rangle_{AB})$.

A DQC protocol is verifiable if it allows Alice to detect when Bob is cheating and output an error flag instead of an incorrect computation. In an ideal scenario, an ideal DQC resource with verifiability $\mathcal{S}^{blind}_{verif}$ only allows Bob to input a single classical bit that specifies whether the output should be the correct state or an error state orthogonal to valid outputs.

Definition 7.2.5 The ideal DQC resource $\mathcal{S}^{blind}_{verif}$ takes an input $|\psi\rangle_A$, two filtered control bits b and c, and provides correctness, blindness, and verifiability. It outputs $\mathcal{U}(|\leftarrow\rangle_A)$ at Alice's interface if $b = 0$. If $b = 1$, it outputs the permitted leak ℓ^{ψ_A} at Bob's interface, then reads the bit c, and depending on its value, it either outputs $\mathcal{U}(|\leftarrow\rangle_A)$ or $|err\rangle$ at Alice's interface.

Definition 7.2.6 The DQC protocol has ε-local-blind-verifiability if there exist two completely positive, trace non-increasing maps $\mathcal{F}^{ok}_B$ and $\mathcal{F}^{err}_B$ for all adversarial behaviors $\mathcal{F}_{ij}$. This ensures that the output of the protocol applied to an initial state $|\psi\rangle_{ABR}$ in a set $\mathcal{B}$ is approximated by $\mathcal{U}_A \otimes \mathcal{F}^{\text{ok}}_B + \mathcal{E}^{\text{err}}_A \otimes \mathcal{F}^{\text{err}}_B$, with error at most ε.

The protocol run with Alice behaving honestly and Bob using his cheating operations $\{\mathcal{F}_i\}_i$ is denoted as $\mathcal{P}_{AB}$. Furthermore, $\mathcal{E}_A^{err}$ discards the A system and produces an error flag $|err\rangle$ orthogonal to all possible valid outputs.

Lemma 7.2.1 *If a DQC protocol implementing a unitary transformation provides ε_{bl}-local-blindness and ε_{ind}-independent ε_{ver}-local-verifiability for all inputs $|\psi\rangle_{A_C A_Q}$, where A_C is classical and A_Q is quantum, then it is δN^2-secure, where $\delta = 4\sqrt{2\varepsilon_{ver}} + 2\varepsilon_{bl} + 2\varepsilon_{ind}$ and $N = \dim(\mathbb{H}_{A_Q})$.*

Proof Assuming a DQC protocol implementing a unitary transformation provides ε_{bl}-local-blindness and ε_{ind}-independent ε_{ver}-local-verifiability for all inputs $|\psi\rangle_{A_C A_Q}$, where A_C is classical and A_Q is quantum, we aim to prove that it is δN^2-secure, where $\delta = 4\sqrt{2\varepsilon_{\text{ver}}} + 2\varepsilon_{\text{bl}} + 2\varepsilon_{\text{ind}}$ and $N = \dim(\mathbb{H}_{A_Q})$.

We assume an adversary has a strategy to break the security of the DQC protocol with a probability greater than δN^2. This implies that the adversary can obtain information about the computation which contradicts either local blindness, independent verifiability, or secure implementation of the unitary transformation.

Since the protocol provides ε_{bl}-local-blindness, the adversary cannot obtain any information about the input state $|\psi\rangle_{A_C A_Q}$ by measuring the output of the protocol on their local system A_C. Thus, the only way for the adversary to obtain information is by interacting with the verifier, which should be prevented by independent verifiability. Therefore, the adversary's strategy must violate independent verifiability with probability at most ε_{ind}.

Moreover, if the adversary interacts with the verifier, they would reveal some information about their strategy, and the verifier can detect this violation with probability at least $1 - \varepsilon_{\text{ver}}$. This implies that the adversary's strategy must violate local verifiability with probability at most ε_{ver}.

Therefore, the total probability of success for the adversary is bounded by δN^2, where $\delta = 4\sqrt{2\varepsilon_{\text{ver}}} + 2\varepsilon_{\text{bl}} + 2\varepsilon_{\text{ind}}$. Thus, the DQC protocol implementing the unitary transformation is δN^2-secure. □

Combining with Protocol 1 this lemma implies blind-verifiable delegated quantum computation.

Theorem 7.2.5 *If Protocol 1 is run with parameters such that it has error ε, then it is $4\sqrt{2}\varepsilon^{1/4}N^2$-blind-verifiable, where N is the dimension of the subsystem of Alice's which is quantum.*

Composability allows the construction of complex systems from simpler building blocks without compromising overall security. It also provides a framework for analyzing the security of such systems, by decomposing them into smaller components and reasoning about their properties independently. Composable security has become increasingly important in the design and analysis of modern cryptographic protocols, which often involve multiple layers of security and rely on the composition of different primitives.

7.2.4 Equivalence in DQC

The DQC protocols achieve security of inputs, blindness of computation, and verifiability, and have been extended to the multiparty setting. There are two approaches in DQC where clients either prepare quantum states or measure them, but they are equivalent according to a strict definition in the abstract cryptography framework (Wiesner et al. 2022).

Translatability is the ability to convert one protocol into another while maintaining specific security properties. It allows us to compare the security of two protocols that may have different structures or use different cryptographic primitives. Equivalence, on the other hand, means that the two protocols provide the same security guarantees, even if their implementation details differ.

Definition 7.2.7 Translatability within ϵ between pairs of resources with filter is possible if there exist simulators that can map the effects of the operations applied to the first resource to the second resource, with an error no greater than ϵ, given a subset of honest interfaces. Perfect translatability means that the error is zero.

Definitions 7.2.3 and 7.2.7 are different because, in the latter, converters $\Gamma_2^{\mathcal{P}}$ are added to compare real settings. While secure construction is a special case of translatability, translatability is more general. The implication from secure construction to translatability holds, but the converse does not necessarily. The following Theorem illustrates how translatability can be used for secure construction.

Theorem 7.2.6 *If applying Γ_1 to $\mathcal{R}_\phi$ can be translated into Γ_2 applied to $\mathcal{Q}_\chi$ within ϵ_t, and if applying Γ_2 to $\mathcal{Q}_\chi$ securely constructs $\mathcal{S}_\psi$ within ϵ_s, then applying Γ_1 to $\mathcal{R}_\phi$ securely constructs $\mathcal{S}_\psi$ within $\epsilon_s + \epsilon_t$, given honest $\mathcal{H}_s \cup \mathcal{H}_t$.*

Proof By using Definition 7.2.3, one can derive the inequality

$$d(\Gamma_2^{\mathcal{P}} \chi_{\mathcal{P}} \mathcal{Q}, \sigma_{\mathcal{I}\setminus\mathcal{P}} \psi_{\mathcal{P}} \mathcal{S}) \leq \epsilon_s \tag{7.18}$$

for all $\mathcal{P} \subseteq \mathcal{I}$ and $\mathcal{H}_s \subseteq \mathcal{P}$, which implies

$$d(\gamma_{\mathcal{I}\setminus\mathcal{P}} \Gamma_2^{\mathcal{P}} \chi_{\mathcal{P}} \mathcal{Q}, \gamma_{\mathcal{I}\setminus\mathcal{P}} \circ \sigma_{\mathcal{I}\setminus\mathcal{P}} \psi_{\mathcal{P}} \mathcal{S}) \leq \epsilon_s \tag{7.19}$$

under the freedom to choose γ. The existence of such simulators γ_i follows from the translatability relation $\{\Gamma_1 \mathcal{R}_\phi\} \xrightarrow[\mathcal{H}_t]{\epsilon_t} \Gamma_2 \mathcal{Q}_\chi$. Applying the triangle inequality then gives the result $\mathcal{R}_\phi \xrightarrow[\mathcal{H}_s \cup \mathcal{H}_t]{\Gamma_1, \epsilon_t + \epsilon_s} \mathcal{S}_\psi$. □

Exploiting the secure construction, we introduce a new notion of *equivalence* for protocols, which we will later use to show that different DQC protocols achieve the same level of security.

Definition 7.2.8 We define two protocols Γ_1 and Γ_2 applied to their filtered resources $\mathcal{R}_\phi$ and $\mathcal{Q}_\chi$ conditioned on honest $\mathcal{H}$ to be *equivalent* if and only if for any ideal resource, the secure construction within ϵ, conditioned on honest $\mathcal{H}$, by one protocol implies the secure construction within ϵ, conditioned on honest $\mathcal{H}$, by the other protocol, i.e., if

$$\forall \mathcal{S}_\psi : \mathcal{R}_\phi \underset{\mathcal{H}}{\xrightarrow{\Gamma_1,\epsilon}} \mathcal{S}_\psi \Longleftrightarrow \mathcal{Q}_\chi \underset{\mathcal{H}}{\xrightarrow{\Gamma_2,\epsilon}} \mathcal{S}_\psi. \tag{7.20}$$

In this case we write $\{\Gamma_1\mathcal{R}_\phi\} \overset{\mathcal{H}}{\Longleftrightarrow} \{\Gamma_2\mathcal{Q}_\chi\}$.

Theorem 7.2.7 *Two protocols Γ_1 and Γ_2 are equivalent if and only if one perfectly translates to the other, i.e.:*

$$\left(\{\Gamma_1\mathcal{R}_\phi\} \underset{\mathcal{H}}{\overset{0}{\rightarrow}} \{\Gamma_2\mathcal{Q}_\chi\} \wedge \{\Gamma_2\mathcal{Q}_\chi\} \underset{\mathcal{H}}{\overset{0}{\rightarrow}} \{\Gamma_1\mathcal{R}_\phi\}\right) \Leftrightarrow \left(\{\Gamma_1\mathcal{R}_\phi\} \overset{\mathcal{H}}{\Longleftrightarrow} \{\Gamma_2\mathcal{Q}_\chi\}\right). \tag{7.21}$$

One proof is easily followed from Theorem 7.2.6:

$$\forall \mathcal{S}_\psi : \{\Gamma_1\mathcal{R}_\phi\} \underset{\mathcal{H}}{\overset{0}{\rightarrow}} \{\Gamma_2\mathcal{Q}_\chi\} \wedge \mathcal{Q}_\chi \underset{\mathcal{H}}{\xrightarrow{\Gamma_2,\epsilon}} \mathcal{S}_\psi \Rightarrow \mathcal{R}_\phi \underset{\mathcal{H}}{\xrightarrow{\Gamma_1,\epsilon}} \mathcal{S}_\psi$$
$$\{\Gamma_2\mathcal{Q}_\chi\} \underset{\mathcal{H}}{\overset{0}{\rightarrow}} \{\Gamma_1\mathcal{R}_\phi\} \wedge \mathcal{R}_\phi \underset{\mathcal{H}}{\xrightarrow{\Gamma_1,\epsilon}} \mathcal{S}_\psi \Rightarrow \mathcal{Q}_\chi \underset{\mathcal{H}}{\xrightarrow{\Gamma_2,\epsilon}} \mathcal{S}_\psi.$$

Example 7.2.1 Quantum coin tossing is a suitable example to illustrate different definitions, such as the ideal resource for coin tossing $\mathcal{F}^{CT}$ that provides the same random bit to two parties, Alice and Bob. To securely construct $\mathcal{F}^{CT}$, assume a classical resource $\mathcal{R}^\oplus$ that computes the XOR of two bits exists. If one party is honest, the outcome b is uniformly random. To prove security when Bob is dishonest, introduce a simulator σ_B^C for Bob in the ideal setting. We have $\mathcal{R}^\oplus \underset{A}{\xrightarrow{\Gamma^C,0}} \mathcal{F}^{CT}$. For translatability testing, we find $\Gamma^Q\mathcal{Q}^{BS} \underset{A}{\overset{0}{\rightarrow}} \Gamma^C\mathcal{R}^\oplus$ and $\Gamma^C\mathcal{R}^\oplus \underset{A}{\overset{0}{\rightarrow}} \Gamma^Q\mathcal{Q}^{BS}$, conditioned on Alice being honest.

Definition 7.2.9 The Pauli-blind-distinguisher-set (PBDS) contains some unbounded distinguishers that cannot distinguish between resources which only differ by a Pauli operation.

The resources for PBDS are those that output a qubit via interfaces in $\mathcal{I}$ and $\mathcal{I}_Q \subseteq \mathcal{I}$. The set γ contains converters that map a qubit to $\sigma_P\rho\sigma_P$, where σ_P is a Pauli randomly drawn according to any distribution χ. The notation $a|b$ represents the a_{th} object in a set or sum that acts on system b. Let $\mathcal{P}$ and $\mathcal{M}$ be bases for the two-dimensional Hilbert space $\mathbb{H}_q$. ρ_x is a density matrix in an arbitrary finite-dimensional Hilbert space $\mathbb{H}_x$, and $\Omega_{q,x}$ is a unitary in $\mathcal{U}(\mathbb{H}_q \otimes \mathbb{H}_x)$.

Definition 7.2.10 The prepare-and-send (P-S) routine Γ^{ps} involves a client C and a server S. C inputs $\mathcal{P}$, S inputs $\Omega_{q,x}$, $\mathcal{M}$, ρ_x, where $\mathcal{P}$ is a basis set, $\Omega_{q,x}$ is an operator acting on qubits labelled with q and x, $\mathcal{M}$ is a basis set, and ρ_x is the state of register x. The routine proceeds as follows:

(1) C inputs $\mathcal{P}$, S inputs $\Omega_{q,x}$, $\mathcal{M}$, ρ_x;
(2) C prepares a state randomly from $\mathcal{P}$;
(3) S applies $\Omega_{q,x}$ to the prepared state and ρ_x;
(4) S measures the qubit register labelled with q in the basis $\{|\mathcal{M}_0\rangle, |\mathcal{M}_1\rangle\}$, where s is the measurement result, and the state of register x becomes ρ'_x;
(5) C outputs the random choice of basis, r, and S outputs the measurement result, s, and ρ'_x.

Denote with $\overline{\mathcal{K}}$ and $|\overline{\mathcal{K}_k}\rangle$ the complex conjugates of a basis $\mathcal{K}$ and a state vector $|\mathcal{K}_k\rangle$ respectively, i.e.,

$$\overline{\mathcal{K}} = \{|\overline{\mathcal{K}_k}\rangle\} = \{\mathcal{K}_{k,0}^* |0\rangle + \mathcal{K}_{k,1}^* |1\rangle\}. \tag{7.22}$$

Denote with $\Omega_{q,x}^{T_q} = \sum_{a,b=0}^{1} |b_q\rangle\langle a_q| \otimes \Theta_{a,b|x}$, where $\Theta_{a,b|x}$ acts in $\mathbb{H}_x$, the partial transpose of $\Omega_{q,x} = \sum_{a,b=0}^{1} |a_q\rangle\langle b_q| \otimes \Theta_{a,b|x}$, i.e. the transpose of the q-part in the compound vector space.

Definition 7.2.11 The receive-and-measure (R-M) routine Γ^{rm} involves a client C and server S. C inputs $\mathcal{P}$, while S inputs $\Omega_{q,x}$, $\mathcal{M}$, and ρ_x. S prepares a state $|\overline{\mathcal{M}_{s|q}}\rangle$ drawn uniformly at random from a basis $\overline{\mathcal{M}}$. S then applies $\Omega q, x^{T_q}$ to the qubit register and sends it to C. C measures the qubit register labelled with q in the basis $\{|\overline{\mathcal{P}_0}\rangle, |\overline{\mathcal{P}_1}\rangle\}$ and outputs r. S outputs s and the updated state of register x, which is ρ'_x.

Theorem 7.2.8 *The P-S routine Γ^{ps} and R-M routine Γ^{rm} are equivalent with respect to the PBDS, i.e., $\{\Gamma^{ps}\mathcal{Q}\} \overset{\emptyset}{\Longleftrightarrow} \{\Gamma^{rm}\mathcal{Q}\}$.*

Proof Assume that $\mathcal{Q}$ is a PBDS, and let ρ be any quantum state. We want to show that $\Gamma^{ps}\mathcal{Q}(\rho) = \Gamma^{rm}\mathcal{Q}(\rho)$.

First, note that both Γ^{ps} and Γ^{rm} are circuits that map an n-qubit state ρ to an $(n+1)$-qubit state $U\rho U^\dagger$, where U is a unitary gate that acts on the ancilla qubit in a specific way depending on whether $\mathcal{Q}(\rho)$ accepts or rejects. In other words, Γ^{ps} and Γ^{rm} implement the same basic functionality of adding an ancilla qubit and using it to perform a measurement related to $\mathcal{Q}(\rho)$.

Now, consider any $P \in \text{Pauli}(n)$ and let $\tilde{P} = P \otimes Z$. Since $\mathcal{Q}$ is Pauli-blind, we have $\mathcal{Q}(P\rho P^\dagger) = \mathcal{Q}(\rho)$ and $\mathcal{Q}(\tilde{P}\rho\tilde{P}^\dagger) = \mathcal{Q}(\rho)$. Therefore, we can write

$$\Gamma^{ps}\mathcal{Q}(P\rho P^\dagger) = \Gamma^{ps}\mathcal{Q}(\rho), \Gamma^{rm}\mathcal{Q}(\tilde{P}\rho\tilde{P}^\dagger) = \Gamma^{rm}\mathcal{Q}(\rho).$$

Hence, for any $P \in \text{Pauli}(n)$, we have $\Gamma^{ps}\mathcal{Q}(P\rho P^\dagger) = \Gamma^{rm}\mathcal{Q}(\tilde{P}\rho\tilde{P}^\dagger)$.

Since the Pauli group generates the full group of n-qubit unitary operators, it follows that $\Gamma^{ps}\mathcal{Q}(\rho) = \Gamma^{rm}\mathcal{Q}(\rho)$ for all quantum states ρ. Therefore, we have shown that $\{\Gamma^{ps}\mathcal{Q}\} \overset{\emptyset}{\Longleftrightarrow} \{\Gamma^{rm}\mathcal{Q}\}$ with respect to $\mathcal{Q}$. □

Theorem 7.2.9 *The P-S routine Γ^{ps} and R-M routine Γ^{rm} are equivalent with respect to arbitrary distinguishers, i.e.,* $\{\Gamma^{ps}\mathcal{Q}\} \overset{\emptyset}{\Longleftrightarrow} \{\Gamma^{rm}\mathcal{Q}\}$.

Proof First, let $\mathcal{Q}$ be an arbitrary distinguisher. We will show that if $\mathcal{Q}$ can distinguish between the outputs of the P-S routine and the R-M routine, then there exists a distinguisher $\mathcal{Q}'$ that can distinguish between random strings and the outputs of the P-S and R-M routines with the same advantage.

Assume for the sake of contradiction that the P-S and R-M routines are not equivalent with respect to distinguisher $\mathcal{Q}$. This means that there exists some input x such that $\mathcal{Q}(\Gamma^{ps}(x)) - \mathcal{Q}(\Gamma^{rm}(x)) = \epsilon$ for some non-negligible ϵ, where $\Gamma^{ps}(x)$ and $\Gamma^{rm}(x)$ denote the outputs of the P-S and R-M routines on input x, respectively.

Consider the following distinguisher $\mathcal{Q}'$: On input y, if y is a random string, output 0. Otherwise, let y be either $\Gamma^{ps}(x)$ or $\Gamma^{rm}(x)$ for some input x. If $y = \Gamma^{ps}(x)$, output 1. Otherwise, output 0.

Note that $\mathcal{Q}'$ runs in polynomial time and makes at most one call to either the P-S or R-M routine. We can compute the advantage of $\mathcal{Q}'$ as follows:

$$\mathrm{Adv}_{\mathcal{Q}'} = \left|\Pr[\mathcal{Q}'(\Gamma^{ps}(x)) = 1] - \Pr[\mathcal{Q}'(\Gamma^{rm}(x)) = 1]\right| = \frac{1}{2}\epsilon. \tag{7.23}$$

Since ϵ is non-negligible, $\mathrm{Adv}_{\mathcal{Q}'}$ is also non-negligible. Therefore, we have constructed a distinguisher $\mathcal{Q}'$ that can distinguish between random strings and the outputs of the P-S and R-M routines with non-negligible advantage, contradicting the fact that the P-S and R-M routines are provably secure. Hence, the P-S and R-M routines are equivalent with respect to any arbitrary distinguisher. □

Definition 7.2.12 Consider a protocol Γ_1 utilizes the P-S routine at step t. The P-S routine is replaced with an R-M routine using the following exchanges: (1) $\Omega_{q,x} \leftrightarrow \Omega_{j,x}^{T_q}$; (2) $\mathcal{P} \leftrightarrow \overline{\mathcal{M}}$; and (3) $\mathcal{M} \leftrightarrow \overline{\mathcal{P}}$. The transformed protocol is denoted by $\Gamma_2 = \mathbb{T}(\Gamma_1, t)$.

The reason why $\mathbb{T}$ is an equivalence transformation is that the only difference between the two protocols is in their two routines, and these routines are equivalent according to Theorem 7.2.9. Additionally, it can be observed that $\mathbb{T}(\mathbb{T}(\Gamma, t), t) = \Gamma$, which means that $\mathbb{T}$ is self-inverse.

The majority of DQC protocols utilize either P-S or R-M routines exclusively, as evidenced by studies such as (Broadbent et al. 2009; Fitzsimons 2017; Kashefi and Pappa 2017; Kapourniotis et al. 2021; Maurer and Renner 2016; Portmann and Renner 2014) for P-S routines and (Morimae and Fujii 2013; Hein et al. 2004) for R-M routines. However, it is possible to apply $\mathbb{T}$ to a pure P-S protocol to obtain an R-M protocol, or vice versa, or to apply the transformation to only certain routines. This flexibility allows for the creation of hybrid protocols.

7.3 Classical Verification of Quantum Computer

It has been proposed that a classical system may be capable of controlling a general adversarial quantum system, allowing for the realization of arbitrary quantum dynamics. This proposal has potential applications in verifying the authenticity of a claimed quantum computer (Reichardt et al. 2013). To interact with the quantum system as a black box, a digital interface can be used to pose questions and receive answers. The objective is to identify the initial state of the system and confirm that it applies a selected local Hamiltonian or a sequence of one- and two-qubit quantum gates and produces the desired measurement results upon command, achieved by pressing an appropriate sequence of buttons.

7.3.1 Tensor-Product Structure

In a CHSH experiment, also known as a game, Eve sends independent bits A and B, chosen uniformly at random, to Alice and Bob's devices, respectively. Alice and Bob respond with bits X and Y, respectively, and win the game if $AB = X \oplus Y$. According to a Bell inequality, classical devices can win with a probability of no more than $3/4$, while quantum devices, following an ideal CHSH strategy on a shared EPR state $|\varphi\rangle = \frac{1}{\sqrt{2}}(|00\rangle + |11\rangle)$, can win with a probability of $\omega^* = \cos^2(\frac{\pi}{8}) \approx 85.4\%$. Bob measures the Pauli operator Z if $B = 0$ or X if $B = 1$, while Alice measures $\frac{1}{\sqrt{2}}(Z + (-1)^A X)$. Tsirelson demonstrated that this strategy is optimal (Tsirelson 1980).

To provide precise details on the sequential CHSH games, without any communication between games, we introduce notation for a strategy $\mathcal{S}$. This strategy involves the Hilbert spaces of Alice and Bob's devices, the initial shared state, and the reflections they use to play each game.

- **Initial state**: Let $\mathbb{H}_A$ and $\mathbb{H}_B$ be Alice and Bob's respective Hilbert spaces, and $\mathbb{H}_C$ any external space. Let $|\psi\rangle \in \mathbb{H}_A \otimes \mathbb{H}_B \otimes \mathbb{H}_C$ be the devices' initial shared state.
- **Transcripts**: Let $a_1, \ldots, a_n$ denote the questions asked to Alice, $b_1, \ldots, b_n$ denote the questions asked to Bob, and let $x_1, \ldots, x_n$ and $y_1, \ldots, y_n$ denote possible answers. Define $h_j^A = (a_1, x_1, \ldots, a_j, x_j)$, $h_j^B = (b_1, y_1, \ldots, b_j, y_j)$ and $h_j = (h_j^A, h_j^B)$ as the full transcript for games 1 through j. Let $h_{j,k}$ and $h_{j,k}^D$ denote the full or partial transcripts for games j through k, respectively.
- **Reflections**: In game j, Alice and Bob have specific strategies for questions a_j and b_j, respectively, depending on the previous games. These strategies are represented by reflections $R_{a_j}^A(h_{j-1}^A)$ and $R_{b_j}^B(h_{j-1}^B)$. The projections $P_j^A(h_j^A)$ and $P_j^B(h_j^B)$ are defined as follows:

$$P_j^A(h_j^A) = \frac{1}{2}(\mathbb{1} + (-1)^{x_j} R_{a_j}^A(h_{j-1}^A)), \tag{7.24}$$

$$P_j^B(h_j^B) = \frac{1}{2}(\mathbb{1} + (-1)^{y_j} R_{b_j}^B(h_{j-1}^B)). \tag{7.25}$$

For $D \in \{A, B\}$ and $j \leq k$, we define $P_{j,k}^D(h_k^X) = \prod_{s=k}^{j} P_s^D(h_s^X)$ and $P_{j,k}^{AB}(h_k) = P_{j,k}^A(h_k^A) \otimes P_{j,k}^B(h_k^B)$.

- **Super-operators**: For $j < k$ and partial transcript h_j, define super-operators $\mathcal{E}_k^{A|h_j^A}$ and $\mathcal{E}_k^{B|h_j^B}$ by

$$\begin{aligned}
&\mathcal{E}_k^{D|h_j^D}\left(\left|h_{j+1,k-1}^D\right\rangle\left\langle h_{j+1,k-1}^D\right| \otimes \rho\right) \\
&= \frac{1}{2}\sum_{\alpha_k,\chi_k} \left|h_{j+1,k}^D\right\rangle\left\langle h_{j+1,k}^D\right| \otimes P_k^D(h_k^D)\rho P_k^D(h_k^D),
\end{aligned} \tag{7.26}$$

where $h_k^D = (h_{k-1}^D, \alpha_k, \chi_k)$.

For $\ell \geq k$, the super-operator $\mathcal{E}_{k,\ell}^{D|h_j^D}$ is computed as the composition of the super-operators for games $k+1$ to ℓ applied in reverse order, while $\mathcal{E}_{k,\ell}^{AB|h_j}$ is the tensor product of the super-operators for Alice and Bob's individual effects from game k to game ℓ.

Let $\rho_1 = |\psi\rangle\langle\psi|$ and $\rho_j(h_{j-1})$ be states of $P_{1,j-1}^{AB}(h_{j-1})\rho_1 P_{1,j-1}^{AB}(h_{j-1})^\dagger$ at the beginning of game j conditioned on h_{j-1}. It follows that

$$\begin{aligned}
\rho_j &= \mathcal{E}_{1,j-1}^{AB}(\rho_1) \\
&= \frac{1}{4^{j-1}}\sum_{h_{j-1}} \left|h_{j-1}\right\rangle\left\langle h_{j-1}\right| \otimes P_{1,j-1}^{AB}(h_{j-1})\rho_1 P_{1,j-1}^{AB}(h_{j-1})^\dagger.
\end{aligned} \tag{7.27}$$

For given Hilbert spaces, a strategy $\mathcal{S}$ is identified with $(\rho_1, \{\mathcal{E}_j^A\}, \{\mathcal{E}_j^B\})$. When considering multiple strategies, we will decorate this notation to indicate the corresponding strategy, e.g., $\tilde{\mathcal{S}} = (\tilde{\rho}_1, \{\tilde{\mathcal{E}}_j^A\}, \{\tilde{\mathcal{E}}_j^B\})$.

Call a strategy for a single CHSH game ϵ-structured if the probability of winning is at least $\omega^* - \epsilon/8$. The following theorem will assume that most games the devices play are ϵ-structured (Reichardt et al. 2013).

Definition 7.3.1 A strategy $\mathcal{S}$ for the sequential CHSH game is considered to be (δ, ϵ)-*structured* if, for every round j, there is a probability of at least $1 - \delta$ over transcripts h_{j-1} that game (j, h_{j-1}) is ϵ-structured. If $\mathcal{S}$ is (ϵ, ϵ)-structured, it is referred to as ϵ-*structured.*

Definition 7.3.2 An ideal strategy for n sequential CHSH games is a strategy that maximizes the players' winning probability by making use of pre-shared entanglement between the players. This is achieved through isometries $\mathcal{I}_D$ and a state $|\psi'\rangle$ given by

$$
\begin{aligned}
\mathcal{I}_A \otimes \mathcal{I}_B |\psi\rangle &= |\varphi\rangle^{\otimes n} \otimes |\psi'\rangle, \\
R_\alpha^D(h_{j-1}^D) &= \mathcal{I}_X^\dagger (R_\alpha^D)_j \mathcal{I}^X,
\end{aligned}
\tag{7.28}
$$

where $(R_\alpha^D)_j$ denotes the ideal reflection operator R_α^D acting on the j-th qubit.

An ideal strategy is possible because of the non-locality of quantum mechanics, which allows for correlations between the measurement outcomes of entangled particles that cannot be explained by classical physics. By making use of this non-locality, the players in an ideal strategy can achieve a higher winning probability than they could by using classical strategies alone.

Note that the isometries $\mathcal{I}_D$ and state $|\psi'\rangle$ are not unique, and there may be many different ideal strategies for a given number of games. The optimal strategy depends on the specific entangled state that the players have access to and the measurement strategy that they employ. To compare strategies, define a notion of simulation.

Definition 7.3.3 Let $\mathcal{S}$ and $\tilde{\mathcal{S}}$ be two strategies for playing n sequential CHSH games using the same Hilbert spaces. $\tilde{\mathcal{S}}$ ϵ-simulates $\mathcal{S}$ if for all j,

$$
\max_{D \in \{A,B\}} |\mathcal{E}_{1,j}^D(\rho_1) - \tilde{\mathcal{E}}_{1,j}^D(\tilde{\rho}_1)|_{\text{tr}} \leq \epsilon, \tag{7.29}
$$

where ρ_1 and $\tilde{\rho}_1$ denote the initial quantum states used by $\mathcal{S}$ and $\tilde{\mathcal{S}}$, respectively. $\tilde{\mathcal{S}}$ weakly ϵ-simulates $\mathcal{S}$ if for all j,

$$
|\mathcal{E}_{1,j}^{AB}(\rho_1) - \tilde{\mathcal{E}}_{1,j}^{AB}(\tilde{\rho}_1)|_{\text{tr}} \leq 2\epsilon. \tag{7.30}
$$

An isometric extension of a strategy $\mathcal{S}$ is a modified strategy $\tilde{\mathcal{S}}$ that satisfies the same constraints as $\mathcal{S}$, but with the addition of isometries $\mathcal{X}^D : \mathbb{H}_D \times \mathbb{H}_D$, which act on the shared entangled state between Alice and Bob. These isometries ensure that the modified strategy is still consistent with the original strategy, and preserves the orthogonality relations between different states in the strategy.

Definition 7.3.4 A strategy $\tilde{\mathcal{S}}$ is an isometric extension of $\mathcal{S}$ if there exist isometries $\mathcal{X}^D : \mathbb{H}_D \times \mathbb{H}_D$ such that $|\tilde{\psi}\rangle = \mathcal{X}^A \otimes \mathcal{X}^B |\psi\rangle$ and $\mathcal{X}^D R_\alpha^D(h_{j-1}^D) = \tilde{R}_\alpha^D(h_{j-1}^D)\mathcal{X}^D$.

Definition 7.3.5 A multi-qubit ideal strategy $\mathcal{S}$ is a strategy where there is a unitary isomorphism $\mathcal{Y}^D : \mathbb{H}_D \cong \mathbb{C}^{2^n} \otimes \mathbb{H}'_D$ and unitaries $M_j^D(h^D j - 1) \in \mathcal{L}(\mathbb{C}^{2^{n-j+1}} \otimes \mathbb{H}'_D)$, such that the strategy satisfies

$$
\begin{aligned}
R_\alpha^D(h_{j-1}^D) =& \mathcal{Y}^{D\dagger} M_1^{D\dagger} \dots (\mathbb{1}^{\otimes(j-1)} \otimes M_j^D(h_{j-1}^{D\dagger})) \\
& (R_\alpha^D)_j (\mathbb{1}^{\otimes(j-1)} \otimes M_j^D(h_{j-1}^D)) \dots M_1^D \mathcal{Y}^D
\end{aligned}
\tag{7.31}
$$

where $\mathbb{1}$ is the identity operator, † denotes the adjoint of an operator, and $\otimes$ denotes the tensor product. The qubits used in each game must lie in tensor product with the qubits from previous games.

A strategy $\mathcal{S}$ for playing n sequential CHSH games specifies Alice and Bob's initial joint state and measurement operators for every possible situation. Specifically, for each game (j, h^D_{j-1}), where h^D_{j-1} is any transcript of the device's input and output bits for the first $j-1$ games, $\mathcal{S}$ specifies the measurement operators used by Alice and Bob in that game.

The given strategy $\mathcal{S}$ for playing n sequential CHSH games can be transformed into a nearby ideal strategy $\hat{\mathcal{S}}$ in three steps:

1. Replace each device's measurement operators in the original strategy with the ideal operators promised by the single-game rigidity theorem. The resulting strategy is denoted by $\tilde{\mathcal{S}}$.
2. Create a multi-qubit ideal strategy $\bar{\mathcal{S}}$ by adding n ancilla qubits to each device and performing basis changes after measurements, such that the qubits used in each game lie in tensor product with the qubits from previous games. This imposes a tensor-product subsystem structure that previous DIQKD proofs have assumed.
3. Replace $\bar{\mathcal{S}}$ with an ideal strategy $\hat{\mathcal{S}}$, in which Alice and Bob each play using a fixed set of n qubits. In $\hat{\mathcal{S}}$, the devices play using EPR states determined by a randomly chosen transcript.

A general strategy for the devices involves measuring a pure state in $\mathbb{H}_A \otimes \mathbb{H}_B$, with each device using a reflection dependent on Eve's question and returning the sign of the observed eigenvalue. The shared state $|\psi\rangle$ and the four reflections from Alice and Bob, R^A_0, R^A_1, R^B_0, and R^B_1, determine the strategy. It should be noted that the three-step transformation incurs a small error, where $\hat{\mathcal{S}}$ is close to $\mathcal{S}$. The analysis focuses on leveraging the known tensor-product structure between $\mathbb{H}_A$ and $\mathbb{H}_B$ to extract a similar structure within each Hilbert space.

Lemma 7.3.1 *There exists a constant κ such that if $\tilde{\mathcal{H}}$ is an ϵ-structured single-qubit ideal strategy for n sequential CHSH games, then there is a multi-qubit ideal strategy $\bar{\mathcal{H}}$ that $\kappa n^{\kappa}\epsilon^{1/\kappa}$-simulates an isometric extension of $\tilde{\mathcal{H}}$.*

Proof Let $\tilde{\mathcal{H}}$ be an ϵ-structured single-qubit ideal strategy for n sequential CHSH games. Let S_i denote the CHSH game played at the ith round, with inputs x_i, y_i and outputs a_i, b_i. Recall that an ideal strategy for the CHSH game achieves a success probability of $\frac{3}{4}$.

Let $\mathbb{H}$ be the Hilbert space representing the qubit used in $\tilde{\mathcal{H}}$. Let $\mathbb{H}_1, \mathbb{H}_2$ be two copies of $\mathbb{H}$. We define $\bar{\mathcal{H}}$ to be the following multi-qubit ideal strategy for the sequential CHSH games: 1. At the beginning of the game, the referee prepares the state $\frac{1}{\sqrt{2}}(|00\rangle + |11\rangle)$ on qubits 1 and 2. 2. For each round i, the referee performs the following steps:

- The input (x_i, y_i) is encoded as a unitary U_i acting on qubit 1;
- The referee applies the unitary $U_i \otimes \mathbb{1}$ on qubit 1 and 2 respectively;
- The referee sends qubit 2 to the player who holds the input y_i;
- The player applies an isometry V_i that ϵ-approximates their strategy $\tilde{\mathcal{H}}$ for the CHSH game on qubit 2;

- The player sends back qubit 2 to the referee;
- The referee applies a Bell measurement on qubits 1 and 2 and obtains outcomes a_i, b_i;
- The referee sends the outcome a_i to the player holding input x_i.

Note that $\bar{\mathcal{H}}$ uses a constant number of copies of the qubit used in $\tilde{\mathcal{H}}$, and each copy is used for at most one round. Thus, the total number of qubits used in $\bar{\mathcal{H}}$ is $O(n)$.

Let $|\psi\rangle$ be the state of the qubits after n rounds of the game. We claim that $|\psi\rangle$ is $\kappa n^{\kappa}\epsilon^{1/\kappa}$-close (in trace distance) to the state of $\tilde{\mathcal{H}}$ at the end of n rounds. To see this, let $|\phi\rangle$ be the state of the qubit used in $\tilde{\mathcal{H}}$ at the end of the game. Then we have

$$\begin{aligned} |\langle\psi|\phi\rangle| &= |\langle 0|U_n^\dagger \cdots U_1^\dagger(\bar{V}_n \cdots \bar{V}_1 \otimes \mathbb{1})U_1 \cdots U_n|0\rangle| \\ &= |\langle 0|U_n^\dagger \cdots U_1^\dagger(V_n \cdots V_1 \otimes \mathbb{1})U_1 \cdots U_n|0\rangle| \\ &= |\langle 0|U_n^\dagger \cdots U_1^\dagger(V_n \cdots V_1 \otimes \mathbb{1})U_1 \cdots U_{n-1}V_n^\dagger \cdots V_1^\dagger|0\rangle| \\ &\le \epsilon, \end{aligned} \tag{7.32}$$

where $\bar{V}_i$ is the isometry obtained from the player's isometry V_i by appending a completely mixed state on the qubits that are not used, and we used the fact that $\tilde{\mathcal{H}}$ is an ϵ-structured single-qubit ideal strategy for n sequential CHSH games. The second equality follows from the fact that $\bar{\mathcal{H}}$ applies V_i on qubit 2 for each round, and the third equality follows from the fact that $\bar{V}_i$ acts as V_i on qubit 2 and as a completely mixed state on qubit 1.

By the triangle inequality, we have that the trace distance between the state of $\bar{\mathcal{H}}$ and the isometric extension of $\tilde{\mathcal{H}}$ is at most 2ϵ, which completes the proof. □

Lemma 7.3.2 *There exists a constant κ such that if $\bar{\mathcal{H}}$ is an ϵ-structured multi-qubit ideal strategy for n sequential CHSH games, then there is an ideal strategy that $\kappa n^{\kappa}\epsilon^{1/\kappa}$-simulates $\bar{\mathcal{H}}$.*

Proof Let $\bar{\mathcal{H}}$ be an ϵ-structured multi-qubit ideal strategy for n sequential CHSH games played between Alice and Bob. Let $\mathcal{H}$ be the ideal strategy that $\kappa n^{\kappa}\epsilon^{1/\kappa}$-simulates $\bar{\mathcal{H}}$. We will now construct $\mathcal{H}$.

Consider a sequence of n CHSH games $\{G_i\}_{i=1}^n$, where each game G_i is played with inputs a_i and b_i. Let $\mathcal{H}_i$ be the ideal strategy for G_i. We will construct $\mathcal{H}$ by concatenating these ideal strategies $\mathcal{H}_i$.

Let m be the total number of qubits used by $\bar{\mathcal{H}}$. Since $\bar{\mathcal{H}}$ is ϵ-structured, we can assume that the qubits are divided into κ groups of equal size m/κ, and that $\bar{\mathcal{H}}$ preserves the structure of these groups. Let $\bar{\mathcal{H}}_i$ be the ideal strategy for G_i that ϵ/κ-simulates the part of $\bar{\mathcal{H}}$ that acts on the i-th group of qubits. By the triangle inequality, we have ϵ-similarity between $\bar{\mathcal{H}}$ and the ideal strategy that concatenates these $\bar{\mathcal{H}}_i$.

To construct $\mathcal{H}$, we will κ-simulated each $\bar{\mathcal{H}}_i$ using a number of copies that depends on n and ϵ. Let $\mathcal{H}_i$ be the ideal strategy for G_i that $\kappa n^{\kappa}\epsilon^{1/\kappa}$-simulates $\bar{\mathcal{H}}_i$. By the definition of κ-similarity, we have ϵ/κ-similarity between $\bar{H}_i$ and $\mathcal{H}_i$ using n^{κ}

copies. Concatenating the $\mathcal{H}_i$ gives us $\mathcal{H}$, which $\kappa n^{\kappa}\epsilon^{1/\kappa}$-simulates $\bar{\mathcal{H}}$ by the triangle inequality.

Therefore, we have shown that there exists a constant κ such that if $\bar{\mathcal{H}}$ is an ϵ-structured multi-qubit ideal strategy for n sequential CHSH games, then there is an ideal strategy that $\kappa n^{\kappa}\epsilon^{1/\kappa}$-simulates $\bar{\mathcal{H}}$. □

Combining both Lemmas 7.3.1 and 7.3.2, follows the main theorem states that a structured strategy can be closely simulated by an ideal strategy. This provides a rigidity result for sequential CHSH games.

Theorem 7.3.1 *For any ϵ-structured strategy $\mathcal{S}$ in n sequential CHSH games, there exists an ideal strategy $\hat{\mathcal{S}}$ that can simulate an isometric extension of $\mathcal{S}$ with error at most $\kappa n^{\kappa}\epsilon^{1/\kappa}$, where κ is a constant.*

The Theorem 7.3.1 suggests that for any strategy used to play a sequential CHSH game, there is an ideal strategy that can closely replicate the results of the original strategy with a small error rate (Reichardt et al. 2013). This provides insights into how different strategies can be compared and optimized, potentially leading to better performance in these games.

7.3.2 Verifying Quantum Dynamics

Verifying quantum dynamics uses computation by teleportation, based on preparing resource states and applying Bell measurements, to achieve verified quantum dynamics (Gottesman and Chuang 1999; Luo and Wang 2015). Charlie asks Bob to prepare copies of a state $|0\rangle \otimes (\mathbb{1} \otimes H)|\varphi\rangle \otimes (\mathbb{1} \otimes G)|\varphi\rangle \otimes \mathrm{CNOT}_{2,4}(|\varphi\rangle \otimes |\varphi\rangle)$ needed to simulate a quantum circuit $\mathcal{C}$ over the gate set $\{H, G, \mathrm{CNOT}\}$, where H is the Hadamard gate and $G = \exp(-i\frac{\pi}{8}Y)$. Bob prepares the state using one-, two- and four-qubit measurements on his EPR shares and reports the results to Charlie. Honest play by Bob leads to desired resource states for Alice, with minor corrections. Each resource state corresponds to a basic operation in $\mathcal{C}$, which Charlie connects by instructing Alice to perform Bell measurements between successive operations. After each G gate, an H correction may be necessary.

To enforce honest play, Charlie runs this protocol only a small fraction of the time and otherwise chooses uniformly between three alternative protocols. Let $m = |\mathcal{C}|^{O(1)}$ and $n = m^{O(1)}$. In the CHSH game, each device has two measurement settings, that in the ideal strategy may be identified with Pauli X and Z operators. For carrying out tomography, however, it is generally necessary to be able to measure on the Y basis as well. It is possible to extend the CHSH game to one in which the ideal strategy also uses Y operators on a shared EPR state. However, this extended game will not satisfy the rigidity property. The problem is that a device that consistently measures using $-Y$ will give indistinguishable statistics from one that uses $+Y$. It is impossible to fix the sign of the Y operator.

Definition 7.3.6 A state $\tau \in \mathcal{K}(\mathbb{H})$ is said to be determined by a set of operators $S \subseteq \mathcal{L}(\mathbb{H})$ with exponent $d > 0$ if there exists a constant $c > 0$ such that for all $\epsilon \geq 0$ and any state $\rho \in \mathcal{L}(\mathbb{H})$,

$$\max_{P \in S} |\mathrm{Tr} P(\rho - \tau)| \leq \epsilon \|\rho - \tau\|_{tr} \leq c\,\epsilon^d. \tag{7.33}$$

For $\mathbb{H} = \mathbb{C}^{2^n}$, a state τ is called *XZ-determined* if it is determined with some exponent $d > 0$ by the Pauli operators ∞, X, $Z^{\otimes n}$. The robustness is important for applications.

Lemma 7.3.3 *For a finite-dimensional Hilbert space* $\mathbb{H}$*, a state* $\tau \in \mathcal{L}(\mathbb{H})$ *is determined by a finite set* $S \subset \mathcal{L}(\mathbb{H})$ *if and only if for any state* $\rho \in \mathcal{L}(\mathbb{H})$*, the implication of Eq. (7.33) holds at* $\epsilon = 0$.

Proof First, assume that a state $\tau \in \mathcal{L}(\mathbb{H})$ is determined by a finite set $S \subset \mathcal{L}(\mathbb{H})$. This means that there exists a unique linear combination of elements in S, say $\sum_{i=1}^n \alpha_i P_i$, where $P_i \in S$ and $\alpha_i \in \mathbb{C}$ such that $\tau = \sum_{i=1}^n \alpha_i P_i$. To show the if and only if statement, it suffices to show that the given implication holds at $\epsilon = 0$.

Suppose $\rho \in \mathcal{L}(\mathbb{H})$ satisfies $\max_{P \in S} |\mathrm{Tr} P(\rho - \tau)| \leq 0$. Then, for any $P \in S$ we have

$$|\mathrm{Tr}(P(\rho - \tau))| \leq |\mathrm{Tr}(P\rho)| + |\mathrm{Tr}(P\tau)| \leq \|P\|_{op}\|\rho\|_{\mathrm{tr}} + \|P\|_{op}\|\tau\|_{\mathrm{tr}}, \tag{7.34}$$

where $\|\cdot\|_{\mathrm{op}}$ denotes the operator norm. By setting $\epsilon = 0$, we have $\|\rho - \tau\|_{\mathrm{tr}} = 0$, which implies that $\rho = \tau$ since the trace norm is a metric.

Conversely, suppose that for any state $\rho \in \mathcal{L}(\mathbb{H})$, $\max_{P \in S} |\mathrm{Tr} P(\rho - \tau)| \leq \epsilon\|\rho - \tau\|_{tr} \leq c\,\epsilon^d$ holds at $\epsilon = 0$. Then, by setting $\epsilon = 0$, we have $\|\rho - \tau\|_{\mathrm{tr}} = 0$ for any state ρ. This implies that τ is a fixed point of the map $T : \mathcal{L}(\mathbb{H}) \to \mathcal{L}(\mathbb{H})$ defined as

$$T(\rho) = \sum_{i=1}^n \alpha_i P_i \rho P_i^\dagger, \tag{7.35}$$

where $P_i \in S$ and $\alpha_i \in \mathbb{C}$. Since τ is a unique fixed point of T, we conclude that τ is determined by the finite set S. □

Recall that a *stabilizer state* is an n-qubit pure state $|\psi\rangle$ for which there exists a set of 2^n distinct and pairwise commuting operators $S \subset \{\pm P : P \in \{\mathbb{1}, X, Y, Z\}^{\otimes n}\}$, the *stabilizer group*, such that $P|\psi\rangle = |\psi\rangle$ for all $P \in S$. Any set of n operators that generate the stabilizer group S are called stabilizer generators for $|\psi\rangle$. The Lemma 7.3.3 implies the XZ-determined for the stabilizer state.

Theorem 7.3.2 *If* $|\psi\rangle \in \mathbb{C}^{2^n}$ *is a stabilizer state that has a set of generators in* $\{\mathbb{1}, X, Z\}^{\otimes n}$*, and if* U *is the tensor product of any* n *single-qubit real unitaries, then* $U|\psi\rangle$ *is* XZ*-determined.*

The ideal state tomography protocol involves Alice and Bob sharing m EPR states, potentially with an additional state, and playing honestly m CHSH games directed by Charlie. Charlie sends Bob a random permutation of the m qubits, and Bob measures consecutive 11-qubit blocks of permuted qubits in the basis specified by

$$\begin{aligned} &P^{(0)}\left|0\right\rangle \otimes (P^{(1)} \otimes \mathbb{1})\left|\varphi\right\rangle \otimes (P^{(2)} \otimes H)\left|\varphi\right\rangle \otimes (P^{(3)} \otimes G)\left|\varphi\right\rangle \\ &\otimes (P_1^{(4)} \otimes P_3^{(5)} \otimes \mathrm{CNOT}_{2,4})(\left|\varphi\right\rangle_{12} \otimes \left|\varphi\right\rangle_{34}), \end{aligned} \tag{7.36}$$

where $P^{(0)} \in \{\mathbb{1}, \sigma_x\}$ and the other $P^{(j)}$ vary over $\{\mathbb{1}, \sigma_x, \sigma_y, \sigma_z\}$. Charlie rejects if the tomography statistics returned by Alice are inconsistent with Bob's reported outcomes, meaning that the fraction of times Bob reports any particular state differs from $1/2^{11}$ by more than $\sqrt{\log m/m}$ or if for any state any of its $\{\mathbb{1}, X, Z\}^{\otimes 11}$ Pauli coefficients differ from the corresponding observed estimates by more than $\sqrt{\log m/m}$. This implies the following result.

Theorem 7.3.3 *In an ideal state tomography protocol, if Alice plays honestly, then:*

- ***Completeness**: If Bob plays honestly, then Eve accepts with high probability;*
- ***Soundness**: If Eve accepts with high probability, then there is a high probability that, after Bob and before Alice's play, for most of the consecutive* 11*-qubit subsystems, Alice's reduced state on the subsystem is close to the state that Bob reported to Eve.*

The statement with high probability means with probability inverse polynomially close to one, i.e., at least $1 - 1/m^c$, where the exact exponents are adjustable.

The process tomography protocol is a technique that ensures the honesty of Alice in quantum communication. It starts with $(K-1)m$ CHSH games played between Alice and Bob. The multi-game rigidity theorem enables us to assume that they share m EPR states and would play the K-th block of games honestly following the ideal CHSH game strategy. However, Eve only interacts with Bob by playing CHSH games. Eve requests Alice, uniformly without replacement from $\{1, \ldots, m\}$, to report the results of Bell measurements on random pairs of qubits. Eve does not draw any statistical conclusions but rejects the protocol if Alice's reported outcome for any pair of qubits is inconsistent with Bob's reported outcomes.

For instance, if Alice reports measuring the Bell state $|\varphi\rangle$ and Bob measures the same two qubits both on the Z basis, then the parity of Bob's measurement results should be 0. This is because $Z \otimes Z$ stabilizes $|\varphi\rangle$: $(Z \otimes Z)\left|\varphi\right\rangle = |\varphi\rangle$. The four Bell states are stabilized by $\pm Z \otimes Z$ and $\pm X \otimes X$ with four distinct sign choices. If Bob measures one of the two qubits on the X basis and the other on the Z basis, his results for that pair do not check Alice's honesty.

Alice's interaction is indistinguishable from the questions she would be asked in actual computation. She is entirely unaware that Bob is playing CHSH games instead of preparing resource states. This implies the following result.

Theorem 7.3.4 *In the process tomography protocol, if after the first $(K-1)m$ CHSH games, Bob plays honestly on his halves of m shared EPR states, then:*

- ***Completeness***: *If Alice plays honestly, Eve accepts with probability one.*
- ***Soundness***: *If Eve accepts with high probability, then the result of Alice's super-operator applied to the initial state is close to that of applying the ideal Bell measurements super-operator to the initial state.*

The class QMIP consists of polynomial-time decidable languages that can be verified by a quantum verifier exchanging polynomially many messages with a polynomial number of quantum provers sharing entanglement but unable to communicate among themselves. Kempe et al. (2009) showed that any QMIP protocol can be converted into a three-turn protocol where the verifier broadcasts a random coin flip and the provers each send two quantum messages. The present result adds two new provers, Alice and Bob, but it is conjectured that this is unnecessary. Broadbent et al. (2009) suggested converting a k-prover QMIP protocol to a k-prover MIP* protocol that reuses the first two provers to simulate the verifier's quantum computations. However, this can allow the provers to cheat. Converting to a three-turn protocol or using two fresh provers can avoid these attacks.

To create a MIP* protocol $\mathcal{P}''$ that simulates the verifier's quantum computations using the first two provers, traps are used to detect cheating. However, in $\mathcal{P}''$, the provers can reveal the traps to each other during the intercommunication, allowing them to cheat freely. As a result, $\mathcal{P}''$ is unsound. One way to avoid this attack is to convert to a three-turn protocol or use fresh provers to run the verifier's quantum computation. Another option is to refresh the verifier's secrets before revealing any message to a prover, but this may impact the security of the converted MIP* protocol.

7.4 Multiparty DQC

It appears that creating universal quantum computers for regular use is still a challenge due to scalability limitations. A practical approach would be the delegation of computation to powerful quantum servers on the network (Kashefi and Pappa 2017). This was proposed earlier in studies of blind quantum computation and offers assurances for input and computation secrecy. In this section, we present a multi-party delegated quantum computing protocol within the measurement-based quantum computing framework.

7.4.1 Delegated Multiparty Quantum Computing

This section defines multiparty delegated quantum computing protocols, where multiple clients $\mathsf{C}_1, \ldots, \mathsf{C}_n$ delegate the computation of a unitary U on input ρ_{in} to a Server S. The Server has register $\mathcal{S}$ and no legal input to the computation. The input state is denoted as ρ_{in}, which belongs to the set of all possible quantum states in registers $\mathcal{C}_1 \otimes \cdots \otimes \mathcal{C}_n \otimes \mathcal{R}$, where $\mathcal{R}$ denotes the environment.

The input registers of the participants include classical information necessary for computation. $\mathcal{L}(\mathcal{A})$ is the set of linear mappings from $\mathcal{A}$ to itself. A superoperator $\Phi : \mathcal{L}(\mathcal{A}) \rightarrow \mathcal{L}(\mathcal{B})$ that is completely positive and trace-preserving is called a quantum operation. $\mathbf{I}_{\mathcal{A}}$ and $\mathbb{1}_{\mathcal{A}}$ denote the mixed state and the identity operator respectively, in register $\mathcal{A}$.

A protocol can be modeled as a sequence of local quantum operations on participants' registers and joint quantum operations on the registers for Oracle calls. In delegated communication protocols, n clients delegate computation to a server. References for delegated communication protocols are (Dupuis et al. 2010, 2012; Ben-Or et al. 2006).

Definition 7.4.1 A t-step delegated protocol with oracle calls between n clients C_k and a Server S is denoted by $\pi\mathcal{O} = (\{\pi_k\}_{k=1}^n, \pi_S, \mathcal{O})$.

(1) The strategy π_k for each client C_k consists of a set of t local quantum operators $(L_1^k, \ldots, L_t^k)$ that act on the register $\mathcal{C}_k$.
(2) The Server's strategy π_S consists of a set of t local quantum operators $(L_1^S, \ldots, L_t^S)$ that act on the register $\mathcal{S}$.

The oracle $\mathcal{O}$ includes a set of t global quantum operators $(\mathcal{O}_1, \ldots, \mathcal{O}_t)$ that act on the joint register $\mathcal{C}_1 \otimes \cdots \otimes \mathcal{C}_n \otimes \mathcal{S}$.

Consider an abstract system that takes ρ_{in} and computation instructions for implementing U in MBQC, i.e., measurement angles $\{\phi_j\}$ as input and outputs a state ρ_{out}. This resource is called an Ideal Functionality. The Ideal Functionality can accept input from Server to allow it to act dishonestly, dictating desired deviations in quantum input and classical information. If the Server acts honestly, the Ideal Functionality outputs the correct output $\rho_{\text{out}} = (U \otimes \mathbb{1}_{\mathcal{R}}) \cdot \rho_{\text{in}}$. To simplify, we use $U \cdot \rho$ instead of $U\rho U^\dagger$ when talking about applying a unitary operation U to a quantum state ρ.

Definition 7.4.2 In a multiparty quantum computation (MQC) delegated resource, n clients provide input ρ_{in} and computation instructions, and the size of the computation q is leaked to the server. The server can choose to input a quantum map and a quantum input. The output of the computation is $\rho_{\text{out}} = (U \otimes \mathbb{1}_{\mathcal{R}}) \cdot \rho_{\text{in}}$ if the server follows the protocol, otherwise the output is determined by the quantum deviation on the inputs of the server and the clients.

In delegated communication protocols, a server is entrusted with performing the computation on behalf of multiple clients. This can be modeled by a sequence of local quantum operations performed by each client on their register, as well as joint quantum operations between the clients' registers and the server's register through oracle calls. The overall goal is to execute the protocol in a way that preserves the security and privacy of the client's data.

Definition 7.4.3 We denote a t-step delegated protocol with oracle calls between clients $\mathsf{C}_k (k \in [n])$ and a Server S, with $\Gamma_{\mathcal{O}} = (\{\pi_k\}_{k=1}^n, \Gamma_S, \mathcal{O})$. If we denote by $\mathcal{C}_k$ and $\mathcal{S}$ the registers of the clients and the Server respectively, then:

(1) Each client's strategy $\Gamma_k, k \in [n]$ consists of a set of local quantum operators $(L_1^k, \ldots, L_t^k)$ such that $L_i^k : \mathcal{L}(\mathcal{C}_k) \to \mathcal{L}(\mathcal{C}_k)$ for $1 \le i \le t$.
(2) The Server's strategy Γ_S consists of a set of local quantum operators $(L_1^S, \ldots, L_t^S)$ such that $L_i^S : \mathcal{L}(\mathcal{S}) \to \mathcal{L}(\mathcal{S})$ for $1 \le i \le t$.
(3) The oracle $\mathcal{O}$ is a set of global quantum operators $\mathcal{O} = (\mathcal{O}_1, \ldots, \mathcal{O}_t)$ such that $\mathcal{O}_i : \mathcal{L}(\mathcal{C}_1 \otimes \cdots \otimes \mathcal{C}_n \otimes \mathcal{S}) \to \mathcal{L}(\mathcal{C}_1 \otimes \cdots \otimes \mathcal{C}_n \otimes \mathcal{S})$ for $1 \le i \le t$.

The first procedure is to prepare the state for $j \in I$. The server stores the states it receives from clients C_k in distinct registers $\mathcal{S}_k \subset \mathcal{S}$ where $k \in [n]$. For each $k = j - 1$, a CNOT operation is performed on $\mathcal{S}_k \otimes \mathcal{S}_{k+2}$; Otherwise, a CNOT operation is performed on $\mathcal{S}_k \otimes \mathcal{S}_{k+1}$. The state is $\mathcal{S}_k$ is then measured and the outcome is denoted as t_j^k. When $j = n$, a CNOT operation is performed on $\mathcal{S}_{n-1} \otimes \mathcal{S}_n$, the state in $\mathcal{S}_{n-1}$ is measured and the outcome is denoted as t_n^{n-1}; otherwise, a CNOT operation is performed on $(\mathcal{S}_n \otimes \mathcal{S}_j)$ and the state in $\mathcal{S}_n$ is measured and the outcome is denoted as t_j^n.

The second procedure is to prepare the state for $j \in O^c \setminus I$. The server stores the states received from clients C_k in distinct registers $\mathcal{S}_k \subset \mathcal{S}$ where $k = 1, \ldots, n$. For $k = 1, \ldots, n-1$, a CNOT operation is performed on $\mathcal{S}_k \otimes \mathcal{S}_{k+1}$, and the state in $\mathcal{S}_k$ is measured to obtain the outcome denoted as t_j^k.

The multiparty DQC protocol begins with remote state preparation protocol which involves clients sending quantum states to a server, which entangles and measures them except for one (Dunjko et al. 2012). One client with a quantum input sends it one-time padded to the Server, while others send Z-rotated $|+\rangle$ states using the first prepare procedure. For operational qubits in $O^c \setminus I$, clients also send rotated $|+\rangle$ states using the first prepare procedure.

The proof of security against a malicious Server that is allowed to deviate from the protocol, by applying operations on the data he receives, is based on quantum teleportation. For all the quantum states sent to the Server, there exists an equivalent circuit that uses an EPR pair and teleports via measurement the quantum state to one of the entangled qubits of the EPR, which is then sent to the Server. This implies a secure delegated multiparty quantum computing protocol (Kashefi and Pappa 2017).

Theorem 7.4.1 *The present delegated multiparty quantum computing protocol is blind against a malicious Server and secures against a coalition of malicious clients.*

To demonstrate the security of the protocol against a malicious server, we will show that the set of honest clients and a simulator for dishonest clients are closely related. The proof relies on teleportation techniques and the equivalence of circuits and delays of operations on different wires. We need to ensure that the present protocol and the cloud implement the same map. We can rewrite the one-time pad in the present protocol as a delayed teleportation of the input to the EPR halves that the Server acts upon, and similarly for the rotated states. The random values chosen follow a uniform distribution in the measurements done by the MPQC resource due to the properties of entanglement in the teleportation process. Instead of choosing variables uniformly at random, the simulator chooses δ_j uniformly at random, and a_j

and r_j are defined from the randomness inherent when measuring the EPR pairs. The security of the simulated protocol ensures the security of the real protocol because the Server has no access to the quantum input or measurement angles in the simulated version.

It is possible to show the security of a quantum computation protocol against a coalition of malicious clients under the given conditions, where the malicious clients never receive quantum information from other clients and only share classical information through secure multiparty computation protocols. The quantum outcome received by the malicious clients is predetermined by an honest server and encoded using the information chosen by the malicious clients in a previous step.

7.4.2 Multiparty DQC with Dishonest Majority

This subsection is to develop a composable, multiparty quantum computation protocol that allows relatively powerless clients to delegate tasks to a powerful quantum server. The protocol should be secure against collusion between clients and the server and require only one honest client for secure operation (Dulek et al. 2020). To achieve these goals, we first deconstructed and analyzed an existing single-client VBQC protocol. We identified the necessary modifications required for a multi-client setting and established the conditions that replacement steps must satisfy. Using the AC framework's composability property, we then constructed a multiparty DQC protocol that meets these requirements (Kapourniotis et al. 2021).

The security analysis involves using distinguishers, which are converters with input interfaces matching resources and outputting a single bit, to discriminate between two resources $\mathcal{R}_1$ and $\mathcal{R}_2$ by using pseudometric on the space of resources. The goal is to determine if the resources are indistinguishable, meaning that no distinguisher can accurately guess which resource it is interacting with.

Definition 7.4.4 Let $\epsilon(\eta)$ be a function of the security parameter η, and let $\mathcal{R}_1$ and $\mathcal{R}_2$ be two resources with the same input and output interfaces. We say that the resources are ϵ-statistically indistinguishable if, for all distinguishers $\mathcal{D}$, the following inequality holds:

$$\left|\Pr[b = 1 \mid b \leftarrow \mathcal{D}\mathcal{R}_1] - \Pr[b = 1 \mid b \leftarrow \mathcal{D}\mathcal{R}_2]\right| \leq \epsilon. \tag{7.37}$$

We then write $\mathcal{R}_1 \underset{stat,\epsilon}{\approx} \mathcal{R}_2$.

Definition 7.4.5 Let $\epsilon(\eta)$ be a function of the security parameter η. We say that an N-party protocol Γ ϵ-statistically constructs resource $\mathcal{S}$ from resource $\mathcal{R}$ against adversarial patterns $\mathsf{P} \subseteq \wp([N])$ if:

1. Γ is correct: $\Gamma\mathcal{R} \underset{stat,\epsilon}{\approx} \mathcal{S}$;

2. Γ is secure for all subsets of corrupted parties in the pattern $M \in \mathsf{P}$: there exists a simulator (converter) σ_M such that $\Gamma_{M^c}\mathcal{R} \underset{stat,\epsilon}{\approx} \mathcal{S}\sigma_M$.

These definitions can be extended to quantum polynomial time (BQP) parties for computational security. If the simulator for a statistically secure protocol runs in expected polynomial time, it can be composed with a computationally-secure protocol. However, the resulting protocol is only computationally-secure.

The multiparty quantum computation Protocol 7.1 has $N + 1$ interfaces, with one for each of the N parties and an additional interface for an eavesdropper. Its purpose is to allow N clients to perform a collectively defined computation U over their private quantum inputs, ensuring either proper execution or complete abortion. Leakage of information by the eavesdropper is limited to l_ρ. The resource does not guarantee the fairness of output distribution in cases where a dishonest majority exists. Each party receives a filtered bit o_j indicating when it receives the output in comparison to others.

Protocol 7.1 Multi-Party Quantum Computation

Inputs:

- The N players send their quantum states $\mathcal{X}_j$ and a classical description of a joint unitary U they wish to apply to the collectively possessed ρ_{inp}. They can each input two bits o_j and c_j as a filtered interface;
- Eavesdropper can input two bits e and c as a filtered interface.

Computation by the cloud:

1. If $e = 1$, Cloud sends the leakage l_ρ to eavesdropper's interface;
2. If $c = 1$ or there exists j such that $c_j = 1$, Cloud sends Abort to all players;
3. Otherwise, it computes $U(\rho_{inp})$. If there exists $j \in [N]$ such that $o_j = 1$, it sends qubit j to the interface of player j. It sends the output qubits to all other players in a similar fashion.

Another improvement is to allow a single client to run a quantum computation on a server so that server cannot corrupt the computation and doesn't learn anything besides the leakage l_ρ (Dunjko et al. 2014; Kashefi and Wallden 2017). This implies verifiable delegated quantum computation Protocol 7.2.

The single-client VBQC protocol can be transformed into a multi-party protocol by modifying each step so that they can be performed collectively by several clients without compromising the blindness, even in the situation where some clients and the server collude. In this modified setup, a Trusted Third Party receives classical parameters from the clients and drives the VBQC computation while verifying the traps have been measured correctly. The only operations remaining for the client are encrypting inputs, preparing dummies and encrypted $|+\rangle$ states, sending encryption keys to the Trusted Third Party, sending the state to the server, and recovering the output state from the server and decrypting to get the final result.

The third improvement to the quantum computation protocol allows multiple clients to submit their parts of a quantum input state, which is transformed by an

Protocol 7.2 Verifiable DQC

Inputs:

- The client inputs a quantum state ρ and the classical description of a unitary U;
- The server chooses whether or not to deviate. This interface is filtered by two control bits (e, c) (set to 0 by default for honest behavior).

Computation by the cloud:

1. If $e = 1$, the Resource sends the leakage l_ρ to the server's interface and awaits further input from the server; if it receives $c = 1$, the Resource outputs $|\mathsf{Abort}\rangle$ at the client's output interface;
2. Otherwise, it outputs $U\rho U^\dagger$ at the client's output interface.

orchestrator using a specified unitary transformation. However, malicious parties can manipulate the input state using a CPTP map and an ancillary quantum state, resulting in an encrypted output state that is protected with Q-OTP encryption and a randomly chosen key for blindness. The DBQC protocol has two phases: a state preparation step where each client sends qubits for the VBQC Dotted-Triple Graph and auxiliary qubits for the DBQC Protocol; and a computation step where UBQC is executed on the collaboratively-generated state from the state-preparation step.

Now, we design the delegated multi-party quantum computation (MPQC) protocol. It consists of the following steps: (1) Clients send qubits for the DBQC state preparation Protocol 7.3, with corresponding secret parameters sent to classical secure multipartite computation (SMPC). (2) Server performs the rest of the DBQC Protocol with classical SMPC and returns measurement results. (3) Classical SMPC instructs the server to perform the VBQC Protocol on the prepared state with encrypted bridge operations for computation qubits for input edges. (4) Clients and classical SMPCs perform key-release steps to either abort or receive decryption keys for their output.

Protocol 7.3 DBQC State Preparation

Client j: For each non-output qubit location $i \in O^c$:

- If $i \in I$ and $\Omega(i) = j$, randomly choose $a(i) \in_R \{0, 1\}$ and $\theta_{\Omega(i)}(i) \in_R \Theta$, apply $Z(\theta_{\Omega(i)}(i))X^{a(i)}$ to the input register $\mathcal{X}_j$, and send the qubit to the server while sending classical values $a(i)$ and $\theta_{\Omega(i)}(i)$ to the independent party through a classical secure channel.
- Otherwise, randomly choose $\theta_j(i) \in_R \Theta$ and prepare $\left|+_{\theta_j(i)}\right\rangle$, and send the qubit to the server while sending the classical value $\theta_j(i)$ to the independent party through a classical secure channel.

Server: prepares a $|+\rangle$ state for each output qubit location i in O, and for all other locations i not in O, it performs a CNOT operation between the state received from client $\Omega(i)$ and the qubits received from all other clients $j \neq \Omega(i)$, where the qubits are in state $\left|+_{\theta_j(i)}\right\rangle$. It measures the target qubits received from all clients $j \neq \Omega(i)$ and defines $t(i)$ as the vector of outcomes of the measurements. The protocol publicly broadcasts $t(i)$.

Protocol 7.4 performs a DBQC on a graph state defined by graph $\mathcal{G}$ and flow f, using Protocol 7.3 for state preparation, followed by entanglement given by the edges of GG. Clients privately send random binary values $r_j(i)$ for $i \in O^c$ to the orchestrator, which is combined into $r(i) = \oplus_j r_j(i)$ to OTP the measurement outcomes obtained by the server. The orchestrator then instructs the server to perform single-qubit measurements using given $\theta(i)$ and $a(i)$, resulting in a Q-OTP version of the quantum output of the computation, with encryption key s determined by the values of $s'_X(i)$ and $s'_Z(i)$. Finally, the termination step involves the orchestrator setting its key k equal to s.

Protocol 7.4 Double Blind Quantum Computation

- The clients and server perform the DBQC State Preparation.
- Server entangles the qubits obtained at the end of the State Preparation according to the edges of the computation graph G with CZ gates.
- For $i \in O^c$, each client j sends $r_j(i) \in_R \{0, 1\}$ to the independent party, who recombines them as $r(i) = \oplus_j r_j(i)$. The independent party computes the measurement angle $\delta(i) = \phi'(i) + \theta(i) + r(i)\pi$ and sends this angle to Server, who measures location i along the $\{\left|+_{\delta(i)}\right\rangle, \left|-_{\delta(i)}\right\rangle\}$ basis and returns the measured outcome $b(i)$ to the independent party.
- The independent party sets $k = (s'_X(i), s'_Z(i))_{i \in O}$.

Theorem 7.4.2 *Under the assumption that at least the client is honest and that the independent party is honest, Protocol 7.4 ϵ_D-constructs DBQC protocol for $\epsilon_D = 0$, i.e., perfect security.*

The proof aims to describe simulators attached to malicious parties' interfaces of the ideal cloud so that the real-world implementation is indistinguishable from the ideal-world version with the simulator. We initially establish the correctness of the present protocol for all honest parties, followed by the computation step, which remains identical to the single-client UBQC Protocol, ensuring the protocol's accuracy after measurements. The correctness is proven by its ability to produce identical quantum and classical data.

In the case of at least one honest client, the protocol ensures perfect blindness from malicious clients to honest parties. In the worst-case scenario with a malicious Server and possibly some malicious clients, a single honest client and the independent party can handle it while the corresponding Simulator connects to malicious clients and servers on one side and the corresponding interfaces of the ideal cloud on the other. Both the ideal cloud and the simulator are indistinguishable from the real world by observing the classical messages and quantum states exchanged. For more details, see Ref. (Kapourniotis et al. 2021).

7.4.3 Full Delegated Multiparty Quantum Computation

The security of a delegated multiparty quantum computation protocol requires the DBQC protocol to produce a client-encrypted state that deviates up to a certain limit, independently of the state's secret parameters. This state must be plugged into the VBQC protocol to preserve its original security properties. The DBQC protocol can generate such a state using the MBQC pattern defined in Theorem 7.4.3. This section contributes a fully delegated multiparty quantum computation.

A quantum state is a valid VBQC execution state if it has the form $\mathcal{E}(\rho_V \otimes \rho_{aux})$, where $\mathcal{E}$ is a CPTP map dependent solely on the public and dishonest party parameters, ρ_{aux} is an adversarially-chosen auxiliary state, and ρ_V is the state of quantum systems consisting of input, computation, trap, dummy, and rotation qubits. The quantum inputs are encrypted using the $Z(\theta(i))X^{a(i)}$ operator, and the computation and trap qubits are initialized in $|+_{\theta(i)}\rangle$, the dummy qubits are of the form $|d(i)\rangle$, and the rotation qubits are in classical states represented as computational basis states on three qubits. The traps, computation, input, and dummy qubits are placed according to a verifiable graph $\mathcal{G}$ such that the probability of the server guessing that any qubit in the graph is a trap is lower-bounded by a constant value p_t, and trap vertices are linked only to dummy vertices. The rotation qubits have the form $|\delta(i)\rangle$, where $\delta(i) = (-1)^{a(i)}\phi'(i) + \theta(i) + r(i)\pi + \oplus_{\tilde{i}\in N_G(i)\cap D} d(\tilde{i})\pi$, $r(i) \in_R \{0, 1\}$, $\phi'(i)$ is the flow-corrected angle on qubit i for a branch of computation corresponding to the server's measurements b, D is the set of dummy qubits in $\mathcal{G}$, and $N_G(i)$ is the set of neighbors of qubit i in $\mathcal{G}$.

The state prepared by the DBQC Protocol 7.4 is a valid VBQC execution state. Additionally, it provides a framework for DBQC computations that utilize Clifford operations to produce suitable VBQC execution states. This can be formalized as the following theorem.

Theorem 7.4.3 *The output of DBQC Protocol 7.4 using an MBQC pattern defined by a fixed graph $\mathcal{G} = (V, E, I, O)$ with flow $(f, \preceq)$ and base angles $\{\phi(i)\}_{i\in O^c}$ is a valid VBQC execution state for the clients' desired computation if all the following hold:*

- *For honest input to the DBQC Protocol, correct VBQC client-encrypted state $\rho_{DT(G)}$ is produced through the honest application of the measurement pattern with measurement angles $\delta(i)$ that correspond to the desired computation for any fixed computation branch.*
- *All DBQC measurement angles are Clifford with $\phi(i) \in \{k\pi/2\}_{0\leq k\leq 3}$.*
- *The graph $\mathcal{G} = (V, E, I, O)$, flow $(f, \preceq)$, and all angles associated with vertices having an X-dependency according to the selected flow are invariant under a change of coloring in the Dotted-Triple Graph.*

The full delegated MPQC is given as follows.

A group of N clients has jointly selected a unitary operator U, which serves as a measurement pattern over the graph $\mathcal{G}$, including measurement angles $\{\phi(i)\}_{i\in O^c}$ for non-output qubits and a flow. Each honest client j possesses a quantum register $\mathcal{X}_j$ that contains their portion of a collectively owned state ρ_{inp} consisting of M qubits.

1. The clients select the positions of their input qubits p_i such that $\Omega(i) = j$. They send these values to classical secure multiparty computation (SMPC). Two $|+\rangle$ states are then prepared at positions $\{1, 2, 3\} \setminus p_i$ for each input qubit. Client N prepares $|+\rangle$ states for the non-input qubits of the dotted-triple graph. Finally, Classical SMPC randomly chooses a compatible coloring of the dotted-triple graph based on these inputs and applies Hadamard gates to dummies.
2. The clients and server perform the DBQC Protocol 7.4, with classical SMPC serving as an independent party using the transformation above as DBQC unitary.
3. The server entangles the qubits obtained at the end of the DBQC Protocol using CZ gates according to the edges of the dotted-triple graph $DT(G)$.
4. The classical SMPC involves instructing the server to execute bridge operations on every qubit in the base locations of the edges connected to the input layer. This is achieved by measuring Traps and dummies as in the VBQC Protocol, with random angles for dummies and correct angles for traps up to a random π rotation. The computation qubit is utilized as a bridge, with the angle δ sent to the server utilizing the base angle $\phi = \pi/2$. Let b represent the measurement outcome; the primary computational qubits' θ angles in the two base locations associated with this edge are modified by adding $(-1)^{1+b}\pi/2$.
5. The classical SMPC instructs the server to measure each remaining qubit $i \in O^c$ with an angle $\delta(i)$ computed according to the VBQC Protocol. The server measures the qubit i along the $\{|+_{\delta(i)}\rangle, |-_{\delta(i)}\rangle\}$ basis and returns the measurement result $b(i)$ to Classical SMPC.

Meanwhile, the quantum output key-release procedure is defined as follows.

1. The server provides each client with three qubits from the Dotted-Triple-Graph that correspond to their output base locations. Additionally, through the classical SMPC, each client obtains information regarding the position of the computation, trap, and dummy qubit, along with the measurement angle $\delta(i)$ for the trap. The client measures its output trap qubit in the basis $\{|+_{\delta(i)}\rangle, |-_{\delta(i)}\rangle\}$ and reports the outcome to the classical SMPC.
2. To ensure the accuracy of the measurements, the classical SMPC verifies that all traps have been measured correctly using a verification procedure similar to the VBQC protocol. If the verification fails, the classical SMPC sends an Abort message to all clients. Conversely, it sends the Q-OTP key for the output computation qubit to each client if the verification succeeds.
3. Each client reverses the Q-OTP on its output computation qubit and assigns this quantum register as its output.

Assume that the base graph chosen to perform an MBQC computation through VBQC has degree 1 on all input vertices. Apply the DBQC Protocol using the MBQC pattern for each qubit of the final $DT(G)$ as (i) MBQC input states and $|+\rangle$ states for non-input positions in the $DT(G)$ and (ii) the position of the MBQC input of honest players are randomly chosen in its associated input base-location. The final stage then consists of $\mathcal{E}(\rho_V \otimes \rho_{aux})$, where ρ_V is a correctly-prepared VBQC client-encrypted state, ρ_{aux} is an arbitrary auxiliary state prepared by the Adversary and the deviation $\mathcal{E}$ is independent of ρ_V.

We assume that the base graph chosen by the clients for performing the initial MBQC computation of the joint unitary has degree 1 on all input vertices of set I. This allows us to reuse the security analyses (Kashefi and Wallden 2017; Kapourniotis et al. 2021) for our composed protocol, therefore proving the full security of the DMPQC Protocol.

7.5 Conclusion

The purpose of cryptographic protocols is to allow secure communication between participants without the need for a trusted third party. This typically requires multiple rounds of communication even when all parties are assumed to be honest. This chapter has been devoted to the blind protocols that allow multiple clients to collaboratively prepare a valid VBQC execution state, which can be used for fully verifiable computations as long as the input and unitary meet certain conditions. This protocol effectively bootstraps the verifiability of any computation using blindness alone, breaking the principle that all sub-components must be verifiable for the full protocol to be considered verifiable. The model is different from previous models for cloud data security assisted by QKD for providing cloud storage security and manage with data dynamics (Sundar et al. 2022; Bibak et al. 2022).

Delegated quantum computation presents considerable obstacles, such as controlling device interactions, and managing quantum noise and errors. This chapter has discussed various topics related to delegated quantum computation including measurement-based quantum computing (Raussendorf and Briegel 2001; Raussendorf et al. 2003; Hein et al. 2006; Luo and Wang 2014), secure delegated quantum computation (Childs 2001; Reichardt et al. 2013; Abadi et al. 1989; Wiesner et al. 2022), composable security (Broadbent et al. 2009; Maurer and Renner 2016; Portmann and Renner 2014), equivalence in delegated quantum computing (Wiesner et al. 2022; Broadbent et al. 2009; Fitzsimons 2017; Kashefi and Pappa 2017; Kapourniotis et al. 2021; Morimae and Fujii 2013; Hein et al. 2004), classical verification of quantum computer (Reichardt et al. 2013), verified quantum dynamics (Gottesman and Chuang 1999; Broadbent et al. 2009; Kempe et al. 2009), delegated multiparty quantum computing (Dupuis et al. 2010, 2012; Ben-Or et al. 2006; Dunjko et al. 2012), delegated MPQC protocol with dishonest majority (Dunjko et al. 2014; Kashefi and Wallden 2017; Dulek et al. 2020; Kapourniotis et al. 2021), and full delegated multiparty quantum computation (Kashefi and Wallden 2017; Kapourniotis et al. 2021). This has shown the power of distributed quantum devices for solving large-scale tasks.

References

M. Abadi, J. Feigenbaum, J. Kilian, On hiding information from an oracle. J. Comput. Syst. Sci. **39**, 21–50 (1989)

P. Arrighi, L. Salvail, Blind quantum computation. Int. J. Quantum Inf. **4**, 883–898 (2006)

H. Barnum, C. Crépeau, D. Gottesman, A. Smith, A. Tapp, Authentication of quantum messages, in *Proceedings of the 43rd Annual IEEE Symposium on Foundations of Computer Science (FOCS 2002)* (2002), pp. 449–458

M. Ben-Or, C. Crépeau, D. Gottesman, A. Hassidim, A.D. Smith, Secure multiparty quantum computation with (only) a strict honest majority, in *Proceedings of FOCS* (IEEE Press, 2006), pp. 249–260

K. Bibak, B.M. Kapron, V. Srinivasan, Authentication of variable length messages in quantum key distribution. EPJ Quantum Technol. **9**, 8 (2022)

G. Brennen, D. OLeary, S.S. Bullock, Criteria for exact qudit universality. Phys. Rev. A **71**, 052318 (2005)

A. Broadbent, J.F. Fitzsimons, E. Kashefi, Universal blind quantum computation, in *Proceedings of the 50th IEEE FOCS* (2009), pp. 517–526

A. Broadbent, J. Fitzsimons, E. Kashefi, Universal blind quantum computation, in *Proceedings of the 50th Symposium on Foundations of Computer Science*, IEEE Computer Society (2009), pp. 517–526

A.M. Childs, Secure assisted quantum computation, arXiv: quant-ph/0111046 (2001)

V. Danos, E. Kashefi, Determinism in the one-way model. Phys. Rev. A **74**, 052310 (2006)

D. Deutsch, Quantum computational networks. Proc. R. Soc. A **425**, 99 (1989)

Y. Dulek, A.B. Grilo, S. Jeffery, C. Majenz, C. Schaffner, Secure multi-party quantum computation with a dishonest majority, in *Advances in Cryptology: 39th Annual International Conference on the Theory and Applications of Cryptographic Techniques* (Springer, Zagreb, Croatia, 2020), pp. 729–758

V. Dunjko, J.F. Fitzsimons, C. Portmann, R. Renner, Composable security of delegated quantum computation, in *Advances in Cryptology: 20th International Conference on the Theory and Application of Cryptology and Information Security* (Springer, Berlin, Heidelberg, 2014), pp. 406–425

V. Dunjko, E. Kashefi, A. Leverrier, Universal blind quantum computing with weak coherent pulses. Phys. Rev. Lett. **108**, 200502 (2012)

F. Dupuis, J.B. Nielsen, L. Salvail, Actively secure two-party evaluation of any quantum operation, in *Proceedings of the CRYPTO* (Springer, 2012), pp. 794–811

F. Dupuis, J.B. Nielsen, L. Salvail, Secure two-party quantum evaluation of unitaries against specious adversaries, in *Annual Cryptology Conference* (2010), pp. 685–706

F. Dupuis, J.B. Nielsen, L. Salvail, Secure two-party quantum evaluation of unitaries against specious adversaries, in *Proceedings of the CRYPTO*, LNCS, vol. 6223 (2010), pp. 685–706

J.F. Fitzsimons, Private quantum computation: an introduction to blind quantum computing and related protocols. npj Quantum Inf. **3**, 23 (2017)

D. Gottesman, The Heisenberg representation of quantum computers, arXiv: quant-ph/9807006 (1998)

D. Gottesman, I.L. Chuang, Demonstrating the viability of universal quantum computation using teleportation and single-qubit operations. Nature **402**, 390–393 (1999)

M. Hayashi, T. Morimae, Verifiable measurement-only blind quantum computing with stabilizer testing. Phys. Rev. Lett. **115**, 220502 (2015)

M. Hein, W. Dür, J. Eisert, R. Raussendorf, M. Van den Nest, H.-J. Briegel, Entanglement in graph states and its applications, arXiv: quant-ph/0602096 (2006)

M. Hein, J. Eisert, H.J. Briegel, Multiparty entanglement in graph states. Phys. Rev. A **69**, 062311 (2004)

T. Kapourniotis, E. Kashefi, L. Music, H. Ollivier, Delegating multi-party quantum computations vs. dishonest majority in two quantum rounds, arXiv:2102.12949 (2021)

E. Kashefi, A. Pappa, Multiparty delegated quantum computing. Cryptography **1**, 12 (2017)

E. Kashefi, P. Wallden, Optimised resource construction for verifiable quantum computation. J. Phys. A. Math. Theore. **50**, 145306 (2017)

J. Kempe, H. Kobayashi, K. Matsumoto, T. Vidick, Using entanglement in quantum multi-prover interactive proofs. Comput. Compl. **18**, 273–307 (2009)

M.-X. Luo, X. Wang, Universal quantum computation based on qudits. Sci. China Phys. Mech. Astron. **57**, 1712–1717 (2014)

M.-X. Luo, X. Wang, Universal remote quantum computation assisted by the Cavity input-output process. Proc. R. Soc. London A **471**, 20150274 (2015)

U. Maurer, R. Renner, From indifferentiability to constructive cryptography (and back), in *Theory of Cryptography: 14th International Conference (TCC 2016)*, Beijing, China (2016), pp. 3–24

T. Morimae, K. Fujii, Blind quantum computation protocol in which Alice only makes measurements. Phys. Rev. A **87**, 050301 (2013)

M.A. Nielsen, I.L. Chuang, *Quantum Computation and Quantum Information* (Cambridge University Press, Cambridge, 2000)

C. Portmann, R. Renner, Cryptographic security of quantum key distribution, arXiv:1409.3525 (2014)

R. Raussendorf, H.J. Briegel, A one-way quantum computer. Phys. Rev. Lett. **86**, 5188 (2001)

R. Raussendorf, D.E. Browne, H.J. Briegel, Measurement-based quantum computation on cluster states. Phys. Rev. A **68**, 022312 (2003)

B.W. Reichardt, F. Unger, U. Vazirani, Classical command of quantum systems. Nature **496**, 456–460 (2013)

K. Sundar, S. Sasikumar, C. Jayakumar, Enhanced cloud security model using QKDP (ECSM-QKDP) for advanced data security over cloud. Quantum Inf. Process. **21**, 115 (2022)

B.S. Tsirelson, Quantum generalizations of Bell's inequality. Lett. Math. Phys. **4**, 93–100 (1980)

F. Wiesner, J. Eisert, A. Pappa, Equivalence in delegated quantum computing, arXiv:2206.07469 (2022)

Chapter 8
Nonlocality in Quantum Network

In this section, we present the nonlocality (Bell 1964) in quantum network scenarios. There are at least two motivations for extending the nonlocality in networks. One is similar to Bell's theorem that aims to understand quantum theory and its relationship to other models. The other is from applications of quantum information. There is increasing interest in quantum networks because of their flexibility and efficiency in various tasks. Bell nonlocality in networks may bring novel forms of quantumness. When many independent entangled states are distributed in a network, some parties distribute entanglement among distant, initially independent systems for space-separate parties with the help of local measurements. This fundamental quantumness of so-called entanglement swapping (Żukowski et al. 1993) has no counterpart in traditional Bell tests. This inspires new ideas and tools for exploring the nonlocality in quantum network scenarios beyond standard Bell inequality experiments.

8.1 Bipartite Nonlocality

The CHSH inequality is one of the most important results in quantum mechanics. It is a mathematical expression that describes the behavior of two entangled particles that are measured in different ways. The inequality states that the correlation between the measurements of these particles cannot exceed a certain value, which is known as the CHSH bound. This bound has important implications for the foundations of quantum mechanics and the nature of reality itself.

The CHSH inequality is named after Clauser, Horne, Shimony, and Holt (Clauser et al. 1969). The inequality is derived from the Bell inequalities, which were first proposed by John Bell (1964). Bell's inequalities describe the behavior of entangled particles in a way that is not consistent with classical physics. They show that the

M.-X. Luo, *Quantum Networks*,
https://doi.org/10.1007/978-981-97-6226-2_8

correlations between the measurements of entangled particles cannot be explained by any hidden variables that would make the particles act like classical objects.

The CHSH inequality is a special case of Bell's inequalities that is easier to understand and apply. It involves two particles that are entangled in such a way that their spin states are correlated. The particles are separated and sent to two distant observers who measure their spins in different directions. The CHSH inequality states that the sum of the correlations between the measurements of the two observers cannot exceed a certain value. This value depends on the angles at which the measurements are made and is given by the formula:

$$|E(a, c) + E(a, d) + E(b, c) - E(b, d)| \leq 2, \tag{8.1}$$

where a, b, c, and d are angles at which the measurements are made, and $E(a, b)$ is the correlation between the measurements made at angles a and b. The correlation is defined as the difference between the number of times the two particles are measured in the same state and the number of times they are measured in different states. The CHSH bound of 2 is the maximum value that can be obtained in any experiment that obeys the laws of quantum mechanics.

The CHSH inequality has been tested in many experiments, and the results have consistently shown violations of the inequality. These violations are strong evidence that the behavior of entangled particles cannot be explained by classical physics and that quantum mechanics is the correct theory. The CHSH inequality has also been used to test the concept of local realism, which states that the properties of particles are determined by their local environment and cannot be influenced by distant measurements. The violation of the CHSH inequality shows that local realism is not a valid assumption and that entangled particles can be affected by distant measurements.

The CHSH inequality has important implications for the nature of reality. It shows that the correlations between entangled particles are not the result of any hidden variables that would make the particles act like classical objects. Instead, the correlations are a fundamental feature of quantum mechanics and cannot be explained by any classical theory.

The Bell theory has resolved the argument of Einstein-Podolsky-Rosen (EPR) about the entanglement steering (Einstein et al. 1935), which states that one party (the steerer) can manipulate local measurement settings to remotely influence the state of the other party (the steered). Now, the so-called EPR-steering is shown to be a genuine quantum effect, with applications in quantum communication and computation. The idea of steering suggests that there is a fundamental non-locality in the quantum world, where the act of measuring one part of an entangled system can instantaneously affect the state of the other part, even if they are separated by a large distance.

Another method is to verify entanglement beyond all separable states by using an entanglement witness (Horodecki et al. 2009). An entanglement witness can detect entanglement by measuring the expectation value of the operator on the two parts of the system. If the expected value is less than or equal to zero, then the system is separable, while if it is greater than zero, the system is entangled. This provides a powerful tool for characterizing and manipulating quantum states and is essential for the development of practical quantum technologies.

In summary, the quantum nonlocality (Wiseman et al. 2007) has been classified into three distinct types: quantum entanglement, Einstein-Podolsky-Rosen steering, and Bell nonlocality for bipartite entanglement in Fig. 8.1. All three phenomena are tied to the fundamental non-locality of quantum mechanics, which allows for instantaneous correlations between distant particles. While entanglement and EPR steering are both forms of nonlocality, Bell's nonlocality is a stronger form that cannot be explained by any classical theory. These results have recently been extended for quantum networks. This chapter contributes quantum nonlocality in different network local models.

The CHSH inequality has also been used to develop new technologies such as quantum key distribution, which allows secure communication over long distances. The violation of the CHSH inequality can be used to generate random numbers that can be used to encrypt messages. This is because the violation implies that the correlations between entangled particles are truly random and cannot be predicted by any classical theory. Quantum key distribution has important applications in fields such as banking, military communication, and e-commerce.

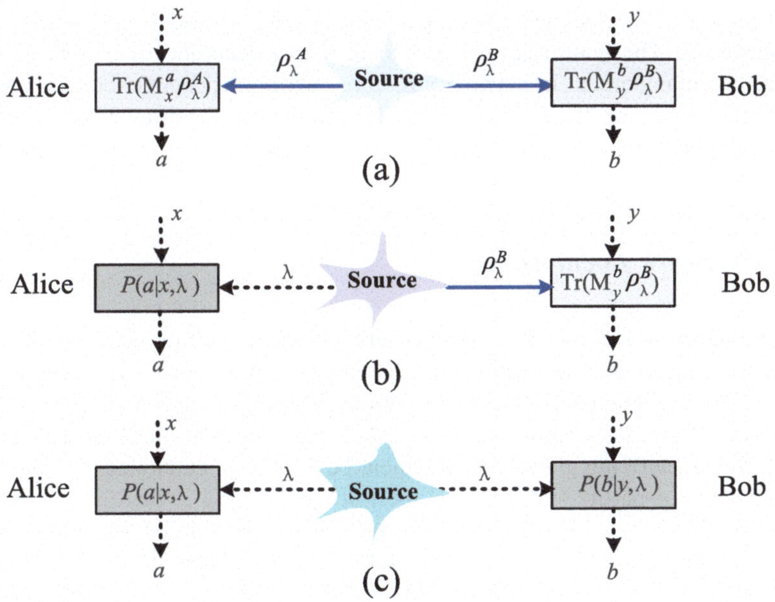

Fig. 8.1 Schematic entanglement, Einstein-Podolsky-Rosen (EPR) steering and nonlocality of bipartite states. **a** The entanglement in the separable model. **c** The EPR steering model in the local hidden state model. **a** The Bell nonlocality in the local hidden variable model. The black box denotes an untrusted measurement device. The white box represents a trusted device. All the sources with different colors are unknown

8.2 Quantum Network Entanglement

In this section, we present a new kind of entanglement in network scenarios. Entanglement is at the heart of quantum information processing. The standard definition of single entanglement is to explore the separability of systems with no less than two particles. In network scenarios, we only consider multipartite systems. This section contributes to the entanglement in quantum networks.

The standard definition of genuine multipartite entanglement generally requires the total state cannot be written as a convex mixture of partially entangled states (Svetlichny 1987; Seevinck and Uffink 2001).

Definition 8.2.1 An n-particle state on Hilbert space $\otimes_{i=1}^n \mathbb{H}_{A_i}$ is genuinely n-partite entanglement if it cannot be decomposed into

$$\rho_{bs} = \sum_S \sum_j p_{j;S} \rho_j^{(S)} \otimes \rho_j^{(\overline{S})}, \tag{8.2}$$

where S and $\overline{S}$ are any bipartition of $\{A_1, \ldots, A_n\}$, $\{p_{j;S}\}$ is a probability distribution, $\rho_j^{(S)}$ and $\rho_j^{(\overline{S})}$ are respective states of the particles in S and $\overline{S}$. Here, the state (8.2) is named a biseparable state.

This notion presents some unpredicted features when considering systems in network scenarios. The main reason is the local tensor decomposition. Investigating multipartite entanglement in network structures with independent sources can lead to novel questions and challenges in entanglement theory, quantum information, and fundamental physics.

8.2.1 Network Local Model

While the standard method for multipartite entanglement requires the decomposition (8.2) of joint systems. The local tensor decomposition is more natural in network scenarios from different network configurations. This allows new models of network local model (Luo 2021; Hansenne et al. 2022; Navascues et al. 2020; Kraft et al. 2020) for characterizing global entanglement beyond local entangled sources.

Consider an n-partite state ρ on Hilbert space $\otimes_{i=1}^n \mathbb{H}_{A_i}$, where $\mathbb{H}_{A_i}$ has local dimension $d_i \geq 2$, $i = 1, \ldots, n$. A network local state is defined according to partially entangled sources, i.e., no entangled source has been shared by n parties (Fig. 8.2).

Definition 8.2.2 An n-partite state is network local if it allows the following decomposition

$$\rho_{1\cdots n} = \sum_\lambda p(\lambda) \mathcal{E}_{1;\lambda} \otimes \cdots \otimes \mathcal{E}_{n;\lambda}(\rho_1 \otimes \cdots \otimes \rho_m)$$

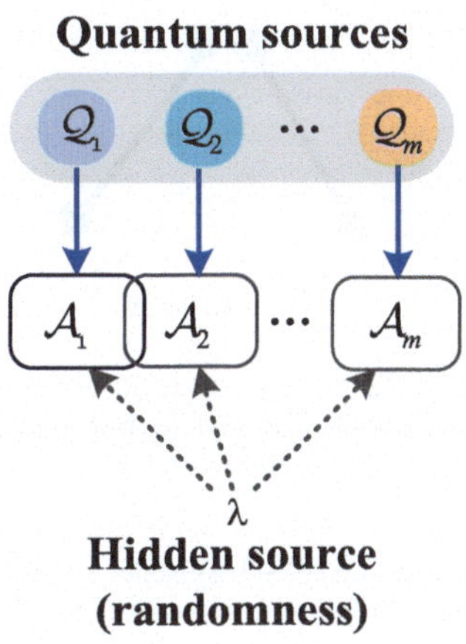

Fig. 8.2 (Color online) Schematic network local model. Each entangled source $\mathcal{Q}_i$ is shared at most $n-1$ parties $\mathcal{A}_i = \{\mathsf{A}_{s_1}, \ldots, \mathsf{A}_{s_i}\}$, $i = 1, \ldots, m$. All parties may share a variable λ

$$= \sum_{\lambda} p(\lambda) \sum_{i} \otimes_{i=1}^{n} U_{i,j;\lambda}(\rho_1 \otimes \cdots \otimes \rho_m)(\otimes_{j=1}^{n} U_{i,j;\lambda}^{\dagger}) \tag{8.3}$$

where ρ_i's are entangled states shared by at most $n-1$ parties, $\{p(\lambda)\}$ is the probability distribution of λ, and $\mathcal{E}_{i;\lambda}$ is a CPTP mapping that may depend on λ.

The channel $\mathcal{E}_{i;\lambda}$ can be represented by using Kraus operators $M_{i,j;\lambda}$ which satisfies $\sum_j M_{i,j;\lambda} M_{i,j;\lambda}^{\dagger} = \mathbb{1}$ with the identity operator $\mathbb{1}$.

While the genuinely multipartite entanglement rules out any biseparable state decomposition (Seevinck and Uffink 2001). Distributing bipartite entangled states can generate genuinely multipartite entanglement such as chain networks, star networks, and other resource states of cluster states (Raussendorf and Briegel 2001) or graph states (Hein et al. 2004). Take a tripartite chain network as an example. The joint state is deterministically transformed into a genuine entanglement of the GHZ state by LOCC. Instead, this kind of network allows network decomposition (8.3). This suggests that classical communication in the standard entanglement model is not allowed in the present model. Meanwhile, Definition 8.2.2 implies another way for verifying genuinely multipartite entanglement in an adversary manner, where an entangled source may be provided by an adversarial factory.

8.2.2 Network-Based Genuine Multipartite Entanglement

All the local states of the network consist of a convex set. This implies a natural definition of a new kind of entanglement, namely, a state is considered as network-based genuine multipartite entanglement if it is not network local. The network of local states can rule out special entanglement with local tensor decomposition. This suggests a stronger entanglement model beyond the biseperable model (Seevinck and Uffink 2001).

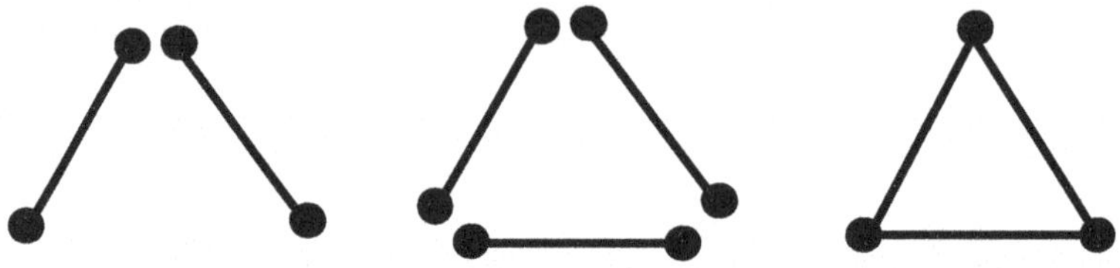

Fig. 8.3 (Color online) Schematic tripartite quantum networks. Chain and triangle networks consist of bipartite states while the last one consists of one tripartite state such as GHZ or W state

For a given number of network nodes, there are entangled networks with exponential configurations. The tripartite case contains a chain network consisting of two bipartite states, a triangle network consisting of three bipartite states, or one tripartite entanglement, as shown in Fig. 8.3. Verifying these different networks may be difficult for general entangled states. Instead, lacking local symmetry is the essential feature of the network of local states. This provides an efficient way of distinguishing network-based genuine multipartite entanglement. An n-partite state is permutationally symmetric if its density operator is invariant under any permutation operation in $\mathbb{S}_n$, where $\mathbb{S}_n$ denotes the permutation group associated with a set of n different elements.

Example 8.2.1 Consider a generalized d-dimensional GHZ state defined by

$$|GHZ\rangle = \sum_{i=0}^{d-1} a_i |i, \ldots, i\rangle, \tag{8.4}$$

where all the Hilbert spaces $\mathbb{H}_i$'s have the same dimension d with $d \geq 2$, and a_i's satisfy $\sum_{i=0}^{d-1} a_i^2 = 1$. This state is permutationally symmetric with respect to all permutation operations.

Example 8.2.2 Consider an n-particle Dicke state (Toth 2007) given by

$$|D_{k,n}\rangle = \frac{1}{\sqrt{N_{k,n}}} \sum_{i_1+\cdots+i_n=k} |i_1, \ldots, i_n\rangle, \tag{8.5}$$

where $N_{k,n} = C_{n-1}^{k+n-1}$ denotes the combination number of choosing $n-1$ balls from a box with $k+n-1$ balls, $k = 1, \ldots, nd - d - 1$. It has special form of the W state $\frac{1}{\sqrt{3}}(|001\rangle + |010\rangle + |100\rangle)$ for $k = 1$. These states are permutationally symmetric.

Both the maximally entangled GHZ and Dicke states are genuinely multipartite entangled in the separable model (Usha Devi et al. 2007). Different from GHZ states, Dicke states are robust against various noises, including particle loss, global dephasing, and bit flip noise. In general, a permutationally symmetric pure state can be represented by the superposition of Dicke states. These kinds of states, except for the product state of single-particle states, are network-based genuinely entangled (Luo 2021).

Theorem 8.2.1 *Any n-partite ($n \geq 3$) permutationally symmetric entangled pure state in the biseparable model is network-based genuine entanglement.*

Proof To show the main idea of proof, consider a 4-dimensional permutationally symmetric GHZ state

$$|GHZ\rangle = \frac{1}{2}\sum_{i=0}^{3}|iii\rangle_{ABC} \tag{8.6}$$

The basis states of particle A (B, or C) can be represented by two qubits A_1 and A_2 (B_1 and B_2, or C_1 and C_2), where $\mathbb{H}_A = \mathbb{H}_{A_1} \otimes \mathbb{H}_{A_2}$, $\mathbb{H}_B = \mathbb{H}_{B_1} \otimes \mathbb{H}_{B_2}$ and $\mathbb{H}_C = \mathbb{H}_{C_1} \otimes \mathbb{H}_{C_2}$. Similar to Eq. (8.3), if $|GHZ\rangle$ can be decomposed into two states under the local operations W_i, it is equivalent to the tensor of two GHZ states (up to local unitary operations), i.e.,

$$(W_1 \otimes W_2 \otimes W_3)|GHZ\rangle = |GHZ_1\rangle|GHZ_2\rangle, \tag{8.7}$$

where $|GHZ_i\rangle = \frac{1}{\sqrt{2}}(|000\rangle + |111\rangle)_{A_iB_iC_i}$. This means the symmetry is not changed under local decomposition (up to local unitary operations). Equation (8.7) can be proved by the contradiction from the symmetry. The main reason is the 2-dimensional Hilbert space cannot be further decomposed. This means GHZ state is network-based entanglement.

A similar proof can be extended for any n-partite (with $n \geq 3$) permutationally symmetric entangled pure state. This means that the symmetric state cannot be generated by some partially entangled states under local operations and shared randomness, where no one state is an n-partite entanglement in the biseparable model. This completes the proof. □

Any permutationally symmetric entangled pure state exhibits a new kind of multipartite nonlocality. This feature highlights the importance of classical communication in distinguishing network-based entanglement from those generated by distributed settings without classical communication. Another example can be found in general systems with small local spaces, where non-separable entanglement also plays an essential role in quantum communication tasks.

Theorem 8.2.2 *Any n-partite ($n \geq 3$) entangled pure state is network-based entanglement if the dimension of local state space is no more than 3.*

The proof is completed by contradiction. If any entangled state satisfies the condition in Theorem 8.2.2, it allows for a network decomposition as in Eq. (8.3). The proof is conducted through induction on the number of decomposed states, and this process relies on three facts. Firstly, the connectedness of entanglement in the chain network under local operations is necessary. For example, in a tripartite chain network, the total joint system state must be entangled. Secondly, entanglement is invariant under local unitary operations. Lastly, each Hilbert space with a dimension no greater than 3 cannot be decomposed into the tensor product of two Hilbert spaces with at least two dimensions.

Example 8.2.3 Consider the following state

$$|\Phi\rangle = \frac{1}{\sqrt{3}}(|000\rangle + |111\rangle + |122\rangle) \tag{8.8}$$

on Hilbert space $\mathbb{H}_A \otimes \mathbb{H}_B \otimes \mathbb{H}_C$, where $\mathbb{H}_A$ is 2 dimensional space while $\mathbb{H}_B$ and $\mathbb{H}_C$ are both 3 dimensional spaces. This state is tripartite entanglement and then network-based entanglement.

Theorem 8.2.2 offers an efficient method for detecting hybrid systems in which the local particles have different state spaces. Both Theorems 8.2.1 and 8.2.2 provide fascinating examples of network-based entanglement. As a result of these theorems, any pure state can be represented as an equivalent entangled state with the use of local unitary operations and auxiliary states. The present symmetries provide a versatile tool for the analysis of correlations in quantum networks (Hansenne et al. 2022).

8.2.3 Noisy Robustness

The present the network-based multipartite entanglement is robust against general noises. In fact, for an n-partite pure state $|\Phi\rangle$ on Hilbert space $\otimes_{i=1}^n \mathbb{H}_i$, denote $D(|\Phi\rangle)$ as the maximum distance from all network local states ρ_{net} as

$$D(|\Phi\rangle) = \sup_{\rho_{net}} F(\rho_{net}, |\Phi\rangle), \tag{8.9}$$

where F denotes the fidelity between ρ and ρ_Φ (Jozsa 1994). We get that

$$F(\rho_{net}, \rho_\Phi) \leq \max_i |\langle\Phi|\Psi_i\rangle|^2 \tag{8.10}$$

if ρ_{net} has the decomposition of $\rho_{net} = \sum_i p_i |\Psi_i\rangle\langle\Psi_i|$. This means that

$$D(|\Phi\rangle) = \sup_{\otimes_{i=1}^m |\Psi_i\rangle} |\langle\Phi| \otimes_{j=1}^n U_j (\otimes_{i=1}^m |\Psi_i\rangle)|^2, \tag{8.11}$$

where $|\Psi_i\rangle$ is any n-partite network local state, U_i is an arbitrary local unitary operation performed by the i-th party, $i = 1, \ldots, n$. The compactness of the Bloch sphere implies that $D(|\Phi\rangle) < 1$ for any network-based entanglement $|\Phi\rangle$. This provides a sufficient condition for witnessing a network-based entanglement ρ if we have

$$F(\rho, \rho_\Phi) > D(|\Phi\rangle). \tag{8.12}$$

From the convexity of all fully separable states (or biseparable states), we can construct an entanglement witness to distinguish a given bipartite entanglement (or each genuinely multipartite entanglement) (Horodecki et al. 2009). A similar result holds for all network-based entanglement. Using Eq. (8.12), we can construct a witness operator as

$$\mathcal{W} = F(\rho, \rho_\Phi)\mathbb{1} - |\Phi\rangle\langle\Phi|. \tag{8.13}$$

Combining with Eqs. (8.11) and (8.12) it follows that

$$\mathrm{Tr}[\mathcal{W}\rho_{net}] \geq 0, \mathrm{Tr}[\mathcal{W}\varrho] < 0, \tag{8.14}$$

for any network local state ρ_{net} and a given network-based entanglement ϱ.

In applications, evaluating the optimal distance for general entangled states can be difficult. Choosing a specific multipartite entangled pure state may be sufficient. Another approach is to use Theorems 8.2.1 and 8.2.2 to inspire some sufficient conditions. We can define a generalized permutationally symmetric entanglement as

$$|\Phi_{ps}\rangle = \sum_{i=0}^{nd-n-1} \alpha_i |D_{i,n}\rangle, \tag{8.15}$$

where $|D_{0,n}\rangle = \beta_0|0, \ldots, 0\rangle + \beta_1|d-1, \ldots, d-1\rangle$ with $\beta_0^2 + \beta_1^2 = 1$, and $|D_{k,n}\rangle$ are Dicke states (8.110). Proper restrictions on α_i are imposed to exclude symmetric product states.

Theorem 8.2.3 *For any n-partite ($n \geq 3$) state ρ, it is network-based entanglement if one of the following conditions holds*

(i) It satisfies

$$F(\rho, \rho_{GHZ}) > \max\{a_0^2, \ldots, a_{d-1}^2\} \tag{8.16}$$

for generalized GHZ states (8.4).

(ii) It satisfies

$$F(\rho, \rho_{D_{k,n}}) = F(\rho, \rho_{D_{nd-n-k-1,n}}) > \frac{n-1}{n+k-1} \tag{8.17}$$

for Dicke state (8.110), where $\lfloor x \rfloor$ denotes the maximal integer no larger than x, $k = 1, \ldots, \lfloor \frac{nd-n-1}{2} \rfloor$.

(iii) *For the state (8.15) it satisfies*

$$F(\rho, \rho_{\Phi_{ps}}) > \sum_{i=1}^{\lfloor\frac{nd-n-1}{2}\rfloor} \frac{n-1}{n+i-1}(\alpha_i^2 + \alpha_{nd-n-1-i}^2) + \alpha_0^2\beta^2 \tag{8.18}$$

for an odd $nd - n - 1$*; or*

$$F(\rho, \rho_{\Phi_{ps}}) > \sum_{i=1}^{\lfloor\frac{nd-n-1}{2}\rfloor} \frac{n-1}{n+i-1}(\alpha_i^2 + \alpha_{n-i}^2) - \frac{2n-2}{nd+n-3}\alpha^2_{\frac{nd-n-1}{2}} + \alpha_0^2\beta^2 \tag{8.19}$$

for an even $nd - n - 1$*, where* β *is defined by* $\beta = \max\{\beta_0, \beta_1\}$.

(iv) *For a network-based entangled qubit state* $|\Phi\rangle$ *it satisfies*

$$F(\rho, \rho_\Phi) > \max_{\mathcal{A}\subset\{A_1,\ldots,A_n\}} \max \lambda(\rho_\mathcal{A}), \tag{8.20}$$

where $\lambda(\rho_\mathcal{A})$ *denotes all the spectrums of* $\rho_\mathcal{A}$*, and* $\rho_\mathcal{A}$ *is the reduced density matrix of systems in* $\mathcal{A}$ *with* $\mathcal{A} \subset \{A_1, \ldots, A_n\}$.

Theorem 8.2.3 provides a useful method to verify network-based multipartite entanglement with general noises. The proof is based on the detailed evaluations of $D(|\Phi\rangle)$ for each specific network-based entanglement (Luo 2021).

Example 8.2.4 Consider an n-partite GHZ state (8.4) with white noise, given by

$$\rho_v = v|GHZ\rangle\langle GHZ| + \frac{1-v}{d^n}\mathbb{1}, \tag{8.21}$$

where $\mathbb{1}$ denotes the identity operator. The weight v may operationally represent the interferometric contrast observed in experiments (Werner 1989). For the maximally entangled GHZ state, ρ_v is network-based entangled for $v > \frac{2^{n-1}-1}{2^n-1}$. The noise visibility is consistent with its biseparable model. This implies an entanglement witness operator of the form

$$\mathcal{W}_{GHZ} = a\mathbb{1} - |GHZ\rangle\langle GHZ|, \tag{8.22}$$

which can be used to witness ρ_v, when $v \geq \frac{d^n a-1}{d^n-1}$, with $a = \max\{a_0^2, \ldots, a_{d-1}^2\}$.

Example 8.2.5 Consider a Dicke state (8.5) with white noise given by

$$\varrho_v = v|D_{k,n}\rangle\langle D_{k,n}| + \frac{1-v}{d^n}\mathbb{1}, \tag{8.23}$$

where $v \in [0, 1]$ and $k \leq \lfloor \frac{nd-n-1}{2} \rfloor$. An entanglement witness operator is given by

$$\mathcal{W}_{D_{k,n}} = \frac{n-1}{n+k-1}\mathbb{1} - |D_{k,n}\rangle\langle D_{k,n}|, \tag{8.24}$$

which can be used to witness the network-based entanglement of ϱ_v with the noise visibility of $v \geq \frac{(n-1)2^n-n-k+1}{(2^n-1)(n+k-1)}$.

Similarly, one may witness a general permutationally symmetric noisy state given by

$$\varrho_v = \sum_{k=1}^{nd-n-1} v_k|D_{k,n}\rangle\langle D_{k,n}| + \frac{v_0}{d^n}\mathbb{1} \tag{8.25}$$

where $v_i \geq 0$ and $\sum_{i=0}^{nd-n-1} v_i = 1$. One method is from Eq. (8.24). It follows from Eq. (8.13) a witness operator

$$\mathcal{W}_{D,n} = \sum_{k=1}^{nd-n-1} \mathcal{W}_{D_{k,n}}, \tag{8.26}$$

when v_0 satisfies $v_0 < \frac{d^n}{d^n-nd+n+1}(1 - \sum_{k=1}^{nd-n-1} \ell_k)$ with $\ell_k = \ell_{nd-n-1-k} = \frac{n-1}{n+k-1}$ for $k = 1, \ldots, \lfloor \frac{nd-n-1}{2} \rfloor$.

The network local model is stronger than the biseparable model, and some entangled states with network decompositions are also network local. Therefore, all the witness operators shown in Eqs. (8.22), (8.24), and (8.26) can be used to verify genuinely multipartite entanglement in the biseparable model.

Example 8.2.6 Consider a network-based entangled pure state $|\Phi\rangle$ with a general noisy operator ϱ, given by

$$\rho_v = v|\Phi\rangle\langle\Phi| + (1-v)\varrho. \tag{8.27}$$

Theorem 8.2.3 provides an efficient method to witness the noisy state. One approach is to use the symmetric space spanned by Dicke states. In particular, the witness operator (8.26) implies a noise visibility of

$$v > \frac{\sum_{k=1}^{nd-n-1} \ell_k}{\sum_{k=1}^{nd-n-1} \beta_k}, \tag{8.28}$$

where $\beta_i = |\langle D_{i,n}|\Phi\rangle|^2$ for $i = 1, \ldots, nd-n-1$, and $\ell_k = \ell_{nd-n-1-k} = \frac{n-1}{n+k-1}$ for $k = 1, \ldots, \lfloor \frac{nd-n-1}{2} \rfloor$.

Another example is cluster states or graph states associated with specific graphs, which are of interest in quantum information processing and measurement-based quantum computations. It is known that single-connected graph states have genuinely multipartite entanglement in the biseparable model. Theorems 8.2.2 and 8.2.3 imply new generic entanglement in the present model. The network local model is stronger than the biseparable model, and some entangled states with network decompositions are also network local. This raises the natural problem of how to detect its Bell nonlocality.

8.3 Robust Multipartite Entanglement

In this section, we present a new kind of entanglement that is robust against particle loss in both single entanglement and network scenarios. In experimental settings, states based on atomic ensembles provide attractive systems for storing or transferring quantum information. However, a noisy evolution may induce an unstable isotope for the nucleus, and the particle loss channel may break the entanglement feature for multipartite systems. However, there are states like Dicke states that maintain their entangling character even after losing any particle. This inspires a new way of characterizing multipartite quantum systems in terms of particle loss channels (Luo and Fei 2021).

8.3.1 *Particle-Lose Channels*

Denote S as a subset of $\{A_1, \ldots, A_n\}$, and $\overline{S}$ as the complement set of S.

Definition 8.3.1 A CPTP mapping $\mathcal{E}_S(\cdot)$ is a particle-lose channel associated with S if it can be decomposed into

$$\mathcal{E}_S(\rho) = \sum_{E_j \in K_S} E_j \rho E_j^\dagger, \tag{8.29}$$

where K_S denotes the Kraus operators of $\mathcal{E}_S(\cdot) = \mathrm{Tr}_{A_j \in S}(\cdot)$ and $\mathrm{Tr}_{A_j \in S}(\cdot)$ denotes partial trace operator over all particles in S.

When $S = \{A_1\}$, we have $E_j = \langle j|_{A_1} \otimes \mathbb{1}_{A_2 \cdots A_n}$, where $\mathbb{1}_{A_2 \cdots A_n}$ is the identity operator on the particles $A_2, \ldots, A_n$. Let $\rho_{(S)}$ denote the output after the particle-lose channel $\mathcal{E}_S(\cdot)$, that is, $\rho_{(S)} = \mathcal{E}_S(\rho)$. In what follows, we explore new features of multipartite entangled systems both in a single source and network configurations beyond the biseparable model or network local model.

Definition 8.3.2 An n-partite state ρ is particle-lose separable if $\rho_{(S)}$ is biseparable for some set S with no more than $n - 2$ particles.

Example 8.3.1 Consider the particle-lose separable (8.4). The reduced state is separable. Thus the GHZ state is particle-lose separable.

Example 8.3.2 Consider the Dicke state (8.5). The remained state is entangled after losing one qubit.

Definition 8.3.2 presents a stronger condition than the k-uniform entanglement because all the k-uniform entangled states require its reductions to be maximally mixed. This model may lose any particles beyond the specific losing particles (Barnea et al. 2015; Sohbi et al. 2015).

Example 8.3.3 Consider the generalized GHZ state given by

$$|GHZ\rangle = (\cos\theta|0,\ldots,0\rangle + \sin\theta|1,\ldots,1\rangle)_{A_1\cdots A_n}, \tag{8.30}$$

where $\theta \in (0, \frac{\pi}{2})$. This state is genuinely multipartite entanglement for $\theta \in (0, \frac{\pi}{2})$. However, it reduces to a fully separable state after losing any one qubit.

All the generalized GHZ states are fragile with respect to particle-loss noises. This kind of state is unique under local unitary operations (Luo and Fei 2021).

Theorem 8.3.1 *A genuinely entangled pure qubit state $|\Phi\rangle$ is equivalent to the generalized GHZ state if and only if the reduced state after losing any one qubit is fully separable, that is, there are local unitary operations U_j such that*

$$(\otimes_{i=1}^n U_i)|\Phi\rangle_{A_1\cdots A_n} = |GHZ\rangle_{A_1\cdots A_n}. \tag{8.31}$$

Proof The sufficient condition is easily proved by a straight evaluation. The necessary condition will be proved by induction. To show the main idea, we consider $n = 3$ as follows. For a genuinely entangled pure qubit state $|\Phi\rangle_{A_1A_2A_3}$, if its reduced state of $\rho_{(A_i)} = \mathcal{E}_{A_i}(|\Phi\rangle\langle\Phi|)$ is fully separable for any $i \in \{1, 2, 3\}$ there are local unitary operations U_i satisfying (8.31). For $i = 1$ the Schmidt decomposition is given by

$$|\Phi\rangle = \sum_{i=0}^{1} \sqrt{\lambda_i}|\psi_i\rangle_{A_1}|\Phi_i\rangle_{A_2A_3}, \tag{8.32}$$

where $|\psi_i\rangle$'s are orthogonal states of qubit A_1, and $|\Phi_i\rangle$'s are orthogonal states of qubits A_2 and A_3, and λ_i are positive Schmidt coefficients.

For $i = 2$ there two qubits A_1 and A_2 are symmetric up to local unitary operations, i.e.,

$$(U_1 \otimes U_2 \otimes \mathbb{1})|\Phi\rangle = \sum_{s=0}^{1} \sqrt{p_s}|ss\rangle_{A_1A_2}|\Psi_s\rangle_{A_3}, \tag{8.33}$$

for some unitary operations U_i, where $|\Psi_s\rangle_{A_3}$'s are orthogonal states. Note that $\mathcal{E}_{A_1}(|\Phi\rangle\langle\Phi|)$ is fully separable. Combined with Eq. (9.45) the reduced state of $\varrho_{(A_1)}$ is given by

$$\varrho_{(A_1)} = \mathcal{E}_{A_1}(U_1 \otimes \mathbb{1})|\Phi\rangle\langle\Phi|(U_1^\dagger \otimes \mathbb{1}) \tag{8.34}$$

$$= \sum_{s=0}^{1} p_s |\Phi_s\rangle\langle\Phi_s|_{A_2A_3}, \tag{8.35}$$

Equation (8.34) is followed from the invariance of $\mathcal{E}_{A_1}(\cdot)$ under local unitary operations. There are two cases. One is at least one state $|\Phi_i\rangle$ is fully separable. The state in Eq. (8.32) is equivalent to a generalized GHZ state under local unitary operations. The other is both states are entangled. In this case, two qubits state in Eq. (8.32) are symmetric under local unitary operations. It follows the results. A similar proof can be extended to general cases. □

Theorem 8.3.1 has a noteworthy implication regarding the uniqueness of fragile entangled qubit systems, which distinguishes this result from those obtained in the biseparable model or the network local model. Specifically, the theorem implies that if a qubit system is genuinely entangled and fragile, then it is unique, up to local unitary operations. This result holds not only for qubits but also for high-dimensional entangled pure states, including absolute maximal entangled states (Helwig et al. 2012) and symmetric states.

Theorem 8.3.2 *A permutationally symmetric entangled pure state is equivalent to the generalized GHZ state in Eq. (8.4) if and only if the reduced state after losing any one state is fully separable.*

Proof The sufficient condition is easily followed from Eq. (8.4). The necessary condition is followed by the symmetry of all particles. The fact is that $\{|GHZ_d\rangle, |D_{k,n}\rangle, \forall k\}$ consists of an orthogonal basis of permutationally symmetric subspace of $\otimes_{i=1}^n \mathbb{H}_{A_i}$, where $|D_{k,n}\rangle$ denotes the maximally entangled n-particle Dicke state in Eq. (8.5). This means a permutationally symmetric state can be represented by

$$|\Phi_{sy}\rangle = \sum_k \beta_k |D_{k,n}\rangle + \beta|GHZ_d\rangle, \tag{8.36}$$

where β_k's and β satisfy $\sum_k |\beta_k|^2 + \beta^2 = 1$.

Note that

$$\sum_k \beta_k |D_{k,n}\rangle = \sum_{j_1=0}^{d-1} |j_1\rangle_{A_1} \sum_k \frac{\beta_k}{\sqrt{N_{k-j_1,n}}} |D_{k-j_1}\rangle_{A_2\cdots A_n}, \tag{8.37}$$

where $|D_{k-j_1,n}\rangle$ denotes $n-1$-particle Dicke state with excitations $k-j_1$. The reduced state in Eq. (8.36) after losing A_1 is given by

$$\rho_{(A_1)} = \mathcal{E}_{A_1}(|\Phi_{sy}\rangle\langle\Phi_{sy}|)$$

$$= \sum_k \frac{\beta_k^2}{N_{k-j_1,n}} |D_{k-j_1,n-1}\rangle\langle D_{k-j_1,n-1}| + \beta^2 \sum_{j=0}^{d-1} \alpha_j^2 |j,\ldots,j\rangle\langle j,\ldots,j|. \tag{8.38}$$

The states $|D_{k_1-j_1,n-1}\rangle$ and $|D_{k_2-j_1,n-1}\rangle$ have different excitations for any $k_1 \neq k_2$. So, all the states of $\{|D_{k-j_1,n-1}\rangle, \forall k\}$ are on different orthogonal subspaces $\mathbb{H}_{k_1}$ and $\mathbb{H}_{k_2}$ of $\otimes_{i=2}^{n}\mathbb{H}_{A_i}$, where $\mathbb{H}_{k_s}$ is spanned by all the states $\{|j_2, \ldots, j_n\rangle, \sum_{t=2}^{n} j_t = k_s - j_1, s = 1, 2\}$. This means $\{|D_{k-j_1,n-1}\rangle, \forall k\}$ are orthogonal states. Moreover, the state $|D_{k-j_1,n-1}\rangle$ is genuinely entangled in the biseparable model. It follows that

$$\varrho_{A_2\cdots A_n} = \sum_{k} \frac{\beta_k^2}{N_{k-j_1,n}} |D_{k-j_1,n-1}\rangle\langle D_{k-j_1,n-1}| \tag{8.39}$$

is genuinely entangled if $\sum_k \beta_k^2 \neq 0$. Note that $\sum_{j=0}^{d-1} \alpha_j^2 |j, \ldots, j\rangle\langle j, \ldots, j|_{A_2\cdots A_n}$ is fully separable. This implies that $\sum_k \beta_k^2 = 0$, that is, $\beta_k = 0$ for all k's. From Eq. (8.38) we obtain

$$\rho_{(A_1)} = \beta^2 \sum_{j=0}^{d-1} \alpha_j^2 |j, \ldots, j\rangle_{A_2\cdots A_n}\langle j, \ldots, j|. \tag{8.40}$$

The symmetry is then used to complete the proof. □

Theorem 8.3.2 is for any permutationally symmetric entangled pure state. A simple extension is for general high-dimensional states. However, it is a challenge for general high-dimensional scenarios which may be not permutationally symmetric. One example is $|\Phi\rangle = a^2|000\rangle + ab|011\rangle + ab|120\rangle + b^2|131\rangle$ with $a^2 + b^2 = 1$. Only losing the particle A_2 will result in a separable.

8.3.2 *Robust Multipartite Entanglement*

Definition 8.3.2 implies a new feature of multipartite entanglement as follows.

Definition 8.3.3 For a multipartite state, it is robust entanglement if it is not a particle-lose separable.

Let $\mathcal{B}_d$ consist of all the particle-lose separable states. Definition 8.3.2 shows that for any state ρ that is not in the set $\mathcal{B}_d$ of particle-lose separable states, there exists strong robustness against particle-lose noise. It is worth noting that $\mathcal{B}_d$ is only star-convex but not convex. This means that any state in $\mathcal{B}_d$ can be connected to a center $\rho_0 \in \mathcal{B}_d$, but not all states in the convex hull of $\mathcal{B}_d$ belong to $\mathcal{B}_d$. As a consequence, a general linear entanglement witness (Horodecki et al. 2009) is ruled out for witnessing robust entanglement, in contrast to entanglement theories such as Bell nonlocality and EPR-steering (Bell 1964; Svetlichny 1987; Luo 2021), where the corresponding sets of separable states are convex.

Consider a generalized Dicke state given by

$$|D_{k,n}\rangle_{A_1\cdots A_n} = \sum_{i_1+\cdots+i_n=k} \alpha_{i_1\cdots i_n}|i_1,\cdots,i_n\rangle, \tag{8.41}$$

where k denotes the number of excitations, n denotes the number of particles, $\alpha_{i_1\cdots i_n}$'s satisfy $\sum_{i_1,\ldots,i_n}\alpha^2_{i_1\cdots i_n}=1$ and $\alpha_{i_1\cdots i_n}\neq 0$ for any $i_1,\ldots,i_n$.

Theorem 8.3.3 *Any generalized Dicke state is robust entanglement.*

Proof The state in Eq. (8.41) is a generalized symmetric state, which means that after any permutation of particles, the final state has a similar form Eq. (8.41) but with different amplitudes. It suffices to prove the genuineness of the reduced state $\rho_{(A_{i+1},\ldots,A_n)}$ for $i=2,\ldots,n-1$.

In what follows, we prove the result with $n=3$. Similar proof holds for general cases. Note that

$$|D_{k,3}\rangle_{A_1A_2A_3} = \sum_{i_3}\alpha_{i_3}|i_3\rangle_{A_3}|D_{k-i_3,2}\rangle_{A_1A_2}, \tag{8.42}$$

where $|D_{k-j_3,2}\rangle$ is defined as $|D_{k-j_3,2}\rangle=\sum_{j_1+j_2=k-j_3}\alpha_{j_1j_2}|j_1j_2\rangle$. The reduced density matrix of $\rho_{(A_3)}$ is given by

$$\rho_{(A_3)} = \sum_{i_3}\alpha^2_{i_3}|D_{k-i_3,2}\rangle_{A_1A_2}\langle D_{k-i_3,2}|. \tag{8.43}$$

For any two different integers i_3 and i_3', the states $|D_{k-i_3,2}\rangle$ and $|D_{k-i_3',2}\rangle$ are orthogonal. This means the state in Eq. (8.43) is a bipartite entanglement if someone of $|D_{k-i_3,2}\rangle$ is a bipartite entanglement. Similar proof holds for general integers of n and k. □

Example 8.3.4 Consider a superposition of generalized Dicke states as

$$|\Phi\rangle = \sum_k \beta_k|D_{k,n}\rangle, \tag{8.44}$$

where β_k's satisfy $\sum_k\beta_k^2=1$. This state exhibits robust entanglement for certain choices of β_k if it is a genuinely n-partite entangled state in the biseparable model. It is worth noting that the Dicke states in Eq. (8.5) are a special case of the generalized Dicke states, obtained when all the amplitudes are equal. These Dicke states are known to be genuinely multipartite entangled. Therefore, generalized Dicke states can be seen as a network-based entanglement, characterized by the amplitudes β_k.

8.3.3 Robust Entanglement of Quantum Networks

A multipartite quantum network is a collection of independent entangled states that are distributed among different parties. Such quantum resources are highly convenient for large-scale tasks with short-range settings. In this subsection, we discuss the robust entanglement of quantum networks. We assume that all the particles owned by one party are considered as a single entity in a large space. Definition 8.3.2 can then be applied to analyze the scenario where all the network nodes are lost or offline during the application of the quantum network.

Consider a general n-partite network $\mathcal{N}_q$ consisting of parties A_i's. $\mathcal{N}_q$ is said to be k-independent if there exist k parties (e.g., $\mathsf{A}_1, \ldots, \mathsf{A}_k$) who do not share any entanglement. Examples of bipartite states that exhibit this property are shown in Fig. 8.4. Note that the independence of the parties cannot be altered without the help of other parties. If all the particles shared by A_j with $j \geq k+1$ are lost, the resulting state becomes fully separable.

Theorem 8.3.4 *Any k-independent quantum network with $k \geq 2$ is particle-lose separable.*

Example 8.3.5 Consider an n-partite chain network shown in Fig. 8.4a, where each pair of two adjacent parties A_j and A_{j+1} share a single bipartite state. In this network, all parties A_i with even i (or all parties A_i with odd i) are independent of each other. If a particle-lose channel $\mathcal{E}_{\mathsf{A}_i}(\cdot)$ is applied to the particles owned by A_i, the resulting reduced state becomes separable.

Example 8.3.6 Consider an $n+1$-partite star network shown in Fig. 8.4b, where each A_j shares one bipartite states with B. The total state is given by

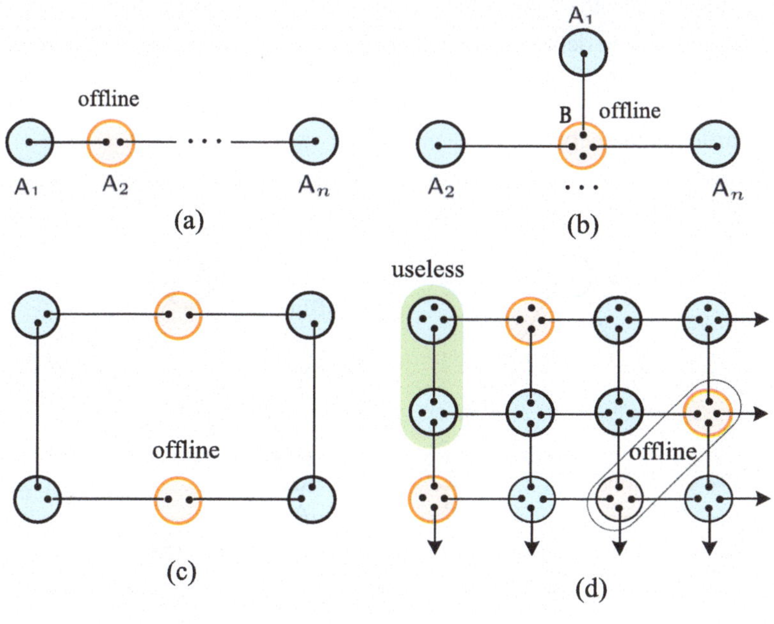

Fig. 8.4 (Color online) Schematic k-independent quantum networks consisting of bipartite entangled states. **a** Chain network. **b** Star network. **c** Cyclic network. **d** Planer network. These networks are particle-lose separable

$$|\Omega\rangle = \frac{1}{\sqrt{2^n}} \sum_{j_1,\ldots,j_n=0}^{1} |j_1, \ldots, j_n\rangle_{\mathsf{A}_1\cdots\mathsf{A}_n} |j_1, \ldots, j_n\rangle_{\mathsf{B}}. \tag{8.45}$$

All the parties A_j's are independent. Passing through the particle-lose channel $\mathcal{E}_{\mathsf{B}}(\cdot)$ will lead to a maximally mixed state.

The robustness of any k-independent network depends solely on the independence of the parties, and not on the shared entangled states. If a network is not robust against particle loss, it may not be suitable for certain applications, such as measurement-based quantum computation.

Different from k-independent networks other networks show stronger connectedness.

Definition 8.3.4 A multipartite quantum network is a *completely connected* if each pair of two parties shares at least one entanglement.

An example of a quantum network that is not robust against particle-loss is shown in Fig. 8.5. Let $\mathcal{N}_q$ consist of bipartite entangled pure states $|\phi\rangle = \sum_j \beta_j |jj\rangle_{AB}$ and generalized Dicke states (8.41), where β_j's satisfy $\sum_j \beta_j^2 = 1$ and $\beta_j \neq 0$ for at least two integers j.

Theorem 8.3.5 *Any completely connected quantum network $\mathcal{N}_q$ is robust entanglement.*

Proof The first step to demonstrate that the total state is an n-partite genuine entanglement is to show that the network is connected. Specifically, this means that each pair of parties can share at least one bipartite entanglement, assisted by the LOCC of other parties in the network.

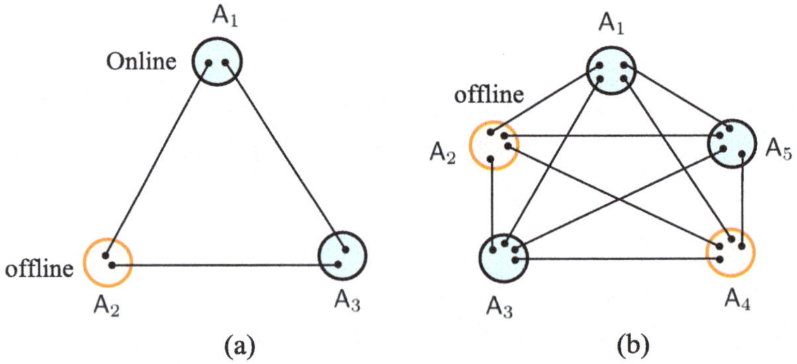

Fig. 8.5 (Color online) Schematic completely connected quantum networks consisting of bipartite entangled states. **a** Triangle network. **b** 5-partite network

Another is to demonstrate that the extended network $\mathcal{N}_q \cup \{|\phi\rangle\}$ is a robust entanglement, assuming that $\mathcal{N}_q$ is a robust entanglement and $|\phi\rangle$ is a bipartite entanglement shared by parties A_i and A_j. Let $|\phi_{i;j}\rangle = \sum_k \alpha_{k;ij}|kk\rangle$ be the shared entanglement between A_i and A_j, where $\alpha_{k;ij}$'s satisfy $\sum_k \alpha_{k;ij}^2 = 1$. The total system is given by

$$|\Phi\rangle_{\mathsf{A}_1\cdots\mathsf{A}_n} = \otimes_{1\leq i\neq j\leq n}|\phi_{i;j}\rangle. \tag{8.46}$$

Now, consider a proper subset $S = \{\mathsf{A}_{i_1}, \ldots, \mathsf{A}_{i_k}\} \subset \{\mathsf{A}_1, \ldots, \mathsf{A}_n\}$ and let $\overline{S}$ be the complement set of S. The joint state after passing through the particle-loss channel $\mathcal{E}_{\overline{S}}(\cdot)$, where $\mathcal{E}_{\overline{S}}(\cdot) := \mathcal{E}_{\mathsf{A}_i\in\overline{S}}(\cdot)$, is then given by

$$\rho(\overline{S}) = (\otimes_{s\neq t, s,t\in\{i_1,\ldots,i_k\}}|\phi_{s;t}\rangle\langle\phi_{s;t}|) \otimes_{j=1}^k \rho_{i_j}, \tag{8.47}$$

where ρ_{i_j} denotes the local state shared by A_{i_j} after losing all the particles entangled with A_{i_j}.

The final step is to prove $\otimes_{s\neq t, s,t\in\{i_1,\ldots,i_k\}}|\phi_{s;t}\rangle\langle\phi_{s;t}|$ is fully separable. This can be established based on the assumption of completely connected networks. □

A similar result holds for any completely connected quantum network that consists of other pure entangled states or mixed entangled states. These quantum networks are called "network local" if each multipartite state is shared among at most $n-1$ parties. This shows another useful feature of completely connected quantum networks.

Example 8.3.7 Consider a tripartite completely connected quantum network consisting of three EPR states in Fig. 8.3a. The total state is given by

$$\begin{aligned} |\Omega\rangle = \frac{1}{2\sqrt{2}}(&|00\rangle_{\mathsf{A}_1}|00\rangle_{\mathsf{A}_2}|00\rangle_{\mathsf{A}_3} + |10\rangle_{\mathsf{A}_1}|00\rangle_{\mathsf{A}_2}|01\rangle_{\mathsf{A}_3} \\ &+|00\rangle_{\mathsf{A}_1}|01\rangle_{\mathsf{A}_2}|10\rangle_{\mathsf{A}_3} + |10\rangle_{\mathsf{A}_1}|01\rangle_{\mathsf{A}_2}|11\rangle_{\mathsf{A}_3} \\ &+|01\rangle_{\mathsf{A}_1}|10\rangle_{\mathsf{A}_2}|00\rangle_{\mathsf{A}_3} + |11\rangle_{\mathsf{A}_1}|10\rangle_{\mathsf{A}_2}|01\rangle_{\mathsf{A}_3} \\ &+|01\rangle_{\mathsf{A}_1}|11\rangle_{\mathsf{A}_2}|10\rangle_{\mathsf{A}_3} + |11\rangle_{\mathsf{A}_1}|11\rangle_{\mathsf{A}_2}|11\rangle_{\mathsf{A}_3}) \end{aligned} \tag{8.48}$$

which can be regarded as a pure state on Hilbert space $\otimes_{j=1}^3 \mathbb{H}_{A_j}$, where $\mathbb{H}_{A_j}$ are 4-dimensional spaces. The reduced state includes an EPR state $|\phi\rangle_{\mathsf{A}_2\mathsf{A}_3}$. Similar result holds for the other reduced state $\rho_{(\mathsf{A}_2)}$ or $\rho_{(\mathsf{A}_3)}$. This implies the total state is robust entanglement.

8.3.4 Robustness-Depth of Multipartite Entanglement

Denote $\mathcal{B}_b$ as the set of all biseparable states on a Hilbert space $\otimes_{j=1}^n \mathbb{H}_{A_j}$. It follows that $\mathcal{B}_b \subset \mathcal{B}_d$. Let $\mathcal{B}_n$ consist of all network local states. We can then establish that $\mathcal{B}_b \subseteq \mathcal{B}_d \cap \mathcal{B}_n$, meaning there exist genuinely entangled states that are neither network-based entanglement nor robust entanglement. This further implies $\mathcal{B}_b \subset \mathcal{B}_d \cap \mathcal{B}_n$. These facts inspire a natural problem of verifying the robustness-depth.

Definition 8.3.5 An n-partite state has robustness depth k if the reduced state is genuinely entangled after passing through any particle-losing channel $\mathcal{E}_S$ for any S with $|S| \leq k - 1$.

Let $\mathcal{B}_d^{(k)}$ as the set of all states with a robustness depth less than k. It follows that $\mathcal{B}_d^{(k)}$ is star convex. Any state not belonging to $\mathcal{B}_d^{(k)}$ has a robustness depth that is greater than or equal to k.

Definition 8.3.6 An n-partite quantum network is k-connected if each party shares at most k bipartite entangled states with other parties.

Theorem 8.3.6 *The robustness-depth of any n-partite k-connected quantum network is at most $k - 1$.*

The main idea of the proof is based on the k-connectedness of networks. Any n-partite k-connected quantum network is genuinely entangled. This can be established by considering the activation ability of each pair, aided by other parties' LOCC. These networks are inherently nonlocal due to their connectedness (Contreras-Tejada et al. 2021). However, they are not robust against the loss of k particles since there are $n - k$ parties whose total states are biseparable.

This theorem implies a general hierarchy of multipartite states in terms of the robustness-depth as

$$\mathcal{B}_s \subset \mathcal{B}_b \subset \mathcal{B}_d^{(1)} \subset \mathcal{B}_d^{(2)} \subset \cdots \subset \mathcal{B}_d^{(n-1)}, \tag{8.49}$$

where $\mathcal{B}_s$ consists of all fully separable states.

Due to the non-convexity of $\mathcal{B}_d^{(k)}$, verifying the robustness-depth for $\rho \notin \mathcal{B}_d^{(k)}$ can be a challenging task. Interestingly, quantum networks provide an easy example. Suppose that $\mathcal{N}_q$ is an n-partite network consisting of bipartite entangled states. The robustness-depth of $\mathcal{N}_q$ is then determined by its minimum degree, which can be expressed as $\deg(\mathcal{N}_q) = \min\{\deg(\mathsf{A}_j), \forall j\}$.

Theorem 8.3.7 *The robustness depth of $\mathcal{N}_q$ is given by $d = \deg(\mathcal{N}_q) - 1$, where $\deg(\mathcal{N}_q)$ denotes its minimum degree.*

The proof relies on the following fact: for every pair of two parties in $\mathcal{N}_q$, there exist at least k distinct chain subnetworks that connect them. This is useful in verifying the final entanglement state after passing through the particle-lose channel.

Cluster states play a crucial role in quantum computations and are often decomposed into networks based on their building blocks. Linear cluster states, represented by a chain network as shown in Fig. 8.4a, are associated with a linear graph with a degree no greater than 2. This limited connectedness makes them vulnerable to the loss of a single particle, hence they are not robust. For 2D cluster states, as shown in Fig. 8.4d, we can easily determine their minimum degree to verify their robustness depth. Other examples include multipartite states such as GHZ states (Luo 2018).

8.4 Quantum Network Steering

The concept of quantum steering was first introduced by Einstein, Podolsky, and Rosen in their paper (Einstein et al. 1935). It refers to a scenario in which one party in a bipartite system can influence the state of the other party through local measurements. This phenomenon is formalized as steering, which arises from the incompatibility of quantum predictions with a classical-quantum model (Wiseman et al. 2007). The observation of quantum steering is considered entanglement detection under the assumption of one uncharacterized measurement, as illustrated in Fig. 8.1b. This provides a new form of quantum nonlocality that lies between standard entanglement tests and Bell nonlocality tests. In this section, we discuss the concept of steering in network scenarios.

8.4.1 Quantum EPR Steering

Consider a bipartite scenario, as depicted in Fig. 8.1b, where Alice and Bob share an unknown state ρ_{AB}. Alice performs m_A measurements on her subsystem labeled by $x = 0, \ldots, m_A - 1$, each having n_A outcomes $a = 0, \ldots, n_A - 1$. Conditional on the measurement setup x and outcome a, Bob's state is transformed into ρ_x^a with a probability $p(a|x)$. While Bob can fully control his measurements and the conditional states ρ_x^a, there is no assumption made about Alice's measurements. This scenario is called a one-sided device-independent scenario, which is different from the two-sided device-independent Bell nonlocality test.

In EPR steering, the available information is all the conditional states and probabilities, i.e., $\{\rho_x^a, p(a|x)\}_{a,x}$. This information can be summarized by unnormalized quantum states $\{\varrho_x^a\}_{a,x}$, where $\varrho_x^a = p(a|x)\rho_x^a$. These states are often referred to as an assemblage.

According to quantum theory, the conditional states are given by

$$\varrho_x^a = \mathrm{Tr}_A[(M_x^a \otimes \mathbb{1})\rho_{AB}], \tag{8.50}$$

where $\{M_x^a\}$ are POVM operators satisfying $\sum_a M_x^a = \mathbb{1}$. Here, the state ρ_{AB} and the measurements $\{M_x^a\}_{a,x}$ are completely unknown in the steering scenario. A more involved assumption is the unknown space dimension for Alice. This gives a quantum assemblage as

$$\begin{aligned}
&\sum_a \varrho_x^a = \sum_a \varrho_{x'}^a = \rho_B, \forall x, x' \\
&\mathrm{Tr} \sum_a \varrho_x^a = 1, \forall x, \\
&\rho_B = \mathrm{Tr}_A \rho_{AB}.
\end{aligned} \tag{8.51}$$

Equation (8.51) is from no-signaling constrictions.

Quantum steering provides a new type of impossibility with the local hidden state (LHS) model. The LHS model entails a source transmitting a classical variable λ to one party, Alice, and a corresponding state ρ_λ to the other party, Bob. The outcome a, conditional on Alice's measurement x, is contingent on the variable λ with a probability of $p(a|x, \lambda)$.

Definition 8.4.1 An LHS admits the following decomposition for joint measurement probability density as

$$\rho(x^A, y^B) = \int d\lambda \mu(\lambda) p(x^A|\lambda) \rho_Q(y^B|\lambda), \tag{8.52}$$

where $\rho_Q(y^B|\lambda)$ is the probability density of measurement outcome y^B conditional on the variable λ.

In an LHS model, the final state after one party's local measurement is given by

$$\varrho_x^a = \int d\lambda \mu(\lambda) p(a|x, \lambda) \rho_\lambda \tag{8.53}$$

with a normalisation of $\mathrm{Tr}[\varrho_x^a] = \int d\lambda \mu(\lambda) p(a|x, \lambda) = p(a|x)$. In what follows, denote LHS decomposition (8.53) by $\varrho_{a|x}^{LHS}$.

A quantum state is considered steerable from A to B (B to A) if Alice (Bob) can perform measurements that yield an assemblage that is not LHS. If an LHS assemblage is produced based on the other party's local measurements, the state is deemed unsteerable.

The steering experiment can provide valuable entanglement verification. Typically, steerable states are stronger than entanglement separability. In fact, suppose that two parties share a separable state, $\rho_{sep} = \int d\lambda \mu(\lambda) \rho_\lambda^A \otimes \rho_\lambda^B$. This follows an assemblage similar to Eq. (8.53) as

$$\begin{aligned} \varrho_x^a &= \mathrm{Tr}_A[(M_x^a \otimes \mathbb{1}) \rho_{sep}] \\ &= \int d\lambda \mu(\lambda) \mathrm{Tr}[M_x^a \rho_\lambda^A] \rho_\lambda^B \\ &= \int d\lambda \mu(\lambda) p(a|x, \lambda) \rho_\lambda^B. \end{aligned} \tag{8.54}$$

Alice's compatible measurements means that there are POVM operators $\{N_\lambda\}_\lambda$, and probability distributions $p(a|x, \lambda)$, such that $M_x^a = \int d\lambda p(a|x, \lambda) N_\lambda$. This implies the same structure (8.53) as

$$\begin{aligned} \varrho_x^a &= \mathrm{Tr}_A[(M_x^a \otimes \mathbb{1}) \rho_{AB}] \\ &= \int d\lambda p(a|x, \lambda) \mathrm{tr}_A[(N_\lambda \otimes \mathbb{1}) \rho_{AB}] \end{aligned}$$

$$= \int d\lambda \mu(\lambda) p(a|x, \lambda) \rho_\lambda^B, \tag{8.55}$$

where $\mu(\lambda) := \mathrm{Tr}[N_\lambda \otimes \mathbb{1}\rho_{AB}]$ and $\rho_\lambda^B := \mathrm{Tr}_A[(N_\lambda \otimes \mathbb{1})\rho_{AB}]/\mathrm{Tr}[(N_\lambda \otimes \mathbb{1})\rho_{AB}$.

In comparison to the correlation functions generated from variables on two sides in a Bell test, EPR-steering involves variables for one party and quantum measurements for the other. Linear EPR-steering inequalities with n settings per party can be used, as described in Wiseman et al. (2007), Saunders et al. (2010). For qubit cases, Bob's k-th measurement setups correspond to the Pauli observable $\hat{\sigma}_k^B$, along some axis u_k. Denoting Alice's corresponding result by the variable $A_k \in \{-1, 1\}$ for all k, the EPR-steering inequality is of the following form:

$$\mathcal{S}_n := \frac{1}{n} \sum_{k=1}^{n} \langle A_k \hat{\sigma}_k^B \rangle \leq \max\{|A_k|, \lambda_{\max}(W)\} \tag{8.56}$$

if Bob has a pre-existing state known to Alice, where $\lambda_{\max}$ denotes the largest eigenvalue and $W = \frac{1}{n}\sum_{k=1}^{n} A_k \hat{\sigma}_k^B$. Consider the Werner state:

$$\rho_{AB} = v|\Phi\rangle\langle\Phi| + \frac{1-v}{4}\mathbb{1} \tag{8.57}$$

with $v \in [0, 1]$. For $n = 2$, projective measurements imply a steerable bound of $v \geq \frac{1}{\sqrt{2}}$, similar to the CHSH inequality. The asymptotic bound is $v > 1/2$ for $n \to \infty$.

Another way is making use of entropic steering inequalities (Walborn et al. 2011). It generally gives a nonlinear LHS constraint for observables as

$$H(X^B|X^A) \geq \sum_\lambda p(\lambda) H_Q(X^B|\lambda), \tag{8.58}$$

where $H_Q(X^B|\lambda)$ is the Shannon entropy of the probability distribution $P_Q(X^B|\lambda)$, and the subscript Q means a corresponding quantum state. All observables of systems compatible with an LHS model should satisfy the inequality. By the entropic uncertainty relation, it is further rewritten into

$$H(X^B|X^A) + H(Y^B|Y^A) \geq \log(\varpi^B), \tag{8.59}$$

where $\varpi^B = \min_{i,j}\{|\langle X_i|Y_j\rangle|^{-2}\}$.

The bipartite EPR steering can be easily extended for multipartite systems by the bipartition of correlations (He and Reid 2013; Li et al. 2015). A different way is a one-sided device-independent scenario (Cavalcanti et al. 2011). Similar to the entanglement (Coffman et al. 2000) and Bell nonlocality, EPR steering satisfies monogamy constraints which limits the free distribution of correlation over different subsystems. This provides fundamental features of multipartite correlations. Take a tripartite system shared by Alice, Bob, and Charlie as an example. Monogamy means

Alice and Bob cannot steer the state of Charlie simultaneously. This has recently been studied in both theory and experiment with high-dimensional systems (Zhen et al. 2019).

8.4.2 *Quantum Network Steering*

In this subsection, we extend EPR steering to quantum network scenarios, using a new model of quantum network steering. In quantum networks, we generally assume a set of independent sources that distribute states to different parties. Similar to EPR steering, we assume that a subset of the parties in the network steering scenario is untrusted. In this case, we are interested in the ability of the trusted parties to steer the network, using the local measurements of the untrusted parties. This is generally referred to as network steering (Jones et al. 2021).

We first consider a simple scenario of an n-partite chain network. In this network, the endpoint parties are trusted, while all the other parties are untrusted. An example is of a tripartite scenario, consisting of two sources β and γ. Its quantum realizations with ρ_{AB} and $\rho_{B'C}$ allow for swapping entanglement for two endpoint parties, Alice and Charlie. Especially, conditional on Bob's measurement $M_b^{BB'}$, we obtain sub-normalized states steered for Alice and Bob as

$$\varrho_b^{AC} = \mathrm{Tr}_{BB'}((\mathbb{1}_{AC} \otimes M_b^{BB'})\rho_{AB} \otimes \rho_{B'C}) \tag{8.60}$$

with probability $p(b) = \mathrm{Tr}\varrho_b^{AC}$. Here, we refer to $\{\varrho_b\}_b$ as a network assemblage.

To determine the network steering ability, we consider the following network local hidden state model (NLHS).

Definition 8.4.2 A tripartite network local hidden state admits the following decomposition for joint probability density as

$$\rho(x^A, y^b, z^C) = \int d\mu(\beta)\mu(\gamma)p(y^b|\beta, \gamma)\rho_Q(x^A|\beta) \otimes \rho_Q(z^C|\gamma), \tag{8.61}$$

where $\rho_Q(x^A|\beta)$ is probability density of measurement outcome x^A conditional on the variable β. Similar definition holds for $\rho_Q(z^C|\gamma)$.

In a tripartite NLHS model, the steered state of Alice and Bob conditional on the outcome b of Bob can be written into

$$\varrho_b^{AC} = \sum_{\beta,\gamma} p(\beta)p(\gamma)p(b|\beta, \gamma)\varrho_\beta^A \otimes \varrho_\gamma^C, \tag{8.62}$$

where β and ϱ_β^A are variable and corresponding states, γ and ϱ_γ^C be those of the second source, and $p(b|\beta, \gamma)$ denotes the local response function of Bob.

A quantum network is considered steering if the generated network assemblage σ_b^{AC} does not have the decomposition in Eq. (8.62). While EPR steering requires multiple measurement setups by the untrusted party, a fixed measurement may be useful for verifying network steering.

In fact, generalized entanglement swapping will activate a bipartite entanglement ϱ_b for Alice and Charlie, conditional on Bob's local measurement. This rules out an NLHS model with a separable state in Eq. (8.62). A similar result can be easily extended to the n-partite chain network, where all parties A_i is assumed to be untrusted.

Theorem 8.4.1 *For any chain network with trusted endpoints, the network steering is verified by a generalized entanglement swapping, i.e., the single entanglement* $\varrho_{b_2,\ldots,b_{n-1}}$ *rules out an NLHS model.*

We will now show how to verify network steering. Consider the tripartite network. If one source distributes a steerable state, one can demonstrate network steering even if the second source distributes separable states. This can be proved as follows.

Consider the case with two untrusted parties without inputs steering a third. This can yield a network assemblage ϱ_x^a in the NLHS model as follows:

$$\varrho_x^a = \sum_{\beta,\gamma} p(\beta)p(\gamma)p(x|\beta)p(a|\beta,\gamma)\varrho_\gamma. \tag{8.63}$$

Note that

$$\varrho_x^a = p(x)\sum_{\gamma} p(\gamma)p(a|x,\gamma)\varrho_\gamma \tag{8.64}$$

where the probability $p(x)$ is defined by $p(x) = \mathrm{Tr}[\sum_a \varrho_x^a] = \sum_\beta p(\gamma)p(x|\gamma)$ and $p(a|x,\gamma) = \frac{1}{p(x)}\sum_\beta p(\beta)p(x|\beta)p(a|\beta,\gamma)$. This means if ϱ_x^a has an NLHS model, then $\hat{\varrho}_x^a := \varrho_x^a/p(x)$ has an LHS model.

Furthermore, if ϱ_b has an NLHS model, then $\varrho_x^a := \mathrm{Tr}_A(M_x^A \otimes \mathbb{1}^C\sigma_b)$ also has an NLHS model for any measurement M_x. This can be proved similar to a Bell test with joint statistics $P(a,b|x,y)$. A simple explanation is that both endpoints are trusted. If ϱ_b has an NLHS model, then $\varrho_x^b := \mathrm{Tr}_A(M_x^A \otimes \mathbb{1}_{\varrho_b}^C)$ also has an NLHS model for any M_x. It follows that

$$\varrho_{b,x} = \sum_{\beta,\gamma} p(\beta)p(\gamma)\mathrm{Tr}(M_x\varrho_\beta)p(b|\beta,\gamma)\varrho_\gamma \tag{8.65}$$

which is an NLHS model with $p(x|\beta) = \mathrm{Tr}(M_x\sigma_\beta)$.

To sum up, suppose that $\rho_{B'C}$ is steerable, i.e., $\varrho_x^b := \mathrm{Tr}((M_x^b \otimes \mathbb{1})\rho_{B'C})$ shows steering for some M_x^b. Denote

$$\rho_{AB} = \sum_x \frac{1}{d}|x\rangle\langle x| \otimes |x\rangle\langle x|, \tag{8.66}$$

where d is the number of measurements, and $\{|x\rangle\}_x$ is an orthogonal basis, and $M_b = \sum_{x'} |x'\rangle\langle x'| \otimes M^b_{x'}$. We get network assemblage ϱ_b as

$$\varrho_b = \sum_x \frac{1}{d} |x\rangle\langle x| \otimes \varrho^b_x. \tag{8.67}$$

These two facts mean network steering.

Similar results hold for 4-partite chain networks if the central state is nonlocal while the others are separable states. That is, if $\varrho_{b,c}$ has an NLHS model, then it follows an NLHV model for two trusted parties associated probability distributions $p(b, c, x, y)$.

For general networks, we use undirected graphs, where each node is either untrusted or trusted, and the edges denote independent sources. If all parties are untrusted, it reduces to the hidden variable model by considering the joint statistics $P(\vec{a}|\vec{x})$. Instead, the outcome is replaced by the network assemblage $\varrho^{\vec{a}}_{\vec{x}}$ when at least one party is trusted. This can be used to verify network steering by generalized entanglement swapping. Here, the input assumption of untrusted parties may be replaced by additional sources and parties without inputs.

Especially, for any network with an untrusted party A who has an input x with probability $p(x)$, and outcome a, it is equivalent to a new network by adding untrusted party A_0 who shares an additional source with A. In the new network, both A_0 and A have an input. Let x be the outcome of A_0. We get for the network assemblages in both networks $p(x)\sigma^{A\cdots}_{a,\ldots|x,\ldots} = \sigma^{AA'\cdots}_{a,x,\ldots}$.

Furthermore, quantum mechanics admits local tomography. This means that a trusted party who shared n independent sources can be equivalently replaced by n trusted parties, each of them connected to a single source. With this transformation, we can verify network steering from generalized entanglement swapping.

8.4.3 *Quantum Network Activation*

The above network steering is based on the steering or nonlocality in standard scenarios. Instead, it may be possible to use only (two-way) unsteerable states. This can be regarded as a form of activation. This complements previous examples of activation of steering in the standard bipartite scenario (Quintino et al. 2016).

Consider a two-qubit Werner state: $\rho_\omega = \omega|\psi^-\rangle\langle\psi^-| + (1-\omega)\frac{\mathbb{1}}{4}$, which passes through two-side identical erasure channel as:

$$\rho_{DEW}(\eta, \omega) = \Lambda_\eta \otimes \Lambda_\eta(\rho_\omega), \tag{8.68}$$

where $|\psi^-\rangle = \frac{1}{\sqrt{2}}(|01\rangle - |10\rangle)$ and $\Lambda_\eta(\rho) = \eta\rho + (1-\eta)|2\rangle\langle 2|$, $|2\rangle$ represents the loss of the system. $\rho_{DEW}(\eta, \omega)$ is entangled when $\omega > \frac{1}{3}$ (and $\eta \neq 0$), and is unsteerable for $\eta \leq \frac{2}{3}(1-\omega)$ (Tischler et al. 2018).

As for any state ρ_{AB} if it is unsteerable from Alice to Bob, the state $\mathbb{1}_A \otimes \mathcal{E}_B(\rho_{AB})$ is also unsteerable for any channel $\mathcal{E}$ (Quintino et al. 2015).

Instead, in terms of entanglement swapping, we can project $\rho_{DEW}(\eta, \omega) \otimes \rho_{DEW}(\eta, \omega)$ onto a Bell state $|\psi^-\rangle$. This will result in $\rho_{DEW}(\eta, \omega^2)$ with probability $\eta^2/4$ and squared visibility.

8.4.4 Simple NLHS Models

In this subsection, we consider what properties of quantum states affect an NLHS model. We regard a source as separable, unsteerable, or local if it can only generate separable, unsteerable, or local states, respectively.

Consider the tripartite network, where one is separable while the other is unsteerable towards the trusted party. For any fixed measurement for the central party, the network assemblage ϱ_b will always be NLHS. Indeed, by taking $\rho_{AB} = \sum_\gamma p(\gamma)\varrho_\gamma^A \otimes \varrho_\gamma^B$, Eq. (8.60) implies

$$\varrho_b = \sum_\gamma p(\gamma)\varrho_\gamma^A \otimes \mathrm{Tr}_{BB'}(M_b \otimes \mathbb{1}_C(\varrho_\gamma^B \otimes \rho_{B'C})). \tag{8.69}$$

Let $M_\gamma^b := \mathrm{Tr}_B(M_b\varrho_\gamma^B \otimes \mathbb{1}_C)$ be measurement operators. It follows that

$$\varrho_b = \sum_\gamma p(\gamma)\varrho_\gamma^A \otimes \mathrm{Tr}_{B'}(M_\gamma^b \otimes \mathbb{1}_C(\rho_{B'C})). \tag{8.70}$$

If the state $\rho_{B'C}$ is unsteerable from B' to C, it yields to an NLHS decomposition as

$$\varrho_b = \sum_{\gamma,\lambda} p(\gamma)p(\lambda)p(b|\lambda,\gamma)\varrho_\gamma^A \otimes \varrho_\lambda^C, \tag{8.71}$$

This means that the combination of a separable and unsteerable source cannot lead to network steering. The main concept behind the result is similar to generalized entanglement swapping, which cannot be completed by one separable state and one entanglement. Similar results hold for general scenarios.

8.5 Bell Nonlocality of Quantum Networks

In this section, we present Bell nonlocality in quantum network scenarios. There are at least two motivations for extending Bell nonlocality in networks. One is similar to Bell's theorem, which aims to explore new insights into quantum theory and its relationship to other models. The other is for quantum applications, as Bell nonlocality in networks may bring novel forms of quantumness.

8.5.1 Single Entanglement

A standard Bell scenario contains several parties who perform local measurements on the shares from a single source. The measurement processes are space-like separated and independent. Each party can freely and independently choose measurement setups denoted by $x_1, \ldots, x_n$, respectively. The outcome of their measurements is respectively labeled by $a_1, \ldots, a_n$. This admits a probability distribution $P(\vec{a}|\vec{x})$ with $\vec{a} = (a_1, \ldots, a_n)$ and $\vec{x} = (x_1, \ldots, x_n)$. The only assumption for these correlations comes from the so-called no-signaling principle, given by

$$\sum_{a_j, j \neq i} P(\vec{a}|\vec{x}) = p(a_i|\vec{x}), \forall a_i, x_i. \tag{8.72}$$

This allows us to describe these correlations with a shared variable.

Definition 8.5.1 A local hidden variable (LHV) model is to explain the correlations by sharing a common variable λ with unknown probability density $\mu(\lambda)$. In statistics, the correlations admitting a local model can be decomposed into

$$P(\vec{a}|\vec{x}) = \int d\lambda \mu(\lambda) \prod_{i=1}^{n} p(a_i|x_i, \lambda), \tag{8.73}$$

where the measurement settings of the i-th party are independent from the local variable, i.e., $\mu(\lambda|\vec{x}) = \mu(\lambda)$.

Any correlation consistent with the no-signaling principle is Bell nonlocal if it does not admit the local model in Eq. (8.73).

In the LHV model, any randomness in the response functions can be absorbed in its probability density $\mu(\lambda)$. This results in deterministic response functions for each party, i.e., $p(a_i|x_i, \lambda), \in, 0, 1$. Each choice of deterministic response functions constitutes a vertex of a high-dimensional set in correlation space. For finite settings per party, there are only finitely many deterministic responses. In this case, the LHV model admits a simple characterization for all the correlations by a local polytope-a high-dimensional convex polyhedron.

A directive result from the local polytope is to solve the compatibility of given correlations and the LHV model by the linear program. An alternative way is to determine all the facets of the local polytope. These facet inequalities provide an optimal way to verify Bell nonlocality, known as Bell inequalities.

One well-known bipartite Bell inequality is the Clauser-Horne-Shimony-Holt (CHSH) Bell inequality (Clauser et al. 1969). Here, each party has two measurements, $x, y \in \{0, 1\}$, and binary outcomes, $a, b \in \{0, 1\}$. The binary settings imply only one facet given by Eq. (8.74):

$$\mathcal{S}_{\text{CHSH}} = \langle A_0 B_0 \rangle + \langle A_0 B_1 \rangle + \langle A_1 B_0 \rangle - \langle A_1 B_1 \rangle \leq 2, \tag{8.74}$$

where $\langle A_x B_y \rangle = \sum_{a,b} (-1)^{a+b} P(a, b|x, y)$ is the joint expectation value of the observations of the two parties.

In quantum scenarios, given a shared state ρ, we obtain correlations from the Born rule:

$$P(a, b|x, y) = \text{Tr}(A_x^a \otimes B_y^b \rho), \tag{8.75}$$

where $\{A_x^a\}$ and $\{B_y^b\}$ are the POVM operators of both parties, i.e., A_x^a and B_y^b are non-negative and sum to identity. The remarkable result of Bell's theorem has ruled out LHV models for all quantum correlations (Bell 1964).

Example 8.5.1 Let Alice and Bob share a generalized EPR state

$$|\Phi\rangle = \cos\theta|00\rangle + \sin\theta|11\rangle \tag{8.76}$$

with $\theta \in (0, \frac{\pi}{2})$. Let $A_0 = Z$, $A_1 = X$, $B_0 = \cos\vartheta Z + \sin\vartheta X$ and $B_1 = \cos\vartheta$ $Z - \sin\vartheta X$. It follows that

$$\mathcal{S}_{CHSH} = 2\cos\vartheta + 2\sin\vartheta\sin 2\theta = 2\sqrt{1 + \sin^2 2\theta} > 2 \tag{8.77}$$

for any $\theta \in (0, \frac{\pi}{2})$, where ϑ satisfies $\sin\vartheta = 1/\sqrt{1 + \sin^2 2\theta}$. This means the total quantum expectation violates the CHSH inequality. This proves the Bell nonlocality of quantum correlations (Cirel'son 1980).

The linear Bell inequality is useful for testing the incompatibility of quantum theory and the LHV model (Bell 1964). This feature is generic for entangled two spin-$\frac{1}{2}$ particles (Gisin 1991) or multipartite entangled states (Popescu and Rohrlich 1994; Clauser et al. 1969), using the CHSH inequality.

Although the Bell method may be easily extended to multipartite scenarios, it is more involved in featuring genuine correlations. Specific motivations follow different local models. The LHV model (8.73) gives the weakest nonlocality for featuring bipartite correlations. The second is the so-called biseparable model (Svetlichny 1987) for exploring genuinely multipartite nonlocality. This multipartite nonlocality rules out any mixing of local and partial nonlocal correlations. Another model is the network-based nonlocality (Navascues et al. 2020) or robust nonlocality (Luo and Fei 2021), which will be discussed in the following subsections.

8.5.2 Source Independent

Inspired by the bipartite Bell inequality (Bell 1964), we follow the multilocality of correlations in networks from a causal network (Pearl 2009). Formally, all parties share variables of $\Lambda = (\lambda_1, \ldots, \lambda_m)$, where Λ can be arbitrary and may exist before the measurement choices, and m is the number of hidden variables. The dichotomic output a_i of any particular system can depend arbitrarily on Λ and the measurement type x_i.

Definition 8.5.2 A generalized local hidden variable (GLHV) model suggests a joint distribution of the measurement outcomes as

$$P(\vec{a}|\vec{x}) = \int_{\Omega} d\mu(\Lambda) \prod_{i=1}^{n} p(a_i|x_i, \Lambda), \tag{8.78}$$

where (Ω, Σ, μ) denotes the measure space of Λ. $\mu(\Lambda)$ is the normalized probability density of Λ, $p(a_i|x_i, \Lambda)$ is the probability of a_i for the i-th party conditional on input x_i and Λ.

Consider an n-partite network with finite settings. Each party A_i shares variables $\Lambda_i = \{\lambda_{j_1}, \ldots, \lambda_{j_{\ell_i}}\}$. Let $\mu_i(\lambda_i)$ be the normalized probability density of λ_i, and $(\Omega_i, \Sigma_i, \mu_i)$ be the measure space of λ_i, $i = 1, \ldots, m$. We obtain a generalized network local model from Eq. (8.78) as follows.

Definition 8.5.3 A generalized network local model admits the following decomposition for joint probability as

$$P(\vec{a}|\vec{x}) = \int_{\Omega} \prod_{i=1}^{m} d\mu_i(\lambda_i) \prod_{j=1}^{n} p(a_j|x_j, \Lambda_j). \tag{8.79}$$

When compared with the LHV model of one source (Bell 1964), there is no network local polytope that can feature a generalized network local model with independent variables. The statistical correlations in entanglement swapping imply a non-convex set (Branciard et al. 2010; Sangouard et al. 2011), which may require new methods such as nonlinear Bell-type inequalities (Geiger and Meek 1999), linear programming techniques (Chaves 2016), or network expansion (Rosset et al. 2016).

Definition 8.5.4 A network is k-independent if there are k separated parties who do not share sources.

The k-independence can be equivalently represented by the k-locality as follows. There are k subsets $\Lambda_{i_1}, \ldots, \Lambda_{i_k}$ consisting of variables such that

$$\cup_{j=1}^{k} \Lambda_{i_j} \subseteq \Lambda, \Lambda_{i_s} \cap \Lambda_{i_t} = \emptyset, 1 \leq s < t \leq k. \tag{8.80}$$

Denote $\mathcal{I} = \{i_1, \ldots, i_k\}$ and its complement set as $\overline{\mathcal{I}}$ of $\{1, \ldots, n\}$. For given measurements A_{x_1} of parties A_i, the correlator is given by $\langle A_{x_1} \cdots A_{x_n}\rangle = \sum_{\vec{a}}(-1)^{\sum_{i=1}^n a_i} P(\vec{a}|\vec{x})$. We define one quantity $I_{n,k}$ of multipartite correlations as

$$I_{n,k} = \frac{1}{2^k} \sum_{x_i, i\in\mathcal{I}} \langle A_{x_1} \cdots A_{x_n}\rangle, \tag{8.81}$$

for any given measurements A_{x_i} of parties A_i, $i \in \overline{\mathcal{I}}$. Similarly, using another collection of measurements by all parties A_i with $i \in \overline{\mathcal{I}}$, we define

$$J_{n,k} = \frac{1}{2^k} \sum_{x_i, i\in\mathcal{I}} (-1)^{\sum_{j\in\mathcal{I}} x_j} \langle A_{x_1} \cdots A_{x_n}\rangle. \tag{8.82}$$

The k-independence means

$$|I_{n,k}| \leq \frac{1}{2^k} \prod_{i\in\mathcal{I}} \Delta_i^+, \tag{8.83}$$

$$|J_{n,k}| \leq \frac{1}{2^k} \prod_{i\in\mathcal{I}} \Delta_i^-, \tag{8.84}$$

where $\Delta_i^+ = A_{x_i=0} + A_{x_i=1}$ and $\Delta_i^- = A_{x_i=0} - A_{x_i=1}$. This implies a general result for featuring the correlations by $I_{n,k}$ and $J_{n,k}$ (Luo 2018).

Theorem 8.5.1 *For any k-independent network the correlations from the network local model satisfy the following nonlinear Bell inequality as*

$$|I_{n,k}|^{\frac{1}{k}} + |J_{n,k}|^{\frac{1}{k}} \leq 1. \tag{8.85}$$

In quantum scenarios, denote $\rho = \otimes_{i=1}^n \rho_i$ as the total quantum resource.

Theorem 8.5.2 *The correlations from quantum networks satisfy the following inequality as*

$$|I_{n,k}^q|^{\frac{1}{k}} + |J_{n,k}^q|^{\frac{1}{k}} \leq \sqrt{2}, \tag{8.86}$$

where $I_{n,k}^q$ and $J_{n,k}^q$ are the corresponding quantities of $I_{n,k}$ and $J_{n,k}$ in terms of quantum correlations.

The proof is based on the following triangle geometric inequality

$$(\prod_{i=1}^n \sin\theta_i)^{\frac{1}{n}} \leq \sin(\frac{1}{n}\sum_{i=1}^n \theta_i) \tag{8.87}$$

for any $\theta_i \in [0, \pi]$ and $n \geq 2$. The equality holds if and only if all θ_i's are equal to each other.

Given a network configuration, different Bell inequalities may be constructed from (8.85) by choosing $I_{n,k}$ and $J_{n,k}$. The number k is another important network feature. A larger k implies more multipartite correlations being involved. However, finding the maximum k (i.e., $k_{\max}$) and the corresponding independent parties is generally NP-hard. Instead, one may find suboptimal k according to the maximal matching of the graph (Hopcroft and Karp 1973; Ausiello et al. 2003; Luo 2018).

In what follows, we resolve the inconsistency problem for general quantum networks. Assume that a quantum network $\mathcal{G}_q$ is k-independent, where $\mathsf{A}_1, \ldots, \mathsf{A}_k$ denote independent observers, while all the others are regarded as one party B. Assume that the quantum resources consist of bipartite entangled pure states and generalized GHZ states. We have the following result.

Theorem 8.5.3 *For any k-independent ($k \geq 2$) quantum network $\mathcal{G}_q$, there are observables for all observers such that the generated quantum correlations are inconsistent with generalized local realism. The maximal violation of the inequality (8.85) is achievable by maximally entangled EPR states and GHZ states.*

Theorem 8.5.3 provides an efficient method for verifying the nonlocality of quantum networks, such as the chain or star network consisting of EPR states (Branciard et al. 2010; Gisin et al. 2017), acyclic or cyclic networks consisting of bipartite entangled pure states and generalized GHZ states.

Consider the noisy quantum resources of Werner states:

$$\rho_w = \otimes_{i=1}^{m_1} \otimes_{j=1}^{m_2} (v_i|\Phi_i\rangle\langle\Phi_i| + \frac{1-v_i}{4}\mathbb{1}_4) \\ \otimes(w_j|\Psi_j\rangle\langle\Psi_j| + \frac{1-w_j}{2^{s_j}}\mathbb{1}_{2^{s_j}}), \tag{8.88}$$

where $|\Phi_i\rangle$ are generalized EPR states, $|\Psi_j\rangle = \hat{a}_j|0,\ldots,0\rangle + \hat{b}_j|1,\ldots,1\rangle$ are s_j-particle GHZ states with $s_j \geq 3$, m_1 and m_2 denote the numbers of the respective generalized EPR states and GHZ states, $\mathbb{1}_{2^{s_j}}$ is 2^{s_j} square identity matrix, and $0 \leq v_i, w_j \leq 1$. By following the proof of Theorem 8.5.3, we can prove the following noise viabilities.

Theorem 8.5.4 *Assume that a k-independent ($k \geq 2$) quantum network $\mathcal{G}_q$ consists of Werner states ρ_w. The product of the noise visibilities v_j^*, w_j^* satisfies*

$$\prod_{i=1}^{m_1}\prod_{j=1}^{m_2} v_i^* w_j^* \leq (1 + \prod_{i=1}^{m_1}\prod_{j=1}^{m_2}(4a_i b_i \hat{a}_j \hat{b}_j)^{\frac{2}{k}})^{-\frac{k}{2}} \tag{8.89}$$

Example 8.5.2 Consider a long-distance chain network. There are k independent parties, where $k = \lceil n/2 \rceil$ and $\lceil x \rceil$ denote the smallest integer not less than x. According to Theorem 8.5.3, quantum correlations violate the inequality (8.85) for any bipartite entangled pure states. The maximal violation is achieved for maximally entangled EPR states.

Example 8.5.3 Consider an $n+1$-partite star network, where each pair of parties A_i and B shares one bipartite entangled state. Theorem 8.5.3 implies that multipartite quantum correlations violate the inequality (8.85). This violation can be extended to hybrid networks that consist of generalized EPR and GHZ states.

Example 8.5.4 Consider an n-partite cyclic network where any adjacent two parties share one bipartite entangled pure state. Theorem 8.5.3 shows that multipartite quantum correlations violate the inequality (8.85) with $k = \lfloor \frac{n}{2} \rfloor$ when $n \geq 4$.

So far, we have constructed nonlinear Bell inequalities to verify the nonlocality in networks with binary setups. This mainly depends on identifying the largest set of mutually unconnected parties. A recent generalization is achieved through the stabilizer formalism (Hsu and Chen 2021). The second method is the network extension method (Rosset et al. 2016), which allows for the addition of independent sources one by one. The third method is variable elimination (Chaves 2016; Mukherjee et al. 2020).

The chain quantum network has recently been utilized to verify the use of complex numbers in quantum theory (Renou et al. 2021). Although real quantum theory can produce similar outcomes in any multipartite experiment, there are differences between real and complex quantum theories in network scenarios consisting of independent states and measurements. Particularly, for a real tripartite chain network, the joint correlation $P(a,b,c|x,z)$ satisfies the following wired inequality, derived from three CHSH inequalities:

$$
\begin{aligned}
&\mathcal{S}_{CHSH}(A_1, A_2; B_1, B_2) + \mathcal{S}_{CHSH}(A_1, A_3; B_3, B_4) \\
&+\mathcal{S}_{CHSH}(A_2, A_3; B_5, B_6) \leq 7.66,
\end{aligned} \tag{8.90}
$$

where $\mathcal{S}_{CHSH}(X_1, X_2; Y_1, Y_2)$ denotes the CHSH operator (Clauser et al. 1969) with two measurement setups, and Alice and Bob perform three and six measurements, respectively. This inequality is further improved to $6\sqrt{2}$ by complex quantum systems. This provides a Bell-like experiment for disproving a real quantum theory.

8.5.3 Without Source Independence

The network local model defined in Definition 8.5.3 is useful for featuring quantum correlations in multiple-source networks. Unfortunately, it has limited power when it comes to featuring the genuinely multipartite nonlocality similar to the single-entanglement case (Svetlichny 1987). The main reason for this limitation is the local setups. Instead, we consider the stronger nonlocality inspired by the biseparable model (Svetlichny 1987) in this subsection.

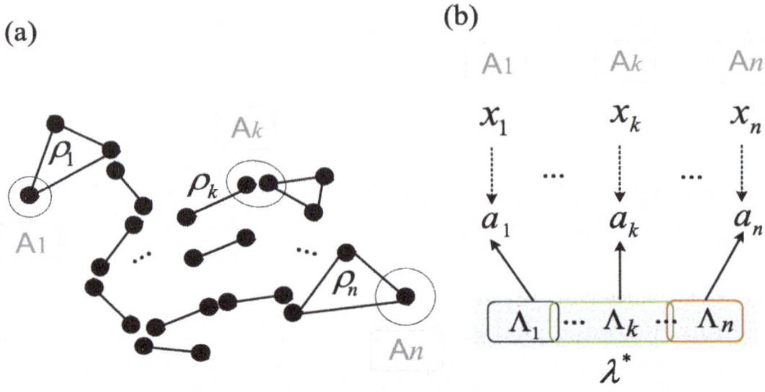

Fig. 8.6 Schematic multiple-source networks. **a** An n-partite quantum network $\mathcal{N}_q$ consisting of m entangled states $\rho_1, \ldots, \rho_m$. **b** A directed acyclic graph (DAG) of $\mathcal{N}_q$ consisting of m independent variables $\lambda_1, \ldots, \lambda_m$. In device-independent applications, an eavesdropper may correlate variables λ_i's into a new variable λ_* (red dot square)

Consider an n-partite quantum network $\mathcal{N}_q$, as shown in Fig. 8.6a, consisting of m entangled states. $\mathcal{N}_q$ is represented by a directed acyclic graph (DAG) as shown in Fig. 8.6b. Unlike the local model in Definition 8.5.3, in some secure applications such as quantum secret sharing, a strong attacker may correlate all distributed variables λ_i's into a new variable λ^*, as shown in Fig. 8.6b. The network local model with source independence is then insufficient for characterizing device-independent applications.

Definition 8.5.5 An n-particle correlation $P(\vec{a}|\vec{x})$ is genuinely multipartite nonlocality (Svetlichny 1987) if it does not admit the following decomposition

$$P(\vec{a}|\vec{x}) = \sum_S \sum_j p_{j;S} P_j(a_i \in S|x_i \in S) P_j(a_i \in \overline{S}|x_i \in \overline{S}), \tag{8.91}$$

where S and $\overline{S}$ are any bipartition of $\{A_1, \ldots, A_n\}$, $\{p_{j;S}\}$ is a probability distribution, $P_j(a_i \in S|x_i \in S)$ and $P_j(a_i \in \overline{S}|x_i \in \overline{S})$ are respective local or quantum correlations of all parties in S and $\overline{S}$.

The post-quantum correlations of $P_j(a_i \in S|x_i \in S)$ and $P_j(a_i \in \overline{S}|x_i \in \overline{S})$ are allowed for verifying genuinely multipartite nonlocality (Svetlichny 1987). However, we do not consider this case in this subsection. Here, we focus on the genuinely multipartite nonlocality (Svetlichny 1987) of quantum networks without the source independence assumption by ruling out all biseparable correlations.

Let $\mathcal{I} = \{s_1, \ldots, s_i\}$ and its complement set $\overline{\mathcal{I}}$ be a bipartition of $\{1, \ldots, n\}$. Let $M_{x_k(\mathcal{I})}$ denote the dichotomic observable of the observer A_k for the input index $x_k(\mathcal{I})$, where x_k depends on the set $\mathcal{I}$. Denote $M_{\mathcal{I}} = \prod_{k\in\mathcal{I}} M_{x_k(I)}$ and $M_{\overline{\mathcal{I}}} = \prod_{k\in\overline{\mathcal{I}}} M_{x_k(\overline{\mathcal{I}})}$. Inspired by the Mermin inequality (Mermin 1990), we can derive a linear Bell inequality with multiple setups for verifying genuinely multipartite nonlocality.

Theorem 8.5.5 *Any biseparable correlations satisfy the following Bell inequality:*

$$\sum_{i=1}^{\lfloor\frac{n}{2}\rfloor}\mathcal{L}_i \leq 2^n\sqrt{2}-4\sqrt{2}+2, \tag{8.92}$$

where the correlator $\mathcal{L}_i$ denotes the summation of multipartite CHSH-type quantities, that is, $\mathcal{L}_i = \sum_{|\mathcal{I}|=i}\mathcal{S}_\mathcal{I}$ with $\mathcal{S}_\mathcal{I} = (M_\mathcal{I}+\hat{M}_\mathcal{I})M_{\overline{\mathcal{I}}}+(M_\mathcal{I}-\hat{M}_\mathcal{I})\hat{M}_{\overline{\mathcal{I}}}$, M_S and $\hat{M}_S$ are observables depending on the set $S=\mathcal{I}$ or $\overline{\mathcal{I}}$. Quantum correlations from networks can achieve the largest bound of $2^n\sqrt{2}-2\sqrt{2}$.

Proof Consider a network $\mathcal{N}_q$ consisting of n parties $\mathsf{A}_1,\ldots,\mathsf{A}_n$. Suppose the total correlation from local measurements is biseparable and has the following decomposition as

$$P(\vec{a}|\vec{x}) = \sum_{\mathcal{I}} p_\mathcal{I} P_{\mathcal{I},\overline{\mathcal{I}}}, \tag{8.93}$$

where $\{p_\mathcal{I}\}$ is a probability distribution, $P_{\mathcal{I},\overline{\mathcal{I}}}$ denotes biseparable correlations respect to the bipartition $\mathcal{I}$ and $\overline{\mathcal{I}}$. Here, the summation is over all the bipartitions of $\{1,\ldots,n\}$.

In what follows, each party makes use of multiple observables. For each bipartition $\mathcal{I}$ and $\overline{\mathcal{I}}$, we define $M_{x_j(\mathcal{I})}$ as the dichotomic observable of A_j for the input x_j depending on $\mathcal{I}$, that is, the input x_j depends on $\mathcal{I}$. For simplicity, we denote

$$M_\mathcal{I} = \prod_{j\in I} M_{x_j(\mathcal{I})},\ M_{\overline{\mathcal{I}}} = \prod_{j\in\overline{\mathcal{I}}} M_{x_j(\overline{\mathcal{I}})}. \tag{8.94}$$

Since the left side of the inequality (8.92) is linear for all the local observables, all biseparable correlations have consisted of a convex set. The maximal bound is achievable at the vertex of product correlations. In fact, $\mathcal{S}_\mathcal{I}$ is a multipartite Mermin operator given by

$$\mathcal{S}_\mathcal{I} = (M_\mathcal{I}+\hat{M}_\mathcal{I})M_{\overline{\mathcal{I}}}+(M_\mathcal{I}-\hat{M}_\mathcal{I})\hat{M}_{\overline{\mathcal{I}}}, \tag{8.95}$$

where $|M_\mathcal{I}|$, $|\hat{M}_\mathcal{I}|$, $|M_{\overline{\mathcal{I}}}|$, $|\hat{M}_{\overline{\mathcal{I}}}| \leq 1$. For each bipartition $\mathcal{I}_0$ and $\overline{\mathcal{I}}_0$, there are two subcases for any product correlations $P(\vec{a}|\vec{x}) = P(a_i\in\mathcal{I}|x_i\in\mathcal{I})P(a_j\in\overline{\mathcal{I}}|x_j \in\overline{\mathcal{I}})$.

(i) The correlation $P(\vec{a}|\vec{x})$ is separable in terms of the bipartition $\mathcal{I}_0$ and $\overline{\mathcal{I}}_0$. In this case, all the parties A_j with $j\in\mathcal{I}$ can be regarded as one party, while all the parties A_t with $t\in\overline{\mathcal{I}}$ as the other. The CHSH inequality implies

$$\mathcal{S}_{\mathcal{I}_0} \leq 2. \tag{8.96}$$

(ii) The correlation $P(\vec{a}|\vec{x})$ is quantum nonlocal in terms of the bipartition $\mathcal{I}_0$ and $\overline{\mathcal{I}}_0$. In this case, there are two parties A_s and A_t with $s,t\in\mathcal{I}$, or $\overline{\mathcal{I}}$ such that they are nonlocal in the bipartition $\mathcal{I}_0$ and $\overline{\mathcal{I}}_0$, and $s\in\mathcal{I}_0$, $t\in\overline{\mathcal{I}}_0$. This implies that

$$\mathcal{S}_{\mathcal{I}_0} \leq 2\sqrt{2}. \tag{8.97}$$

There is at least one term $\mathcal{S}_{\mathcal{I}}$ in the summation of $\sum_{i=1}^{\lfloor \frac{n}{2} \rfloor} \mathcal{L}_i$ satisfies that $\mathcal{S}_{\mathcal{I}} \leq 2$. Combined with the linearity of $\mathcal{L}_i$, it follows that

$$\sum_{i=1}^{\lfloor \frac{n}{2} \rfloor} \mathcal{L}_i \leq 2(N-1)\sqrt{2} + 2 \tag{8.98}$$

for any product correlations, where N denotes the total number of $\mathcal{S}_{\mathcal{I}}$'s.

The number N is evaluated as $N = \lfloor \frac{n}{2} \rfloor$ by using the standard combination theory (Luo 2022; Diestel 2017). This has completed the proof. □

In quantum scenarios, it is easy to shown $\mathcal{S}_{\mathcal{I}} \leq 2\sqrt{2}$ for each bipartition $\mathcal{I}$ and $\overline{\mathcal{I}}$. Equation (8.98) implies the following inequality as

$$\sum_{i=1}^{\lfloor \frac{n}{2} \rfloor} \mathcal{L}_i \leq (2^n - 2)\sqrt{2}. \tag{8.99}$$

Here, the upper bound is achievable for quantum networks consisting of maximally entangled EPR states.

The Bell inequality (8.92) provides a way to verify the genuinely multipartite nonlocality of quantum networks. This provides the first Bell test for verifying general cyclic networks, including triangle networks. These results can inspire interesting applications on large-scale quantum networks in a device-independent manner.

The inequality (8.92) may be trivial if all the observables of $M_{\mathcal{I}}$, $\hat{M}_{\mathcal{I}}$, $M_{\overline{\mathcal{I}}}$, and $\hat{M}_{\overline{\mathcal{I}}}$ are different for each bipartition I and $\overline{\mathcal{I}}$, where each party has 2 inputs and outputs for each bipartition $\mathcal{I}$ and $\overline{\mathcal{I}}$. In this case, the inequality (8.92) is a summation of different CHSH inequalities (Clauser et al. 1969), where each party has $2N$ inputs and two outputs. Generally, the number of inputs will depend on the involved state or quantum network. However, we can prove that the total number of inputs is no more than 8.

Example 8.5.5 Consider a triangle network $\mathcal{N}_q$ shared by three parties A_1, A_2, and A_3. A specific linear inequality for genuinely tripartite nonlocality is given by

$$\begin{aligned} \mathcal{B}_3 := & A_1B_1C_1 + A_2B_1C_1 + A_1B_2C_2 \\ & -A_2B_2C_2 + A_3B_3C_1 + A_3B_4C_1 \\ & +A_4B_3C_2 - A_4B_4C_2 + A_3B_5C_3 \\ & +A_3B_5C_4 + A_4B_6C_3 - A_4B_6C_4 \leq 4\sqrt{2} + 2 \end{aligned} \tag{8.100}$$

Here, A_i, B_j, and C_k are dichotomic observable of A_1, A_2, and A_3, respectively, $i, k = 1, \ldots, 4$; $j = 1, \ldots, 6$. It means that A_1 and A_3 have 4 inputs while A_2 has 6 inputs.

In quantum scenarios, each pair shares one EPR state: $|\Phi_i\rangle_{2i-1,2i} = \cos\theta_i|00\rangle + \sin\theta_i|11\rangle$ with $\theta_i \in (0, \pi)$ and $i = 1, 2$, and 3. The total system is denoted as $|\Phi_1\rangle_{12}|\Phi_2\rangle_{34}|\Phi_3\rangle_{56}$. Here, $|\Phi_1\rangle$ and $|\Phi_3\rangle$ will be respectively changed into four-particle GHZ states $|\Phi_1\rangle = \cos\theta_1|0000\rangle + \sin\theta_1|1111\rangle$ and $|\Phi_4\rangle = \cos\theta_3|0000\rangle + \sin\theta_3|1111\rangle$, with local operations and auxiliary qubits. Define local observables as

$$\begin{aligned}
&A_1 = B_3 = C_3 = \cos\theta Z^{\otimes 3} + \sin\theta X^{\otimes 3}, \\
&A_2 = B_4 = C_4 = \cos\theta Z^{\otimes 3} - \sin\theta X^{\otimes 3}, \\
&A_3 = B_1 = Z^{\otimes 3}, A_4 = B_2 = X^{\otimes 3}, \\
&B_5 = Z^{\otimes 4}, B_6 = X^{\otimes 4}, \\
&C_1 = Z \otimes Z, C_2 = X \otimes X.
\end{aligned} \tag{8.101}$$

From a forward evaluation, we get

$$\mathcal{B}_3 = 6\cos\theta + 6\sin\theta\sin 2\theta_1 \sin 2\theta_2 \sin 2\theta_3 > 4\sqrt{2} + 2 \tag{8.102}$$

which violates the inequality (8.100) when θ_i's satisfy $\sin 2\theta_1 \sin 2\theta_2 \sin 2\theta_3 > 0.7928$.

For noisy networks consisting of Werner state (Werner 1989) given by $\rho_i = v|\phi_i\rangle\langle\phi_i| + \frac{1-v}{4}\mathbb{1}$ with $\theta_i = \frac{\pi}{4}$, and $v \in (0, 1]$, the noise visibility is given by $v = 0.966$.

Instead of the inequality (8.100), another Bell inequality with fewer inputs is given by

$$\begin{aligned}
\mathcal{B}_3' := &A_1B_1C_1 + A_2B_1C_1 + A_1B_2C_2 \\
&-A_2B_2C_2 + A_3B_3C_1 + A_3B_4C_1 \\
&+A_4B_3C_2 - A_4B_4C_2 + A_3B_1C_3 \\
&+A_3B_1C_4 + A_4B_2C_3 - A_4B_2C_4 \le 4\sqrt{2} + 2.
\end{aligned} \tag{8.103}$$

Here, each party has 4 inputs. For verifying the genuinely tripartite nonlocality the two-qubit entanglement will be locally transformed into a four-particle GHZ state for the inputs 3 and 4. For special case of $\theta_1 = \theta_2 = \theta_3$, it follows that $\theta > 0.55$, where θ satisfies $\cos\theta \approx 0.59$. If $\theta_1 = \theta_2 = \frac{\pi}{4}$, it follows that $\theta_3 > 0.38$. The noise visibility is the same as the inequality (8.100).

The inequality (8.92) is a relaxed version of the Svetlichny-type inequality (Svetlichny 1987), which considers all quantum nonlocal correlations but excludes post-quantum correlations. This linear inequality can be useful in detecting genuinely multipartite nonlocality in networks that include EPR and GHZ states or noisy quantum networks. Such multipartite nonlocality is stronger than the one assumed under source independence.

8.5.4 Full Quantum Network Nonlocality

In this subsection, we discuss full network nonlocality, which describes correlations that require all links in a network to distribute nonlocal resources. The basic idea is similar to that for genuinely multipartite entanglement in networks, where at least one nonlocal source can be replaced by a local source while the others can be either nonlocal or post-quantum (Pozas-Kerstjens et al. 2022).

Definition 8.5.6 In a given network the correlations $P(\vec{a}|\vec{x})$ are fully network nonlocality if they cannot be simulated by allowing at least one source in the network to be of a local variable, while all other sources may be general independent nonlocal resources.

Here, the fully network nonlocality of certain networks, including chain or star networks, requires all sources to be nonlocal resources, as shown in Fig. 8.7. These sources may distribute entangled states or some post-quantum resources under non-signaling constraints (Barrett 2007). In this case, it can be defined independently of quantum theory, similar to the biseparable model (Svetlichny 1987).

The full network nonlocality is stronger than genuinely multipartite nonlocality. Similar to nonlocality in the generalized network local model, this type of nonlocality exhibits specific properties relative to a given network configuration. This means

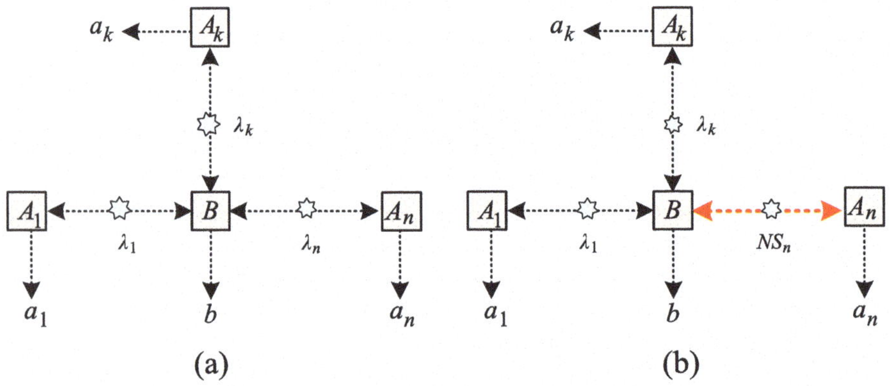

Fig. 8.7 Star network scenario. **a** Classical sources. **b** Hybrid sources. At least one pair of two parties A_i and B share a nonlocal resource (red line) while other pairs share classical variable λ_j's. This kind of hybrid network can generate nonlocal correlations by violating the nonlinear Bell inequality (8.85)

that fully network nonlocal correlations can be simulated by relaxing the network configurations. This property is different from standard Bell nonlocality.

Consider a tripartite chain network, where Alice and Bob share a variable while Bob and Charlie share an entangled state. Here, Alice and Charlie are independent and can share one entanglement from entanglement swapping in quantum realizations. Assume Alice and Charlie have binary inputs, $x, z \in \{0, 1\}$, and produce binary outcomes, $a, c \in \{0, 1\}$, while Bob has a fixed measurement with four possible outcomes, $b \equiv (b_0, b_1) \in \{0, 1\}^2$. In the bilocal scenario, models of the form (8.79) respect the network Bell inequality (Branciard et al. 2012):

$$\mathcal{S}_2 := \sqrt{|I_0|} + \sqrt{|I_1|} \leq 1, \tag{8.104}$$

where $I_t = \frac{1}{4}\sum_{a,b,c,x,z}(-1)^{a+b_t+c+t(x+z)}P(a,b,c|x,z)$ for $t \in 0, 1$. The network nonlocality will be verified by violating Eq. (8.104). The maximal violation is $\sqrt{2}$ obtained by two maximally entangled EPR states. Unfortunately, all of these violations may be simulated by one post-quantum source, i.e., a Popescu-Rohrlich (PR) box (Popescu and Rohrlich 1994), which implies $I_0 = I_1 = \frac{1}{2}$ (Pozas-Kerstjens et al. 2022). Thus, the quantum violation of the inequality (8.104) can be simulated in a bilocal network with one local-variable source and one post-quantum resource.

Another linear no-signaling correlation can be expressed as $|I_0| + |I_1| \leq 1$. This inequality provides a tight constraint on non-full network nonlocal correlations. Although the projection of the correlations in the (I_0, I_1)-plane is a convex set, all the non-full network nonlocal correlations are generally non-convex. A simple example is the uniform mixture of two distributions $p_k(a,b,c) = \delta_{a,k}\delta_{b,k}\delta_{c,k}$ with $k \in \{0, 1\}$, which is compatible with the maximal violation of the inequality (8.104) and is fully network nonlocal (Pozas-Kerstjens et al. 2022).

This inspires a problem for characterizing full network nonlocal correlations from quantum theory. One method is through nonlinear inequalities. Consider a star network consisting of n independent parties $\mathsf{A}_1, \ldots, \mathsf{A}_n$, with binary inputs $(x_1, \ldots, x_n)$ and binary outputs $(a_1, \ldots, a_n)$. Each party A_i is connected to the center party B who performs a fixed measurement with an n-bit outcome $b \equiv (b_1, \ldots, b_n) \in \{0, 1\}^n$. Generalizing the inequality (8.104), it follows a network Bell inequality as:

$$\mathcal{S}_n = \frac{1}{2^{n-2}}\sum_{t=1}^{2^{n-1}} |I_t|^{1/n} \leq 1, \tag{8.105}$$

where $I_t = \frac{1}{2^n}\sum_{\vec{a},\vec{x},b}(-1)^{a_1+\ldots+a_n+\tilde{t}\cdot b+\hat{t}\cdot\vec{x}}P(\vec{a},b|\vec{x})$, the n-bit strings $\tilde{t}$ and $\hat{t}$ are $\tilde{t} = (1, t_1, t_2, \ldots)$ and $\hat{t} = (\oplus_{k=1}^{n-1}t_k, t_1, t_2, \ldots)$, and $(t_1, t_2, \ldots)$ is the string of $n-1$ bits representing t. The best-known quantum protocol achieves $\mathcal{S}_n = \sqrt{2}$ by having each source emitting maximally entangled EPR states (Einstein et al. 1935).

Taking $n = 3$ as an example, the linear constraint of $|I_1| + |I_2| + |I_3| + |I_4| \leq 1$ is a tight constraint for all non-full network nonlocal correlations. In particular, it implies the full network nonlocality witness:

$$\mathcal{S}_3 \leq 2^{1/3}, \tag{8.106}$$

The quantum correlations with $\mathcal{S}_3 \in (1, 2^{1/3}]$ imply verification of network nonlocality but not fully network nonlocality. Instead, the quantum correlations with $\mathcal{S}_3 \in (2^{1/3}, \sqrt{2}]$ exhibit both network nonlocality and fully network nonlocality. This provides a visibility of $v \approx 89.1\%$ for verifying noisy states $\rho_v = v|EPR\rangle\langle EPR| + \frac{1-v}{4}\mathbb{1}$.

The star networks provide basic examples for exploring the relationship between network nonlocality and full network nonlocality. On one hand, a single bipartite entanglement shared by some A_i and B can exhibit standard network nonlocality but not fully network nonlocality. On the other hand, all n pairs of bipartite entanglement may exhibit maximal violation of network nonlocality and imply full network nonlocality. Notably, for $n = 3$, all quantum violations can be simulated using one post-quantum source. For $n \geq 4$, the full network nonlocality witness is the same as its quantum violation.

Now, we return to the bilocal scenario of $n = 2$ by violating the inequality (8.104). A good setup is an elegant joint measurement for the center party (Tavakoli et al. 2021). Here, Bob projects two qubits onto a partially entangled basis that forms a regular tetrahedron inside the Bloch sphere with Bell state measurement at the vertex. Fully network nonlocal correlations may be witnessed by inflating networks (Wolfe et al. 2019; Navascués and Wolfe 2020; Wolfe et al. 2021; Pironio et al. 2010), which will be discussed in the next chapter. In short, network inflation is a way to copy sources and measurement devices in a given network and then arrange these copied resources in different configurations. By exploring the correlation constraints arising from new networks, nontrivial constraints on the correlations from the original network can be imposed. Here, we consider another method by the simultaneous violation of two inequalities:

$$-A_1B_2C_3 - A_2B_2 + C_3[A_1B_2 + A_2B_2C_3 + C_3] \leq 1, \tag{8.107}$$
$$-A_1B_2C_3 + B_2C_2 + A_1[B_2C_3 - A_1B_2C_2 + A_1] \leq 1. \tag{8.108}$$

Here, the expectation values are defined as $A_xB_yC_z = \sum_{a,b,c} ab_ycP(a,b,c|x,z)$, where $a, c \in \{\pm 1\}$ and $(b_1, b_2, b_3) \in \{\pm 1\}^3$ satisfies $b_1b_2b_3 = 1$. A violation of Eq. (8.107) implies non-bilocality, while a simultaneous violation of Eq. (8.108) implies full network nonlocality, ruling out any combination of local variables and nonlocal resources. The critical visibility per source is given by

$$v \geq \frac{4}{\cos\theta + \sqrt{8 + 8\sin\theta + \cos^2\theta}}, \tag{8.109}$$

for noisy sources.

The present method provides a simple primitive for characterizing fully network nonlocality. Unfortunately, it cannot be extended to verifying general quantum networks, even for entangled pure states. Such resources can, in principle, activate nonlocal correlations in terms of generalized entanglement swapping, while any hybrid realizations of classical variables and nonlocal sources cannot. This generally requires new methods for characterizing all entanglement-swapping experiments.

8.5.5 Nonlocality of Robust Entanglement

It is generally difficult to verify the Bell nonlocality of robust entanglement due to the non-convexity of $\mathcal{B}_d$ in Sect. 8.3.2. We may generally make use of the following set of Bell inequalities

$$\begin{cases} \mathcal{L}_n(\rho_{A_1,\ldots A_n}) \leq c_n, \\ \mathcal{L}_k(\mathcal{E}_S(\rho)) \leq c_k, \forall |S| \leq k, \end{cases} \tag{8.110}$$

where $\mathcal{L}_m(\cdot)$ is an m-partite Bell operator for verifying the genuinely m-partite nonlocality in the biseparable model (Svetlichny 1987), c_n and c_k are constants. It is necessary to violate all these inequalities to verify the nonlocality of the robust entanglement.

Example 8.5.6 Consider a triangle network consisting of three maximally entangled EPR states. The genuinely tripartite nonlocality in the biseparable model can be verified by using LOCC or linear inequalities with multiple setups. Moreover, the final state after passing through a single particle-loss channel $\mathcal{E}_{\mathsf{A}_i}(\cdot)$ can be verified by violating the CHSH inequality (Clauser et al. 1969).

Example 8.5.7 Consider a generalized W state $|D_{1,3}\rangle$ in (8.41). Its genuinely tripartite nonlocality can be verified by using the Svetlichny inequality (Svetlichny 1987). Surprisingly, all the linear inequalities with two measurement setups such as the CHSH inequality are useless for verifying the final states. Instead, it may be witnessed by violating the following nonlinear inequality as

$$(\langle X_1X_2\rangle - \langle Y_1Y_2\rangle)^2 + (\langle X_1Y_2\rangle + \langle Y_1X_2\rangle)^2 - (1 - \langle Z_1Z_2\rangle)^2 \leq 0, \tag{8.111}$$

where X_j, Y_j and Z_j are Pauli observables on the j-th qubit, $j = 1, 2$. The maximal quantum bound is 4.

The inequality (8.111) can be further extended in multipartite scenarios. A general method may require verifying all the final states. This is impossible because of the exponential number. One may consider special symmetry to further reduce the computation complexity.

8.6 Nonlocality Without Input Choice

In this subsection, we discuss the quantum nonlocality of networks without input choice. In standard Bell theory, quantum correlations can always be simulated by a classical model under the assumption of one measurement setup per party (Bell 1964). However, this result cannot be extended to network scenarios. The independent source assumption may inspire different correlations even with one group of measurements, which is the nonlocality without input choice.

8.6.1 Rigidity-Based Arguments

Consider a triangle quantum network as shown in Fig. 8.3. Here, each pair of two parties shares a bipartite source, classical variable, or quantum state. The main ingredient is the independence of these three sources, meaning that the three parties share no common information. Now, each party generates an output (a, b, and c, respectively) depending on their shares without any input assumption, contrary to standard Bell tests (Bell 1964).

By classical realization with three variables $\lambda_1, \lambda_2, \lambda_3$, the joint probability distribution admit the following decomposition as

$$P_C(a,b,c) = \int d\mu(\lambda_1)d\mu(\lambda_2)d\mu(\lambda_3)p(a|\lambda_2,\lambda_3)p(b|\lambda_1,\lambda_3)p(c|\lambda_1,\lambda_2), \quad (8.112)$$

where $p(a|\lambda_2,\lambda_3)$, $p(b|\lambda_1,\lambda_3)$, $p(c|\lambda_1,\lambda_2)$ denotes deterministic response functions for three parties. The assumption of source independence makes all the distributions $P(a,b,c)$ non-convex. This forbids the extension of standard methods in Bell nonlocality.

In quantum scenarios, each party has two qubits from two entangled EPR states as $|\Phi_i\rangle|\Phi_j\rangle$, where $|\Phi_i\rangle = \cos\theta_i|00\rangle + \sin\theta_i|11\rangle$ with $\theta_i \in (0, \frac{\pi}{4})$. The total state is regarded as a high-dimensional state of

$$\begin{aligned}|\Phi\rangle = (&a_1|000\rangle + a_2|012\rangle + a_3|120\rangle + a_4|132\rangle \\ &+a_5|201\rangle + a_6|213\rangle + a_7|321\rangle + a_8|333\rangle)_{61,23,45}\end{aligned} \quad (8.113)$$

on Hilbert space $\mathbb{C}^4\otimes\mathbb{C}^4\otimes\mathbb{C}^4$, where a_i are defined by $a_1 = \cos\theta_1\cos\theta_2\cos\theta_3$, $a_2 = \cos\theta_1\sin\theta_2\cos\theta_3$, $a_3 = \sin\theta_1\cos\theta_2\cos\theta_3$, $a_4 = \sin\theta_1\sin\theta_2\cos\theta_3$, $a_5 = \cos\theta_1\cos\theta_2\sin\theta_3$, $a_6 = \cos\theta_1\sin\theta_2\sin\theta_3$, $a_7 = \sin\theta_1\cos\theta_2\sin\theta_3$, and $a_8 = \sin\theta_1\sin\theta_2\sin\theta_3$. By performing local projective measurements given by

$$\mathcal{Q} := \{|01\rangle, \cos\theta|00\rangle + \sin\theta|11\rangle, \sin\theta|10\rangle - \cos\theta|11\rangle, |10\rangle\} \quad (8.114)$$

with $\theta \in (0, \frac{\pi}{4})$, the Born's rule implies a joint quantum correlation given by $P_Q(a, b, c) = \mathrm{Tr}(M_a \otimes M_b \otimes M_c|\Phi\rangle\langle\Phi|)$ with $M_a, M_b, M_c \in \mathcal{Q}$. From the symmetry of local measurements, it will give rise to new symmetry for quantum correlations while it cannot be realized by classical sources (Renou et al. 2019).

Theorem 8.6.1 *The quantum distribution $P_Q(a, b, c)$ cannot be reproduced by any classical trilocal model (8.112) when $1 < \theta < \theta_{min}$, where $\theta_{min} = \arccos 0.785$.*

The main idea is that the quantum distribution $P_Q(a, b, c)$ features a certain number of specific constraints under the local measurement (8.114). One example is $\Pr(x = y = 01) = \Pr(x = y = 10) = 0$ for any $x, y \in a, b, c$. These symmetry relations for measurement outcomes are key to proving this theorem. For some specific choice of the measurement angle θ, there is no trilocal model compatible with all these constraints at once (Renou et al. 2019). This provides new network nonlocality beyond the standard Bell nonlocality (Bell 1964).

Symmetry plays an important role in general cyclic networks, but it cannot be extended to general quantum networks. This issue can be resolved by new methods (Renou and Beigi 2022). One such method is the token-counting strategy, which considers the maximally entangled EPR state $\frac{1}{\sqrt{2}}(|01\rangle + |10\rangle)$ as a quantum superposed token, with $|01\rangle$ and $|10\rangle$ being sent to different parties. The local measurement operators are defined by the number of tokens received, as

$$\mathcal{Q}_t := \{\bar{0} := |00\rangle\langle 00|, \bar{1} := |01\rangle\langle 01| + |10\rangle\langle 10|, \bar{2} := |11\rangle\langle 11|\} \tag{8.115}$$

where $\bar{n}$ counts the number of received tokens. These local measurements generate a joint distribution given by

$$P_{token} = \mathrm{Tr}(M_a \otimes M_b \otimes M_c|\Phi\rangle\langle\Phi|), M_a, M_b, M_c \in \mathcal{Q}_t \tag{8.116}$$

with $\theta = \frac{\pi}{4}$. For the outcome $\bar{1}$, the party performs the second dichotomic measurement $\{|\bar{1}_0\rangle := \cos\vartheta|01\rangle + \sin\vartheta|10\rangle, |\bar{1}_1\rangle := \sin\vartheta|01\rangle - \cos\vartheta|10\rangle\}$. The joint distribution P_{token} is simulated by classical strategy because $\{\bar{0}, \bar{1}, \bar{2}\}$ are diagonal in the token number basis.

In the quantum strategy, when all parties measure the $\bar{1}$ token, the global state is a coherent superposition of the tokens turning clockwise and anticlockwise. On the other hand, classical token counting results in a mixed state of $\frac{1}{2}(|\circlearrowleft\rangle\langle\circlearrowleft| + |\circlearrowright\rangle\langle\circlearrowright|)$, while quantum token counting is a pure state of $\frac{1}{\sqrt{2}}(|\circlearrowleft\rangle + |\circlearrowright\rangle)$. This implies a rigidity-based method for identifying nonlocality in general networks without input assumptions (Renou and Beigi 2022). For example,

$$q(a, b, c, t) = \Pr(a, b, c, t|\mathrm{N} \text{ for all outcomes}), \tag{8.117}$$

where $q(a, b, c, t)$ denotes the probability of measurement outcomes a, b, c, t when all parties measure the $\bar{1}$ token. These marginals are incompatible with any correlations from classical realizations of triangle networks under specific measurement choices of $\{\bar{1}_0, \bar{1}_1\}$. This result can be extended to general classical token-counting strategies by distributing an arbitrary fixed number of tokens. Such strategies are rigid in all networks where any pair of parties shares at most one source.

A similar approach is to color a pre-established set in superposition. A d-dimensional bipartite source of $\frac{1}{\sqrt{d}}(|0, 0\rangle + \cdots + |d-1, d-1\rangle)$ can be represented by d different colors. Each party first checks the consistency of the set obtained from neighboring parties and outputs the same color. In a triangle network, each party can implement local projections with $\{|00\rangle, |11\rangle, |22\rangle\}$ to generate a color-matching distribution P_{match}. The subsequent steps are similar to the token-counting procedure, leading to rigid strategies for identifying network non-locality without input assumptions (Renou and Beigi 2022).

The rigidity in Finner inequalities, which will be discussed in the following chapter, allows for finding network nonlocal quantum color-matching strategies in the n-partite complete networks, comprising all bipartite sources, or a combination of tripartite and bipartite sources. This refers to methods for achieving nonlocality in quantum systems without making assumptions about the inputs. The rigidity of classical strategies is used as a form of self-testing, verifying the nonlocality present in the system.

8.6.2 Disguised Nonlocality on Networks

So far, we have discussed various versions of network nonlocality, each with a specific motivation. However, the central question of how to characterize genuine network nonlocality has not been completely resolved. This may be considered a genuine feature of the network configuration. Although all network nonlocalities comply with the decomposition (8.78), some manifestations may be linked to standard Bell nonlocality (Fritzm 2012).

Consider the bilocal scenario of a tripartite chain network, where three parties generate a joint correlation $P(a, b, c|x, y, z)$. Suppose Alice and Bob want to test the CHSH inequality by the marginal distribution $p(a, b|x, y)$. This follows from the network local model (8.78) that

$$\begin{aligned} p(a, b|x, y) &= \sum_c \Big[\int d\lambda_1 d\lambda_2 \mu_1(\lambda_1)\mu_2(\lambda_2) p(a|x, \lambda_1) \\ &\quad \times p(b|y, \lambda_1, \lambda_2) p(c|z, \lambda_2) \Big] \\ &= \int d\lambda_1 \mu(\lambda_1) p(a|x, \lambda_1) p(b|y, \lambda_1). \end{aligned} \tag{8.118}$$

This is precisely the standard LHV model. This means if $P(a, b, c|x, y, z)$ satisfies the conditions for network locality, then the bipartite marginal $p(a, b|x, y)$ admits a local hidden variable (LHV) model. Conversely, in a network local model, nonlocal correlations in $p(a, b|x, y)$ will result in the nonlocality of $p(a, b, c|x, y, z)$. This illustrates how the hybrid network in Fig. 8.7 can simulate nonlocal behavior using local resources.

Consider a network consisting of sources $\rho_1, \ldots, \rho_m$, which are shared by n parties $\mathsf{A}_1, \ldots, \mathsf{A}_n$. Local measurements imply a joint distribution $P(\vec{a}|\vec{x})$. For a given network local model $\mathcal{M}_1$, we may find some constraints as

$$\mathcal{L}(P(\vec{a}|\vec{x}))_{\mathcal{M}_1} \leq c_1 \tag{8.119}$$

with a constant c_1. The violation of the inequality (8.119) implies specific network nonlocality in the local model $\mathcal{M}1$. To explore other possible models, we can use the correlations $P(\vec{a}|\vec{x})$. There are at least two methods for this. One is to relax or enhance the assumption of sources, such as LHV or NLHV models, and verify the violation of new constraints as

$$\mathcal{L}(P(\vec{a}|\vec{x}))_{\mathcal{M}_2} \leq c_2 \tag{8.120}$$

with a constant c_2. Simple extensions of tripartite networks include chain networks or star networks. Another method is to relax the measurement setups. For example, we may consider multiple variables in network scenarios, unlike the LHV model with a single variable. The source independence of these variables may inspire new network nonlocality (Renou et al. 2019). One application of this is a network experiment without measurement choices, which highlights an important difference between network nonlocality and standard Bell nonlocality in terms of measurement inputs.

Disguising network nonlocality is difficult for general networks because it requires exploring the correlations from any realizations with hybrid sources, in local, quantum, or post-quantum theory. In some cases, the disguised nonlocality cannot be easily detected by relaxing sources or evaluating marginal distributions. The main problem is from the local joint measurements for each party, which may be difficult to analyze for general networks. One possibility is to regard all the inputs as the outputs of one additional party (added one). This leads to an extension of the original network, which may help to identify disguised network nonlocality. This kind of disguising may rule out some weak models such as network local models (Supic et al. 2020), which are not able to capture the full extent of the nonlocality present in the system. Disguised network nonlocality is an active area of research, as it has important implications for our understanding of the limits of local and quantum theories.

8.6.3 Classical Network Inflation

Network inflation is an efficient way to explore constraints on the correlations generated from any network (Wolfe et al. 2019). The main idea is to analyze hypothetical situations by copying the sources and measurement devices in the original network.

The new network is arranged in arbitrary configurations, except for preserving the invariant of the original network. Constraints on the correlations from the new network configurations are then mapped back to the original network, implying special conditions on the correlations. This process allows for a more detailed analysis of the underlying correlations and can reveal new insights into the nonlocal behavior of the network. Examples of such insights include the discovery of new hidden-variable models for quantum correlations and the identification of novel forms of nonlocality that are not captured by standard Bell inequalities.

Classical inflation was initially developed for solving the problem of causal compatibility for a given distribution $P(\vec{a}|\vec{x})$ and network configuration. Causal compatibility refers to the consistency of the observed correlations with a specified causal structure, and classical inflation is a method for testing this consistency. In terms of network nonlocality, classical inflation is equivalent to discerning whether the probability distribution $P(\vec{a}|\vec{x})$ admits a specific network local model. One example of a network local model is the LHV model (8.78), which assumes that the observed correlations can be explained by a shared, local hidden variable that influences the outcomes of the local measurements. The ability to discern whether a given distribution admits a network local model is an important tool for detecting and characterizing network nonlocality, as it allows us to rule out certain types of classical explanations for the observed correlations.

The main procedure is as follows. We first inflate a given network by adding copies of the sources and parties. The new arrangement of all sources and parties should satisfy that each copied party is connected to copied sources that are connected to the original party. The second is to explore necessary constraints on probability distributions compatible with the inflated network. The third is to translate these constrict into original networks by considering certain marginals. One example is to show a given joint distribution $P_{abc} = \frac{1}{2}[000] + \frac{1}{2}[111]$ which is incompatible with a chain network and implies a separability condition of $P(a, c) = p(a)p(c)$ for any outcomes x, z. Here, random variables A, B, C are assigned to a, b, c, respectively. $P(a, c)$, $p(a)$ and $p(c)$ are marginal distributions.

In addition to identifying specific distribution, the main idea is from the intrinsic symmetries of the inflation network. This requires a strong understanding of inflation. The general inflation requires solving some optimizations. This provides an efficient way for verifying Bell nonlocality in specific networks. The further extensions of high-order or quantum inflations will be discussed in the next chapter.

8.7 Conclusion

In this chapter, we have extended Bell nonlocality for quantum networks. The standard Bell nonlocality or non-multilocality is weak and cannot show the network features of multiple independent sources. Instead, the full network nonlocality shows some genuine features of quantum networks similar to the biseparable model of single entanglement (Svetlichny 1987). This implies strong motivations for exploring the essential ingredient of quantum tasks on networks such as entanglement swapping.

Different disguising models may be combined in network scenarios. This emphasizes the importance of identifying genuine features of network nonlocality. The reasonable new model should be combined with disguised explorations with well-known nonlocalities. The network entanglement is based on the results (Luo 2021; Luo and Fei 2021). The entanglement model may be extended by using shallow quantum circuits (Luo 2022). The network steering is defined in Ref. (Jones et al. 2021). The fully network nonlocality is from Ref. (Pozas-Kerstjens et al. 2022), while the network nonlocality is without input choice from Ref. (Renou et al. 2019; Renou and Beigi 2022). For further discussion, see the recent review (Tavakoli et al. 2022).

References

G. Ausiello, P. Crescenzi, G. Gambosi, V. Kann, A. Marchetti-Spaccamela, M. Protasi, *Complexity and Approximation: Combinatorial Optimization Problems and Their Approximability Properties* (Springer, 2003)

T.J. Barnea, G. Pütz, J.B. Brask, N. Brunner, N. Gisin, Y.-C. Liang, Nonlocality of W and Dicke states subject to losses. Phys. Rev. A **91**, 032108 (2015)

J. Barrett, Information processing in generalized probabilistic theories. Phys. Rev. A **75**, 032304 (2007)

J.S. Bell, On the Einstein Podolsky Rosen paradox. Phys. Physique Fizika **1**, 195 (1964)

C. Branciard, N. Gisin, S. Pironio, Characterizing the nonlocal correlations created via entanglement swapping. Phys. Rev. Lett. **104**, 170401 (2010)

C. Branciard, D. Rosset, N. Gisin, S. Pironio, Bilocal versus nonbilocal correlations in entanglement-swapping experiments. Phys. Rev. A **85**, 032119 (2012)

E.G. Cavalcanti, Q.Y. He, M.D. Reid, H.M. Wiseman, Unified criteria for multipartite quantum nonlocality. Phys. Rev. A **84**, 032115 (2011)

R. Chaves, Nonlinear bell inequalities tailored for quantum networks. Phys. Rev. Lett. **116**, 010402 (2016)

B.S. Cirel'son, Quantum generalizations of Bell's inequality. Lett. Math. Phys. **4**, 93 (1980)

J.F. Clauser, M.A. Horne, A. Shimony, R.A. Holt, Proposed experiment to test local hidden-variable theories. Phys. Rev. Lett. **23**, 880–884 (1969)

V. Coffman, J. Kundu, W.K. Wootters, Distributed entanglement. Phys. Rev. A **61**, 052306 (2000)

P. Contreras-Tejada, C. Palazuelos, J.I. de Vicente, Genuine multipartite nonlocality is intrinsic to quantum network. Phys. Rev. Lett. **126**, 40501 (2021)

R. Diestel, *Graph Theory* (Springer, Berlin Heidelberg, 2017)

A. Einstein, B. Podolsky, N. Rosen, Can quantum-mechanical description of physical reality be considered complete? Phys. Rev. **47**, 777–780 (1935)

T. Fritzm, Beyond Bell's theorem I: correlation scenarios. New J. Phys. **14**, 103001 (2012)

D. Geiger, C. Meek, in *Proceedings of the 15th Conference on Uncertainty in Artificial Intelligence* (Morgan Kaufmann Publishers Inc., 1999), p. 226

N. Gisin, Bell's inequality holds for all non-product states. Phys. Lett. A **154**, 201 (1991)

N. Gisin, Q. Mei, A. Tavakoli, M.O. Renou, N. Brunner, All entangled pure quantum states violate the bilocality inequality. Phys. Rev. A **96**, 020304(R) (2017)

K. Hansenne, Z.-P. Xu, T. Kraft, O. Gúhne, Symmetries in quantum networks lead to no-go theorems for entanglement distribution and to verification techniques. Nat. Commun. **13**, 496 (2022)

Q.Y. He, M.D. Reid, Genuine multipartite Einstein-Podolsky-Rosen steering. Phys. Rev. Lett. **111**, 250403 (2013)
M. Hein, J. Eisert, H.J. Briegel, Multiparty entanglement in graph states. Phys. Rev. A **69**, 062311 (2004)
W. Helwig, W. Cui, J.I. Latorre, A. Riera, H.K. Lo, Absolute maximal entanglement and quantum secret sharing. Phys. Rev. A **86**, 52335 (2012)
H.E. Hopcroft, R.M. Karp, An $n^5/2$ algorithm for maximum matchings in bipartite graphs. SIAM J. Comput. **2**, 225 (1973)
R. Horodecki, P. Horodecki, M. Horodecki, K. Horodecki, Quantum entanglement. Rev. Mod. Phys. **81**, 865 (2009)
L.-Y. Hsu, C.-H. Chen, Exploring Bell nonlocality of quantum networks with stabilizing and logical operators. Phys. Rev. Res. **3**, 023139 (2021)
B.D.M. Jones, I. Supic, R. Uola, N. Brunner, P. Skrzypczyk, Network quantum steering. Phys. Rev. Lett. **127**, 170405 (2021)
R. Jozsa, Fidelity for mixed quantum states. J. Mod. Opt. **41**, 2315–2323 (1994)
T. Kraft, S. Designolle, C. Ritz, N. Brunner, O. Gühne, M. Huber, Quantum entanglement in the triangle network, arXiv:2002.03970 (2020)
C.-M. Li, K. Chen, Y.-N. Chen, Q. Zhang, Y.-A. Chen, J.-W. Pan, Genuine high-order Einstein-Podolsky-Rosen steering. Phys. Rev. Lett. **115**, 010402 (2015)
M.X. Luo, Computationally efficient nonlinear bell inequalities for quantum networks. Phys. Rev. Lett. **120**, 140402 (2018)
M.X. Luo, New genuine multipartite entanglement. Adv. Quantum Technol. **4**, 2000123 (2021)
M.X. Luo, Fully device-independent model on quantum networks. Phys. Rev. Res. **4**, 013203 (2022)
M.X. Luo, S.-M. Fei, Robust multipartite entanglement without entanglement breaking. Phys. Rev. Res. **3**, 043120 (2021)
N.D. Mermin, Extreme quantum entanglement in a superposition of macroscopically distinct states. Phys. Rev. Lett. **65**, 1838 (1990)
K. Mukherjee, B. Paul, A. Roy, Characterizing quantum correlations in a fixed-input n-local network scenario. Phys. Rev. A **101**, 032328 (2020)
M. Navascués, E. Wolfe, The inflation technique completely solves the causal compatibility problem. J. Causal Infer. **8**, 70–91 (2020)
M. Navascues, E. Wolfe, D. Rosset, A. Pozas-Kerstjens, Genuine network multipartite entanglement. Phys. Rev. Lett. **125**, 240505 (2020)
J. Pearl, *Causality* (Cambridge University Press, 2009)
S. Pironio, M. Navascues, A. Acín, Convergent relaxations of polynomial optimization problems with noncommuting variables. SIAM J. Optim. **20**, 2157 (2010)
S. Popescu, D. Rohrlich, Quantum nonlocality as an axiom. Found. Phys. **24**, 379–385 (1994)
A. Pozas-Kerstjens, N. Gisin, A. Tavakoli, Full network nonlocality. Phys. Rev. Lett. **128**, 010403 (2022)
M.T. Quintino, T. Vértesi, D. Cavalcanti, R. Augusiak, M. Demianowicz, A. Acin, N. Brunner, Phys. Rev. A **92**, 032107 (2015)
M.T. Quintino, N. Brunner, M. Huber, Superactivation of quantum steering. Phys. Rev. A **94**, 062123 (2016)
R. Raussendorf, H.J. Briegel, A one-way quantum computer. Phys. Rev. Lett. **86**, 5188 (2001)
M.-O. Renou, S. Beigi, Network nonlocality via rigidity of token counting and color matching. Phys. Rev. Lett. **128**, 060401 (2022)
M.-O. Renou, E. Bäumer, S. Boreiri, N. Brunner, N. Gisin, S. Beigi, Genuine quantum nonlocality in the triangle network. Phys. Rev. Lett. **123**, 140401 (2019)
M.-O. Renou, D. Trillo, M. Weilenmann, T.P. Le, A. Tavakoli, N. Gisin, A. Acín, M. Navascues, Quantum theory based on real numbers can be experimentally falsified. Nature **600**, 625–629 (2021)
D. Rosset, C. Branciard, T.J. Barnea, G. Pütz, N. Brunner, N. Gisin, Nonlinear bell inequalities tailored for quantum networks. Phys. Rev. Lett. **116**, 010403 (2016)

N. Sangouard, C. Simon, H. de Riedmatten, N. Gisin, Quantum repeaters based on atomic ensembles and linear optics. Rev. Mod. Phys. **83**, 33 (2011)

D.J. Saunders, S.J. Jones, H.M. Wiseman, G.J. Pryde, Experimental EPR-steering using Bell-local states. Nat. Phys. **6**, 845–849 (2010)

M. Seevinck, J. Uffink, Sufficient conditions for three-particle entanglement and their tests in recent experiments. Phys. Rev. A **65**, 012107 (2001)

A. Sohbi, I. Zaquine, E. Diamanti, D. Markham, Decoherence effects on the non-locality of symmetric states. Phys. Rev. A **91**, 022101 (2015)

I. Supic, J.-D. Bancal, N. Brunner, Quantum nonlocality in networks can be demonstrated with an arbitrarily small level of independence between the sources. Phys. Rev. Lett. **125**, 240403 (2020)

G. Svetlichny, Distinguishing three-body from two-body nonseparability by a Bell-type inequality. Phys. Rev. D **35**, 3066 (1987)

A. Tavakoli, N. Gisin, C. Branciard, Bilocal bell inequalities violated by the quantum elegant joint measurement. Phys. Rev. Lett. **126**, 220401 (2021)

A. Tavakoli, A. Pozas-Kerstjens, M.-X. Luo, M.-O. Renou, Bell nonlocality in networks. Rep. Prog. Phys. **85**, 056001 (2022)

N. Tischler, F. Ghafari, T.J. Baker, S. Slussarenko, R.B. Patel, M.M. Weston, S. Wollmann, L.K. Shalm, V.B. Verma, S.W. Nam, H.C. Nguyen, H.M. Wiseman, G.J. Pryde, Conclusive experimental demonstration of one-way Einstein-Podolsky-Rosen steering. Phys. Rev. Lett. **121**, 100401 (2018)

G. Toth, Detection of multipartite entanglement in the vicinity of symmetric Dicke states. J. Opt. Soc. Am. B **24**, 275–282 (2007)

A.R. Usha Devi, R. Prabhu, A.K. Rajagopal, Characterizing multiparticle entanglement in symmetric N-qubit states via negativity of covariance matrices. Phys. Rev. Lett. **98**, 060501 (2007)

S.P. Walborn, A. Salles, R.M. Gomes, F. Toscano, P.H. Souto Ribeiro, Revealing Hidden Einstein-Podolsky-Rosen Nonlocality. Phys. Rev. Lett. **106**, 130402 (2011)

R.F. Werner, Quantum states with Einstein-Podolsky-Rosen correlations admitting a hidden-variable model. Phys. Rev. A **40**, 4277 (1989)

H.M. Wiseman, S.J. Jones, A.C. Doherty, Steering, entanglement, nonlocality, and the Einstein-Podolsky-Rosen paradox. Phys. Rev. Lett. **98**, 140402 (2007)

E. Wolfe, R.W. Spekkens, T. Fritz, The inflation technique for causal inference with latent variables. J. Causal Infer. **7**, 20170020 (2019)

E. Wolfe, A. Pozas-Kerstjens, M. Grinberg, D. Rosset, A. Acín, M. Navascués, Quantum inflation: a general approach to quantum causal compatibility. Phys. Rev. X **11**, 021043 (2021)

Y.-Z. Zhen, X.-Y. Xu, L. Li, N.-Le Liu, K. Chen, The Einstein-Podolsky-Rosen steering and its certification. Entropy **21**, 422 (2019)

M. Żukowski, A. Zeilinger, M.A. Horne, A.K. Ekert, Event-ready-detectors Bell experiment via entanglement swapping. Phys. Rev. Lett. **71**, 4287–4290 (1993)

Chapter 9
Compatibility of Quantum Networks

Network configuration for Internet communication is the process of configuring network control, flow, and operation for a specific agent. This term encompasses multiple processes of configuration and setup on all network components and is also known as network setup. On the other hand, the quantum Internet requires a different type of network processing, as it involves special forms of entangled quantum networks that are essential for achieving quantum supremacy beyond classical networks. In this chapter, we focus on a more restricted concept of quantum network configuration, particularly on network topology and compatible correlations.

9.1 Network Inflation

We consider a general network consisting of several independent sources, which can be modeled by Bayesian networks (Pearl 2009). Each party in the network has some shared sources and performs local measurements on them, resulting in a joint distribution of all outcomes that are conditional on their measurement setups. Bell theory (Bell 1964) offers a useful method for exploring the quantumness of networks beyond classical realizations by examining the restrictions of all correlations, as discussed in Chap. 8.

A standard Bell test involves at least two sets of local measurements for each party. Alternatively, another method involves one set of local measurements that produce a joint distribution $P(\vec{x})$ of all measurement outcomes. With this setup, the incompatibility of networks can be divided into two parts. One arises from the sources, which is similar to standard quantum nonlocality and aims to verify the quantumness beyond other theories. The other arises from the network topology. Thus, the fundamental problem is to determine the compatibility between a given distribution and a network's topology.

M.-X. Luo, *Quantum Networks*,
https://doi.org/10.1007/978-981-97-6226-2_9

Definition 9.1.1 For a given probability distribution $P(\vec{x})$ on some finite measurable space $\times_{i=1}^{n}\mathcal{X}_i$, it is compatible with a given network $\mathcal{N}$ consisting of independent sources if it can be generated by specific local measurements. Otherwise, it is incompatible.

In classical realizations of Bayesian networks, each network is associated with a directed acyclic graph (DAG) (Pearl 2009). Here, each edge is represented by a measurable variable on a proper measurable space, and each outcome depends on all the related variables. The joint distribution is a multivariate function of all outcomes, which allows us to decompose it into a product of conditional distributions, each depending on a single latent variable λ (Bell 1964):

$$P(\vec{x}) = \int_{\Omega} p(x_1|\lambda)\cdots p(x_n|\lambda)\mu(\lambda)d\lambda, \tag{9.1}$$

where $(\Omega, \mu(\lambda))$ denotes the measurable space of the unobservable latent variable λ, $\mu(\lambda)$ denotes the probability of λ, and $p(x_i|\lambda)$ denotes the characteristic function of outcome x_i conditional on the variable λ. Bayesian networks with latent variables exhibit non-trivial constraints on their correlations, which can be explored using methods from Bell's theory.

In general network, it allows decomposing

$$P(\vec{x}) = \int_{\Omega} d\mu(\Lambda)\prod_{i=1}^{n} P(x_i|\Lambda_i), \tag{9.2}$$

where the output a_i depends on all patent nodes in $\Lambda_i \subset \Lambda$. The network compatibility is defined as follows.

Definition 9.1.2 (Pearl 2009) (*Markov Compatibility*) If a probability function $P_{\vec{x}}$ admits the factorization of (9.2) relative to DAG $\mathcal{G}$, then $\mathcal{G}$ and $P_{\vec{x}}$ are compatible.

9.1.1 Classical Network Inflation

The concept of network inflation is a powerful tool for setting constraints on the correlations that can be generated by local measurements on any network (Wolfe et al. 2019). The basic idea is to create a hypothetical network by inflating the given network with multiple copies. These inflated networks should conform to the original network, and the copied resources can be arranged in arbitrary configurations.

In the second step, the correlations from the hypothetical network are analyzed using multiple copies of the sources and measurement devices. It is important to note that any conclusions obtained in the inflated network can be pulled back to the original network since the inflated network is a faithful representation of the original.

This provides a useful method for investigating the compatible correlations under the assumption of one measurement set per party. By analyzing the inflated network, one can set constraints on the possible correlations that can be generated by local measurements, which can help in understanding the fundamental limitations of quantum mechanics and the nature of entanglement.

An example of network inflation is shown in Fig. 9.1. It involves a triangle network consisting of three variables λ, γ, and η for three variables X, Y, and Z. The inflation is shown in Fig. 9.1b by copying three variables λ_1, γ_1, and η_1 for the other three variables X_1, Y_1, and Z_1. This is known as first-order inflation.

In Fig. 9.1b, X_1, Y_1, and Z_1 are connected to both the sources and the copied sources. The inflation provides correlations $P(x, y, z, x_1, y_1, z_1)$. It is important to note that all the outputs of X, Y, and Z remain unchanged during this inflation. This means that the new correlations $P(x, y, z, x_1, y_1, z_1)$ from Fig. 9.1b can be used to obtain information about the correlations $P(x, y, z)$ from Fig. 9.1a. This implies a general method for characterizing the correlations $P(x, y, z)$ by exploring the restrictions of $P(x, y, z, x_1, y_1, z_1)$. Similar extensions can be made by copying other variables in the network.

Classical inflation was originally developed for solving the problem of causal compatibility for a given distribution and network. In the context of Bell nonlocality in networks, the compatible problem is equivalent to determining whether an observed probability distribution admits a classical realization of the network.

A general procedure for solving compatible problems with classical inflations is as follows. Suppose that a probability distribution $P(\vec{x})$ is generated by local measurements on a given classical network $\mathcal{N}$. This may be equivalently rephrased by the existence of variables and deterministic functions to generate $P(\vec{x})$. And then, the inflation of $\mathcal{N}$ is completed by adding multiple copies of sources and parties. Here, one should connect all the nodes in a compatible way such that each copy of a given party is only connected to copied sources that are connected to this party in the

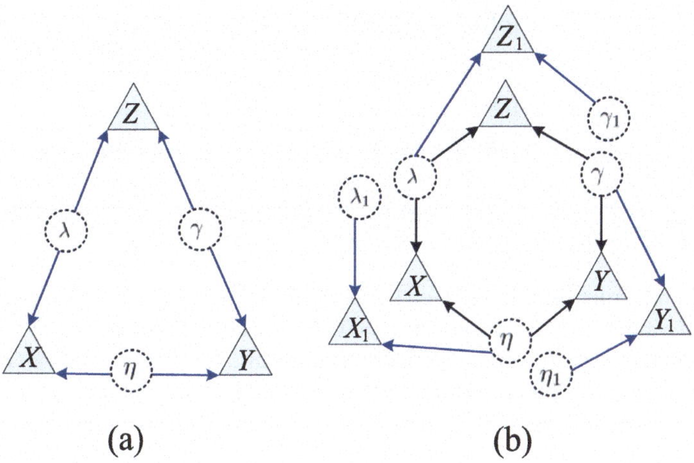

Fig. 9.1 (Color online) **a** Triangle network consisting of three variables λ, γ and η. **b** Inflation of triangle network by copying three variables λ_1, γ_1 and η_1

original network. The following step is to explore special constraints on probability distributions $P_{\rm inf}(\vec{x})$ compatible with the inflation network $\mathcal{N}_{\rm inf}$ as

$$\mathcal{S}_i(P_{\rm inf}(\vec{x})) \leq c_i, i \in I \tag{9.3}$$

with general functions $\mathcal{S}_i$ and constants c_i. All the constraints are further translated back to the original network as

$$\mathcal{S}_i(P(\vec{x})) \leq \hat{c}_i, i \in I \tag{9.4}$$

where the constant $\hat{c}_i$ may be different from c_i.

Note that certain marginal distributions of $P_{\rm inf}(\vec{x})$ compatible with the inflated networks should match it from the original distribution $P(\vec{x})$. This ensures the possibility of the back-forward translation. One example is the inflation of a triangle network in a web manner (Wolfe et al. 2019). Instead, specific distributions (eg. the intrinsic symmetries) from the inflated network may result in incompatible marginal distributions of the original distribution. All the constrictions for these distributions may be interesting for special goals of network nonlocality without the assumption of inputs. Finally, one needs to determine whether a distribution $P_{\rm inf}$ satisfies all the constraints stated above. This may be formulated as a linear program, which allows efficient numeric solutions. The present method in principle can solve the causal compatibility problem by using high-order inflations. We illustrate this idea by using generalized W-type distributions (Luo 2022).

Theorem 9.1.1 *Consider the following W-type distributions*

$$P_{xyz} = p_1[001] + p_2[010] + p_3[100] + p_4[011], \tag{9.5}$$
$$P_{xyz} = p_1[001] + p_2[010] + p_3[100] + p_4[101], \tag{9.6}$$
$$P_{xyz} = p_1[001] + p_2[010] + p_3[100] + p_4[110], \tag{9.7}$$
$$P_{xyz} = p_1[010] + p_2[011] + p_3[100] + p_4[101], \tag{9.8}$$
$$P_{xyz} = p_1[001] + p_2[010] + p_3[101] + p_4[110], \tag{9.9}$$
$$P_{xyz} = p_1[001] + p_2[011] + p_3[100] + p_4[110]. \tag{9.10}$$

All distributions are compatible with the tripartite network realized by three independent variables for any $p_i \in (0, 1)$ *and* $\sum_i p_i = 1$.

Proof Second-order inflation is used to prove this theorem while first-order inflation can be used for the W distribution with $d = 0$ (Wolfe et al. 2019). In fact, it is sufficient to prove results for both distributions (9.5) and (9.8). A similar proof can be proved for other distributions by permuting outcome variables.

From Eq. (9.5) it is easy to get marginal distributions of two outcomes as

$$P_{xy} = p_1[00] + p_3[10] + (p_2 + p_4)[01], \tag{9.11}$$
$$P_{yz} = p_3[00] + p_2[10] + p_1[01] + p_4[11], \tag{9.12}$$
$$P_{xz} = p_3[10] + p_2[00] + (p_1 + p_4)[01], \tag{9.13}$$
$$P_x = (p_2 + p_1 + p_4)[0] + p_3[1], \tag{9.14}$$
$$P_y = (p_1 + p_3)[0] + (p_2 + p_4)[1], \tag{9.15}$$
$$P_z = (p_2 + p_3)[0] + (p_1 + p_4)[1] \tag{9.16}$$

for any $p_i \in (0, 1)$ and $\sum_i p_i = 1$.

For a distribution P_{xyz} compatible with the triangle network consisting of three variables X, Y and Z, we get second-order spiral inflation. The first-order inflation consists of X_1, Y_1 and Z_1. From Wolfe et al. (2019, Lemma 4), all the marginal distributions of variables $\{A, B, C\}$, $\{X, Z_1\}$, $\{Y_1, Z\}$, $\{X_1, Y\}$, $\{X_1\}$, $\{Y_1\}$ and $\{Z_1\}$ should have the same distributions with these in Eqs. (9.11)–(9.16). This implies that

$$P_{xz_1} = p_3[10] + p_2[00] + (p_1 + p_4)[01], \tag{9.17}$$
$$P_{y_1z} = p_3[00] + p_2[10] + p_1[01] + d[11], \tag{9.18}$$
$$P_{x_1y} = p_1[00] + p_3[10] + (p_2 + p_4)[01], \tag{9.19}$$
$$P_{x_1} = P_x, P_{y_2} = P_y, P_{z_1} = P_z. \tag{9.20}$$

Equation (9.17) implies a deterministic outcome of $X = 0$ whenever $Z_1 = 1$. Similarly, from Eq. (9.19) it follows $Y = 0$ whenever $X_1 = 1$. This means that

$$\Pr(X = 0|Z_1 = 1) = \Pr(Y = 0|X_1 = 1) = 1 \tag{9.21}$$

As for the second-order inflation network consisting of X_2, Y_2 and Z_2, all the marginal distributions of variables $\{X, Y, Z\}$, $\{Y, Z_2\}$, $\{X, Y_2\}$, $\{X_2, Z\}$, $\{X_2\}$, $\{Y_2\}$ and $\{Z_2\}$ should have the same distributions with these in Eqs. (9.11)–(9.16). This implies the following constrictions as

$$P_{yz_2} = p_3[00] + p_2[10] + p_1[01] + p_4[11], \tag{9.22}$$
$$P_{xy_2} = p_1[00] + p_3[10] + (p_2 + p_4)[01], \tag{9.23}$$
$$P_{x_2z} = p_3[10] + p_2[00] + (p_1 + p_4)[01], \tag{9.24}$$
$$P_{x_2} = P_x, P_{y_2} = P_y, P_{z_2} = P_z. \tag{9.25}$$

From Eq. (9.23) we get a deterministic outcome $X = 0$ whenever $Y_2 = 1$ while $Z = 0$ whenever $X_2 = 1$ from Eq. (9.24). This implies that

$$\Pr(X = 0|Y_2 = 1) = \Pr(Z = 0|X_2 = 1) = 1. \tag{9.26}$$

From Fig. 9.1b all the variables of X_i, Y_j, Z_k are marginally independent in any compatible distribution. Equations (9.16) and (9.20) implies that

$$P[X_1 = Z_1 = X_2 = Y_2 = 1] \neq 0. \tag{9.27}$$

Combined with Eqs. (9.21) and (9.26), it follows that $P[X = Y = Z = 0] \neq 0$, which is contradicted to Eq. (9.5). This has proved the result for the distribution (9.5). Similar proof holds for the distribution in Eq. (9.8). □

9.1.2 Quantum Network Inflation

The inflation technique for quantum networks poses a challenge since classical inflation relies on the ability to clone classical information, which is impossible for faithful cloning of local quantum states in a distributed manner, as dictated by the quantum no-go theorem. However, the problem of local cloning has been recently resolved by cloning the total sources. This involves distributing multiple copies of joint quantum states among parties who then perform local measurements on their respective shares.

The local cloning problem is recently resolved by cloning the total sources (Wolfe et al. 2021). In this case, each party receives multiple copies of joint quantum states and performs local measurements on its shares. For any probability distribution $P(\vec{x})$ that is compatible with a quantum network, joint state, and local measurements can be found such that the statistical outcome satisfies two conditions. Firstly, it is invariant under the permutation of all the state spaces that correspond to different copies of the same source. Secondly, it reproduces the original distribution when all parties perform local measurements on the original network. One method of achieving this is by using the tensor product of copied sources coupled with the original measurements on different subspaces.

Although finding a quantum realization in a given inflation is as challenging as finding one in the original scenario, the easier problem of finding a quantum state and measurements whose statistics are invariant under suitable permutations can be solved using non-commuting polynomial optimization (Pironio et al. 2010). If there is no solution to this relaxed problem, then the original distribution cannot be realized in the original quantum network.

Example 9.1.1 Consider the quantum network $\mathcal{N}$ consisting of three bipartite quantum states $\rho_{AB}, \rho_{BC}, \rho_{CD}$, and ρ_{AD}. The local bipartite measurements will generate three variables A, B, and C which take the values a, b, and c, respectively. For a given distribution $P_{obs}(a, b, c, d)$, the problem is to determine the compatibility with the network $\mathcal{N}$. This can be realized in four steps as follows.

(1) Consider an inflated network $\mathcal{N}_{\text{inf}}$ with copied bipartite states of systems A_2B_1, B_2C_1, C_2D_1 and A_1D_2. Suppose the correlations $P_{obs}(a, b, c, d)$ is generated by performing local measurements with commuting measurement observables E_a, F_b, G_c and H_d, on local systems A_1A_2, B_1B_2, and C_1C_2, respectively. Now, consider n copies $\rho_{AB}, \rho_{BC}, \rho_{CD}$ and $\rho_{AD}, i = 1, \ldots, n$, being distributed. Denote ρ as the total state.
(2) For any $i, j = 1, \ldots, n$, let $\{E_a^{i,j}\}$ be the measurement observables on the i-th copy of ρ_{AD} and the j-th copy of ρ_{AB}. Similar definitions are for $\{F_b^{i,j}\}$, $\{G_c^{i,j}\}$

and $\{H_d^{i,j}\}$. This is used to extend the original measurements on all the copies of each system.

(3) This step is to evaluate the correlations related to P_{obs}. In fact, we have

$$\langle X\rangle_\rho := \mathrm{Tr}(\rho X), \tag{9.28}$$

which may satisfy nontrivial constrictions for $X \in \{E, F, G, H\}$. Especially, for $i \neq k$, $j \neq l$, $X_-^{i,j}$ and $X_-^{k,l}$ are acted on different Hilbert spaces and then commute, i.e., $[X_-^{i,j}, X_-^{k,l}] = 0$. This implies that

$$\langle E_a^{1,1} E_{a_1}^{1,2} F_b^{2,2}\rangle_\rho = \langle E_a^{1,2} E_{a_1}^{1,1} F_b^{1,2}\rangle_\rho, \tag{9.29}$$

where one can exchange $\rho_{AB}^{(1)}$ with $\rho_{AB}^{(2)}$. This follows a general result

$$\begin{aligned}&\langle Q(\{E_a^{i,j}, F_b^{k,l}, G_c^{m,n}, H_d^{u,v}\})\rangle_\rho\\ &= \langle Q(\{E_a^{g(i),g_1(j)}, F_b^{g_1(k),g_2(l)}, G_c^{g_1(m),g_2(n)}, H_d^{g_2(u),g_2(v)}\})\rangle_\rho\end{aligned} \tag{9.30}$$

for any permutations g, g_1, and g_2 of the indices $1, \ldots, n$, where $Q(\cdot, \cdot, \cdot, \cdot)$ is an analytical function. Finally, the measurement outcomes a^i, b^j, c^k and d^s satisfy

$$\langle \otimes_{i=1}^n E_{a^i}^{i,j} F_{b^i}^{i,j} G_{c^i}^{i,j} H_{d^i}^{i,j}\rangle_\rho = \prod_{i=1}^n P_{obs}(a^i, b^i, c^i, d^i) \tag{9.31}$$

if $P_{obs}(a, b, c, d)$ is compatible with $\mathcal{N}$. This implies $P_{obs}(a, b, c, d)$ admits an n-th-order quantum inflation. Here, the different number n implies hierarchic conditions for any compatible distribution $P_{obs}(a, b, c, d)$. Otherwise, Eq. (9.31) does not hold for any measurement operators $E_{a^i}^{i,j}$, $F_{b^i}^{i,j}$, $G_{c^i}^{i,j}$ and $H_{d^i}^{i,j}$.

(4) All the constraints may be explored by using NPO theory (Pironio et al. 2010).

Similar to the scalar extension method, quantum inflation characterizes correlations generated from local models. Quantum inflation provides a relaxed version of classical inflation in terms of sources. The program derived from inflation at a specific inflation level is semidefinite relaxation, which can be used to characterize network local models.

Finding an efficient way for inflating networks is a challenging problem. Exhaustive searching may require a hierarchy of inflation tests, which is a general NP-hard problem. Therefore, inflating special networks may be a more efficient approach, which can provide constraints for solving the compatibility problem. In principle, higher-order inflation can lead to a conclusive result (Navascues and Wolfe 2020).

These inflation techniques associate to any causal structure a hierarchy of increasingly strict compatibility tests, where each test can be formulated as a computationally efficient convex optimization problem (Navascues and Wolfe 2020). Recently, an inflation hierarchy has also been formulated for causal structures that allow for the observed classical random variables to arise from measurements on quantum states, however, no proof of completeness of this quantum inflation hierarchy has been supplied (Ligthart et al. 2023). The modified quantum inflation hierarchy introduced in this paper makes use of generators that need not be projections. The basic idea underlying the quantum inflation hierarchy is to relax independence conditions to symmetry conditions which are linear in elements of the algebra and easily incorporated into an SDP.

9.2 Measurable Inequalities

Our goal is to derive configuration limits for joint distributions from general networks using Finner inequalities (Luo 2022; Finner 1992; Renou et al. 2019). The main idea is to prove generalized Finner inequalities for correlations from classical networks consisting of independent sources. This approach is extended for any quantum network with local projection measurements, and similar results hold for any positive quantum network with local entangled measurements.

9.2.1 Classical Networks

We consider an n-partite network $\mathcal{N}_c$ consisting of independent variables $\lambda_1, \ldots, \lambda_m$. Each party A_j shares variables $\Lambda_j \subseteq \{\lambda_1, \ldots, \lambda_m\}$ with others, and the outcome a_j of A_j depends on the shared sources Λ_j, $j = 1, \ldots, n$. The local causal model (Pearl 2009) allows us to decompose the joint probability conditional on outcomes as:

$$P(\vec{a}) = \int_{\times_{j=1}^{n}\Omega_j} \prod_{j=1}^{n} p(a_j|\Lambda_j) d\mu_1(\lambda_1) \cdots d\mu_m(\lambda_m), \tag{9.32}$$

where $\vec{a} = (a_1, \ldots, a_n)$, $a_j \in \{0, \ldots, k_j\}$, and $(\Omega_j, \Sigma_j, \mu_j)$ denotes the measurable space of λ_j. $\mu_j(\lambda_j)$ is the probability of λ_j, and $p(a_j|\Lambda_j)$ is the probability of a_j conditional on Λ_j. It satisfies $\sum_{a_j} p(a_j|\Lambda_j) = 1$ for any x_j, Λ_i. We denote $P = \sum_{\vec{a}} P_{\vec{a}}[\vec{a}]$ as a general distribution, where $P_{\vec{a}}$ denotes the probability of the outcome $\vec{a}$.

Definition 9.2.1 The *configuration* of a given network $\mathcal{N}_c$ is a set of $(\mathcal{A}, \Lambda, \mathcal{R})$, where $\mathcal{A} = \{\mathsf{A}_j, \forall j\}$ and $\mathcal{R} = \{\mathsf{A}_j \to \lambda_i, \forall i, j\}$, and $\mathsf{A}_j \to \lambda_i$ means A_j receives one source from λ_i.

The network configuration described here is independent of the local sources per party. Consider a classical tripartite network, as shown in Fig. 9.2a, with a single source λ. In this case, we have:

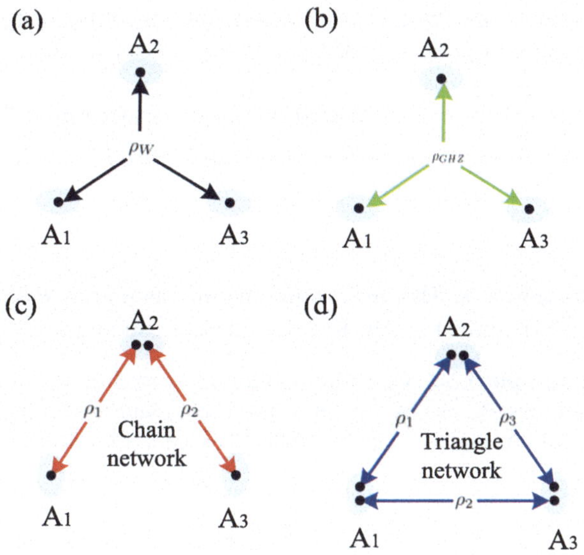

Fig. 9.2 (Color online) Schematically tripartite networks. Here, ρ_W or ρ_{GHZ} are tripartite sources while ρ_i are bipartite sources, $i = 1, 2, 3$. Each agent A_i may receive some states from one or two independent sources

$$\mathcal{R} = \{\mathsf{A}_1 \to \lambda, \mathsf{A}_2 \to \lambda, \mathsf{A}_3 \to \lambda\}. \tag{9.33}$$

Unlike in quantum nonlocality, the network configuration can be verified using the observed correlations without assuming any inputs. These inherent features of correlations may be compatible with a given network, irrespective of the realization of sources.

Example 9.2.1 One example is from triangle network (similar to Fig. 9.2d). The Finner inequality (Finner 1992; Renou et al. 2019) is given by

$$P(a_1, a_2, a_3) \leq \sqrt{p(a_1)p(a_2)p(a_3)}, \tag{9.34}$$

which provides a nontrivial limitation of correlations compatible with triangle networks consisting of bipartite sources.

The inequality (9.34) is useful for verifying the GHZ-type distribution $P_{GHZ} = p_{000}[000] + p_{111}[111]$ from a single source. However, it is not useful for the W-type distribution:

$$P_w(a_1, a_2, a_3) = p_{001}[001] + p_{010}[010] + p_{100}[100]. \tag{9.35}$$

In fact, we have that

$$\begin{aligned} p(a_1 = 0) &= 1 - p_3^2, \; p(a_1 = 1) = p_3^2, \\ p(a_2 = 0) &= 1 - p_2^2, \; p(a_2 = 1) = p_2^2, \\ p(a_3 = 0) &= 1 - p_1^2, \; p(a_3 = 1) = p_1^2, \end{aligned} \tag{9.36}$$

where $p(a_1)$, $p(a_2)$, $p(a_3)$ denote the respective marginal distribution of the three parties. From Eqs. (9.35) and (9.36), we can see that:

$$\begin{aligned}&P(0,0,1)^2 - p(a_1=0)p(a_2=0)p(a_3=1)\\ =&\ p_1^4 - (1-p_3^2)(1-p_2^2)p_1^2\\ =&\ -p_3^2p_2^2p_1^2\\ \leq&\ 0\end{aligned} \tag{9.37}$$

for any α_i's. This means that the distribution of $P(0,0,1)$ satisfies the inequality (9.34). Similar results hold for other distributions.

Example 9.2.2 Consider a chain network Fig. 9.2c consisting of two generalized sources $P_{\lambda_i} = \cos^2\theta_i[00] + \sin^2\theta_i[11]$ and $\theta_i \in (0, \frac{\pi}{2})$. Here, A_1 and A_2 share λ_1, while A_2 and A_3 share λ_2. They can get a joint distribution given by

$$\begin{aligned}P(a_1,a_2,a_3) =& \cos^2\vartheta_1\cos^2\theta_1\cos^2\theta_2[000] + \sin^2\vartheta_1\cos^2\theta_1\cos^2\theta_2[010]\\ &+\cos^2\vartheta_2\cos^2\theta_1\sin^2\theta_2[021] + \sin^2\vartheta_2\cos^2\theta_1\sin^2\theta_2[031]\\ &+\sin^2\vartheta_1\sin^2\theta_1\sin^2\theta_2[101] + \cos^2\vartheta_2\sin^2\theta_1\sin^2\theta_2[111]\\ &+\sin^2\vartheta_2\sin^2\theta_1\cos^2\theta_2[120] + \cos^2\vartheta_2\sin^2\theta_1\cos^2\theta_2[130],\end{aligned} \tag{9.38}$$

where the outcomes of A_2 are encoded into $\{0,\ldots,3\}$. It follows that

$$\begin{aligned}&p(a_1=0) = \cos^2\theta_1,\ p(a_1=1) = \sin^2\theta_1,\\ &p(a_2=0) = \cos^2\vartheta_1\cos^2\theta_1\cos^2\theta_2 + \sin^2\vartheta_1\sin^2\theta_1\sin^2\theta_2,\\ &p(a_2=1) = \sin^2\vartheta_1\cos^2\theta_1\cos^2\theta_2 + \cos^2\vartheta_2\sin^2\theta_1\sin^2\theta_2,\\ &p(a_2=2) = \cos^2\vartheta_2\cos^2\theta_1\sin^2\theta_2 + \sin^2\vartheta_2\sin^2\theta_1\cos^2\theta_2,\\ &p(a_2=3) = \sin^2\vartheta_2\cos^2\theta_1\sin^2\theta_2 + \cos^2\vartheta_2\sin^2\theta_1\cos^2\theta_2,\\ &p(a_3=0) = \cos^2\theta_2,\ p(a_3=1) = \sin^2\theta_2.\end{aligned} \tag{9.39}$$

We have for the distributions (9.38) and (9.39) that

$$\begin{aligned}&P(0,0,0)^2 - p(a_1=0)p(a_2=0)p(a_3=0)\\ =&-\cos^2\vartheta_1\sin^2\vartheta_1\cos^4\theta_1\cos^4\theta_2\\ &-\sin^2\vartheta_1\sin^2\theta_1\cos^2\theta_1\sin^2\theta_2\cos^2\theta_2 \leq 0.\end{aligned} \tag{9.40}$$

Similarly, we have

$$\begin{aligned}&P(0,1,0)^2 - p(a_1=0)p(a_2=1)p(a_3=0)\\ =&-\sin^2\vartheta_1\cos^2\vartheta_1\cos^4\theta_1\cos^4\theta_2\\ &-\cos^2\vartheta_2\sin^2\theta_1\sin^2\theta_2\cos^2\theta_1\cos^2\theta_2 \leq 0.\end{aligned} \tag{9.41}$$

This implies the following inequality

$$P(a_1, a_2, a_3) = \sqrt{p(a_1)p(a_2)p(a_3)} \tag{9.42}$$

for any $a_1, a_3 \in \{0, 1\}$ and $a_2 \in \{0, \ldots, 3\}$. It means that the inequality (9.34) cannot verify chain configuration.

For the chain network, we can prove the following configuration inequality

$$P(a_1, a_2, a_3) \leq p(a_1)^{\frac{m-k}{m}} p(a_2)^{\frac{k}{m}} p(a_3)^{\frac{m-k}{m}} \tag{9.43}$$

for any $m \geq 2$ and $k \leq m-1$. A special case of this inequality gives the separable condition of $P(a_1, a_3) = p(a_1)p(a_3)$. To prove this inequality, we make use of the concept of a generalized fractional independent set of graphs (Schrijver 2003; Diestel 2017).

In what follows, we will explore configuration inequalities for classical networks consisting of any sources.

Definition 9.2.2 For an n-partite network $\mathcal{N}$, there is an associated undirected graph $\mathcal{G} = (V, E)$, where $i \in V$ denotes A_i in $\mathcal{N}$. The edge $e_{ij} \in E$ connected with i and j represents one bipartite source shared by A_i and A_j. The s-hyper edge $e_{i_1 \cdots i_s} \in E$ connected with $i_1, \ldots, i_s$ represents one s-partite source shared by $\mathsf{A}_{i_1}, \ldots, \mathsf{A}_{i_s}$.

Definition 9.2.3 A fractional independent set of a graph $\mathcal{G}$ associated with the network $\mathcal{N}$ is a non-negative vector of $\vec{s} = (s_1, \ldots, s_n)$ satisfying

$$\begin{aligned} s_j &\geq 0, \\ \sum_{j:i\to j} s_j &\leq 1, \forall i, \end{aligned} \tag{9.44}$$

where j is ranging over agents, and i over sources.

In the graph $\mathcal{G} = (V, E)$ associated with an n-partite network $\mathcal{N}$, each node has a nonnegative weight for source j such that the summation of the weights of nodes connected to it is no more than 1. For a simple graph consisting of only edges, all the solutions may be explored using its extreme points, i.e., the optimal fractional clique cover numbers $\frac{1}{2}$. However, it is NP-complete to find the extreme points of a general graph (Diestel 2017; Schrijver 2003). In what follows, we will find suboptimal solutions using this framework, which can also be used to characterize the correlations in quantum networks.

Algorithm 9.1 has a total time complexity of $O(mnN)$, where m denotes the number of sources and N denotes the number of edges. It has polynomial time complexity in terms of sources and edges. Another method is shown in Algorithm 9.2.

Algorithm 9.1 Fractional independent set

Input: An undirected graph $\mathcal{G} = (V, E)$.

- Set $x_j = \frac{1}{m_j}$ for each m_j-hyper edge $e_j \in E$;
- Set $s_i = \min\{x_j | j \to i\}$;
- Set $\vec{s} = (s_1, \ldots, s_n)$.

Output: A fractional independent set $\vec{s}$.

Algorithm 9.2

Input An undirected graph $\mathcal{G} = (V, E)$.

Output A fractional independent set $\vec{s} = (s_1, \ldots, s_n)$

- Decompose $\mathcal{G}$ into a combination of different subgraphs $\mathcal{G}_j$, that is, $\mathcal{G} = \cup_{j=2}^{t} \mathcal{G}_j$, where $\mathcal{G}_2$ denotes the bipartite graph, and $\mathcal{G}_j$ denotes the j-hyper edge associated with j-partite source;
- Set a fractional independent set $\vec{s}_j$ for the subgraph $\mathcal{G}_j$;
- Update s_i for the i-th node as $s_i = \min\{s_k \in \vec{s}_j | k \to i\}$;
- Set $\vec{s} = (s_1, \ldots, s_n)$.

The main difference between Algorithms 9.1 and 9.2 is the covering set for each subgraph. In Algorithm 9.2, one can choose different fractional independent sets depending on the network configurations. Algorithm 9.2 has the total time complexity $O(mLN)$, where N denotes the number of edges and L denotes the maximal time complexity for setting $\vec{s}_j$. Note L is upper bounded by some constant depending on all the sub-graphs. This is because $\mathcal{G}_j \subset \mathcal{G}$. This means Algorithm 9.2 has polynomial time complexity in terms of sources and edges.

Example 9.2.3 Consider the network shown in Fig. 9.3. There are five parties $\mathsf{A}_1, \ldots, \mathsf{A}_5$. A_1, A_2 and A_3 share three bipartite sources. A_3, A_4 and A_5 share a hyper edge. Set $\frac{1}{2}$ for each edge of A_1, A_2 and A_3 in Fig.2 while $\{\frac{1}{2}, \frac{1}{4}, \frac{1}{4}\}$ for three nodes of the hyper edge. And then, we update the covering number as $s_i = \frac{1}{2}$ for $i = 1, 2, 3$, and $s_j = \frac{1}{4}$ for $j = 4, 5$. Finally, a fractional independent set $\vec{s} = (s_1, \ldots, s_6)$ is given by

$$\vec{s} = (\frac{1}{2}, \frac{1}{2}, \frac{1}{2}, \frac{1}{4}, \frac{1}{4}). \tag{9.45}$$

Example 9.2.4 Consider an n-partite chain network shown in Fig. 9.4. Here, each pair of A_j and A_{j+1} are connected by one edge, $j = 1, \ldots, n-1$. The set of $\{\mathsf{A}_{2j} | j = 1, \ldots, \lceil \frac{n}{2} \rceil\}$ is an independent set. We get $s_{2j} = \frac{m-k}{m}$ for A_{2j}, and $s_{2j+1} = \frac{k}{m}$ for A_{2j+1}, $j = 1, \ldots, \lceil \frac{n}{2} \rceil$, where $\lceil x \rceil$ denotes the minimal integer no less than x, $m \geq 2$ and $k \leq m-1$. The fractional independent set of $\vec{s} = (s_1, s_2, \ldots, s_n)$ is given by

$$\vec{s} = (\frac{k}{m}, \frac{m-k}{m}, \frac{k}{m}, \ldots, \frac{m-k}{m}) \tag{9.46}$$

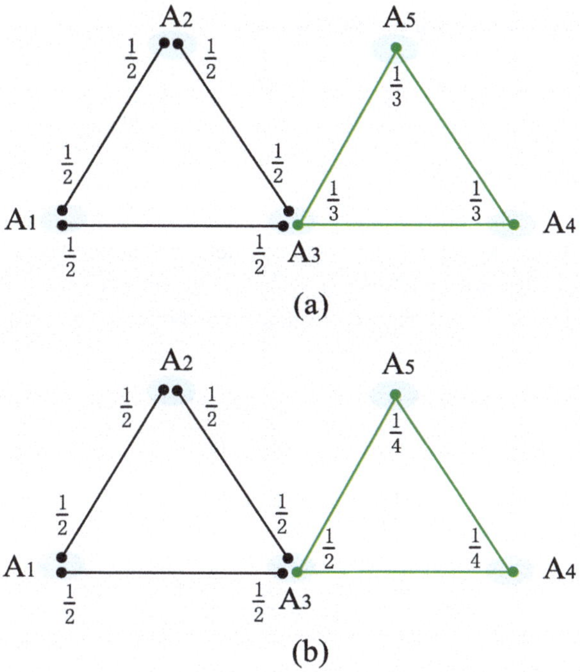

Fig. 9.3 (Color online) 5-partite network. A_1, A_2 and A_3 share a triangle network consisting of three bipartite sources. A_3, A_4 and A_5 share a tripartite source corresponding to the 3-hyper edge in the graph

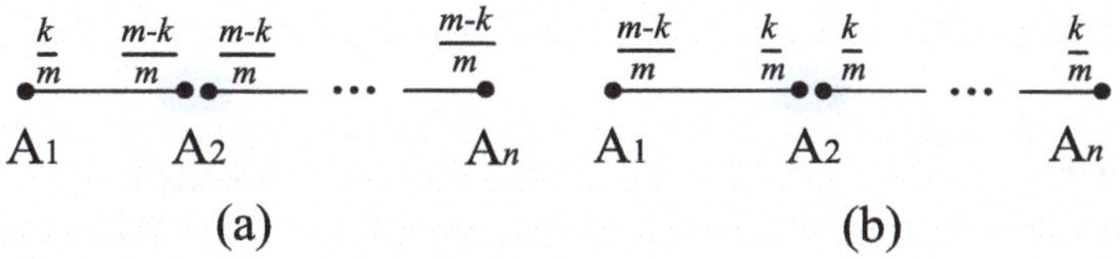

Fig. 9.4 (Color online) An n-partite chain network

for an even integer n, as shown in Fig. 9.4a, and

$$\vec{s} = (\frac{m-k}{m}, \frac{k}{m}, \frac{m-k}{m}, \ldots, \frac{k}{m}, \frac{m-k}{m}) \tag{9.47}$$

for an odd integer n, as shown in Fig. 9.4b.

Another method is $s_{2j} = \frac{k}{m}$ for A_{2j}, and $s_{2j+1} = \frac{m-k}{m}$ for A_{2j+1}, $j = 1, \ldots, \lceil \frac{n}{2} \rceil$. The fractional independent set of $\vec{s}$ is given by

$$\vec{s} = (\frac{m-k}{m}, \frac{k}{m}, \frac{m-k}{m}, \ldots, \frac{k}{m}) \tag{9.48}$$

for an even integer n, as shown in Fig. 9.4b, and

$$\vec{s} = (\frac{m-k}{m}, \frac{k}{m}, \frac{m-k}{m}, \ldots, \frac{k}{m}, \frac{k}{m}) \tag{9.49}$$

for an integer odd n, where $2k \leq m$.

Example 9.2.5 Consider an $n+1$-partite star network, as shown in Fig. 9.5. Here, B has shared one edge with each A_j, $j = 1, \ldots, n$. We get $s_{n+1} = \frac{m-k}{m}$ for B and $s_j = \frac{k}{m}$ for A_j, $j = 1, \ldots, n$. The fractional independent set $\mathbf{s} = (s_1, \ldots, s_n, s_{n+1})$ is given by

$$\vec{s} = (\frac{k}{m}, \ldots, \frac{k}{m}, \frac{m-k}{m}) \tag{9.50}$$

with $m \geq 2$ and $k \leq m-1$, as shown in Fig. 9.5a.

Another method is to get

$$\vec{s} = (\frac{m-k}{m}, \ldots, \frac{m-k}{m}, \frac{k}{m}), \tag{9.51}$$

as shown in Fig. 9.5b.

Example 9.2.6 Consider an n-partite cyclic network, shown as Fig. 9.6. Here, each pair of A_j and A_{j+1} (or A_1 and A_n) shares one bipartite source, $j = 1, \ldots, n-1$. Similar to n-partite chain network shown in Fig. 9.4, we get the fractional independent sets (9.46)–(9.49).

Example 9.2.7 Consider an n-partite complete network (associated with a complete hypergraph), shown in Fig. 9.7, where each m parties $\mathsf{A}_{i_1}, \ldots, \mathsf{A}_{i_m}$ shares one source, $1 \leq i_1, \ldots, i_m \leq n$. We set $s_i = \frac{1}{m}$ for each A_i. The fractional independent set of $\vec{s} = (s_1, s_2, \ldots, s_n)$ is given by

$$\vec{s} = (\frac{1}{m}, \ldots, \frac{1}{m}). \tag{9.52}$$

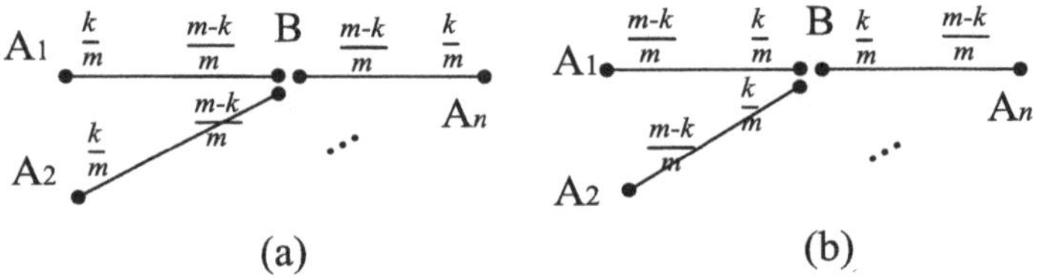

Fig. 9.5 (Color online) An $n+1$-partite star network

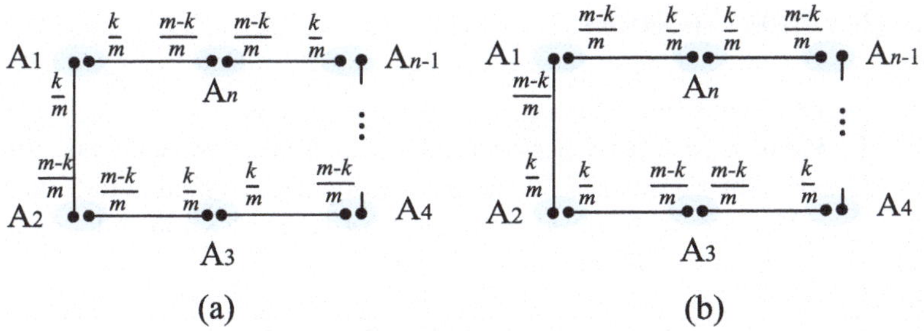

Fig. 9.6 (Color online) An n-partite cyclic network

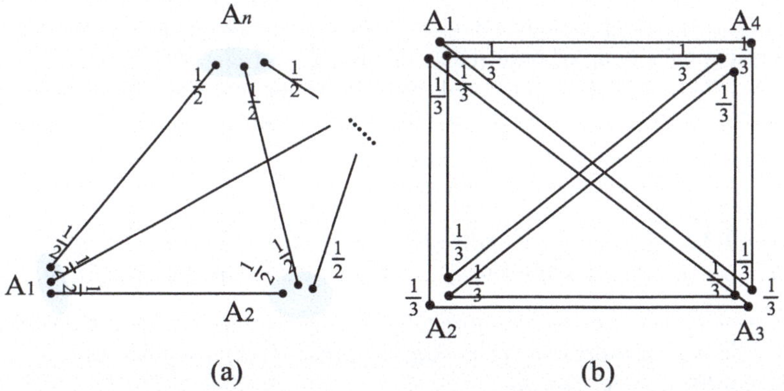

Fig. 9.7 (Color online) **a** An n-partite complete network. **b** A 4-partite complete network

For the special case of $m = 2$, it reduces to the Finner inequality (Finner 1992), as shown in Fig. 9.7a. Another example of $m = 3$ is shown in Fig. 9.7b.

In general, each achievable fractional independent set **s** of given networks can give a nontrivial limitation of compatible correlations. In fact, for a given classical network $\mathcal{N}_c$ associated with graph $\mathcal{G}$, each achievable fractional independent set $\vec{s}$ implies a constriction of compatible correlations as

$$P(\vec{a}) \leq \prod_{j=1}^{n} p(a_j)^{s_j} \tag{9.53}$$

for any outcome $\vec{a}$. The inequality (9.53) provides a computationally efficient method for verifying network configurations of classical networks consisting of any sources beyond results with bipartite sources (Renou et al. 2019).

9.2.2 Quantum Networks

Consider a general quantum network $\mathcal{N}_q$ consisting of states $\rho_1, \ldots, \rho_m$ on Hilbert space $\otimes_j \mathbb{H}_j$. Suppose each party performs a local measurements with measurement operator M_{a_j} on its shared states in $\mathcal{N}_q$, where M_{a_i} are positive semidefinite operators satisfying $\sum_{a_i} M_{a_i} = \mathbb{1}$ with the identity operator $\mathbb{1}$. Born's rule implies a joint probability distribution conditional on all outcomes as

$$P_q(\vec{a}) = \mathrm{Tr}(\otimes_{i=1}^n M_{a_i}(\otimes_{j=1}^m \rho_j)). \tag{9.54}$$

Denote special measurements as projection measurement under computation basis, that is, $M_{a_i} \in \{|j\rangle, \forall j\}$. Denote entangled measurements as projection measurement under orthogonal entangled pure states, that is, $M_{a_i} \in \{|\phi_j\rangle, \forall j\}$.

The distribution $\{P_q(\vec{a})\}$ generated from single quantum entanglement is consistent with its from classical networks (Bell 1964). A similar result holds for quantum networks consisting of bipartite entangled states (Renou et al. 2019). We present an efficient method for general quantum networks.

Theorem 9.2.1 (Quantum generalized Finner inequality) *Let* $\mathbf{s} = \{s_1, \ldots, s_n\}$ *be a fractional independent set of graph* $\mathcal{G}$ *associated with quantum network* $\mathcal{N}_q$.

(i) Suppose A_j *performs special projection measurements, and* f_j *are any nonnegative local post-processing (function) of classical outputs of* A_j. *Any correlation compatible with* $\mathcal{N}_q$ *satisfies*

$$\mathbb{E}_q \prod_{j=1}^n f_j \leq \prod_j \|f_j\|_{\frac{1}{s_j}}, \tag{9.55}$$

where $\|f\|_s = \mathbb{E}[f^{1/s}]^s$ *and* $\mathbb{E}$ *is the expectation with respect to* P. *Especially, if* f_j*'s are indicator functions of* a_j, *it follows that*

$$P_q(\vec{a}) \leq \prod_{j=1}^n p_q(a_j)^{s_j}. \tag{9.56}$$

(ii) Suppose A_j *performs general projection measurements under general orthogonal basis. Any correlation compatible with a positive quantum network (under any local unitary operations)* $\mathcal{N}_q$ *satisfies the inequality (9.56).*

A well-known result is the consistence of correlations from both classical and quantum networks consisting of a single source.

Lemma 9.2.1 (Bell 1964) *Any correlation* $P(\vec{a})$ *achievable in n-partite states can be generated by n-partite variable.*

Proof of Theorem 9.2.1 Different from the proof (Renou et al. 2019) for all bipartite sources, we consider a general case. This is essential for extensions in the no-signaling model. In fact, consider a general quantum network $\mathcal{N}_q$ consisting of states $\rho_1, \ldots, \rho_m$ on Hilbert space $\otimes_j \mathbb{H}_j$. For each state ρ_j, define $\overline{s}_j := \{s_{i\to j}\}$ for all A_i's who share ρ_j, where $s_{i\to j} \in [0, 1]$ and satisfies $\sum_i s_{i\to j} \le 1$. Consider a set $\vec{s} = \cup_j \overline{s}_j = (s_1, \ldots, s_n)$ be a total fractional independent set of the associated graph $\mathcal{G}$, where $s_j = \min\{s_{i\to j}\}$.

For $s_j = 0$ we get

$$\|f_j\|_{\frac{1}{s_j}} = \|f_j\|_\infty = \max_{a_j : P_{A_j}(a_j)\neq 0} f(a_j), \tag{9.57}$$

and

$$\mathbb{E}_q \prod_{i=1}^n f_i \le \|f_j\|_\infty \mathbb{E}_q[\prod_{i\neq j} f_i]. \tag{9.58}$$

If the generalized Finner inequality (9.55) holds by ignoring the n-th partite (for the joint distribution $P(\vec{a})$), it also holds including the same party for $s_j = 0$. This means that all the weights can be supposed to be $\mathbf{s} = (s_1, \ldots, s_n)$ such that $s_j \in (0, 1]$ for all j's.

Moreover, for $s_j = 1$ the party A_j cannot share any source with others. We get

$$\mathbb{E}_q \prod_{i=1}^n f_i \le \mathbb{E}_q \prod_{i\neq j} f_j \mathbb{E}_q f_j, \tag{9.59}$$

where f_j satisfies $\|f_j\|_{\frac{1}{s_j}} = \|f_j\| = \mathbb{E}_q[f_j]$. As a result, all the parties with weights equal to 1 can be ignored.

In summary, assume $\vec{s} = (s_1, \ldots, s_n)$ such that $s_j \in (0, 1)$ for all j's. In what follows, it is sufficient to prove

$$\mathbb{E}_q \prod_{j=1}^n f_j \le \prod_j \|f_j\|_{\frac{1}{s_j}}. \tag{9.60}$$

Note that f_j is a positive function on the output of A_j. Composing the projection measurements of A_j with the classical post-processing, assume f_j is the outcome of observable X_j. Suppose that X_j is decomposed into the superposition of complete orthogonal projection operators. Assume that

$$X_j = \sum_{a_j} f_j(a_j) M_{a_j}, \tag{9.61}$$

where $\{M_{a_j} = |a_j\rangle\langle a_j|\}$ are projection bases which satisfy $M_{a_j=x_1} M_{a_j=x_2} = \delta_{x_1x_2} M_{a_j=x_1}$ with the delta function $\delta_{x_1x_2}$, that is, $\delta_{x_1x_2} = 1$ for $x_1 = x_2$ and $\delta_{x_1x_2} = 0$ for $x_1 \neq x_2$.

According to the Born rule, we get

$$\mathbb{E}_q \prod_{j=1}^{n} f_j = \mathrm{Tr}(\otimes_j X_j \rho), \tag{9.62}$$

where $\rho = \otimes_i \rho_i$ and ρ_i are pure states. Otherwise, consider the purification $|\Phi\rangle_{A_1\cdots A_n;B}$ of ρ with an auxiliary particle B. It is easy to check

$$\mathbb{E}_q \prod_{j=1}^{n} f_j = \mathrm{Tr}(\mathbb{1}_B \otimes (\otimes_{j=1}^{n} X_j)|\Phi\rangle\langle\Phi|), \tag{9.63}$$

where $\mathbb{1}_B$ denotes the identity operator on the particle B.

Case 1. The inequality (9.55) holds for local projection measurements under the computation basis, i.e., $\{M_{a_j} = |a_j\rangle\langle a_j|\}$ defined in Eq. (9.61) are computation bases.

Case 2. The inequality (9.56) holds with local projection measurements under general bases for positive quantum networks.

From Case 1, the configuration inequality is invariant under local unitary operations of each particle. For each s-partite entangled state $|\psi_i\rangle$, consider the following normal form under local unitary operations:

$$|\psi_i\rangle = \sum_{j_{1;i}\cdots j_{s;i}} \lambda_{j_{1;i}\cdots j_{s;i}} |j_{1;i} \cdots j_{s;i}\rangle, \tag{9.64}$$

which satisfies that all the reduced states of $\rho_{j_{t;i}}$ have diagonal density matrices given by

$$\rho_{j_{t;i}} = \sum_{\ell=0}^{d-1} \lambda_\ell^2 |\ell\rangle\langle\ell|, \tag{9.65}$$

where $\lambda_\ell \geq 0$, s denotes the largest number of parties who share one entanglement, and all the other states $|\psi_i\rangle_{A_1\cdots A_k}$ with $k < s$ can be regarded as $|\psi_i\rangle_{A_1\cdots A_k}|0\cdots 0\rangle_{A_{k+1}\cdots A_s}$ by introducing axillary systems $A_{k+1}, \ldots, A_s$. The new network has the same configuration with $\mathcal{N}_q$. From Eqs. (9.62) and (9.64), we get

$$\begin{aligned} \mathbb{E}_q \prod_{j=1}^{n} f_j &= \mathrm{Tr}[(\otimes_{j=1}^{n} X_j)(\otimes_{i=1}^{m} |\psi_i\rangle\langle\psi_i|)] \\ &= \sum_{\substack{j_{1;1},\cdots,j_{s;1} \\ \cdots \\ j_{1;m},\cdots,j_{s;m}}} \prod_{i=1}^{m} \lambda_{j_{1;i}\cdots j_{s;i}}^{(\ell_i)} \lambda_{j_{1;i}\cdots j_{s;i}}^{(\ell_i')} \mathrm{Tr}(X_j \otimes_{i:i\to j} |\ell_i\rangle\langle\ell_i'|), \end{aligned} \tag{9.66}$$

where $\otimes_{i:i\to j}|\ell_i\rangle\langle\ell_i'|$ denotes the computation basis of all the systems shared by the agent A_j, $\{|\ell_i\rangle\}\in\{\{|j_{1;i}\rangle\},\ldots,\{|j_{s;i}\rangle\}$ (the system derived from the source $|\psi_i\rangle$), $j=1,\ldots,n$.

For any j define $g_j:\times_{i:i\to j}F_d^{1/s_j}\to\mathbb{R}$ as

$$\begin{aligned} g_j(\times_{i:i\to j}(\ell_i,\ell_i')) &= \prod_{i=1}^m(\lambda^{(\ell_i)}_{j_{1;i}\cdots j_{s;i}}\lambda^{(\ell_i')}_{j_{1;i}\cdots j_{s;i}})^{s_j}\mathrm{Tr}(X_j\otimes_{i:i\to j}|\ell_i\rangle\langle\ell_i'|) \\ &\geq 0 \end{aligned} \tag{9.67}$$

from the assumption of positive quantum network, where $F_d=\{0,\ldots,d-1\}$. Suppose Δ_j is random variable with uniform distribution over $F_d^{\frac{1}{2s_j}}$, that is, $\Delta_j(\times_{i:i\to j}(\ell_i,\ell_i'))=1/d^{2/s_j}$.

From Eqs. (9.66) and (9.67) we get

$$\mathbb{E}_q\prod_{j=1}^n f_j\leq d^{2sm}\mathbb{E}_q[\prod_{j=1}^n g_j((\Delta_i)_{i\to j})]. \tag{9.68}$$

By regarding Δ_j as the classical variables shared by all the agents who have applied local functions g_j on them, we get the result as

$$\mathbb{E}_q\prod_{j=1}^n f_j=d^{2sm}\prod_{j=1}^n\|g_j\|_{\frac{1}{s_j}}. \tag{9.69}$$

Define the operator Y_j as

$$Y_j=\sum_{(\ell_i)_{i:i\to j}}(\prod_{i:i\to j}\lambda^{(\ell_i)}_{j_{1;i}\cdots j_{s;i}})^{\frac{1}{s_j}}\otimes_{i:i\to j}|\ell_i\rangle\langle\ell_i|, \tag{9.70}$$

which is a positive semi-definite matrix from any positive quantum network. This follows that

$$\begin{aligned} d^{\frac{1}{s_j}|\{i:i\to j\}|}\|g_j\|_{\frac{1}{s_j}}^{\frac{1}{s_j}} &= d^{\frac{1}{s_j}|\{i:i\to j\}|}\mathbb{E}g_j^{\frac{1}{s_j}} \\ &= \sum_{\times_{i:i\to j}(\ell_i,\ell_i')}\prod_{i=1}^m\lambda^{(\ell_i)}_{j_{1;i}\cdots j_{s;i}}\lambda^{(\ell_i')}_{j_{1;i}\cdots j_{s;i}}\mathrm{Tr}(X_j\otimes_{i:i\to j}|\ell_i\rangle\langle\ell_i'|)^{\frac{1}{s_j}} \\ &= \mathrm{Tr}((Y_j^{2s_j}X_j)^{\frac{1}{s_j}}) \\ &\leq \mathrm{Tr}(Y_jX_j^{\frac{1}{s_j}}Y_j) && (9.71) \\ &\leq \mathrm{Tr}[(\sum_{(\ell_i)_{i:i\to j}}\prod_{i:i\to j}(\lambda^{(\ell_i)}_{j_{1;i}\cdots j_{s;i}})^2\otimes_{i:i\to j}|\ell_i\rangle\langle\ell_i|)X_j^{\frac{1}{s_j}}] && (9.72) \end{aligned}$$

$$= \|f_j\|_{\frac{1}{s_j}}^{\frac{1}{s_j}}, \tag{9.73}$$

where the inequality (9.71) is by the Araki-Lieb-Thirring inequality. The inequality (9.72) is from the inequality of $x^t \leq x$ with $x \in (0, 1)$ and $t \geq 1$. The summation in Eq. (9.72) denotes the reduced density matrix of A_j. Equation (9.66) is easily proved by using equality (9.73). This has proved the inequality (9.56). □

The correlations generated from local projection measurements on quantum networks are compatible with the inequality (9.53) for the classical networks. This result shows special restrictions for correlations from positive quantum networks consisting of such as EPR, GHZ or W states. Note the local joint measurement in an entangled basis is equivalent to local joint unitary operations and projection measurements, which is essentially changing the configuration of the network. The present result shows the inherent configuration of quantum networks under local unitary operations and then provides non-trivial features of quantum networks.

Theorem 9.2.1 provides an efficient method for solving the *network configuration compatibility problem* (NCCP) by testing the inequality (9.56). Here, NCCP is to determine whether a given correlation $P(\vec{x})$ is generated from local measurements on a given network or not. Take a single-source distribution $P_{GHZ} = \frac{1}{2}[000] + \frac{1}{2}[111]$ as an example. It violates both the inequalities (9.34) and (9.43). This implies P_{GHZ} is incompatible with triangle and chain networks. A general extension can be obtained for general quantum networks as Algorithm 9.3.

Algorithm 9.3 Verifying network configuration

Input: An n-partite network $\mathcal{N}_q$.
- Consider another n-partite network $\mathcal{N}'_q$.
- Find a fractional independent set $\vec{s}$ for $\mathcal{N}'_q$.
- Obtain the configuration inequality from Theorem 1 for $\mathcal{N}'_q$.
- Get the joint distribution $P(\vec{a})$ for $\mathcal{N}_q$.
- Verify whether $P(\vec{a})$ violates the present configuration inequality or not.

Output: The configuration $(\mathcal{A}, \rho, \mathcal{R})$ of $\mathcal{N}_q$.

This algorithm provides a general method to test the configuration of given networks. It depends on the proposed configuration inequality in Theorem 9.2.1. Moreover, for an n-partite network $\mathcal{N}_q$ consisting of any bipartite entangled pure states or generalized GHZ states, the compatible joint distribution may violate specific configuration inequalities from other networks.

The present network configurations in Theorem 9.2.1 are robust against general noises. This provides an efficient method to verify the correlations available in noisy quantum networks.

Example 9.2.8 Consider a noisy GHZ state (Greenberger and Horne 1989; Dür et al. 2000) as

$$\rho_{GHZ} = v|GHZ\rangle\langle GHZ| + \frac{1-v}{8}\mathbb{1}, \tag{9.74}$$

where $|GHZ\rangle = \cos\theta|000\rangle + \sin\theta|111\rangle$ with $\theta \in (0, \frac{\pi}{2})$, and $v \in (0, 1]$. The inequality (9.43) yields an equality of $P(a, c) = p(a)p(c)$, which will be violated by correlations from a noisy GHZ state beyond the inequality (9.34).

Example 9.2.9 Consider a noisy W state (Werner 1989; Dür et al. 2000) as

$$\rho_W = v|W\rangle\langle W| + \frac{1-v}{8}\mathbb{1}, \tag{9.75}$$

where $|W\rangle = \cos\theta\cos\gamma|001\rangle + \sin\theta\cos\gamma|010\rangle + \sin\gamma|100\rangle$ and $v \in (0, 1)$. The joint distribution of P_{ρ_W} derived from local projection measurements violates $P(a, c) = p(a)p(c)$ for any $v \in (0, 1)$ while it is impossible for the inequality (9.34).

Example 9.2.10 Consider a triangle network consisting of three noisy states of (Einstein et al. 1935; Werner 1989):

$$\rho = v|EPR\rangle\langle EPR| + \frac{1-v}{4}\mathbb{1}, \tag{9.76}$$

where $|EPR\rangle = \cos\theta|00\rangle + \sin\theta|11\rangle$ with $\theta \in (0, \frac{\pi}{2})$, and $v \in (0, 1)$. The inequality (9.43) is useful for verifying correlations from local entangled measurements (Renou et al. 2019). A similar result holds for star networks by violating the inequality (9.46).

9.3 Entropic Method

Another method is from entropy theory (Fritz 2012; Chaves and Budroni 2016; Gisin et al. 2020; Cover and Thomas 2012). Different from special limits for each outcome, this method provides average restricts for the whole distribution.

9.3.1 Chain Networks

Consider a chain network as shown in Fig. 9.2c, where three parties Alice, Bob, and Charlie share two independent sources. This kind of network configuration may be explored by using the Shannon entropy of statistical correlations.

For a joint distribution $P(x, y, z)$ which is compatible with the chain network, we prove a correlation constrictions different from the Markov chain. The Markov chain allows the following conditional independent decomposition of $p(x, y|z) = p(x|z)p(y|z)$ (Pearl 2009). This implies a general inequality as

$$I(X;Y;Z) = I(X;Y) - I(X;Y|Z) > 0. \tag{9.77}$$

Here, $I(X;Y|Z)$ is the conditional mutual information defined by $I(X;Y|Z) = H(Y,Z) + H(X,Z) - H(X,Y,Z) - H(Z)$ (Cover and Thomas 2012).

However, for chain networks, the typical application is entanglement swapping for its quantum realizations. This implies a different condition of the so-called ant-Markov chain, i.e., two independent parties may share new correlations conditional on the other party's local measurement. In general, a conditional dependence for $P(x, y, z)$ is compatible with chain network as

$$I(X;Y) = 0, I(X;Y|Z) > 0. \tag{9.78}$$

This means two independent variables X and Y on $\mathcal{X} \times \mathcal{Y}$ can build new correlations conditional on the outcome of variable Z. It can be formulated as

$$p(x, y) = p(x)p(y), p(x, y|z) \neq p(x|z)p(y|z). \tag{9.79}$$

This kind of activated correlation $p(x, y|z)$ shows new features of three variables with the condition given by

$$I(X;Y;Z) \leq 0. \tag{9.80}$$

This intrigues new kinds of non-Shannon-type information inequalities for any independent set beyond the polymatroidal axioms of $H(X), I(X;Y) \geq 0$. It also implies special constraints for correlations dependent on network configurations.

We present a simple example to show the inequality (9.78), where two bipartite sources are $\frac{1}{2}[00] + \frac{1}{2}[11]$. A compatible joint distribution is given by

$$P_{xyz_1z_2} = \frac{1}{4}[0000] + \frac{1}{4}[0101] + \frac{1}{4}[1010] + \frac{1}{4}[1111], \tag{9.81}$$

which is further combined into

$$P_{xyz} = \frac{1}{4}[000] + \frac{1}{4}[011] + \frac{1}{4}[101] + \frac{1}{4}[110] \tag{9.82}$$

by using classical Birkhoff transformation on two variables z_1 and z_2, e.g., a doubly stochastic matrix (a_{ij}) which satisfies each column or each row consists of a probability distribution, i.e., $\sum_i a_{ij} = \sum_j a_{ij} = 1$ and $a_{ij} \geq 0$. A formal statement for this is to use classical encoding: $\mathcal{F} : z_1z_2 \mapsto z = z_1 \oplus z_2$. This means that

$$\begin{aligned} &H(X) = H(Y) = 1, \\ &H(X,Y) = H(X,Y,Z) = 2, \\ &H(X,Z) = H(Y,Z) = 1. \end{aligned} \tag{9.83}$$

The probability distribution $P_{xyz_1z_2}$ satisfies the inequality (9.78). Moreover, the probability distribution in Eq. (9.82) is compatible with the triangle networks.

So, a compatible distribution in Eq. (9.82) can be decomposed into Eq. (9.78). The distribution P_{xyz} allows a chain network decomposition as shown in Fig. 9.2c. This can be represented by

$$P(x, y, z) = \int_{\Omega_1 \otimes \Omega_2} p(x|\lambda)p(y|\gamma)p(z|\lambda, \gamma)\mu(\lambda)\mu(\gamma)d\lambda d\gamma, \tag{9.84}$$

where $(\Omega, \mu(\lambda))$ denotes the measurable space of variable λ, and $\mu(\lambda)$ denotes the probability of λ. It can be extended for a general probability distribution satisfying Eq. (9.78).

Theorem 9.3.1 *Any joint probability distribution satisfying the condition (9.78) is compatible with a chain Bayesian network.*

In Hilbert scenarios, after all the local measurements on a given state ρ on Hilbert space $\otimes_{i=1}^n \mathcal{H}_{A_i}$, according to the Born's rule, it allows a quantum joint probability given by

$$P(x_1, \ldots, x_n) = \mathrm{Tr}(M_{x_1} \otimes \cdots \otimes M_{x_n} \rho) \tag{9.85}$$

for all outcomes x_i's.

Instead of the classical chain network, any joint distribution satisfying the condition (9.78) should be compatible with quantum networks consisting of two generalized Einstein-Podolsky-Rosen (EPR) states (Einstein et al. 1935), as shown in Fig. 9.2c. In fact, Alice and Bob perform respective projection measurements $\{M_x, x \in \{0, 1\}\}$ and $\{M_y, y \in \{0, 1\}\}$, while the other performs local measurement $\{M_z = |z\rangle\langle z|, z \in \{00, \ldots, 11\}\}$. These simple local measurements allow a joint distribution as

$$P_{xyz} = \frac{1}{4}[0000] + \frac{1}{4}[0101] + \frac{1}{4}[1010] + \frac{1}{4}[1111]. \tag{9.86}$$

This provides an experiment way for verifying any negative Shannon information satisfying Eq. (9.78). As a directive result, any joint distribution (9.86) from local measurements on the chain network with any bipartite states ρ_1 and ρ_2 is compatible with the classical distribution (9.84), that is,

$$\mathcal{S}_c = \mathcal{S}_q, \tag{9.87}$$

where $\mathcal{S}_c$ consists of the classical probability distribution (9.84) with respect to any two variables λ and γ, or it can be equivalently represented by $\mathcal{S}_c = \{P_{xyz}^{(c)} | I(X; Y) = 0, I(X; Y|Z) > 0\}$. $\mathcal{S}_q$ consists of all distributions (9.86) from any two states ρ_1 and ρ_2, that is,

$$\mathcal{S}_q = \{P_{xyz}^{(q)} | I(X; Y) = 0, I(X; Y|Z) > 0\}. \tag{9.88}$$

This means there is no quantum nonlocality beyond classical networks under the assumption of one measurement per party.

9.3.2 *Triangle Network*

For a triangle network consisting of three independent variables X, Y and Z on finite sample spaces, as shown Fig. 9.2d, we prove the generated joint distribution satisfies the following inequality (Fritz 2012):

$$H(X) + H(Y) + H(Z) \leq H(X, Y) + H(Y, Z). \tag{9.89}$$

Define the tripartite mutual information of X, Y and Z as

$$\begin{aligned} I(X; Y; Z) = & \; H(X) + H(Y) + H(Z) + H(X, Y, Z) \\ & -H(X, Y) - H(Y, Z) - H(X, Z) \end{aligned} \tag{9.90}$$

This further implies that

$$\begin{aligned} I(X; Y; Z) \leq & \; H(X, Y, Z) - H(X, Z) \\ = & \; H(Y|X, Z). \end{aligned} \tag{9.91}$$

Similarly, permutating random variables in the inequality (9.89) yields to other two inequalities as

$$I(X; Y; Z) \leq H(X|Y, Z), \tag{9.92}$$

$$I(X; Y; Z) \leq H(Z|X, Y). \tag{9.93}$$

These implies a general inequality (Luo 2022) for characterizing triangle network as

$$I(X; Y; Z) \leq \min\{H(X|Y, Z), H(Y|X, Z), H(Z|X, Y)\}. \tag{9.94}$$

Another is from special constrictions of joint distribution $P_{xyz} = \widetilde{\Sigma}_{x,y,z} P(x, y, z)[xyz]$ which is compatible with the triangle network. Consider the Finner inequality (Renou et al. 2019) as

$$P(x, y, z) \leq \sqrt{p(x)p(y)p(z)}. \tag{9.95}$$

The marginal distributions are given by

$$P_x = \widetilde{\Sigma}_x p(x)[x],\ P_y = \widetilde{\Sigma}_y p(y)[y],\ P_z = \widetilde{\Sigma}_z p(z)[z]. \tag{9.96}$$

Together with the inequality (9.95), we get that

$$\begin{aligned} H(X, Y, Z) &= \sum_{x,y,z} P(x, y, z) \log \frac{1}{P(x, y, z)} \\ &\geq \sum_{x,y,z} P(x, y, z) \log \frac{1}{\sqrt{p(x)p(y)p(z)}} \\ &= \frac{1}{2}(H(X) + H(Y) + H(Z)) \end{aligned} \tag{9.97}$$

by using the increasing function $f(x) = \log x$. This implies another lower bound for tripartite mutual information as

$$\begin{aligned} I(X; Y; Z) > {} & \tfrac{3}{2}(H(X) + H(Y) + H(Z)) \\ & -H(X, Y) - H(X, Z) - H(Y, Z). \end{aligned} \tag{9.98}$$

Instead, for a triangle network consisting of one variable λ, as shown in Fig. 9.2a, we prove that the compatible joint distribution P_{xyz} satisfies the following inequalities as

$$I(X; Y; Z) \leq 4H(X, Y, Z) - H(X, Y) - H(X, Z) - H(Y, Z), \tag{9.99}$$

$$\begin{aligned} I(X; Y; Z) > {} & \frac{4}{3}(H(X) + H(Y) + H(Z)) \\ & -H(X, Y) - H(X, Z) - H(Y, Z). \end{aligned} \tag{9.100}$$

Moreover, for two dependent variables X and Y on respective finite sample spaces $\mathcal{X}$ and $\mathcal{Y}$, another scenarios for $I(X; Y; Z) \leq 0$ is defined by

$$I(X; Y) > 0, I(X; Y|Z) > 0. \tag{9.101}$$

Example 9.3.1 Consider one example shown as

$$\begin{aligned} P_{xyz} = {} & \frac{1}{8}[000] + \frac{1}{8}[011] + \frac{1}{8}[102] + \frac{1}{8}[113] \\ & +\frac{1}{8}[220] + \frac{1}{8}[231] + \frac{1}{8}[322] + \frac{1}{8}[333]. \end{aligned} \tag{9.102}$$

It follows that $I(X; Y) = 1$ and $I(X; Y|Z) = 2$. By using the binary representation of i as $g : i \mapsto i_1 i_2$, P_{xyz} can be split into two distributions of $P_{x_1y_1} = \frac{1}{2}[00] + \frac{1}{2}[11]$ and $P_{x_2y_2z}$ defined in Eq. (9.81). This means that

$$P_{g(x)g(y)z} = P_{x_1x_2y_1y_2z} = P_{x_1y_1} \times P_{x_2y_2z} \overset{g^{-1}}{\rightarrow} P_{xyz} \quad (9.103)$$

under the inverse mapping g^{-1} of g for x and y. The joint distribution $P_{x_1y_1}$ satisfies $I(X_1; X_2) = I(X; Y)$, while $P_{x_2y_2z}$ satisfies $I(X_2; Y_2) = 0$ and $I(X_2; Y_2|Z) = 2$. This implies a new compatible network consisting of three variables, as shown in Fig. 9.2d.

In general, for any distribution satisfying Eq. (9.101), suppose that there is a transformation $g : x \mapsto x_1x_2$ satisfying $P_{g(x)g(y)z} = P_{x_1x_2y_1y_2z} = P_{x_1y_1} \times P_{x_2y_2z}$. Under this assumption, we obtain the following result.

Theorem 9.3.2 *For any joint distribution satisfying the condition (9.101), it is compatible with the triangle network if the decomposition (9.103) holds.*

For a general network, the main method is from so-called multipartite independent networks (Luo 2018), that is, some nodes have not shared any entanglement.

Consider an $n + m$-partite quantum network $\mathcal{N}_n$ consisting of generalized EPR states (Einstein et al. 1935): $|\phi_1\rangle = \cos\theta_i|00\rangle + \sin\theta_i|11\rangle$ with $\theta_i \in (0, \frac{\pi}{2})$, $i = 1, \ldots, N$. Assume that there are n nodes $\mathsf{A}_1, \ldots, \mathsf{A}_n$ in $\mathcal{N}_n$ with $n \geq 2$ such that each pair of them has not shared any entanglement.

Denote $X_i \in \mathcal{X}_i$ and $Y_j \in \mathcal{Y}_j$ as the respective outcomes of A_i and B_j under local projection measurements, $i = 1, \ldots, n$; $j = 1, \ldots, m$. By using $H(X_1, \ldots, X_j|X_{j+1}) = H(X_1, \ldots, X_j, X_{j+1}) - H(X_{j+1})$, the mutual information is defined as

$$I(X_1; \cdots; X_n; Y) = I(X_1; \cdots; X_n) - I(X_1; \cdots; X_n|Y), \quad (9.104)$$

where $I(X_1; \cdots; X_n)$ denotes the mutual information of the variables $X_1, \ldots, X_n$ given by

$$I(X_1; \cdots; X_n) = \sum_{\text{odd } i, i \leq n} \sum_{1 \leq j_1 < \cdots < j_i \leq n} H(X_{j_1}, \ldots, X_{j_i}) - \sum_{\text{even } s, s \leq n} \sum_{1 \leq \ell_1 < \cdots < \ell_i \leq n} H(X_{\ell_1}, \ldots, X_{\ell_i}), \quad (9.105)$$

and $I(X_1; \cdots; X_n|Y)$ denotes the mutual information of $X_1, \ldots, X_n$ conditional on the outcomes of $Y = Y_1 \cdots Y_m$ and is given by

$$I(X_1; \cdots; X_n) = \sum_{\text{odd } i, i \leq n} \sum_{1 \leq j_1 < \cdots < j_i \leq n} H(X_{j_1}, \ldots, X_{j_i}|Y) - \sum_{\text{even } s, s \leq n} \sum_{1 \leq \ell_1 < \cdots < \ell_i \leq n} H(X_{\ell_1}, \ldots, X_{\ell_i}|Y). \quad (9.106)$$

Any n-independent quantum network $\mathcal{N}_n$ with $n \geq 3$ may show different features beyond the chain network. Specifically, both the negative and positive Shannon mutual information may be generated from $\mathcal{N}_n$ with $n \geq 3$.

One example is star network, which is compatible with the joint distribution given by

$$P_{x_1 \cdots x_n y} = \tilde{\Sigma}_{i_1,\ldots,i_n} p_{i_1} \cdots p_{i_n} [x_{i_1} \cdots x_{i_n} y_{i_1 \cdots i_n}], \tag{9.107}$$

with $n+1$ finite-dimensional variables $X_1, \ldots, X_n, Y$, where $\{p_{i_j}\}$ are probability distributions, $j = 1, \ldots, n$.

9.4 Bell Compatibility of Networks

In this subsection, we consider the compatibility problem of causal networks with multiple measurement setups per party. This is related to the Bell nonlocality (Bell 1964).

9.4.1 *Bell Compatibility*

Consider a multisource network $\mathcal{N}$ consisting of n parties $\mathsf{A}_1, \mathsf{A}_2, \ldots, \mathsf{A}_n$. Under the assumption similar to local hidden variable model (Bell 1964; Luo 2020), each A_i shares some independent sources Λ_i of $\lambda_1, \lambda_2, \ldots, \lambda_m$. x_i and a_i denote the respective measurement input and output of A_i. Each output is conditional on the shared variables and measurement setups. This implies a Bell-type DAG in an input-output manner.

In what follows, each correlation is regarded as a vector with components $P_{\vec{a},\vec{x}} := P(\vec{a}|\vec{x})$ in the real space $\mathbb{R}^d$ with the dimension $d = \prod_{i=1}^n |a_i| \times |x_i|$. Given inputs $\{\mathbf{I}_1, \ldots, \mathbf{I}_n\}$ for all parties, denote $R_{\Lambda_i}^{(i)}$ as local deterministic response function of a_i by A_i conditional on the inputs $\mathbf{I}_i$ and sources Λ_i, $i = 1, \ldots, n$.

Definition 9.4.1 A conditional distribution $P(\vec{a}|\vec{x})$ is compatible with a Bell DAG if it allows the following decomposition as

$$P(\vec{a}|\vec{x}) = \sum_{\lambda_1,\ldots,\lambda_m} \prod_{i=1}^{n} R_{\Lambda_i}^{(i)}(a_i|\mathbf{I}_i) \prod_{j=1}^{m} \mu_i(\lambda_j), \tag{9.108}$$

where $|\mathbf{I}_i|$ denotes the number of the inputs of the i-th output, and μ_i is the probability distribution of the source λ_i.

The local strategy function of $R_{\Lambda_i}^{(i)}$ may be further represented by a characteristic function as

$$R_{\Lambda_i}^{(i)}(a_i|\mathbf{I}_i) := \delta_{a_i, f_{\Lambda_i}^{(i)}}(\mathbf{I}_i), \tag{9.109}$$

where δ denotes the Kronecker delta and $f_{\Lambda_i}^{(i)}$ is the local deterministic assignment of $\mathbf{I}_i$ into a_i. The global response is then defined by

$$\mathbf{R}_\Lambda := R_{\Lambda_1}^{(1)} \times \cdots \times R_{\Lambda_n}^{(n)}, \tag{9.110}$$

where $\mathbf{R}_\Lambda$ can be represented by a vector in $\mathbb{R}^d$, with components $R_{\Lambda,\vec{a},\vec{x}} := R_\Lambda(\vec{a}|\vec{x}) = (R_{\Lambda_1}^{(1)}(a_1|\mathbf{I}_1), \ldots, R_{\Lambda_n}^{(n)}(a_n|\mathbf{I}_n))$.

Similar to Bell theory (Bell 1964), if all the sets Λ_i are non-intersect, P consist of a polytope in $\mathbb{R}^d$. This kind of polytope may be featured by the convex hull of a finite number of external points given by R_{Λ_i} for different sources. This implies a simple way to solve the compatibility problem for given correlations. In fact, it may be used to determine whether a given correlation is compatible with a Bayesian network with respect to the inputs $\{\mathbf{I}_1, \ldots, \mathbf{I}_n\}$ by using linear programs (Chaves et al. 2017).

Nevertheless, all the sets of Λ_i's may intersect for general networks. For a general k-independent network, denote $\mathcal{H} = \{i_1, i_2, \ldots, i_k\}$ as all the indexes of independent parties, $\overline{\mathcal{H}} = \{1, 2, \ldots, n\} \setminus \mathcal{H}$. Denote $I_{n,k}$ and $J_{n,k}$ as two quantities given by

$$I_{n,k} = \frac{1}{2^k} \sum_{x_i, i \in \mathcal{H}} \langle a_{x_1} a_{x_2} \cdots a_{x_n} \rangle_{\mathcal{I}}^0, \tag{9.111}$$

$$J_{n,k} = \frac{1}{2^k} \sum_{x_i, i \in \mathcal{H}} (-1)^{\sum_{j \in \mathcal{I}} x_j} \langle a_{x_1} a_{x_2} \cdots a_{x_n} \rangle_{\mathcal{H}}^0, \tag{9.112}$$

where they expect of measurement outcome is given by

$$\langle a_{x_1} \cdots a_{x_n} \rangle_{\mathcal{I}}^0 = \sum_{\vec{a}} (-1)^{\sum_{i=1}^n a_i} P(\vec{a}|\vec{x}_{\mathcal{H}}; x_s = 0, s \in \overline{\mathcal{I}}), \tag{9.113}$$

$$\langle a_{x_1} \cdots a_{x_n} \rangle_{\mathcal{H}}^1 = \sum_{\vec{a}} (-1)^{\sum_{i=1}^n a_i} P(\vec{a}|\vec{x}_{\mathcal{H}}; x_s = 1, s \in \overline{\mathcal{H}}). \tag{9.114}$$

With straightforward evaluations, all the compatible correlations under the no-signaling restrictions satisfy the following inequality:

$$\mathcal{R}_{ns} := |I_{n,k}|^{\frac{1}{k}} + |J_{n,k}|^{\frac{1}{k}} \leq 2, \tag{9.115}$$

where the non-signaling principle means the local actions will not be affected by other's local actions, i.e., $p(a_i|\vec{x}) = p(a_i|x_i)$.

These multisource networks imply a general non-convex causal set, instead a star-convex set in $\mathbb{R}^d$ with the center point at the origin. In the classical variable model, it has $\mathcal{R}_c \leq 1$, while quantum correlations satisfy $\mathcal{R}_q \leq \sqrt{2}$ (Luo 2018). This implies a hierarchic relationship as

$$\{P|\mathcal{R}_c\} \subseteq \{P|\mathcal{R}_q\} \subseteq \{P|\mathcal{R}_{ns}\}. \tag{9.116}$$

For cyclic networks without independent agents, all the non-signaling correlations satisfy a linear restriction of

$$|I_{n,k}| + |J_{n,k}| \leq 2. \tag{9.117}$$

This restriction is impossible to distinguish two correlation sets from the classical causal model or quantum model. Other methods such as partial correlations or post-selection may be applied.

In what follows, we consider a general problem of determining whether a given non-signaling correlation is compatible with a Bell DAG with multiple sources (acyclic networks). By contracting Λ into a vector, Eq. (9.108) is rewritten into

$$P = \mathbf{R}_\Lambda \mu, \tag{9.118}$$

where $\mathbf{R}_\Lambda$ is a contracted matrix, while $\mu = (\mu_1, \ldots, \mu_m)^T$ is a contracted vector over the contracted vector Λ. Given inputs $\{\mathbf{I}_1, \ldots, \mathbf{I}_n\}$, the compatibility problem of a given correlation and Bayesian network is equivalent to determining $\mu_1, \ldots, \mu_m$ which satisfy the inequality (9.117) and Eq. (9.118). This follows an optimization problem as

$$\begin{aligned} &\min_{\forall \mu_i \geq 0, \|\mu\|=1} \quad \mathbf{X}\mu, \\ &s.t., \quad \mathbf{R}_\lambda \mu = P, |I_{n,k}|^{\frac{1}{k}} + |J_{n,k}|^{\frac{1}{k}} \leq \alpha, P(\vec{a}_i|\vec{x}) = P(\vec{a}_i|\vec{x}_i), \end{aligned} \tag{9.119}$$

where $\mathbf{X}$ is a matrix function. Choosing a proper parameter α with $1 \leq \alpha \leq 2$ is depending on specific goals from different causal models. This kind of multi-variable star-convex optimization may be solved by using the ellipsoid method in a polynomial-time complexity (Lee and Valiant 2016). This implies an efficient way for searching new causal correlations and Bell inequalities.

9.4.2 Bell Compatibility Problems

This section provides a detailed classification of correlations that are compatible with tripartite networks.

Definition 9.4.2 A $\mathcal{G}_1$ is non-signaling implying a DAG $\mathcal{G}_2$, if every non-signaling correlation that are compatible with $\mathcal{G}_1$ is compatible with $\mathcal{G}_2$.

Here, the correlations in $\mathcal{G}_2$ can be regarded as causal relaxations of its in $\mathcal{G}_1$. $\mathcal{G}_1$ and $\mathcal{G}_2$ are non-signaling equivalent if they mutually are non-signaling implication. Similar to the network consisting of one source, the multipartite correlations may be useful for information-theoretic protocols (Chaves et al. 2017) when two associated Bell DAGs are non-signaling equivalent.

In what follows, a Bell DAG is named an input-output (IO) DAG if causal relaxations are from the Bell locality. One example is a tripartite chain network. Let $\mathcal{G}_{gen}$ and $\mathcal{G}_{io}$ be two Bell DAGs, where $\{a_j, x_j\} \subseteq S_{x_i}$ and $\{a_j, x_j\} \subseteq S_{a_i}$ for $\mathcal{G}_{gen}$, while $\{x_j\} \subseteq S_{a_i}$ for $\mathcal{G}_{io}$ for $1 \leq i \neq j \leq n$. These two graphs are non-signaling equivalent. One example is the chain network consisting of three parties sharing two variables λ_1 and λ_2. By using locality relaxation from two independent inputs x_1 and x_3, the fan out of x_1 and x_3 being connected to all outputs a_i's can generate the same non-signaling correlations from the case of all inputs x_i's to x_2, while x_1 to a_1 and x_3 to a_3. Another case is to consider the fan out of x_1 and x_2. Moreover, consider the connection from λ_i to x_i, this is equivalent to x_i to λ_i from the relaxations of measurement independence (Luo 2020). This graph can be then described by using input sets in terms of its incoming nodes. With these definitions, we can classify all the tripartite Bell DAGs in terms of entanglement swapping.

Theorem 9.4.1 *Consider a Bell DAG with three nodes and two variables. There are fifteen non-signaling Bell-type classes.*

The proof of Theorem 1 is completed by examining all the possible Bell DAGs (Luo 2020). Generally, for DAGs consisting of multiple sources, it is NP-hard to classify all the external non-signaling correlations using the convex-optimization tools. It might be possible for exploring special correlation constrictions or specific networks.

9.5 Conclusion

In this chapter, we have investigated some problems related to network configuration. The Bell theory (Bell 1964) provides an interesting way to explore local correlation sets under the assumption of multiple measurements per party. This is easy for quantum networks consisting of one source but difficult for general networks or local measurement setups.

For one measurement setup, the multisource quantum network may exhibit distinguished features of quantumness beyond its classical realizations with variables (Renou et al. 2019). This has sparked new interest in non-standard Bell theory for quantum networks (Renou and Beigi 2022). It is also related to another interesting problem of network compatibility in classical statistics that has been recently discussed using network inflation in Sect. 9.1.1 based on the classical model (Wolfe et al. 2019; Navascues and Wolfe 2020; Luo 2022) or in Sect. 9.1.2 based on the quantum model (Wolfe et al. 2021). The Bell compatibility problem is a big problem that may be related to quantumness (Luo 2020; Pironio et al. 2010; Supic et al. 2020) in Sect. 9.4.

References

J.S. Bell, On the Einstein Podolsky Rosen paradox. Phys. Physique Fizika **1**, 195 (1964)

R. Chaves, C. Budroni, Entropic nonsignaling correlations. Phys. Rev. Lett. **116**, 240501 (2016)

R. Chaves, D. Cavalcanti, L. Aolita, Causal hierarchy of multipartite Bell nonlocality. Quantum **1**, 23 (2017)

T.M. Cover, J.A. Thomas, *Elements of Information Theory* (Wiley, 2012)

R. Diestel, *Graph Theory* (Springer, Berlin, Heidelberg, 2017)

W. Dür, G. Vidal, J.I. Cirac, Three qubits can be entangled in two inequivalent ways. Phys. Rev. A **62**, 062314 (2000)

A. Einstein, B. Podolsky, N. Rosen, Can quantum-mechanical description of physical reality be considered complete? Phys. Rev. **47**, 777–780 (1935)

H. Finner, A generalization of Hölder's Inequality and some probability inequalities. Ann. Probab. **20**, 1893–1901 (1992)

T. Fritz, Beyond Bell's theorem I: correlation scenarios. New J. Phys. **14**, 103001 (2012)

N. Gisin, J.-D. Bancal, Y. Cai, P. Remy, A. Tavakoli, E.Z. Cruzeiro, S. Popescu, N. Brunner, Constraints on nonlocality in networks from no-signaling and independence. Nat. Commun. **11**, 2378 (2020)

D.M. Greenberger, M.A. Horne, A. Zeilinger, in *Bell's Theorem, Quantum Theory and Conceptions of the Universe*, ed. by M. Kafatos (Kluwer, Dordrecht, 1989), pp. 69–72

J.C.H. Lee, P. Valiant, Optimizing star-convex functions, in *IEEE 57th Annual Symposium on Foundations of Computer Science* (IEEE, USA, 2016), pp. 603–614

L.T. Ligthart, M. Gachechiladze, D. Gross, A convergent inflation hierarchy for quantum causal structures. Commun. Math. Phys. **401**, 2673–2714 (2023)

M.-X. Luo, Computationally efficient nonlinear bell inequalities for quantum networks. Phys. Rev. Lett. **120**, 140402 (2018)

M.-X. Luo, Nonsignaling causal hierarchy of general multisource networks. Phys. Rev. A **101**, 062317 (2020)

M.-X. Luo, Negative Shannon information hides networks. Adv. Quantum Tech. **10**, 202200069 (2022)

M. Navascues, E. Wolfe, The inflation technique completely solves the causal compatibility problem. J. Causal Infer. **8**, 70–91 (2020)

J. Pearl, *Causality* (Cambridge University Press, 2009)

S. Pironio, M. Navascues, A. Acín, Convergent relaxations of polynomial optimization problems with noncommuting variables. SIAM J. Optim. **20**, 2157–2180 (2010)

M.-O. Renou, S. Beigi, Network nonlocality via rigidity of token counting and color matching. Phys. Rev. Lett. **128**, 060401 (2022)

M.-O. Renou, Y. Wang, S. Boreiri, S. Beigi, N. Gisin, N. Brunner, Limits on correlations in networks for quantum and no-signaling resources. Phys. Rev. Lett. **123**, 070403 (2019)

M.-O. Renou, E. Bäumer, S. Boreiri, N. Brunner, N. Gisin, S. Beigi, Genuine quantum nonlocality in the triangle network. Phys. Rev. Lett. **123**, 140401 (2019)

A. Schrijver, *Combinatorial Optimization: Polyhedra and Efficiency* (Springer, 2003)

I. Supic, J.-D. Bancal, N. Brunner, Quantum nonlocality in networks can be demonstrated with an arbitrarily small level of independence between the sources. Phys. Rev. Lett. **125**, 240403 (2020)

R.F. Werner, Quantum states with Einstein-Podolsky-Rosen correlations admitting a hidden-variable model. Phys. Rev. A **40**, 4277 (1989)

E. Wolfe, R.W. Spekkens, T. Fritz, The inflation technique for causal inference with latent variables. J. Causal Infer. **7**, 20170020 (2019)

E. Wolfe, A. Pozas-Kerstjens, M. Grinberg, D. Rosset, A. Acín, M. Navascues, Quantum inflation: a general approach to quantum causal compatibility. Phys. Rev. X. **11**, 021043 (2021)

M. Zukowski, A. Zeilinger, M.A. Horne, A.K. Ekert, Event-ready-detectors Bell experiment via entanglement swapping. Phys. Rev. Lett. **71**, 4287–4290 (1993)

Index

M.-X. Luo, *Quantum Networks*,
https://doi.org/10.1007/978-981-97-6226-2

SPRINGER NATURE

GPSR Compliance

The European Union's (EU) General Product Safety Regulation (GPSR) is a set of rules that requires consumer products to be safe and our obligations to ensure this.

If you have any concerns about our products, you can contact us on ProductSafety@springernature.com

In case Publisher is established outside the EU, the EU authorized representative is:

Springer Nature Customer Service Center GmbH
Europaplatz 3
69115 Heidelberg, Germany

The manufacturer's authorised representative in the EU is Springer
Nature Customer Service Centre GmbH, Europaplatz 3, 69115 Heidelberg,
Germany. If you have any concerns regarding our products, please
contact ProductSafety@springernature.com

Printed and bound by CPI Group (UK) Ltd, Croydon, CR0 4YY
01/10/2026
02225858-0016